Acoustic Technologies in Biology and Medicine

Acoustic Technologies in Biology and Medicine

Edited by Adem Ozcelik, Ryan Becker, and Tony Jun Huang

WILEY VCH

Editors

Dr. Adem Ozcelik
Adnan Menderes University
Zafer Mahallesi
Muhendislik Fakultesi, Efeler
09010 Aydın
Turkiye

Dr. Ryan Becker
Duke University
144 Hudson Hall
Box 90300
27708 Durham NC
United States

Dr. Tony Jun Huang
Duke University
144 Hudson Hall
Box 90300
NC
United States

Cover Image: © oxygen/Getty Images

Library of Congress Card No.: applied for

British Library Cataloguing-in-Publication Data
A catalogue record for this book is available from the British Library.

Bibliographic information published by the Deutsche Nationalbibliothek
The Deutsche Nationalbibliothek lists this publication in the Deutsche Nationalbibliografie; detailed bibliographic data are available on the Internet at <http://dnb.d-nb.de>.

Print ISBN: 978-3-527-35062-9
ePDF ISBN: 978-3-527-84130-1
ePub ISBN: 978-3-527-84131-8
oBook ISBN: 978-3-527-84132-5

Typesetting Straive, Chennai, India
Printing and Binding: CPI Group (UK) Ltd, Croydon, CR0 4YY

C9783527350629_130524

Contents

Preface

Acoustic technologies have played an increasingly significant role in both biology and medicine, empowering scientists and clinicians to explore and understand the intricate workings of living organisms and develop innovative diagnostic and therapeutic approaches. From ultrasound imaging to acoustic tweezers, from acoustic biosensors to acoustic stimulation, acoustic technologies have become indispensable tools in the arsenal of researchers and practitioners in these fields.

Acoustic Technologies in Biology and Medicine is designed to be a cornerstone reference bridging the rapidly advancing acoustic technologies with state-of-the-art applications in biology and medicine. The research in acoustics for biology and medicine requires a multidisciplinary approach involving fluid mechanics, physics, chemistry, electronics, and the life sciences. The burgeoning success and eminence of this field is fueled by the realization of groundbreaking technological innovations and impactful solutions addressing significant biological problems. Thus, it is imperative to present this field to a diverse audience spanning multiple engineering disciplines and the life sciences, targeting their involvement and education on the basic principles, breadth of technologies, and interdisciplinary applications of the acoustics. In this respect, *Acoustic Technologies in Biology and Medicine* adopts a unique role and context to introduce the graduate students and professionals from a broad range of backgrounds to this active research field, aiming to increase the synergy and interaction between the engineers and life scientists.

The book takes an interdisciplinary approach by covering the mechanistic and physical basics, as well as the applications. First, the book covers the underlying physical and theoretical principles of acoustic technologies. This aims to help the reader grasp the fundamental principles of acoustics. Second, different technologies of acoustic systems, including bulk and surface acoustic wave-based platforms, acoustic imaging, acoustic sensors, and acoustic levitators, are discussed. A clear understanding of the principles, advantages, and disadvantages of different acoustic technologies is crucial for evaluating and determining which acoustic methods are most suitable for particular biological applications. Third, state-of-the-art applications of acoustics in biology and medicine are presented. These applications include single cell and organism manipulation, acoustic biosensing, cancer cell isolation (liquid biopsy), cell/tissue stimulation and ablation, micro-robot actuation, acoustic imaging, and drug delivery.

The book is intended for a broad audience, including researchers and students in physics, biology, and engineering, as well as clinicians and practitioners in medicine. It is our hope that this book will provide a useful reference and inspire further research and innovation in the exciting and rapidly evolving field of acoustic technologies in biology and medicine.

February 28, 2023
Durham, NC, USA

Dr. Adem Ozcelik
Dr. Ryan Becker
Dr. Tony Jun Huang

1

Fundamentals of Acoustic Wave Generation and Propagation

Mehmet A. Sahin[1], Mushtaq Ali[2], Jinsoo Park[2], and Ghulam Destgeer[1]

[1] Technical University of Munich, Central Institute for Translational Cancer Research (TranslaTUM), Control and Manipulation of Microscale Living Objects, School of Computation, Information and Technology (CIT), Department of Electrical Engineering, Einsteinstraße 25, Munich 81675, Germany
[2] Chonnam National University, Multiscale Flow Control Laboratory, Department of Mechanical Engineering, Gwangju 61186, Republic of Korea

1.1 Introduction

1.1.1 Acoustic or Sound Waves

The use of acoustic technologies is widespread, from radio frequency (RF) surface acoustic wave (SAW) filters in most telecommunication devices and nondestructive testing (NDT) of solid parts in expensive machineries to ultrasonic imaging of fetuses during pregnancy and of joints and bones in case of injuries. At the heart of any acoustic technology are sound waves at work – a combination of pressure and velocity fluctuations propagating through a medium. The terms "acoustic" or "sound" waves are interchangeably used. Since "sound" is commonly associated with hearing audible acoustic waves as perceived by the human auditory organs, it would be restrictive to limit the term sound to human perception only, the audible frequency range of 20 Hz to 20 kHz (Figure 1.1). Dogs and cats can even hear much higher-frequency ultrasound waves, whereas elephants via their feet can sense very low-frequency seismic vibrations propagating long distances across the land, i.e. infrasound. The ability of whales and bats to echolocate their targets at much higher frequencies of up to 200 kHz is also well known. Moreover, the use of MHz frequency waves in NDT and GHz frequency waves in SAW filters further broadens the scope of sound/acoustic waves. Therefore, any periodic pressure/velocity fluctuation in a solid, liquid, or gaseous media over a wide range of frequencies (10^0–10^9 Hz) can be safely called a sound/acoustic wave [1].

1.1.2 Dominos Effect

In an acoustic wave, the localized pressure at any given point within the medium oscillates between the maximum and minimum pressure values, linked to regions

Acoustic Technologies in Biology and Medicine, First Edition.
Edited by Adem Ozcelik, Ryan Becker, and Tony Jun Huang.

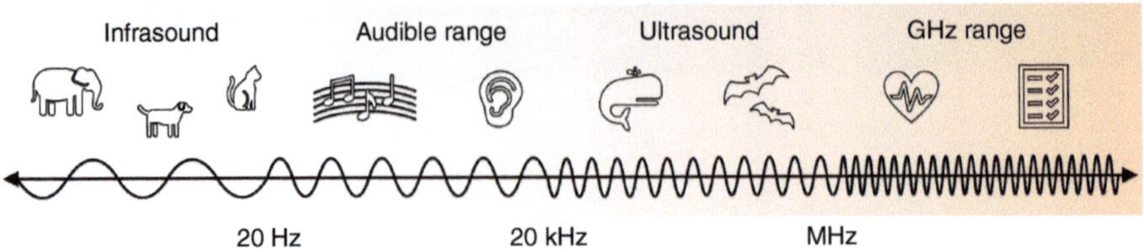

Figure 1.1 Categorization of acoustic waves based on frequency range.

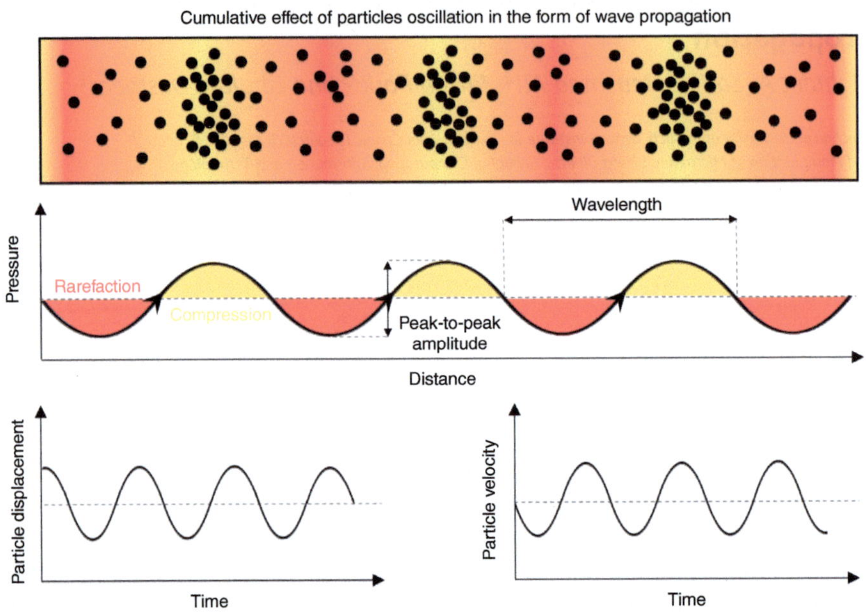

Figure 1.2 Particle oscillation in a medium because of an acoustic wave propagation.

of compression and rarefaction, respectively (Figure 1.2). Any particle of the discretized medium, without moving along the propagating wave, will vibrate about a mean position with an oscillating velocity field. Propagation of an acoustic wave with these oscillating pressure and velocity fields through a medium can be understood by a simple analogy of a stack of falling dominoes, where the disturbance originating from one end of the dominoes' stack reaches the opposite end in no time, even without a single domino piece significantly moving away from its original position. Here, the average velocity of an individual domino is related to the particle velocity of the medium, whereas the disturbance moving across the stack is analogous to wave propagation. It is important to distinguish the oscillating particle velocity from the velocity of wave propagation [2].

1.1.3 Elastic vs Inelastic Waves

Acoustic waves under discussion are broadly categorized as elastic waves since they need an elastic medium to propagate. For an acoustic wave propagation, as the particles of the elastic medium move about an equilibrium position, the velocity of

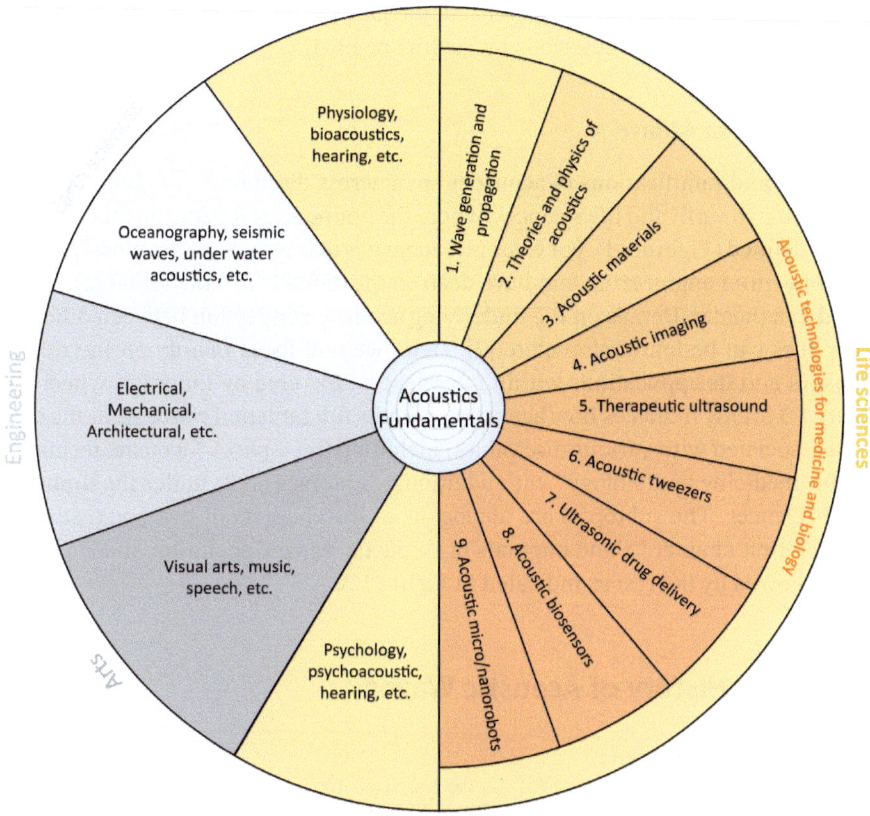

Figure 1.3 Position of "acoustic technologies for medicine and biology" within the scope of acoustics plotted in a circular chart analogous to Lindsay's wheel. Source: Adapted from Lindsay [5].

these particles determines the kinetic energy, whereas the displacement away from the equilibrium position (i.e. wave amplitude) determines the potential energy associated with the acoustic wave. Similarly, mechanical waves, e.g. a transverse wave visualized on a vibrating string, require a medium to propagate that can store kinetic and potential energies. Elastic waves also share a lot in terms of wave propagation and energy with electromagnetic waves that can even travel through vacuum and have an enormous frequency range: $10–10^6$ Hz radio waves, 10^{14} Hz visible light, 10^{20} Hz gamma rays, etc. Since vacuum can store electric and magnetic energies, analogous to potential and kinetic energies in elastic waves, it can sustain electromagnetic waves propagation. Electromagnetic waves (e.g. visible light) can also propagate through medium other than vacuum; however, the waves will be subjected to phenomena like reflection and refraction at interfaces where electromagnetic properties of media (e.g. refractive index) change. Similarly, acoustic waves can be reflected and refracted at an interface where the wave propagation velocity or acoustic impedance is not continuous. Building upon the fundamental assumptions of geometrical optics, where a wave can be represented by a ray directed normal to the wavefront, the paths taken by an acoustic wave in a complex medium can be described by Snell's law [3, 4].

1.1.4 Scope of Acoustics

The scope and ramifications of acoustics span across disciplines, i.e. from arts and engineering to earth and life sciences, where the boundaries separating them are not strictly defined (Figure 1.3). For example, acoustics of large spaces are closely linked to architectural engineering and have deep connections with how visual art is perceived in a theater. Details on the underlying acoustic connection between different disciplines can be found elsewhere [2]. Here, we will focus mainly on the role of acoustics and its applications within life sciences. Inspired by Lindsay's wheel [5], Figure 1.3 briefly indicates how acoustics, with its fundamental concepts at the center, is associated with various disciplines, including the topic of "acoustic technologies for medicine and biology" with additional subtopics (1–9) under the umbrella of life sciences. The subtopics are analogous to the chapters of this book, starting with the first chapter "Fundamentals of Acoustic Wave Generation and Propagation" followed by the rest as indicated in Figure 1.3.

1.2 Brief History of Acoustic Waves

1.2.1 Early History

In the sixth century BCE, Pythagoras discovered the relationship between the length of the vibrating strings and the variable-pitch musical sounds they produced. More than two millennia later, in 1643, Torricelli demonstrated that sound does not propagate in vacuum, thus establishing the need for media for sound propagation. Half a century later, in 1701, Sauveur introduced the term "acoustics" to describe the science of "sound," thereby pitching the two terms as synonyms to each other. A century later, in 1802, Chladni brought acoustics to the spotlight by introducing it as a separate branch of physics. With the development of acoustic resonators in 1859, also known as Helmholtz resonators, which are based on resonance of sound waves within a cavity, acoustics principles later found their applications in sound amplification or silencing in loudspeakers or mufflers, respectively. In 1866, Kundt measured the speed of sound propagating through a tube filled with fine powder to trace the acoustic pressure nodes of standing waves. As the classical acoustics matured by the end of the nineteenth century, Rayleigh culminated the state of the art of that time in his book *The Theory of Sound* (1896), which is still relevant today [2, 6, 7].

1.2.2 History of Acoustic Streaming

A historical advancement of acoustics can be further traced at multiple fronts; however, for the sake of brevity, we will only focus on the development of the following two topics: acoustic streaming flow and acoustic radiation force (ARF). Lord Rayleigh (1884), while explaining observations done by Faraday (patterns on vibrating Chladni plates) [8] and Dvorak (circulation of air in Kundt's tubes experiment), [9] described a "boundary layer driven" flow within the context of standing wave resonators with mean fluid motion outside of the boundary layer, also known as "outer streaming" or "Rayleigh streaming" [10]. A mean flow within the boundary layer is called "Schlichting streaming" after Schlichting provided detailed analysis of this secondary flow within an incompressible boundary layer [11]. Contrary to a standing wave, a progressive or traveling wave dissipating acoustic energy within a viscous media along its path results in a time-averaged flow field in the form of "quartz wind" or "Eckart streaming," first described by Eckart [12]. With the advent of SAW technology (1965), it became possible to produce high-frequency (>10 MHz) acoustic waves in an efficient manner [13]. Shiokawa et al. later (in 1990s) coupled the SAW with small droplet volumes to study phenomena like acoustic streaming flow, jetting, and atomization [14, 15]. Acoustic manipulation of small droplets (2003) and mixing using SAW-induced streaming flow (2006) were further elaborated with potential applications in biomedicine by Wixforth et al. using frequencies as high as 146 MHz [16, 17]. Friend and coworkers demonstrated ultrasonic atomization using a bulk acoustic waves (BAWs) device operated at 1.6 MHz, which highlights the contrast between the BAW- and SAW-based techniques [18]. These earlier works have led to discovery of interesting phenomena and exciting applications in recent years, e.g. unique finger instabilities (2012) and micro-centrifugation (2021) [19, 20].

1.2.3 History of Acoustic Radiation Force

On the other front, with King's [21] theoretical model of the "ARF" on a rigid spherical particle, a foundation was laid to calculate the force on a compressible droplet or bubble [22] or an elastic particle [23] exposed to a traveling or a standing acoustic wave. Gorkov's [24] simplified expression of the ARF model of Yosioka and Kawasima [22] is still a common reference for modern-day theoretical as well as experimental works on the subject. While the aforementioned theoretical models were being validated with corresponding experimental demonstrations of acoustic waves pushing on suspended particles in a fluidic media, it was not until 1993 that Mandralis and Feke [25] used the ARF to fractionate particles in a *continuous flow*. With the work of Coakley and coworkers [26] and Laurell and coworkers [27], a continuous acoustic manipulation of suspended particles was brought to the realm of microfluidics with the use of silicon-based microfabricated channels attached to a BAW transducer. Following suit, Huang et al. demonstrated standing SAW-based focusing (2008) [28] and separation (2009) [29] of microparticles in a continuous flow within a polydimethylsiloxane (PDMS) microchannel. Continuous separation of particles by a traveling SAW was demonstrated only in 2013 by Sung and coworkers [30] and Franke and coworkers [31] Over the past decades, the objects manipulated by the acoustic waves are getting smaller and smaller from cm-scale (1969) [23] elastic spheres to μm-scale (2004) [27] microparticles, droplets, and cells to nm-scale (2022) [32] nanoparticles, extracellular vesicles, etc. For further reading on the historical development of the field, readers may also refer to the following references: [2, 33].

1.3 What Is an Acoustic Wave?

1.3.1 Acoustic Parameters

An acoustic wave is an elastic wave where a medium is needed to carry disturbances within the displacement, velocity, and pressure fields to transfer energy from one point in space to another, where the medium properties play a critical role in regulating energy transport. Ripples on the surface of water, vibrations in a guitar string, and seismic activities resulting in earthquakes are examples of such waves. In contrast, inelastic electromagnetic waves do not depend upon any medium to carry the oscillating electric and magnetic fields to transport energy, and can propagate even in vacuum. For visualization of elastic waves, one can imagine a transverse wave on a vibrating rope in a sinusoidal shape, where one end of the rope is moving up and down while the other is tied to an anchor. The highest and lowest points on the rope are called the wave *crest* and *trough*, respectively. The peak-to-peak wave amplitude is measured vertically from crest to trough, whereas the wavelength is measured horizontally between adjacent crests or troughs (Figure 1.2). The number of crests (or troughs) passing by a specific point per second defines the wave frequency, which is directly related to the wavelength of a given waveform and the speed of that waveform propagating in the media. More specifically, the wave frequency equals the ratio of wave speed to wavelength. Some examples of waves propagating at very different frequencies and wavelengths within very different media include seismic waves (0.01–10 Hz) traveling through the earth, audible sound waves (20–20 000 Hz) traveling through the atmospheric air, and ultrasound (2–20 MHz) traveling through the body for biomedical imaging. In addition to the wave frequency, wave speed, and wavelength, other key parameters that are important to fully characterize an acoustic wave include particle displacement, particle velocity, absolute acoustic pressure, oscillating acoustic pressure, density fluctuations (as a result of oscillating pressure), acoustic attenuation, acoustic intensity and acoustic impedance. Since most of these parameters are interdependent, it is hard to manipulate one without affecting the others. For example, displacement, velocity, and pressure fields have the same waveforms with certain phase differences, where a variation in one field would directly alter the rest. Acoustic intensity, a measure of acoustic energy carried by a wave per unit area per unit time, is defined as the product of acoustic pressure and particle velocity. Similarly, acoustic impedance, a measure of the opposition a material offers to an acoustic wave propagation, is a product of the speed of sound and density of the medium.

1.3.2 Displacement, Velocity, and Pressure Fields

At the heart of acoustic wave propagation through a medium is the particle–particle interaction, where any particle of the medium transfers energy by simply interacting with its adjacent neighbor, triggering a displacement of the neighboring particle from its rest position. Although the acoustic wave causes local oscillations in the medium it travels through, the particles of the medium do not translocate with the wave (Figure 1.2). The particles' to-and-fro displacement over time is represented by a time-dependent particle velocity field depicting the acoustic wave. Oscillation of these particles with a certain velocity field leads to periodic regions of compression (high pressure) and rarefaction (low pressure) within a medium, which constitute a periodic pressure field that is 90° out-of-phase with the displacement field [34]. The phase difference between the pressure field and velocity field is 0° or 180° depending upon the reference coordinates and direction of wave propagation. That means, in a traveling wave, the particles having the highest or the lowest displacement at any given point correspond to the pressure value of zero, and vice versa. A period of displacement, velocity, or pressure field containing exactly one compression and one rarefaction region forms one complete wavelength of the acoustic wave. It is worth noting that an increase in the acoustic pressure or the particle velocity would result in an increased intensity of the acoustic wave, as both of these parameters work in perfect synchrony.

1.3.3 Wave Propagation

Just like an optical beam of light, acoustic waves can also exhibit phenomenon like reflection, refraction, and transmission, where the mechanical properties of the media, e.g. density, compressibility, and speed of sound, play a pivotal role in acoustic wave propagation, in contrast to the important optical properties of media, e.g. refractive indices, for light wave propagation. At an interface between two media of different mechanical properties, i.e. acoustic impedances, a part of the incident acoustic wave is reflected back, and a part is transmitted across the interface (Figure 1.4). The angles, proportions, wavelengths (or frequencies), and velocities of the reflected and transmitted waves depend upon the relative acoustic properties of the two media. The angles and proportions will be decided by Snell's law and acoustic impedances of media, respectively. The acoustic wavelength across an interface would vary based on the relative acoustic wave velocity within the respective medium.

1.3.4 Wave Dissipation

Imagine an acoustic wave at 2 kHz frequency traveling from air (sound velocity of ~330 m/s) to water (~1500 m/s) (Figure 1.4). The acoustic wavelength would change

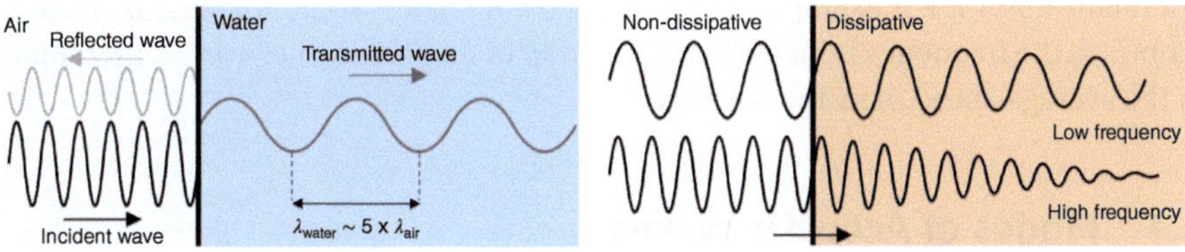

Figure 1.4 (left) Acoustic wave reflection and transmission at an air–water interface. (right) Transmission of an acoustic wave from a non-dissipative medium to a dissipative medium where the wave amplitude attenuates faster for high-frequency waves compared to low-frequency waves.

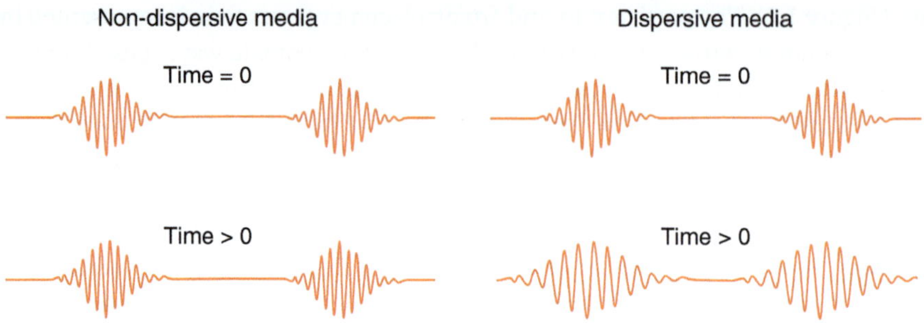

Figure 1.5 Nondispersive vs dispersive media.

from 165 to 750 mm across the air–water interface. A large portion of the wave amplitude will be reflected back into the air, and only a small portion will be transmitted into the water, which, in addition to ~5× longer wavelength, would make it extremely difficult to hear any sound underwater. Acoustic waves traveling through a viscoelastic dissipative medium such as water would also lose their energy due to viscous damping as the wave amplitude gradually attenuates over a length scale that is dependent on the acoustic wave frequency and medium viscosity (Figure 1.4). High-frequency waves attenuate faster in high-viscosity media, leading to shorter attenuation lengths, and vice versa. Since air viscosity is two orders of magnitude lower than that of water, it can sustain acoustic waves for longer distances for similar attenuation and frequency. Therefore, it is possible to hear low-frequency sounds through air over long distances compared to high-frequency sounds that dampen faster over short length scales.

1.3.5 Wave Dispersion

Acoustic wave velocity, or the speed of sound, stays constant for any range of frequencies in a nondispersive medium, but in a dispersive medium, it is frequency dependent (Figure 1.5). For instance, a sound waves packet composed of different frequencies can travel through the nondispersive media while maintaining a constant wave velocity for any given frequency and a consistent pulse shape over time. However, in a dispersive medium, high-frequency waves would travel faster than low-frequency waves; therefore, the acoustic pulse spreads out with time as it travels forward. The change in acoustic wave velocity is generally insignificant for most materials but not totally absent. For example, the speed of sound in air would increase by meager 0.1 m/s as the frequency is increased by an order of magnitude, i.e. <0.05% increment in the average speed of sound ~330 m/s.

1.4 Modes of Acoustic Waves

Acoustic waves can exist in several different modes, each with their own unique characteristics. They can be categorized based on the frequency range, wave propagation mode, or wave configuration.

1.4.1 Categorization Based on Frequency Range

Acoustic waves with frequencies in the audible range of 20 Hz to 20 kHz are known to human ears, which provides a good reference point to categorize the rest based on this frequency range (Figure 1.1). Low-frequency acoustic waves with frequencies below the audible range, i.e. <20 Hz, fall into the infrasound category. Avalanches, volcanoes, meteorites, and earthquakes are a few examples of infrasound-producing natural phenomena. Acoustic waves with frequency above the audible frequency range, i.e. >20 kHz, are usually referred to as ultrasound or ultrasonic. Ultrasonic waves are widely used in industrial settings, e.g. for cleaning parts located in difficult-to-reach regions where high-frequency acoustic waves remove the dirt, grease, and dust particles off the targeted surfaces. Moreover, ultrasonic waves have found a plethora of applications in biomedical sciences, including therapeutics, drug delivery and diagnostics (Figure 1.6). With their characteristic short wavelengths, ultrasonic waves imply a high degree of discrimination and a high concentration of energy; therefore, they can be used as a means of exploration, detection, actuation, etc. Starting beyond the audible frequency limit of 20 kHz, ultrasonic waves cover another three orders of magnitude higher frequencies beyond 10 MHz. These waves, with their broad range of frequencies, have been used for the improved delivery of drugs and genes to organs that are difficult to access, such as brain, for sono-dynamic and gene therapies, for thrombolysis and hemostasis, for healing bone fractures and treating osteoporosis, to mention a few. Acoustic waves having frequencies in GHz can be employed effectively in a variety of applications, including RF filters for 5G communication technologies, ultrawide band filters, and also in the components (such as actuators) of microelectromechanical systems. Bulk and SAW devices have been utilized at the RF front-end for a long time for filtering purposes where their operating frequencies are below 2.6 GHz range [36, 37]. Ultrasonic imaging, also known as ultrasound imaging or sonography, uses high-frequency sound waves to create images of the inside of the body. Also, ultrasonic sensors use sound waves at similar frequencies to measure distance, detect objects, and monitor fluid levels.

1.4.2 Categorization Based on Propagation Mode

Mechanical waves traveling at distances of 100s of kilometers during earthquakes to 10s of micrometers within microfluidic channels and miniaturized substrates demonstrate similar characteristics, which help in their categorization (Figure 1.7). Acoustic waves are categorized into following different types of waves based on particle movement in the medium and direction of wave propagation.

1.4.2.1 Longitudinal Waves

Longitudinal waves are the waves in which the particles of a medium vibrate in the same direction as the wave propagation. Material experiences periodic regions of compression and expansion as the waves travel forward (Figure 1.7). Longitudinal waves are also known as primary (P) compressional waves since they travel faster

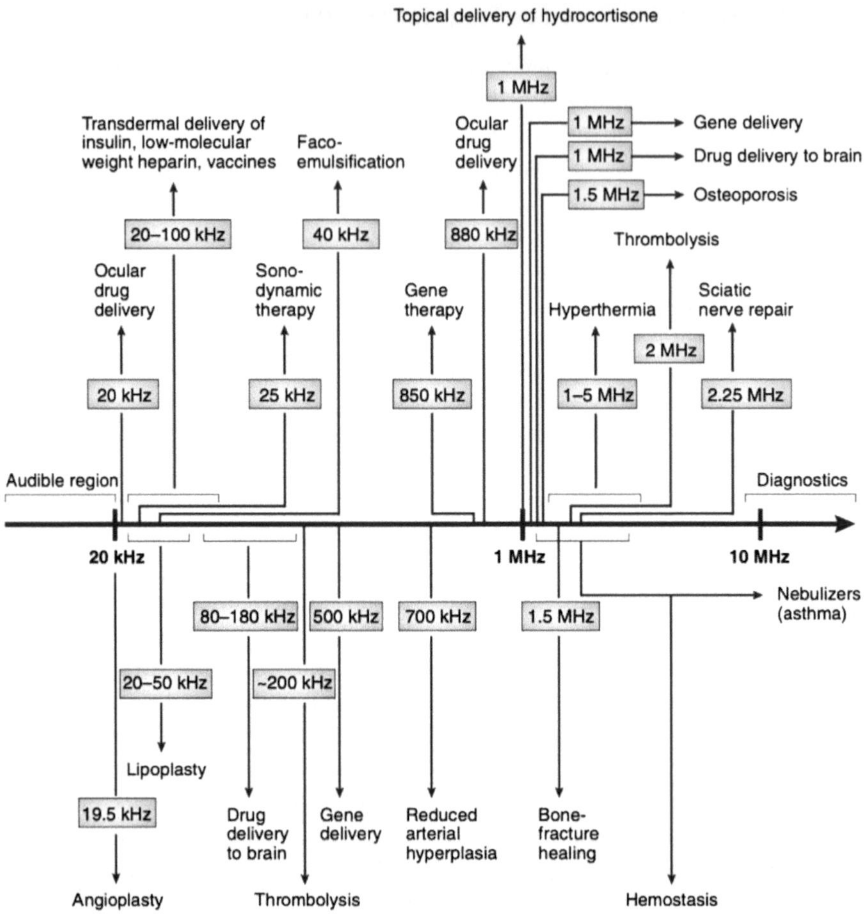

Figure 1.6 Summary of ultrasonic frequencies and their applications. Source: Reproduced from Mitragotri [35].

through the earth's crest than shear waves, which are also known as secondary (S) transverse waves (Figure 1.7a). A piezoelectric substrate sandwiched by a pair of electrodes can produce, on a much smaller scale, similar longitudinal or shear waves depending upon the orientation of piezo crystals, frequency of input current, thickness of the substrate, etc. (Figure 1.7b).

1.4.2.2 Shear Waves

Transverse or shear waves are waves in which the particles of a medium vibrate perpendicular to the direction of the wave propagation. This type of wave can propagate only in solids and not in liquids or gases since fluids cannot sustain a shear stress and only allow longitudinal waves to pass through. During an earthquake, the transverse S waves propagate with a velocity slower than the longitudinal P waves, arriving several seconds later than the longitudinal waves at

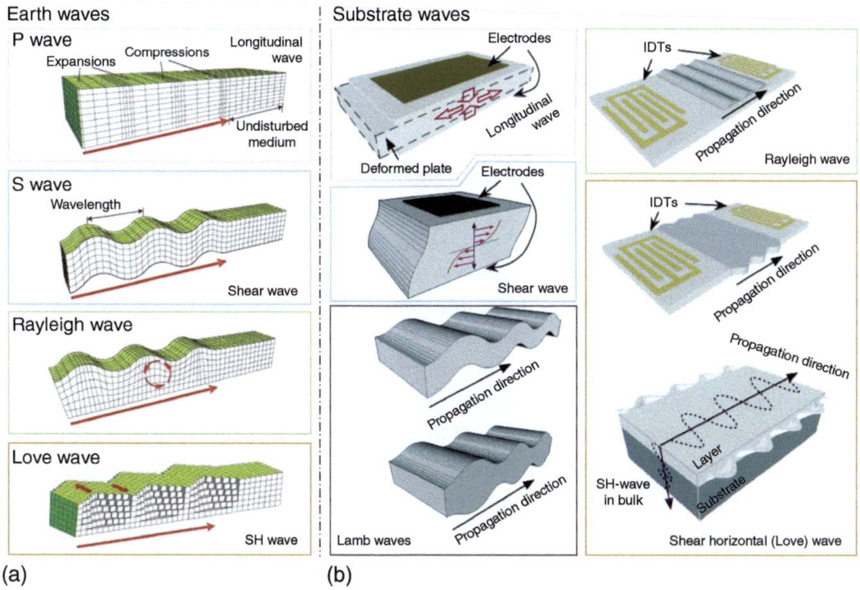

Figure 1.7 Categorization of waves based on propagation. (a) Propagation modes of earth waves. Source: Images courtesy: Copyright, Science Learning Hub - Pokapü Akoranga Pütaiao, University of Waikato [38]. (b) Wave propagation modes at a much smaller scale, i.e. within a substrate. Source: Fu et al. [39]/Reproduced with permission from Royal Society of Chemistry.

any given point. The same is true for a piezoelectric substrate, where longitudinal wave velocity is always greater than shear wave velocity.

1.4.2.3 Rayleigh Waves

Rayleigh waves, also known as surface waves, propagate across the surface of a medium where the particle displacements are both in the longitudinal and transverse directions. This renders a Rayleigh wave as a modal superposition of the longitudinal and shear waves. Due to these characteristics, the medium particles move in an elliptical orbit with velocity components parallel and normal to the wave propagation direction simultaneously. The amplitude of the particle displacement diminishes exponentially from the surface into the medium; therefore, the Rayleigh wave energy is mainly concentrated within a few wavelengths from the surface of the medium. These waves are mostly found in solids with wave velocity solely dependent on the material and the orientation of the crystals in the solid substrate. In liquids, a similar surface wave type is observed, i.e. water waves. However, the medium particles move in circular clockwise orbits in a rightward propagating water wave instead of elliptical orbits in Rayleigh wave, which switch from counterclockwise rotation to clockwise moving from the surface of a substrate to inwards of a substrate [34]. Rayleigh-type SAWs can be produced on specially oriented bulk piezoelectric substrates such as ST-cut and Y-cut quartz, 128° Y-X-cut lithium niobate ($LiNbO_3$), and X-112° Y-cut lithium tantalite ($LiTaO_3$) substrates.

Additionally, films made of zinc oxide, aluminum nitride, lead zirconate titanate, or $LiNbO_3$ that are vertically oriented (commonly referred to as *c*-plane or *z*-plane oriented) are also utilized to generate Rayleigh-type SAWs [39].

1.4.2.4 Love Waves

Love waves are horizontally polarized surface waves, which are named after Augustus Edward Hough Love, who first predicted them mathematically in 1911. These waves are responsible for horizontal shifting of the earth's crust during an earthquake and are only observed when the wave velocity in the upper layer is lower than the underlying sublayer. On a substrate, these shear-horizontal (SH) Love waves propagate through an isotropic thin layer sitting on a relatively thicker substrate and have a lower shear wave velocity compared to that of the substrate [40]. The isotropic layer on the substrate is usually much thinner than the wavelength of the SH Love waves, which are characterized by a unique horizontal shearing motion of the particles perpendicular to the direction of wave propagation and parallel to the device surface [39]. The SH Love waves can be produced on specially oriented bulk piezoelectric substrates, such as ST-cut quartz, 64° Y-X-cut $LiNbO_3$, and 36° Y-X-cut $LiTaO_3$.

1.4.2.5 Lamb Waves

Lamb waves are produced in a wave guide sandwiched between two parallel surfaces, such as the top and bottom surfaces of a plate or a substrate; therefore, they are also known as guided waves or plate waves [41]. The Lamb waves can propagate in a symmetric or an antisymmetric mode, where motion on the top surface is a mirror image of the bottom surface about the midplane of the plate or both surfaces are moving exactly in sync, respectively. Contrary to the Rayleigh waves, Lamb waves produce stresses throughout the plate thickness. The motion of any given particle within the plate vibrating in Lamb wave mode is always in the plane defined both by the surface normal and the wave propagation direction. The number of modes that each form of Lamb wave can achieve is unlimited. Frequency of the wave, angle of incidence, and material are some of the variables that affect the modes. The relation of frequency to wafer thickness determines the Lamb wave velocity. Because Lamb waves are extremely dispersive, their velocities change with frequency. Lamb waves can be used to solve a wide range of problems involving the identification of subsurface discontinuities for NDT since they can flow within thin plates.

1.4.3 Categorization Based on Wave Configuration

1.4.3.1 Traveling Waves

One can observe the crest of the wave moving along with the particles as it travels. After this crest, there is a trough, which is then followed by the next crest. In fact, one would identify a clear wave pattern moving through the medium in the shape of a sine wave. This sine wave pattern moves continuously until it encounters another wave in the same direction or until it changes the medium that have different property. This type of wave form that is observed traveling through a medium is described as a traveling wave (Figure 1.8).

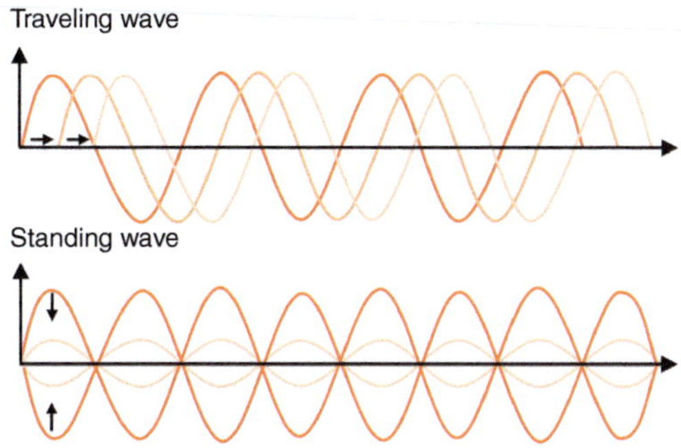

Figure 1.8 Traveling waves vs standing waves.

1.4.3.2 Standing Waves

Another type of wave pattern which is generated when reflected waves from one end of the medium interfere with incident waves from the source is defined as standing waves. Because of this interference, certain locations along the medium seem to be at a standstill. The pattern is commonly referred to as a standing wave pattern because the observed wave pattern is marked by points that appear to be standing still.

1.5 Acoustic Wave Propagation and Interaction

1.5.1 Transmission and Reflection of Acoustic Waves

When an acoustic wave encounters a boundary between two different media, a portion of the incident wave is reflected while the rest is transmitted through the boundary. The transmitted waves propagate through the interface from one medium to another, whereas the reflected waves remain in the same medium as the incident wave but propagate in the opposite direction. The ratio of the transmitted and reflected waves depends on the acoustic impedance values of the two media that form a boundary and the incident angle of the waves with reference to the normal to the boundary. The acoustic impedance (Z) of a medium is one of the fundamental physical properties of any medium and is defined as the product of its density (ρ) and its speed of sound (c), such that $Z = \rho c$. This specific property of the medium plays a crucial role in the reflection and transmission at boundaries between different media. When an acoustic wave travels from one medium with $Z_1 = \rho_1 c_1$ to another medium with $Z_2 = \rho_2 c_2$, the wave intensity and pressure amplitude of the incident, the transmitted and reflected waves, are determined by Z_1 and Z_2. The pressure transmission ($\boldsymbol{T}_P$) and reflection ($\boldsymbol{R}_P$) coefficients can be defined as the ratios of complex pressure amplitude of the transmitted ($\boldsymbol{P}_t$) and reflected ($\boldsymbol{P}_r$) waves to those of the incident wave ($\boldsymbol{P}_i$), such that $\boldsymbol{T}_P = \boldsymbol{P}_t/\boldsymbol{P}_i$ and $\boldsymbol{R}_P = \boldsymbol{P}_r/\boldsymbol{P}_i$, respectively. For harmonic plane traveling waves, the wave intensity (I) can be expressed as $P^2/2Z$, and the intensity transmission (T_I) and reflection (T_R) coefficients are defined as $T_I = (Z_1/Z_2)|\boldsymbol{T}_P|^2$ and $T_R = |\boldsymbol{R}_P|^2$, respectively [42, 43]. For oblique incidence of the waves with an incident angle of θ_i, the transmitted and reflected waves propagate at angles of θ_t and θ_r, respectively (Figure 1.9). The

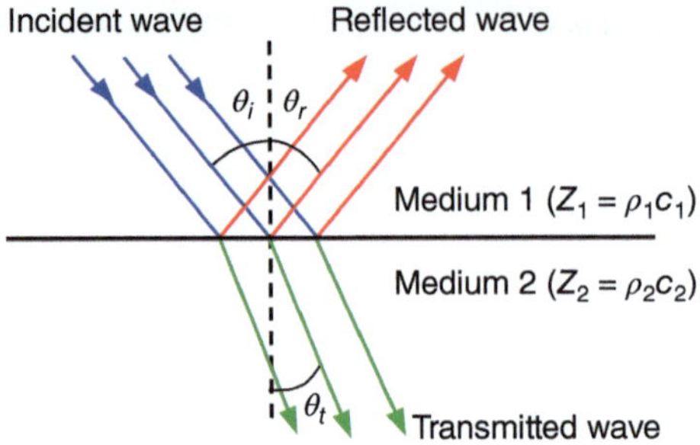

Figure 1.9 Wave reflection and transmission.

reflection and transmission of acoustic waves can be described mathematically using Snell's law, such that $\sin \theta_i / c_1 = \sin \theta_t / c_2$. The refraction angle, at which the transmitted waves travel through the new medium, is determined by the speeds of sound in two media and the incident angle (θ_i). Wave transmission and reflection play an important role in various applications, including acoustic imaging, which will be delineated in Chapters 4 and 5.

1.5.2 Acoustic Scattering

Acoustic scattering refers to the phenomenon in which the incident waves are redirected from the original propagation path when they encounter an obstacle that serves as a scatterer in the wave propagation process. The wave scattering phenomenon is influenced by many physical properties of the object and the acoustic wave, including size, shape, and acoustic properties of the object, and the frequency, amplitude, and polarization of the acoustic wave. Consider an object suspended in a fluid that is subject to an acoustic field, then the incoming waves are scattered off the object. A pressure field forms when vibration is introduced into a fluid medium. Three unique but interrelated scenarios result from the existence of an object in this vibrating medium: first, the interaction of the incident acoustic wave with the object; second, the scattering of the wave from the object; and third, the transmission of the wave into the object. These situations affect the pressure field based on the mechanical characteristics of the object placed in the wave path (i.e. size, shape, density, compressibility, and acoustic impedance). There is a net resultant force from these interactions with the acoustic waves, which is known as ARF. When wave scattering phenomena occur at solid–fluid interface, the wave exponentially decays while propagating in the fluid. In practice, this attenuation is comparatively small for solid–gas boundaries. The wave scattering phenomenon can be described by a nondimensional parameter called the Helmholtz number (ka), which is defined as the product of the wave number (k) and radius (a) when a spherical object (the scatterer) is placed at an infinitely large distance from it (in the far field), as shown in Figure 1.10a. The Helmholtz number can be interpreted as the nondimensional object size compared to the acoustic wavelength, and the three scattering regimes can be categorized depending on ka: Rayleigh scattering ($ka \ll 1$), Mie scattering ($ka \approx 1$), and geometrical scattering ($ka \gg 1$), as shown in Figure 1.10b. When the size of the object is much smaller than the wavelength

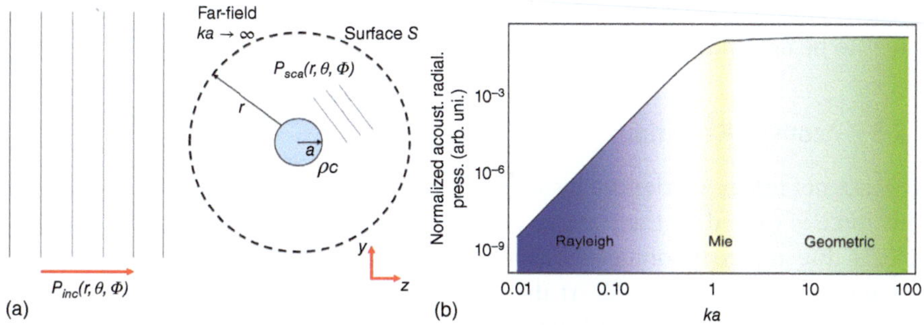

Figure 1.10 (a) Schematics of the acoustic scattering of an incident pressure field (P_{inc}) by a fluid sphere of radius (a), with internal (P_{int}) and external pressure (P_{Scat}) fields. (b) Representation of Rayleigh, Mie-like, and geometric scattering regimes when different values of the size factor ka for the air–water acoustic scattering were presented against normalized acoustic radiation pressure. Source: Pessôa and Neves [44]/Reproduced with permission from Elsevier.

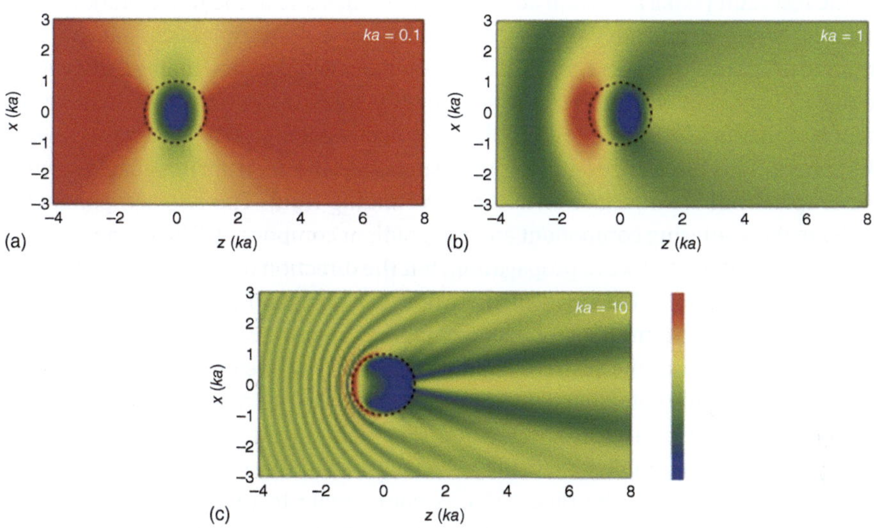

Figure 1.11 Total normalized pressure field. Scattering regimes for different *ka* values (*ka* = 0.01, *ka* = 1, and *ka* = 10). Source: Reproduced with permission from Pessôa and Neves [44], ELSEVIER.

(i.e. $ka < 1$) then the acoustic wave scattering is considered spherically isotropic scattering; however, when $ka \gg 1$, the scattering is no longer considered isotropic, and a net momentum transfer to the sphere occurs as backscattering dominates. The ARF can be assessed using a far-field examination of the scattering.

Figure 1.11 indicates an acoustic scattering of a water droplet of size factor ka in air. For small droplets with $ka = 0.01$, the scattering is in the Rayleigh regime, characterized by a minimum acoustic pressure intensity. For larger droplets with $ka = 1$, the scattering is in the Mie regime and can result in the formation of a high-pressure intensity zone. For even larger droplets with $ka = 10$, the Mie scattering regime continues to be dominant.

1.5.3 Acoustic Radiation

ARF is a net force experienced by an object under the influence of an acoustic field. This force is caused by the transfer of wave momentum to scattering or absorbing objects due to an inhomogeneous wave scattering or reflection off the objects exposed to the acoustic field. At the interface, the acoustic wave can be partially reflected and transmitted, and these reflections and transmissions can contribute to originate the radiation force. The ARF can be discussed in two different ways depending on whether waves are traveling or standing. ARF on the object due to traveling acoustic waves is attributed to the anisotropic scattering of waves, but the force that arises due to standing acoustic waves is subject to the existence of pressure nodes and antinodes. When a solid sphere (elastic) is placed in the path of the wave, the force which is attributed to the scattering of the primary wave mainly depends on the resonant peaks at certain acoustic frequencies due to the free vibration of the spheres. More specifically, an inhomogeneous wave scattering occurs at the solid interface. On the other hand, when the fluid sphere (droplets and bubbles) is exposed to the acoustic scattering waves, the sphere suspended in the immiscible fluid experiences a distinct force due to the inhomogeneous geometrical wave scattering off the sphere. This force arises due to the mismatch in the acoustic impedances across the droplets/bubble interface with the surrounding fluid. The two components of ARF are the scattering component and the gradient component. The former component is directed toward wave propagation, but the direction of latter is perpendicular to wave propagation. The ARF in the Mie, Rayleigh, and geometrical scattering regimes varies by a number of factors, including the acoustic wave frequency, material properties of the scatterer, and the geometry of the object. The strength of the ARF is generally considered to increase in the order of Rayleigh scattering < Mie scattering < geometrical scattering regime. Rayleigh scattering occurs when the object size is much smaller than the wavelength of the acoustic wave, while Mie scattering occurs when the object size is comparable to the wavelength. In geometrical scattering, the object size is much larger than the wavelength, leading to a stronger ARF.

There has been much effort to investigate the ARF both experimentally and theoretically. Some of the milestone studies include King [21] and Maidanik [45] for rigid spheres, Hasegawa and Yosioka [23] and Tahmasebipour et al. [46] for elastic spheres, and Maidanik [45] and Yosioka and Kawasima [22] for fluid spheres. Interestingly, the most widely used polymer microspheres for biomedical applications, made of polystyrene (PS), polymethyl methacrylate (PMMA), polylactic acid (PLA), and polylactide-*co*-glycolide (PLGA), exhibit elastic behaviors when exposed to an acoustic field. The elastic microspheres experience an additional force that can be characterized by resonant peaks at certain acoustic frequencies due to the free vibration of the spheres [47, 48]. The available theories and their connections are summarized in Figure 1.12.

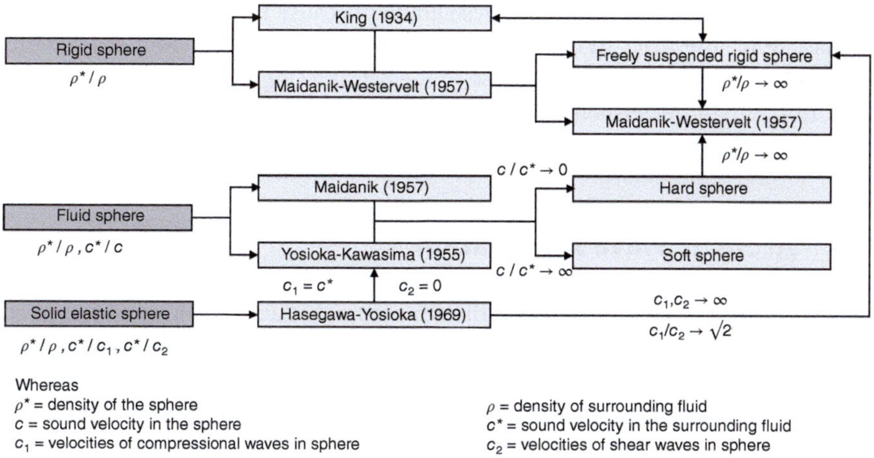

Figure 1.12 Different theories on wave scattering and their connection. Source: Hasegawa [49]/Reproduced with permission from Acoustical Society of America.

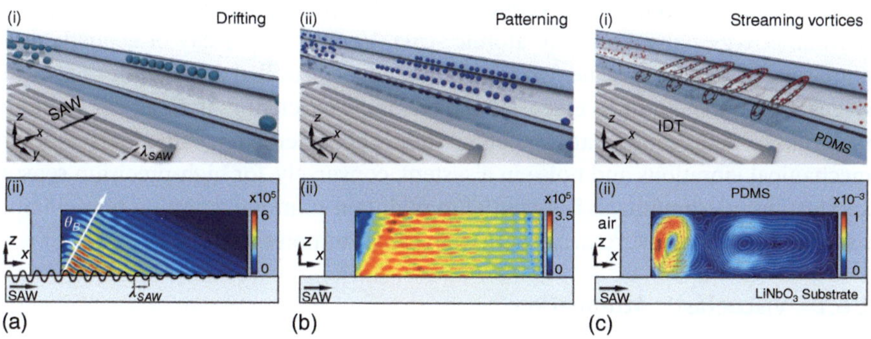

Figure 1.13 Migration, patterning, and swirling behavior of particles in the microfluidic channel. (a) particle drifting mechanism and illustration of acoustic wave propagation in the fluid. (b) particle patterning mechanism and simulated time-averaged absolute pressure field for particle patterning. (c) particle acoustic streaming flow mechanism and simulated streaming filed. Source: Fakhfouri et al. [50]/Reproduced with permission from Royal Society of Chemistry.

Acoustophoresis, defined as the manipulation of objects using acoustic forces, is widely utilized in various microfluidics applications such as manipulation of microparticles [47, 50], cells [51], droplets [52], and bubbles [53]. In a typical acoustically driven system, suspended objects are subject to steady-state acoustic forces that are generated by two main sources. Firstly, the presence of objects in a sound field causes scattering, which in turn produces ARF, which works to drive objects in a specific direction. Secondly, the resulting gradient of acoustic field provides a steady-state flow known as acoustic streaming flow (ASF). As soon as the wave reaches the edge of microfluidic microchannel, the wave refracted into the fluid domain within the microchannel in the form of longitudinal wave at the Rayleigh angle. The phenomenon of ASF results from gradients in the sound field as it travels through a viscous fluid. These gradients cause time-averaged body forces to operate on the fluid. When the two mechanisms, ARF and ASF, arise within the fluid domain, the former give rise to the drifting (Figure 1.13a) and patterning (Figure 1.13b) of the suspended objects, whereas the latter give rise to the swirling behavior of the moving particles (Figure 1.13c) [50].

1.6 Acoustic Wave Attenuation

Loudspeakers in a cinema hall or a theater are not located only around the screen like miniaturized speakers in a cell phone or a tablet but distributed systematically throughout the large space to ensure uniform acoustic pressure for all the seats. This is to avoid uncomfortably high acoustic pressures for the front rows and diminished sound quality and intensity for the seats at the rear side of the hall. Nonuniform sound intensity in a large space can be explained by the attenuation of acoustic waves over long distances.

Attenuation or damping in a vibrating system can be defined as an irreversible reduction in amplitude of the oscillations. A dissipation of vibrational energy into thermal energy happens at the atomic scale, leading to a decay of acoustic wave amplitude at the macroscale. Understanding of a controlled dissipation of acoustic waves within a medium is essential to developing efficient materials. For example, materials with high attenuation coefficients are considered favorable for isolation of mechanical vibrations and noise in a system; conversely, for high-precision sensors and testing instruments, it is essential to maintain minimal damping of oscillating signals to gather measurements with high signal-to-noise ratio [54].

1.6.1 Viscoelastic Attenuation

Acoustic waves can attenuate by losing some of their amplitude at an interface due to reflection, transmission, or refraction of waves as described in Section 1.5.1. Moreover, a continuous attenuation of waves can occur within a medium due to intrinsic wave dissipation because of viscous, elastic, or viscoelastic losses. An elastic material under stress follows Hooke's law, i.e. deformation or strain in the material is directly proportional to the applied stress, and when the stress is removed, the strain also disappears without permanently altering the geometry of the material (Figure 1.14)

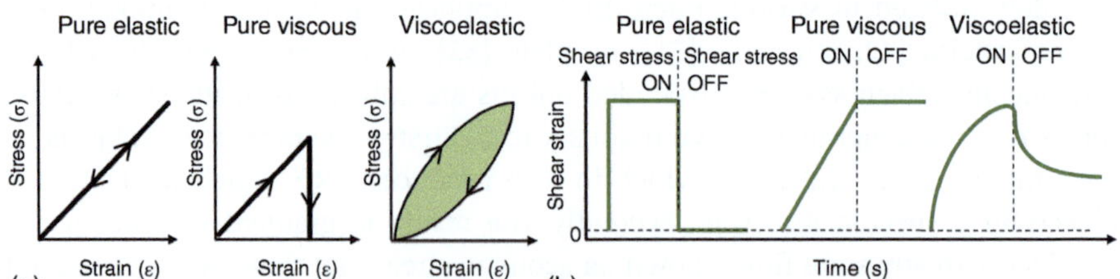

Figure 1.14 (a) Stress–strain diagram for elastic and viscoelastic materials. (b) Shear strain (compliance) over time under the applied shear stress to different material types.

[54]. In an ideal scenario, no energy is lost during the loading and unloading of the material. On the other hand, if the material is purely viscous, all the applied energy will be lost to the viscous dissipation and the shear strain will remain even when the applied stress is removed. Here, the stress–strain ratio determines the viscosity coefficient (μ) of the material, which means the viscous damping of acoustic waves would be higher in high-viscosity fluids. A combination of elastic and viscous behaviors is used to explain the viscoelastic materials, where some of the applied energy to induce strain will be lost during the material loading, and the material will return to its original shape with a hysteresis when the stress is removed. In every loading and unloading cycle, some of the vibration energy will be damped/absorbed by material.

1.6.2 Acousto-Thermal Heating

Under periodic stress in the form of high- frequency acoustic waves or vibrations, 10^3–10^6 of loading/unloading cycles per second can induce significant acousto-thermal heating of the viscoelastic material and high temperatures [54, 55]. The conversion of mechanical energy in ultrasonic waves to heat over time poses a critical challenge in systems where a constant temperature within the medium is essential to the intended operation. For example, viability of biological samples in the form of cells or proteins can be severely compromised within the acousto-microfluidic devices if the thermal energy produced during the acoustic wave dissipation is not timely removed using a constant fluidic flow or an active Peltier cooling element [56, 57]. However, in some cases, instead of actively removing acousto-thermal heating from the system, it can also be used to our advantage for various applications such as droplet sorting (marogami effect), thermochromic display and microheater/reactor [58–63].

1.6.3 Acoustic Streaming Flow

Propagation and attenuation of acoustic waves within a medium can lead to a bulk fluid flow, in the form of Eckart, Rayleigh, or bubble-driven streaming (Figure 1.15), with a non-zero average velocity that depends upon the transducer configuration, wave frequency, wave amplitude, dimensions of the fluidic domain, fluid viscosity, etc.

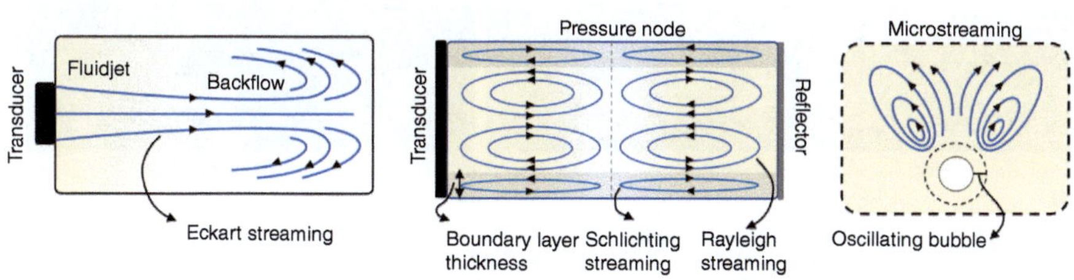

Figure 1.15 Acoustic streaming flows. Source: Wiklund et al. [64]/Reproduced with permission from Royal Society of Chemistry.

1.6.3.1 Eckart Streaming

Acoustic wave attenuation within a viscous fluid leads to a gradual reduction of the wave amplitude and formation of a pressure gradient in the direction of wave propagation, which generates a velocity field in the form of an acoustic streaming flow or Eckart streaming (Figure 1.15(left)) [12]. Eckart streaming is more pronounced when the attenuation length scale within the fluidic medium (L) is significantly larger than the acoustic wavelength (λ), i.e. $L \gg \lambda$. The attenuation length ($L \propto 1/\mu f^2$) is inversely proportional to the medium viscosity (μ) and acoustic frequency (f) squared. A single piezoelectric transducer, producing mainly traveling waves within a medium without significant wave reflection and standing wave formation, is ideal to generate Eckart streaming flow provided enough attenuation length is reached. Standing wave formation within the fluidic medium due to strong wave reflections at the boundaries could suppress the Eckart streaming flow. For example, a single 4 MHz ultrasonic transducer radiating traveling acoustic waves ($\lambda \approx 325$ μm) within a fluidic medium with maximum possible attenuation length of $L \approx 2$ mm (channel width, $L \gg \lambda$) produces Eckart streaming flow when the acoustic wave reflection from the planner PS reflector is minimal (Figure 1.16). However, when the waves are strongly reflected by a molybdenum reflector, they interfere with the incident traveling waves to form a standing acoustic field, which suppresses the Eckart streaming flow formation and concentrates the tracing particles at the standing waves pressure nodes.

1.6.3.2 Rayleigh Streaming

Rayleigh streaming flow is produced within standing waves resonators with acoustic pressure nodes and anti-nodes uniformly spaced at quarter wavelengths across the fluidic domain (Figure 1.15(center)). Due to the viscous attenuation at the boundaries of the acoustic resonator, a momentum flux between the node and anti-nodes of a standing wave results in an inner boundary layer flow also known as "Schlichting streaming" [11]. The Schlichting streaming, confined within a very thin boundary layer, triggers a Rayleigh streaming flow in the form of larger vortices spanning across the bulk of the medium (Figure 1.17) [10]. The boundary layer thickness

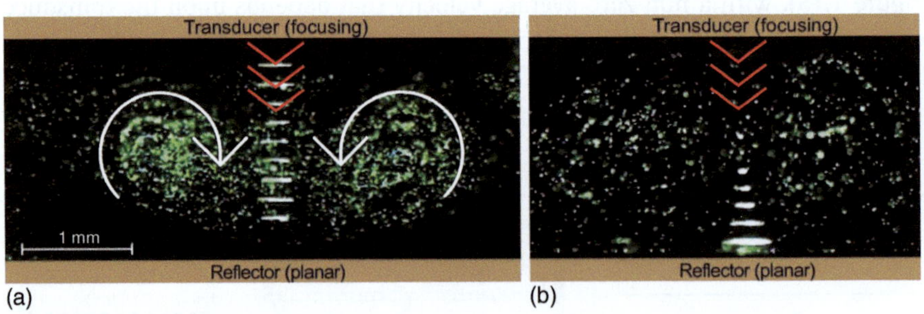

Figure 1.16 (a) Eckart streaming flow in the presence of weak wave reflector (polystyrene). (b) Streaming flow is suppressed, and standing waves are formed in the presence of a strong wave reflector (molybdenum). Source: Reproduced with permission from Wiklund et al. [64], Royal Society of Chemistry.

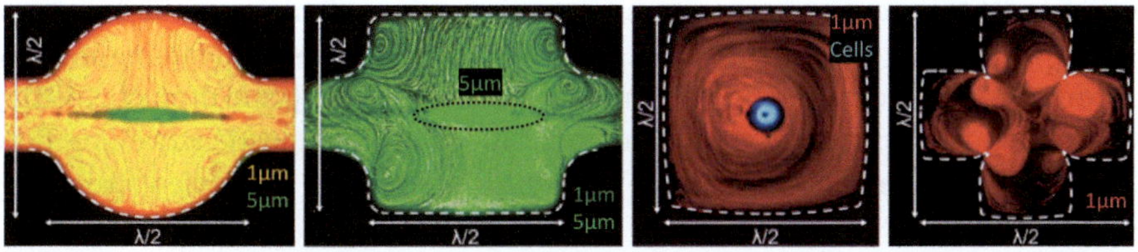

Figure 1.17 Rayleigh streaming patterns within standing wave resonators of different geometries. Source: Reproduced with permission from Wiklund et al. [64], Royal Society of Chemistry.

($\delta = \sqrt{\mu/\pi\rho f}$) provides a key reference dimension to identify Rayleigh streaming within a channel with characteristic length L; the following condition should hold to confirm the Rayleigh streaming flow: $\delta \ll L \ll \lambda$ [64]. For example, acoustic streaming patterns within standing wave resonators of different shapes ($L = \lambda/2$) can be easily traced with 1 µm microparticles, which move with the streaming flow contrary to the larger particles that are concentrated at the pressure nodes in the center due to the direct ARF (Figure 1.17). The boundary layer thickness (δ) for high-frequency O (MHz) acoustic waves is O (1 µm), which fulfills the condition $\delta \ll L(\lambda/2) \ll \lambda$ to confirm the Rayleigh streaming flow.

1.6.3.3 Bubble-Driven Microstreaming

Bubble-driven acoustic microstreaming occurs when a gas bubble trapped under in a liquid is subjected to high-frequency sound waves (Figure 1.15(right)). The bubble experiences a period of compression and rarefaction due to the alternating pressure of the acoustic waves, which forces the bubbles to decrease or expand in size, respectively. This cyclic expansion and contraction of the bubbles leads to the generation of localized flows or microstreaming around the bubbles. The resulting microstreaming depends on the frequency and amplitude of the acoustic waves, the size of the bubble, and the properties (density and viscosity) of the liquid surrounding the bubble. The frequency of the acoustic wave must be tuned for a strong actuation of the bubble for enhanced microstreaming generation, where the optimum frequency relates inversely with the bubble size. Bubble-driven acoustic microstreaming has found important applications in various fields, including microfluidics, drug delivery, and tissue engineering.

1.6.3.4 Applications of Acoustic Streaming Flow

Acoustic streaming flow has a variety of applications, including mixing and pumping fluids within microfluidic platforms, enhanced chemical processing in industrial reactors, cleaning labware, dental equipment and jewelry. For example, acoustic streaming is used to mix fluids within volumes that are difficult to mix otherwise [65]. Acoustic waves are further used to mix and manipulate even femtoliter-volume droplets within nanoslits in controllable manner [66]. Similar acoustic streaming effects have been used for particle or cell aggregation in sessile droplets [67–69]. Beyond acoustofluidic mixing of fluids, a high-power actuation of the sessile droplets could lead to jetting [70], translation [71], splitting [72], and atomization [73] of the droplets. Within confined microchannels, active and rapid mixing of fluids in laminar flow regime [74, 75], sharp-edge and bubble-based acoustic streaming-induced mixing [76, 77] and pumping [78], nanoparticle focusing [79], and droplet pipetting or splitting [80] are other biomedical applications of the acoustic streaming flow. For more examples, the readers are encouraged to refer to the review articles by Connacher et al. [81] and Wu [82].

1.7 Generation and Propagation of Acoustic Waves

1.7.1 Acoustic Waves Generation in Nature

From communication to echolocation and from self-defense to preying on other animals, many living creatures in nature have evolved to generate and sense acoustic or sound waves of varying frequencies and intensities using only their bodily parts. For example, whales produce acoustic waves by moving air between their sinuses in the head, and they use oil-filled channel in their lower jaws to sense the reflected or echoed sound waves and conduct those to the middle ear for hearing (Figure 1.18a) [83]. Similarly, African elephants produce high-amplitude and low-frequency sound waves using their huge vocal cords. The infrasound travels through the ground and air to distant elephants, who hear it using their feet and ears in a way analogous to seeing and hearing lightning and thunder with a short delay time (Figure 1.18b) [84–86].

Similarly, we humans generate and sense sounds for communication, sharing feelings, expressing emotions by talking, singing, laughing, crying, screaming, shouting, humming, yelling, etc. Our bodies have evolved with an advanced mechanism, which incorporates three essential organs, lungs, vocal cords, and articulators, to perform these functions [87]. The lungs are working as a pump creating pressure to flow air through the trachea. Pressurized air flow vibrates the vocal folds to generate audible acoustic waves, where the pitch and tone of the waves can be adjusted by flexing the muscles of the larynx (epiglottis, supraglottic, vocal cord, etc.). The articulators above the larynx, such as tongue, palate, cheek, and lips, convert the sound waves produced by the vocal cords into speech or other sound expressions [88–90].

1.7.2 Generation of Acoustic Waves in Lab

1.7.2.1 Lower-Frequency Acoustic Waves

Acoustic waves are generated by any vibrating system such as vibrating strings of a guitar, diaphragm of a drum, and air within a resonant cavity of a flute. In industrial settings, vibrations are widely produced by an eccentric rotating mass (ERM)-based

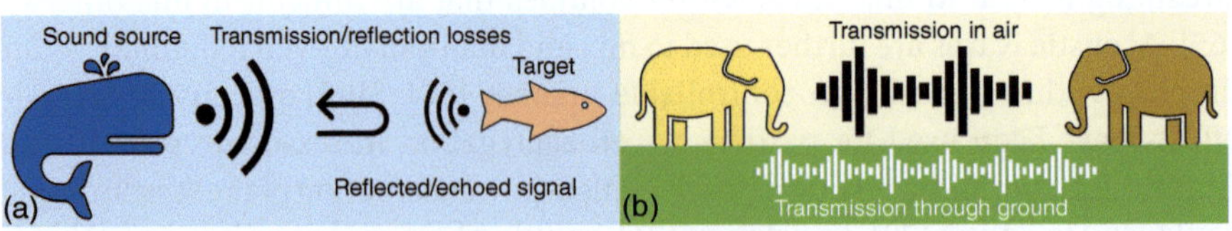

Figure 1.18 Echolocation examples for predator/prey (a) and elephant communication (b). Source: Adapted from Madsen and Surlykke [83].

DC (direct current) motor with an unbalanced mass placed on the motor shaft. Harmonic vibrations with tunable frequency and amplitude are generated by varying the angular speed and the offset of the ERM from the center of rotation. Alternatively, linear resonant actuators (LRAs) can generate an oscillating force across a single axis, resulting in low-frequency vibrations [91]. A moving mass with a "voice coil" wrapped around a magnet on one side and a "wave spring" on the other side for balancing oscillates under resonant-frequency AC excitation. The frequency and amplitude of the vibrations produced by LRA can be adjusted by changing the AC (alternative current) input. The vibrations coming out of the ERM- and LRA-based systems can be construed as acoustic waves with frequencies in hundreds of hertz. These vibratory systems are quite useful in haptic modules on mobile devices or for low-frequency manipulation of living objects by Faraday waves or ripples on a liquid-free surface. However, these low-frequency oscillators cannot generate ultrasonic waves required for most biomedical applications [92].

1.7.2.2 Piezoelectricity and High-Frequency Wave Generation

Since mechanical oscillators cannot reach ultrasonic frequencies, alternate methods have been sought after for decades, where a solution is found in piezoelectric materials to generate high-frequency acoustic waves. Piezoelectric ceramics can convert mechanical stresses (compression, tension, or shear) into electrical potential energy, and vice versa, without the need for any additional moving component. A voltage difference is produced across two electrodes sandwiching a piezo crystal going through a mechanical deformation due to the piezoelectric effect (Figure 1.19). Conversely, an applied voltage across those electrodes can deform the piezo crystal due to inverse piezoelectric effect. If the applied voltage potential is in the form of an alternating current of certain frequency, the deformation of the piezo element will turn in to oscillations of similar frequency. Since AC signals can be provided at a broad range of frequencies using a broad-range RF signal generator, piezoelectric elements can be made to oscillate and produce acoustic waves over several orders of magnitude frequencies (Hz–GHz). In addition to generating acoustic waves, piezoelectric or inverse piezoelectric effects can also be utilized for sensing or energy harvesting applications.

Piezoelectric materials have been used for a long time in a variety of fields such as acoustic detection of submarines [93], ultrasonic imaging of tissues, minimally invasive piezoelectric surgery [94, 95], industrial welding and cleaning by sandwiched

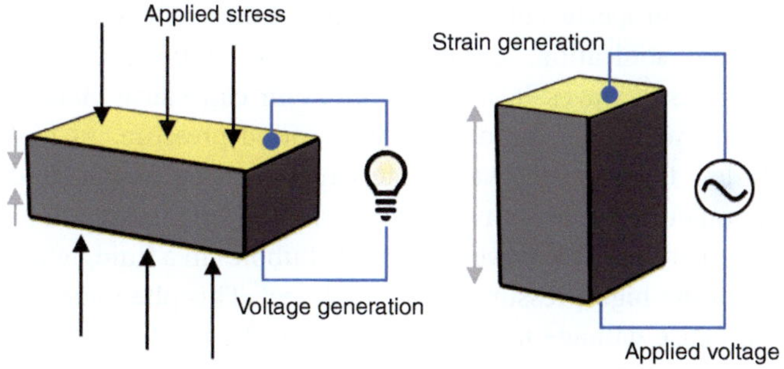

Figure 1.19 Two different ways of working principle for piezoelectric ceramics.

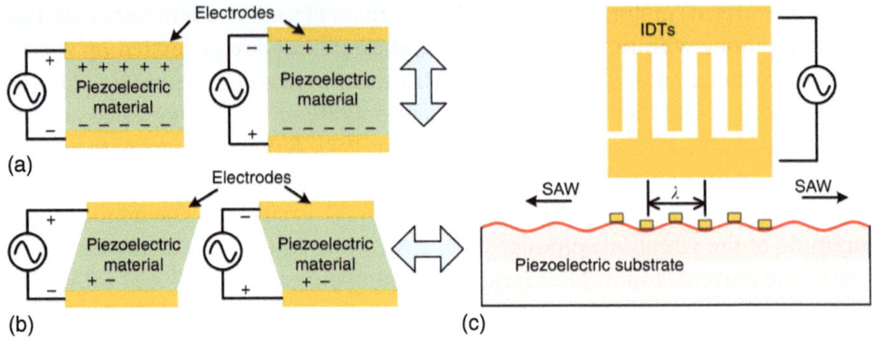

Figure 1.20 (a) Thickness mode vibrations. (b) Shear mode vibrations. (c) Surface acoustic waves (SAWs) produced by an interdigitated transducer (IDT). Source: Wu et al. [99]/Springer Nature/Licensed under CC BY 4.0.

Langevin-type high-power transducers [96], nondestructive and noncontact sensing [97], or energy harvesting system mobilized on movable devices to charge batteries [98]. Most of the early applications were based on BAW-based platforms for relatively low-frequency (kHz–MHz) requirements (Figure 1.20) [13]. The piezoelectric elements can be actuated by a thickness or shear-mode BAW depending on the crystal orientation, whereas the resonance frequency depends on the thickness of the element. It is possible to vibrate bulk of the piezoelectric element at very high frequencies and higher resonance modes; however, it is relatively less energy-efficient method. With the advent of interdigital transducers (IDTs), it became possible to generate acoustic waves at very high frequencies (MHz–GHz) in an energy-efficient manner. The IDTs are composed of thin-film metal electrodes placed periodically on the piezoelectric substrate surface and are able to generate high- frequency SAWs in an energy-efficient fashion as the waves are mainly concentrated at the surface of the substrate. SAW devices are extensively used in electronics industry for signal processing and cover a very broad band of frequencies from 10 MHz to 1 GHz. However, over past two decades, SAW devices have found themselves a new place in microfluidics to actively manipulate fluid and biological samples [100–103].

1.8 Acoustic Waves Effects in Fluidic Media

Acoustic waves in fluidic media can have a variety of effects depending on the frequency, amplitude, and the properties of the fluid. Some of these effects include acoustic streaming, mixing, sonication, cavitation, heating, atomization, and acoustic levitation. The acoustic waves in a fluidic medium can cause changes in the pressure of the fluid, which can be measured as sound pressure. At high frequencies and amplitudes, the acoustic waves can cause heating of the fluid, which can be used for therapeutic applications such as hyperthermia treatment for cancer [104]. The high-intensity acoustic waves can create bubbles in a fluid, which can collapse rapidly, generating high-pressure shocks and heat. This phenomenon, known as cavitation, can cause damage to materials and can be used for various applications such as cleaning, mixing, and sonochemistry [105]. In microfluidic systems, acoustic streaming can be used to generate and control flow patterns, and manipulate particulate matters, including particles, cells, bubbles, and droplets. The acoustic techniques have been also widely utilized for lab-on-a-chip applications for cell sorting and separation, and chemical synthesis [106]. Acoustic waves can be used to levitate small particles and droplets in a fluid by producing an ARF [100]. This effect can be used for various applications, such as material processing and sample handling. Acoustic waves can also be used to mix fluids in laboratory applications by creating vortices and other flow patterns [107].

1.8.1 Vibrating Membranes and Sharp-Edge Structures

Upon submersion of specially made microstructures in the fluid, such as membrane and sharp-edge structures, and subjecting them to acoustic excitation, vortical flows or bubbles can be generated and controlled. The vibration of the microstructures creates pressure waves in the fluid, which can cause microscale fluid flows. The direction and magnitude of the flow can be controlled by manipulating the vibration frequency, amplitude, and geometry of the membrane or structure. Utilizing the vibrations of thin membranes and sharp-edged structures to produce acoustic streaming effects is a widely used method for controlling liquids and droplets in microfluidic systems. As explained in Section 1.6.3, acoustic streaming is a powerful tool for manipulating fluids and droplets at the microscale, where traditional fluid handling methods may not be feasible. The interaction between the acoustic wave and the fluid generates pressure variations that induce fluid motion in the form of acoustic streaming. The streaming effects utilizing these structures are dependent on the sharp-edge geometry, the frequency and amplitude of the applied signal, and the eigenmodes of the glass substrate. This streaming effect results in the formation of a localized vortex. Figure 1.21A,B shows an acoustofluidic device with sharp-edge structures that are utilized to achieve various flow operations. When the piezoelectric transducer is turned on, acoustic streaming patterns form at the tips of the oscillating sharp-edge structures. Figure 1.21C shows droplet manipulation via acoustic streaming generated by a vibrating membrane. Droplets can travel away from the membrane borders by using the vortex flow. Therefore, precise alignment of the membrane in relation to the streamlines results in the displacement of droplets from the edges of the membrane.

1.8.2 Oscillating Bubbles

Oscillating bubbles in microfluidics refer to the phenomenon of gas bubbles that periodically change in size and shape in a confined liquid environment. This is due to the interaction of the bubble with the liquid and the walls of the microfluidic channel. Oscillating bubbles can be generated by subjecting a fluid to a high-frequency vibration, which creates pressure waves in the fluid. These pressure waves can cause a bubble to form and begin to oscillate. The frequency, amplitude, and phase of the applied acoustic force can be used to control the oscillation of

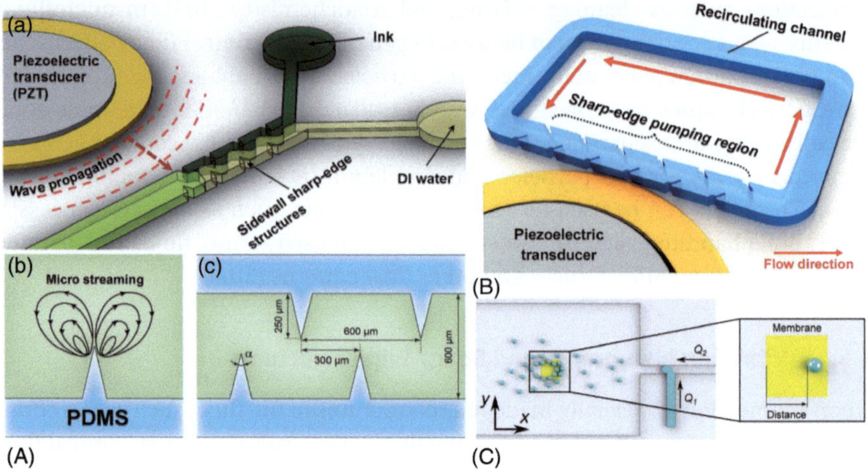

Figure 1.21 (A) Schematic of the sharp-edge-based acoustofluidic pumping device. Source: Nama et al. [108]/Reproduced with permissions from Royal Society of Chemistry. (B) Schematic of the device showing a microfluidic channel with sharp-edge structures on its side walls. Source: Huang et al. [109]/Reproduced with permissions from Royal Society of Chemistry. (C) Effects of the applied voltage on droplet repulsion from a single small membrane. Source: Reproduced with permission from Van Phan et al. [110] ©2016, American Chemical Society.

the bubbles. Oscillating bubbles have a wide range of applications in various fields. One of the most common is in sonochemistry, where the high-energy acoustic fields created by oscillating bubbles are used to promote chemical reactions. In the microfluidic platform, bubbles can be used to generate microstreams, which can be used for fluid mixing, particle manipulation, and droplet control [53]. Bubbles excited by an acoustic field act as an external energy source and induce a driving force at the interface between the bubbles and the surrounding liquid to induce vortical flows in the fluid. This phenomenon, also known as microstreaming, occurs when piezoelectric materials produce an acoustic wave by using an alternating current or radiofrequency. Generally, oscillating bubbles are a versatile tool that can be used to generate high-energy acoustic fields, which can be used to promote chemical reactions, control fluids and droplets, and manipulate particles.

1.8.2.1 Cavitation

Cavitation is becoming increasingly important in therapeutic ultrasonic procedures such as tumor ablation, lithotripsy, and diagnostics [105]. Cavitation refers to the formation and collapse of vapor-filled cavities, or bubbles, in a liquid that is subjected to changes in pressure. The development of bubble cavitation is caused by the rarefaction and compression cycles that an ultrasonic wave-induced pressure oscillation generates in the liquid. These vapor-filled cavities will collapse and produce a powerful shockwave in the form of a micro-jet when they are subjected to a substantially higher pressure; this phenomenon is known as inertial cavitation. In a wide range of industrial applications, ultrasonic cavitation effects can be used. Energy produced by bubble collapse can be used to enhance the reaction rates of chemical processes, such as the solubilization of organic substances and the formation of free radicals [111]. Ultrasonic cavitation is occasionally combined with conventional stirring or other techniques in the production of biodiesel. Microalgae treatment as a feedstock for biodiesel also relies heavily on the mechanism of ultrasound cavitation. The ultrasonic frequency, ultrasonic source power intensity, fluid physical characteristics, initial bubble radius, and extra chemical reactions triggered during cavitation are the few variables that affect the efficiency of ultrasonic cavitation mechanism.

1.8.3 Optoacoustic Imaging

Optoacoustic imaging, also known as photoacoustic imaging or optoacoustic tomography (OAT), is a hybrid imaging technique that combines the high spatial resolution of optical imaging with the deep penetration of ultrasonic imaging. The basic principle of optoacoustic imaging is based on the photoacoustic effect, which is the generation of an acoustic wave in a material as a result of absorption of light. When a short pulse of laser light is absorbed by a biological tissue, it heats up and expands, creating an ultrasonic pressure wave that propagates through the tissue. In general, the optoacoustic technique is used to generate images of the internal structure of biological tissue by measuring the acoustic waves generated by the absorption of short pulses of laser light. Optoacoustic imaging can be used to image a wide range of biological tissues, including skin, breast, brain, and blood vessels. It has several advantages over traditional imaging techniques, including high spatial resolution, high contrast, and the ability to detect both optical absorption and scattering. The technology can be applied in various medical fields such as cancer detection and cardiovascular imaging [112]. When specific tissue components such as hemoglobin or lipids absorb this light, an ultrasound-induced mechanical wave is produced. These signals can be picked up by a transducer or transducer array, and can be transformed into images using different algorithms and other available techniques [112]. As depicted in Figure 1.22, the laser is used to deliver light to the sides of the transducer. Light is scattered by the tissue beneath the transducer array, and waves are produced using the transducer array. Photoacoustic imaging will be further delineated in Chapter 5.

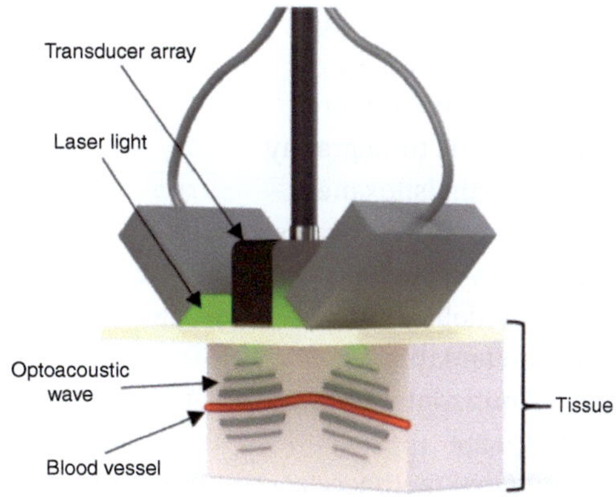

Figure 1.22 Schematic of optoacoustic imaging device. Source: Schellenberg and Hunt [113]/Reproduced with permission from Elsevier.

1.8.4 Manifestations of Acoustic Radiation Force and Acoustic Streaming Flow

ARF and acoustic streaming flow are two different phenomena that can be generated by an applied acoustic wave in a fluid. In microfluidics, these two phenomena are often observed simultaneously, where the ARF can be used to trap and manipulate micro-objects, while the acoustic streaming flow can be used to mix and transport fluids within the microchannel. The acoustic waves are typically applied using piezoelectric transducers or ultrasonic transducers that are placed near the microfluidic channel. The ability to manipulate micro-scale objects is crucial for a wide range of chemical and biological research. For the manipulation of small-scale objects, well-known traditional methods include centrifugation, sedimentation, and mechanical filtering. Later, various microfluidic systems were developed in which the use of optical, magnetic, dielectrophoretic and acoustic forces was of great interest. On-chip micro-object manipulations using acoustophoresis, typically at ultrasonic frequencies in the hundreds of kHz to tens of MHz range, have great potential to manipulate microscale objects. The microfluidic technology in the field of acoustophoresis is relatively straightforward to integrate within the systems. When an acoustic wave field is imposed on a microscale object suspended in the fluid, the former will be affected by ARF originating from wave scattering phenomena, as discussed in earlier section 1.5.3. In the case of BAW manipulation, acoustic waves spread through the bulk of a medium, whereas for SAWs-based devices, waves propagate along an elastic interface and transmit into the surrounding fluid medium. Besides the radiation forces, lateral acoustic streaming happens when an acoustic wave interacts with a fluid–fluid or solid–fluid interface, such as liquid drops in a gas or gas bubbles in a liquid and vibrating solid structure in the liquid. Further details on ARF and acoustic streaming flow can be found in Chapter 2.

List of Abbreviations and Symbols

radio frequency	RF
surface acoustic waves	SAWs
nondestructive testing	NDT
bulk acoustic waves	BAWs
acoustic radiation force	ARF
optoacoustic tomography	OAT
polydimethylsiloxane	PDMS
primary compressional waves	P-wave
secondary transverse waves	S-wave
lithium niobate	$LiNbO_3$
lithium tantalite	$LiTaO_3$
shear-horizontal Love waves	SH-wave
polystyrene	PS
polymethyl methacrylate	PMMA

polylactic acid	PLA
polylactide-*co*-glycolide	PLGA
acoustic streaming flow	ASF
order of magnitude	$O(\ldots)$
eccentric rotating mass based	ERM
direct current	DC
linear resonant actuators	LRA
alternative current	AC
interdigital transducers	IDT
acoustic impedance	Z
density	ρ
speed of sound	c
pressure transmission	$\boldsymbol{T}_P$
pressure reflection	$\boldsymbol{R}_P$
complex pressure amplitude of transmitted wave	$\boldsymbol{P}_t$
complex pressure amplitude of reflected wave	$\boldsymbol{P}_r$
incident angle	θ_i
reflected waves propagate angle	θ_r
wave number	k
spherical object radius	a
complex pressure amplitude of incident wave	P_i
internal pressure fields	P_{int}
external pressure fields	P_{Scat}
intensity transmission	T_I
intensity reflection	T_R
viscosity coefficient	μ
wave intensity	I
length scale within the fluidic medium	L
acoustic wavelength	λ
acoustic frequency	f
boundary layer thickness	δ

References

1 Ginsberg, J.H. (2018). *Acoustics – A Textbook for Engineers and Physicists*. New York: Springer.

2 Rossing, T.D. (ed.) (2007). *Springer Handbook of Acoustics*. Springer Science & Business Media.

3 Hirose, A. and Lonngren, K.E. (2010). *Fundamentals of Wave Phenomena*. IET.

4 Freegarde, T. (2012). *Introduction to the Physics of Waves*. Cambridge: Cambridge University Press. https://doi.org/10.1017/CBO9781139048149.

5 Lindsay, R.B. (1964). Report to the National Science Foundation on the conference on education in acoustics. *The Journal of the Acoustical Society of America* 36: 2241–2243.

6 Xiang, N. and Blauert, J. (2021). *Acoustics for Engineers*. Berlin: Springer.

7 Strutt, J.W. (1877). *The Theory of Sound*, vol. 1. Cambridge: Cambridge University Press. https://doi.org/10.1017/CBO9781139058087.

8 Faraday, M. (1831). XVII. On a peculiar class of acoustical figures; and on certain forms assumed by groups of particles upon vibrating elastic surfaces. *Philosophical Transactions of the Royal Society of London* 121: 299–340. https://doi.org/10.1098/rstl.1831.0018.

9 Dvořák, V. (1876). Ueber die akustische Anziehung und Abstossung. *Annalen der Physik* 42: 157.

10 Rayleigh, L. (1884). On the circulation of air observed in Kundt's tubes, and on some allied acoustical problems. *Philosophical Transactions of the Royal Society of London* 175: 1–21. http://www.jstor.org/stable/109434.

11 Schlichting, H. (1932). Berechnung ebener periodischer Grenzschichtstromungen. *Physikalische Zeit* 33: 327–335.

12 Eckart, C. (1948). Vortices and streams caused by sound waves. *Physical Review* 73 (1): 68–76. https://doi.org/10.1103/PhysRev.73.68.

13 White, R.M. and Voltmer, F.W. (1965). Direct piezoelectric coupling to surface elastic waves. *Applied Physics Letters* 7 (12): 314–316. https://doi.org/10.1063/1.1754276.

14 Shiokawa, S., Matsui, Y., and Ueda, T. (1990). Study on SAW streaming and its application to fluid devices. *Japanese Journal of Applied Physics* 29 (S1): 137. https://doi.org/10.7567/JJAPS.29S1.137.

15 Shiokawa, S. and Matsui, Y. (1994). The dynamics of SAW streaming and its application to fluid devices. *MRS Proceedings* 360: 53. https://doi.org/10.1557/PROC-360-53.

16 Wixforth, A. (2003). Acoustically driven planar microfluidics. *Superlattices and Microstructures* 33 (5, 6): 389–396. https://doi.org/10.1016/j.spmi.2004.02.015.

17 Sritharan, K., Strobl, C.J., Schneider, M.F. et al. (2006). Acoustic mixing at low Reynold's numbers. *Applied Physics Letters* 88 (5): 1–3. https://doi.org/10.1063/1.2171482.

18 Forde, G., Friend, J., and Williamson, T. (2006). Straightforward biodegradable nanoparticle generation through megahertz-order ultrasonic atomization. *Applied Physics Letters* 89 (6): https://doi.org/10.1063/1.2221914.

19 Rezk, A.R., Manor, O., Friend, J.R., and Yeo, L.Y. (2012). Unique fingering instabilities and soliton-like wave propagation in thin acoustowetting films. *Nature Communications* 3: 1167. https://doi.org/10.1038/ncomms2168.

20 Gu, Y., Chen, C., Mao, Z. et al. (2021). Acoustofluidic centrifuge for nanoparticle enrichment and separation. *Science Advances* 7 (1): eabc0467.

21 King, L.V. (1934). On the acoustic radiation pressure on spheres. *Proceedings of the Royal Society of London. Series A: Mathematical and Physical Sciences* 147 (861): 212–240. https://doi.org/10.1098/rspa.1934.0215.

22 Yosioka, K. and Kawasima, Y. (1955). Acoustic radiation pressure on a compressible sphere. *Acta Acustica united with Acustica* 5 (3): 167–173.

23 Hasegawa, T. and Yosioka, K. (1969). Acoustic-radiation force on a solid elastic sphere. *The Journal of the Acoustical Society of America* 46 (5B): 1139–1143. https://doi.org/10.1121/1.1911832.

24 Gorkov, L.P. (1962). On the forces acting on a small particle in an acoustic field in an ideal fluid. *Doklady Akademii Nauk SSSR* 6: 773–776.

25 Mandralis, Z.I. and Feke, D.L. (1993). Continuous suspension fractionation using acoustic and divided-flow fields. *Chemical Engineering Science* 48 (23): 3897–3905. https://doi.org/10.1016/0009-2509(93)80368-Z.

26 Harris, N.R., Hill, M., Beeby, S. et al. (2003). A silicon microfluidic ultrasonic separator. *Sensors and Actuators B: Chemical* 95 (1–3): 425–434. https://doi.org/10.1016/S0925-4005(03)00448-9.

27 Nilsson, A., Petersson, F., Jönsson, H., and Laurell, T. (2004). Acoustic control of suspended particles in micro fluidic chips. *Lab on a Chip* 4 (2): 131–135. https://doi.org/10.1039/b313493h.

28 Shi, J., Mao, X., Ahmed, D. et al. (2008). Focusing microparticles in a microfluidic channel with standing surface acoustic waves (SSAW). *Lab on a Chip* 8 (2): 221–223. https://doi.org/10.1039/b716321e.

29 Shi, J., Huang, H., Stratton, Z. et al. (2009). Continuous particle separation in a microfluidic channel via standing surface acoustic waves (SSAW). *Lab on a Chip* 9: 3354–3359. https://doi.org/10.1039/b915113c.

30 Destgeer, G., Lee, K.H., Jung, J.H. et al. (2013). Continuous separation of particles in a PDMS microfluidic channel via travelling surface acoustic waves (TSAW). *Lab on a Chip* 13 (21): 4210–4216. https://doi.org/10.1039/c3lc50451d.

31 Skowronek, V., Rambach, R.W., Schmid, L. et al. (2013). Particle deflection in a poly(dimethylsiloxane) microchannel using a propagating surface acoustic wave: size and frequency dependence. *Analytical Chemistry* 85 (20): 9955–9959. https://doi.org/10.1021/ac402607p.

32 Zhang, J., Chen, C., Becker, R. et al. (2022). A solution to the biophysical fractionation of extracellular vesicles: Acoustic Nanoscale Separation via Wave-pillar Excitation Resonance (ANSWER). *Science Advances* 8 (47): eade0640.

33 Augustsson, P. (2011). On microchannel acoustophoresis – experimental considerations and life science applications. http://lup.lub.lu.se/record/2203054/file/2203068.pdf.

34 Russell, D.A. (2022). Longitudinal and transverse wave motion. https://www.acs.psu.edu/drussell/Demos/waves/wavemotion.html (accessed 10 May 2022).

35 Mitragotri, S. (2005). Healing sound: the use of ultrasound in drug delivery and other therapeutic applications. *Nature Reviews. Drug Discovery* 4 (3): 255–260. https://doi.org/10.1038/nrd1662.

36 Zou, Y., Gao, C., Zhou, J. et al. (2022). Aluminum scandium nitride thin-film bulk acoustic resonators for 5G wideband applications. *Microsystems & Nanoengineering* 8 (1): 124. https://doi.org/10.1038/s41378-022-00457-0.

37 Dutta, J., Singh, A.V., Singhal, S., and Upadhayay, M.D. (2016). Design and simulation of a zinc oxide thin film bulk acoustic resonator filter for 2.6 GHz band applications. *IETE Journal of Research* 62 (1): 3–8. https://doi.org/10.1080/03772063.2015.1062734.

38 The University of Waikato (2007). Science Learning Hub, 21 July 2007. https://www.sciencelearn.org.nz/images/353-earth-waves (accessed 10 May 2023).

39 Fu, Y.Q., Pang, H.F., Torun, H. et al. (2021). Engineering inclined orientations of piezoelectric films for integrated acoustofluidics and lab-on-a-chip operated in liquid environments. *Lab on a Chip* 21 (2): 254–271. https://doi.org/10.1039/d0lc00887g.

40 Chen, M., Huan, Q., and Li, F. (2019). Excitation of moderate-frequency Love wave in a Plexiglas plate on aluminum semi-space. *The Journal of the Acoustical Society of America* 146 (6): EL482–EL488. https://doi.org/10.1121/1.5139192.

41 Su, Z. and Ye, L. (2009). *Identification of Damage Using Lamb Waves: From Fundamentals to Applications*. Springer Science & Business Media.

42 Kinsler, L.E. and Frey, A.R. (1950). *Fundamentals of Acoustics*. New York: Wiley.

43 Munoz Gutierrez, M.A., Prospero Sanchez, P.L., and do Vale Neto, J.V. (2005). An eigenpath underwater acoustic communication channel simulation. *IEEE Xplore* 1: 355–362.

44 Pessôa, M.A.S. and Neves, A.A.R. (2020). Acoustic scattering and forces on an arbitrarily sized fluid sphere by a general acoustic field. *Journal of Sound and Vibration* 479: 115373. https://doi.org/10.1016/j.jsv.2020.115373.

45 Maidanik, G. (1957). Acoustical radiation pressure due to incident plane progressive waves on spherical objects. *The Journal of the Acoustical Society of America* 29 (6): 738–742. https://doi.org/10.1121/1.1909032.

46 Tahmasebipour, A., Begley, M., and Meinhart, C. (2022). Acoustophoresis of a resonant elastic microparticle in a viscous fluid medium. *The Journal of the Acoustical Society of America* 151 (5): 3083–3093. https://doi.org/10.1121/10.0010418.

47 Destgeer, G., Jung, J.H., Park, J. et al. (2017). Acoustic impedance-based manipulation of elastic microspheres using travelling surface acoustic waves. *RSC Advances* 7 (36): 22524–22530. https://doi.org/10.1039/C7RA01168G.

48 Destgeer, G., Ha, B.H., Park, J. et al. (2015). Microchannel anechoic corner for size-selective separation and medium exchange via traveling surface acoustic waves. *Analytical Chemistry* 87 (9): 4627–4632. https://doi.org/10.1021/acs.analchem.5b00525.

49 Hasegawa, T. (1977). Comparison of two solutions for acoustic radiation pressure on a sphere. *The Journal of the Acoustical Society of America* 61 (6): 1445–1448. https://doi.org/10.1121/1.381460.

50 Fakhfouri, A., Devendran, C., Ahmed, A. et al. (2018). The size dependant behaviour of particles driven by a travelling surface acoustic wave (TSAW). *Lab on a Chip* 18 (24): 3926–3938. https://doi.org/10.1039/C8LC01155A.

51 Wang, K., Zhou, W., Lin, Z. et al. (2018). Sorting of tumour cells in a microfluidic device by multi-stage surface acoustic waves. *Sensors and Actuators B: Chemical* 258: 1174–1183. https://doi.org/10.1016/j.snb.2017.12.013.

52 Xi, H.-D., Zheng, H., Guo, W. et al. (2017). Active droplet sorting in microfluidics: a review. *Lab on a Chip* 17 (5): 751–771. https://doi.org/10.1039/C6LC01435F.

53 Li, Y., Liu, X., Huang, Q. et al. (2021). Bubbles in microfluidics: an all-purpose tool for micromanipulation. *Lab on a Chip* 21 (6): 1016–1035. https://doi.org/10.1039/D0LC01173H.

54 Akay, A. and Carcaterra, A. (2014). Damping mechanisms. In: *CISM International Centre for Mechanical Sciences, Courses and Lectures*, vol. 558, 259–299. Springer International Publishing. https://doi.org/10.1007/978-3-7091-1821-4_6.

55 Zener, C. (1940). Internal friction in solids. *Proceedings of the Physical Society* 52 (1): 152. https://doi.org/10.1088/0959-5309/52/1/322.

56 Augustsson, P., Barnkob, R., Wereley, S.T. et al. (2011). Automated and temperature-controlled micro-PIV measurements enabling long-term-stable microchannel acoustophoresis characterization. *Lab on a Chip* 11 (24): 4152–4164. https://doi.org/10.1039/c1lc20637k.

57 Mulvana, H., Cochran, S., and Hill, M. (2013). Ultrasound assisted particle and cell manipulation on-chip. *Advanced Drug Delivery Reviews* 65 (11, 12): 1600–1610. https://doi.org/10.1016/j.addr.2013.07.016.

58 Park, J., Jung, J.H., Destgeer, G. et al. (2017). Acoustothermal tweezer for droplet sorting in a disposable microfluidic chip. *Lab on a Chip* 17 (6): 1031–1040. https://doi.org/10.1039/c6lc01405d.

59 Park, J., Ha, B.H., Destgeer, G. et al. (2016). Spatiotemporally controllable acoustothermal heating and its application to disposable thermochromic displays. *RSC Advances* 6 (40): 33937–33944. https://doi.org/10.1039/c6ra04075f.

60 Cha, B., Kim, W., Yoon, G. et al. (2021). Enhanced solutal Marangoni flow using ultrasound-induced heating for rapid digital microfluidic mixing. *Frontiers of Physics* 9: https://doi.org/10.3389/fphy.2021.735651.

61 Ha, B.H., Lee, K.S., Destgeer, G. et al. (2015). Acoustothermal heating of polydimethylsiloxane microfluidic system. *Scientific Reports* 5 (1): 11851. https://doi.org/10.1038/srep11851.

62 COMSOL Multiphysics® (2022). *Generic 711 Coupler – An Occluded Ear-Canal Simulator*. Stockholm, Sweden: COMSOL AB.

63 COMSOL Multiphysics® (2022). *Theory of Thermoviscous Acoustics: Thermal and Viscous Losses*. Stockholm, Sweden: COMSOL AB.

64 Wiklund, M., Green, R., and Ohlin, M. (2012). Acoustofluidics 14: applications of acoustic streaming in microfluidic devices. *Lab on a Chip* 12 (14): 2438–2451. https://doi.org/10.1039/c2lc40203c.

65 Wixforth, A., Strobl, C., Gauer, C. et al. (2004). Acoustic manipulation of small droplets. *Analytical and Bioanalytical Chemistry* 379 (7, 8): 982–991. https://doi.org/10.1007/s00216-004-2693-z.

66 Miansari, M. and Friend, J.R. (2016). Acoustic nanofluidics via room-temperature lithium niobate bonding: a platform for actuation and manipulation of nanoconfined fluids and particles. *Advanced Functional Materials* 26 (43): 7861–7872. https://doi.org/10.1002/adfm.201602425.

67 Li, H., Friend, J.R., and Yeo, L.Y. (2007). Surface acoustic wave concentration of particle and bioparticle suspensions. *Biomedical Microdevices* 9 (5): 647–656. https://doi.org/10.1007/s10544-007-9058-2.

68 Destgeer, G., Ha, B., Park, J., and Sung, H.J. (2016). Lamb wave-based acoustic radiation force-driven particle ring formation inside a sessile droplet. *Analytical Chemistry* 88 (7): 3976–3981. https://doi.org/10.1021/acs.analchem.6b00213.

69 Rezk, A.R., Yeo, L.Y., and Friend, J.R. (2014). Poloidal flow and toroidal particle ring formation in a sessile drop driven by megahertz order vibration. *Langmuir* 30 (36): 11243–11247. https://doi.org/10.1021/la502301f.

70 Tan, M.K., Friend, J.R., and Yeo, L.Y. (2009). Interfacial jetting phenomena induced by focused surface vibrations. *Physical Review Letters* 103 (2): https://doi.org/10.1103/PhysRevLett.103.024501.

71 Tan, M.K., Friend, J.R., and Yeo, L.Y. (2007). Microparticle collection and concentration via a miniature surface acoustic wave device. *Lab on a Chip* 7 (5): 618–625. https://doi.org/10.1039/b618044b.

72 Collignon, S., Friend, J., and Yeo, L. (2015). Planar microfluidic drop splitting and merging. *Lab on a Chip* 15 (8): 1942–1951. https://doi.org/10.1039/c4lc01453g.

73 Collins, D.J., Manor, O., Winkler, A. et al. (2012). Atomization off thin water films generated by high-frequency substrate wave vibrations. *Physical Review. E, Statistical, Nonlinear, and Soft Matter Physics* 86 (5): https://doi.org/10.1103/PhysRevE.86.056312.

74 Yaralioglu, G.G., Wygant, I.O., Marentis, T.C., and Khuri-Yakub, B.T. (2004). Ultrasonic mixing in microfluidic channels using integrated transducers. *Analytical Chemistry* 76 (13): 3694–3698. https://doi.org/10.1021/ac035220k.

75 Cui, W., Zhang, H., Zhang, H. et al. (2016). Localized ultrahigh frequency acoustic fields induced micro-vortices for submilliseconds microfluidic mixing. *Applied Physics Letters* 109 (25): 253503. https://doi.org/10.1063/1.4972484.

76 Ahmed, D., Mao, X., Shi, J. et al. (2009). A millisecond micromixer via single-bubble-based acoustic streaming. *Lab on a Chip* 9 (18): 2738–2741. https://doi.org/10.1039/b903687c.

77 Huang, P.H., Xie, Y., Ahmed, D. et al. (2013). An acoustofluidic micromixer based on oscillating sidewall sharp-edges. *Lab on a Chip* 13 (19): 3847–3852. https://doi.org/10.1039/c3lc50568e.

78 Fang, W.-F. and Lee, A.P. (2015). LCAT pump optimization for an integrated microfluidic droplet generator. *Microfluidics and Nanofluidics* 18: 1265–1275.

79 Collins, D.J., Ma, Z., Han, J., and Ai, Y. (2017). Continuous micro-vortex-based nanoparticle manipulation via focused surface acoustic waves. *Lab on a Chip* 17 (1): 91–103. https://doi.org/10.1039/C6LC01142J.

80 Sesen, M., Devendran, C., Malikides, S. et al. (2017). Surface acoustic wave enabled pipette on a chip. *Lab on a Chip* 17 (3): 438–447. https://doi.org/10.1039/c6lc01318j.

81 Connacher, W., Zhang, N., Huang, A. et al. (2018). Micro/nano acoustofluidics: materials, phenomena, design, devices, and applications. *Lab on a Chip* 18 (14): 1952–1996. https://doi.org/10.1039/c8lc00112j.

82 Wu, J. (2018). Acoustic streaming and its applications. *Fluids* 3 (4): 108. https://doi.org/10.3390/fluids3040108.

83 Madsen, P.T. and Surlykke, A. (2013). Functional convergence in bat and toothed whale biosonars. *Physiology* 28 (5): 276–283. https://doi.org/10.1152/physiol.00008.2013.

84 Günther, R.H., O'Connell-Rodwell, C.E., and Klemperer, S.L. (2004). Seismic waves from elephant vocalizations: a possible communication mode? *Geophysical Research Letters* 31 (11): https://doi.org/10.1029/2004GL019671.

85 O'Connell-Rodwell, C.E., Arnason, B.T., and Hart, L.A. (2000). Seismic properties of Asian elephant (*Elephas maximus*) vocalizations and locomotion. *The Journal of the Acoustical Society of America* 108 (6): 3066–3072. https://doi.org/10.1121/1.1323460.

86 Langbauer, W.R. (2000). Elephant communication. *Zoo Biology* 19 (5): 425–445. https://doi.org/10.1002/1098-2361(2000)19:5<425::AID-ZOO11>3.0.CO;2-A.

87 Mayo Clinic Staff (2023). Laryngitis, 22 June 2022. https://www.mayoclinic.org/diseases-conditions/laryngitis/symptoms-causes/syc-20374262 (accessed 10 May 2023).

88 Park, M.C. (2019). Understanding the multi-mass model and sound generation of vocal fold oscillation. *AIP Advances* 9 (10): https://doi.org/10.1063/1.5113911.

89 Stevens, K.N. (2000). *Acoustic Phonetics*. MIT Press.

90 Titze, I.R. (1994). *Principle of Voice Production*. Prentice Hall.

91 Precision Microdrive Using DC motors for energy harvest or LRAs for vibration detection. https://www.precisionmicrodrives.com/eccentric-rotating-mass-vibration-motors-erms (accessed 10 May 2023).

92 Guex, A.G., di Marzio, N., Eglin, D. et al. (2021). The waves that make the pattern: a review on acoustic manipulation in biomedical research. *Materials Today Bio* 10. https://doi.org/10.1016/j.mtbio.2021.100110.

93 Mason, W.P. (1981). Piezoelectricity, its history and applications. *Journal of the Acoustical Society of America* 70 (6): 1561–1566. https://doi.org/10.1121/1.387221.

94 Donald, I. (1974). *Sonar – The Story of an Experiment*. Pergamon Press.

95 Manbachi, A. and Cobbold, R.S.C. (2011). Development and application of piezoelectric materials for ultrasound generation and detection. *Ultrasound* 19 (4): 187–196. https://doi.org/10.1258/ult.2011.011027.

96 Rozenberg, L.D. (1973). The ultrasonic cleaning mechanism. In: *Physical Principles of Ultrasonic Technology* (ed. L.D. Rozenberg), 252–269. Boston, MA: Springer US. https://doi.org/10.1007/978-1-4684-8217-1_13.

97 Hellier, C.J. (2003). *Handbook of Nondestructive Evaluation*. McGraw Hill.

98 Saadon, S. and Sidek, O. (2011). A review of vibration-based MEMS piezoelectric energy harvesters. *Energy Conversion and Management* 52 (1): 500–504. https://doi.org/10.1016/j.enconman.2010.07.024.

99 Wu, M., Ozcelik, A., Rufo, J. et al. (2019). Acoustofluidic separation of cells and particles. In: *Microsystems and Nanoengineering*, vol. 5, no. 1. Nature Publishing Group https://doi.org/10.1038/s41378-019-0064-3.

100 Destgeer, G. and Sung, H.J. (2015). Recent advances in microfluidic actuation and micro-object manipulation via surface acoustic waves. *Lab on a Chip* 15 (13): 2722–2738. https://doi.org/10.1039/c5lc00265f.

101 Yeo, L.Y. and Friend, J.R. (2014). Surface acoustic wave microfluidics. *Annual Review of Fluid Mechanics* 46 (1): 379–406. https://doi.org/10.1146/annurev-fluid-010313-141418.

102 Ding, X., Li, P., Lin, S.C. et al. (2013). Surface acoustic wave microfluidics. *Lab on a Chip* 13 (18): 3626–3649. https://doi.org/10.1039/c3lc50361e.

103 Rufo, J., Cai, F., Friend, J. et al. (2022). Acoustofluidics for biomedical applications. *Nature Reviews Methods Primers* 2 (1): 30. https://doi.org/10.1038/s43586-022-00109-7.

104 Kok, H.P., Cressman, E.N., Ceelen, W. et al. (2020). Heating technology for malignant tumors: a review. *International Journal of Hyperthermia* 37 (1): 711–741. https://doi.org/10.1080/02656736.2020.1779357.

105 Hashmi, A., Yu, G., Reilly-Collette, M. et al. (2012). Oscillating bubbles: a versatile tool for lab on a chip applications. *Lab on a Chip* 12 (21): 4216–4227. https://doi.org/10.1039/C2LC40424A.

106 Wei, W., Wang, Y., Wang, Z., and Duan, X. (2023). Microscale acoustic streaming for biomedical and bioanalytical applications. *TrAC Trends in Analytical Chemistry* 116958. https://doi.org/10.1016/j.trac.2023.116958.

107 Li, Y., Cai, S., Shen, H. et al. (2022). Recent advances in acoustic microfluidics and its exemplary applications. *Biomicrofluidics* 16 (3): 031502. https://doi.org/10.1063/5.0089051.

108 Nama, N., Huang, P.-H., Huang, T.J., and Costanzo, F. (2014). Investigation of acoustic streaming patterns around oscillating sharp edges. *Lab on a Chip* 14 (15): 2824–2836. https://doi.org/10.1039/C4LC00191E.

109 Huang, P.-H., Nama, N., Mao, Z. et al. (2014). A reliable and programmable acoustofluidic pump powered by oscillating sharp-edge structures. *Lab on a Chip* 14 (22): 4319–4323. https://doi.org/10.1039/C4LC00806E.

110 Van Phan, H., Alan, T., and Neild, A. (2016). Droplet manipulation using acoustic streaming induced by a vibrating membrane. *Analytical Chemistry* 88 (11): 5696–5703. https://doi.org/10.1021/acs.analchem.5b04481.

111 Zheng, H., Zheng, Y., and Zhu, J. (2022). Recent developments in hydrodynamic cavitation reactors: cavitation mechanism, reactor design, and applications. *Engineering* https://doi.org/10.1016/j.eng.2022.04.027.

112 Karlas, A., Pleitez, M.A., Aguirre, J., and Ntziachristos, V. (2021). Optoacoustic imaging in endocrinology and metabolism. *Nature Reviews. Endocrinology* 17 (6): 323–335. https://doi.org/10.1038/s41574-021-00482-5.

113 Schellenberg, M.W. and Hunt, H.K. (2018). Hand-held optoacoustic imaging: a review. *Photoacoustics* 11: 14–27. https://doi.org/10.1016/j.pacs.2018.07.001.

2

Basic Theories and Physics of Acoustic Technologies

Khemraj G. Kshetri and Nitesh Nama

University of Nebraska-Lincoln, Department of Mechanical and Materials Engineering, 1400 R St, Lincoln, NE 68588, USA

2.1 Introduction

Acoustic waves refer to mechanical waves that propagate energy through an elastic media via pressure fluctuations. Acoustics is a classical field of physics and has attracted scientific interest from researchers dating as far back as the eighteenth century [1–5]. Specifically, the propagation of acoustic waves in fluid-solid media as well as their interaction with immersed particles (e.g. solid objects, liquid droplets, or gas bubbles) has been the subject of extensive experimental and theoretical research from prominent scientists including Chladni [1], Faraday [2], König [3], Kundt [4], and Rayleigh [5]. Since then, acoustic technologies have been employed for a broad and diverse range of scientific endeavors, with applications, including architectural and musical acoustics [6–8], underwater acoustics [9, 10], physical acoustics [11, 12], acoustic imaging and sensing [13–16], and medical acoustics [14, 17].

Recently, acoustic technologies have attracted renewed interest in the life sciences due to their suitability and potential for biomedical applications [18–21]. These technologies offer many attractive advantages in biomedicine: (i) acoustic technologies are highly biocompatible and are widely employed within medical settings for sensing, imaging, and therapeutic applications, (ii) acoustic technologies provide a contact-less method of energy transfer that can be leveraged to manipulate fluids and particles, (iii) acoustic waves can penetrate deep within the human tissue, making them well-suited for imaging applications, and (iv) acoustic technologies allow label-free manipulation of microparticles compared to magnetic, optical, or electro-kinetic approaches that require the sample or media to have specific dielectric, magnetic, or electrical conductivity properties [18, 20, 22]. The use of acoustics in biomedicine has a long history, dating back to 350 BCE, when Hippocrates used sound emanating from patients' chests to detect fluid in their lungs [23]. Since then, the use of acoustics in biomedicine has steadily increased to enable improved diagnostics and therapeutics via stethoscopes [24], ultrasound

Acoustic Technologies in Biology and Medicine, First Edition.
Edited by Adem Ozcelik, Ryan Becker, and Tony Jun Huang.

imaging [25], photoacoustic imaging [26], and photoacoustic microscopy [27, 28]. Furthermore, the recent progress in microfluidic technologies for chemical, bioengineering, and biomedical applications has also spurred interest in acoustics as a means of overcoming the scaling challenges associated with flow physics at microscales. Indeed, this merger of microscale flow physics and acoustics has resulted in a novel field of study that is often referred to as micro-acoustofluidics or simply acoustofluidics [29]. Acoustics has also benefited from the significant advancements in microfabrication and microelectronics that have enabled the generation and handling of acoustics at microscales. This has led to a broad range of biomedical applications, including cell and particle manipulation [30–33], cell–cell communication [34, 35], three-dimensional (3D) bioprinting [36], microflow cytometry [37–39], fluid manipulation system [40–46], and targeted drug delivery [47–49].

Given the significant interest in acoustics over the past few centuries, the underlying physics of acoustic waves is rather mature, and classical acoustic principles have proven to be extremely useful in achieving a qualitative physical understanding of acoustic systems. However, in recent years, researchers have pushed the limits of classical acoustic theories by exploring ever- expanding range of frequencies, amplitudes, and scales of acoustics. This has resulted in several recent studies that have pointed toward the inadequacy of classical theories in explaining acoustic phenomena in novel acoustic technologies that are governed by a unique combination of microscale hydrodynamics and high-frequency, high-amplitude acoustics, among other factors [50, 51]. While significant theoretical and numerical progress has been made in recent years, a complete understanding of acoustical phenomena, particularly for emerging applications, remains a work in progress.

In this chapter, we describe the basic theory, governing equations, and principles of acoustic systems. Specifically, we first provide an overview of the governing equations for acoustic wave propagation in solid and fluid media. We introduce the phenomenon of acoustic streaming and outline the common modeling approaches to studying fluid response under acoustic actuation. We also provide an overview of the typical forces in acoustic systems, including the primary and secondary acoustic radiation forces and the hydrodynamic drag force. This chapter offers an introduction to the basic theories governing the interactions of acoustics with fluid/solid media; interested readers are referred to references [52–57] for a more detailed discussion on basic theories and applications of acoustics.

2.2 Acoustic Waves in Solids

The propagation of acoustic waves in solid materials has been extensively studied over the last century. As mentioned earlier, this research has attracted renewed interest due to the use of acoustic waves for various diagnostic and biomedical applications. Specifically, emerging biomedical applications of acoustics necessitate a complete understanding of the propagation of acoustic waves in solid materials (e.g. tissue and cells) and the accompanying acoustic-fluid–solid interactions.

In the Sections 2.2.1–2.2.3, we consider acoustic wave propagation in general solid materials. We provide an overview in the case of both piezoelectric and non-piezoelectric solids and derive general expressions for the commonly employed constitutive laws for each case.

2.2.1 Governing Equation

The governing equation for the propagation of acoustic waves in a solid material is given by Newton's second law as:

$$\rho_s \frac{\partial^2 \boldsymbol{u}}{\partial t^2} = \boldsymbol{\nabla} \cdot \boldsymbol{T} \tag{2.1}$$

where ρ_s is the mass density of the solid, t is the time, $\boldsymbol{u}$ is the displacement field, and $\boldsymbol{T}$ is the stress tensor. Here, the stress tensor $\boldsymbol{T}$ is obtained from constitutive models that relate the displacement with the stress tensor for the specific solid material being studied. Next, we describe the typical constitutive models that are used to study acoustic wave propagation in solids.

2.2.2 Acoustic Waves in Non-piezoelectric Solids

Using the index notation and Einstein summation convention, Hooke's law can be used to relate the stress and strain tensors for non-piezoelectric solids and can be expressed as

$$\boldsymbol{T} = \mathbb{C}\boldsymbol{S} \tag{2.2}$$

where $\mathbb{C}$ is the fourth-order elastic stiffness tensor that relates the stress tensor $\boldsymbol{T}$ and the strain tensor $\boldsymbol{S}$. Since both the stress tensor $\boldsymbol{T}$ and the strain tensor $\boldsymbol{S}$ are symmetric second-rank tensors, the elastic stiffness tensor is such that its first two and the last two components are interchangeable, i.e. $\mathbb{C}_{ijkl} = \mathbb{C}_{jikl}$ and $\mathbb{C}_{ijkl} = \mathbb{C}_{ijlk}$. Further, the strain tensor can be expressed in terms of the gradient of displacement as

$$\boldsymbol{S} = \frac{1}{2}[\boldsymbol{\nabla u} + (\boldsymbol{\nabla u})^T] \tag{2.3}$$

where $\nabla \boldsymbol{u}$ denotes the gradient of the displacement field and $(\nabla \boldsymbol{u})^T$ denotes the transpose of displacement gradient. Lastly, the electric displacement, $\mathbf{D}$, is related to the electric field, $\mathbf{E}$, as

$$\mathbf{D} = \boldsymbol{\varepsilon}\mathbf{E} \tag{2.4}$$

where $\boldsymbol{\varepsilon}$ are the dielectric constants; $\mathbf{D}$ and $\mathbf{E}$ are the electric displacement and electric field, respectively. The first three equations describe the relation of mechanical displacement to the stress and strain tensors, while Eq. (2.4) relates the electric displacement to the electric field. Combined, Eqs. (2.1)–(2.4) describe the governing relations between mechanical displacement, electrical displacement, and mechanical stress/strain tensors and can be employed (with appropriate boundary conditions) to study acoustic waves in non-piezoelectric solids.

2.2.3 Acoustic Waves in Piezoelectric Solids

The modeling of acoustic wave propagation in piezoelectric materials is challenging due to their anisotropy as well as their electromechanical coupling efficiency. The electromechanical coupling of piezoelectric materials can be modeled using coupled constitutive relations given as

$$\boldsymbol{T} = \tilde{\mathbb{C}}\boldsymbol{S} - \boldsymbol{e}\mathbf{E} \tag{2.5a}$$

$$\mathbf{D} = \boldsymbol{e}\boldsymbol{S} + \tilde{\varepsilon}\mathbf{E} \tag{2.5b}$$

where $\mathbb{C}$ is the elastic stiffness tensor within a constant electric field, $\boldsymbol{e}$ is the piezoelectric stress tensor, and $\tilde{\varepsilon}$ are the dielectric constants at constant strain. Here, $\boldsymbol{e}$ couples the electrical and mechanical responses of the piezoelectric material. On a typical piezoelectric substrate, the speed of an electromagnetic wave is several orders of magnitude faster than the acoustic wave speed. Consequently, the mechanical dynamics are considerably slower than the electromagnetic radiation, allowing the use of quasi-static assumption

$$\frac{\partial \boldsymbol{D}}{\partial t} = 0 \tag{2.6}$$

Combined, Eqs. (2.1), (2.3), (2.5a), (2.5b), and (2.6) govern the behavior of acoustic waves in piezoelectric materials and can be solved via analytical or numerical approaches [19].

2.3 Acoustic Waves in Fluids

The classical theories for propagation of acoustic waves in fluid media have yielded important insights into the interaction of acoustics with fluid media for simplified geometries. Recent advances in computing and numerical methods have enabled numerical analyses of more general cases where classical theories have been employed for different particle shapes, complex fluid media, and different temporal and spatial profiles of acoustic waves [58–62]. In this section, we will introduce the equations that govern acoustic wave propagation in fluid media and discuss the phenomenon of acoustic streaming. We then discuss common classifications of acoustic streaming and outline the common modeling approaches to studying acoustic streaming.

2.3.1 Governing Equations

The conservation principles that govern a linear, viscous, compressible fluid are given as

$$\frac{\partial \rho}{\partial t} + \nabla \cdot (\rho \boldsymbol{v}) = 0 \tag{2.7}$$

$$\rho\frac{\partial \boldsymbol{v}}{\partial t} + \rho(\nabla \boldsymbol{v})\boldsymbol{v} = -\nabla p + \mu\nabla^2 \boldsymbol{v} + \left(\mu_B + \frac{\mu}{3}\right)\nabla(\nabla \cdot \boldsymbol{v}) \tag{2.8}$$

$$\rho T\left[\frac{\partial s}{\partial t}+\nabla s\cdot \boldsymbol{v}\right]=\mu\left(\nabla \boldsymbol{v}+(\nabla \boldsymbol{v})^{T}-\frac{2}{3}(\nabla\cdot \boldsymbol{v})\boldsymbol{I}\right):\nabla \boldsymbol{v}+\mu_{B}(\nabla.\boldsymbol{v})^{2}+\kappa\nabla^{2}T \tag{2.9}$$

where $\boldsymbol{\rho}$ is the fluid density, $\boldsymbol{\mu}$ and $\boldsymbol{\mu_B}$ are shear and bulk viscosities, respectively, $\boldsymbol{\kappa}$ is the thermal conductivity, $\boldsymbol{p}$ is the pressure, $\boldsymbol{v}$ is the fluid velocity, $\boldsymbol{T}$ is the temperature, $\boldsymbol{s}$ is the energy, and $\boldsymbol{I}$ is the identity tensor. These equations allow the study of the fluid response when combined with appropriate boundary conditions and a thermodynamic relation of pressure to density and energy

$$p-p_0=A\left(\frac{\rho-\rho_0}{\rho_0}\right)+\frac{B}{2}\left(\frac{\rho-\rho_0}{\rho_0}\right)^2+\cdots \tag{2.10}$$

where the subscript "0" denotes the fluid properties at equilibrium, and A and B are

$$A=\rho_0\left(\frac{\partial c_0^2}{\partial\rho}\right)_{\text{constant s with } \rho=\rho_0} \tag{2.11a}$$

$$B=\rho_0^2\left(\frac{\partial c_0^2}{\partial\rho}\right)^2_{\text{constant s with } \rho=\rho_0} \tag{2.11b}$$

where c_0 is the isentropic acoustic wave speed. Equations (2.7)–(2.11b) form a nonlinear system of equations that can be used, with appropriate boundary conditions, to describe acoustic wave propagation in fluids. However, for reasons discussed in the next section, a direct numerical solution of these equations can be challenging, and further simplifications are typically adopted for modeling acoustic waves in fluids.

2.3.2 Acoustic Streaming

The propagation of acoustic waves through a volume element of the fluid can be understood as a state of pressure and velocity fluctuations. For an ideal fluid element, there is no net displacement of the fluid particle over a cycle. In contrast, for real fluids, the viscous dissipations result in a nonzero displacement of the fluid particle in each cycle. Thus, the viscous dissipations result in a time-averaged mean flow that is typically referred to as acoustic streaming. Following Bruus [63, 64], consider the fluid response of a body of typical dimension a, oscillating with a velocity $U_0 \cos \omega t$ in a viscous fluid. If the scaling $U_0/\omega a \ll 1$ (i.e. amplitude of oscillation is small than the particle dimension), then the fluid response contains not only a leading order oscillatory term, but also a nonzero steady component. Mathematically, this can be understood via the fact that $\cos^2\omega t=\frac{1}{2}[1+\cos 2\omega t]$ has a nonzero steady component equal to $\frac{1}{2}$. For modeling purposes, the total fluid response is typically understood to be comprised two components: (i) a harmonic component with the same period as the acoustic actuation, and (ii) a remaining component that represents the mean motion of the fluid in a time-averaged sense. The first component is commonly referred to as the acoustic response of the fluid, while the second component is called the streaming motion.

While streaming flows always arise from viscous losses within the fluid, the specific characteristics of the streaming flow are determined by the mechanisms

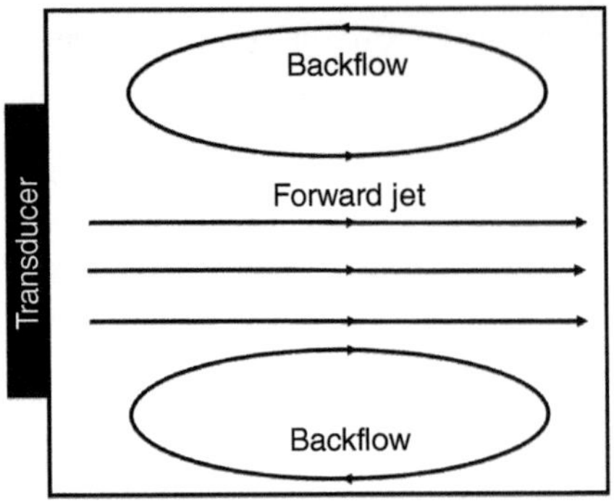

Figure 2.1 Schematic illustration of Eckart streaming flow comprising a fluid jet moving away from the transducer. The confined nature of the geometry results in concomitant backflow. Source: Adapted from Wiklund et al. [67].

of these dissipations. Specifically, typical streaming flows have been observed over several velocity (ranging from few μm/s to several cm/s) and length (ranging from a few μm to several cm) scales. To aid in the understanding of their physical mechanisms, streaming flows have been typically classified either depending on the location of their viscous dissipations or on their velocities.

Firstly, depending on the location of viscous dissipations within the fluid, the streaming can be classified into following categories:

(1) *Quartz wind or Eckart streaming*: This streaming refers to the streaming flows generated by acoustic energy dissipations within the bulk of the fluid [65] (as opposed to near a boundary). Specifically, as the acoustic wave moves through a fluid, some acoustic energy is absorbed by the fluid element, resulting in the attenuation of acoustic pressure and velocity amplitude in propagation direction. This generates a steady, time-averaged net force in the direction of propagation, resulting in a jet of fluid in the same direction [66, 67]. In the case of confined fluid channels, this jet flow is typically accompanied by a corresponding backflow due to the confined nature of fluid (see Figure 2.1). It should be noted that to generate appreciable Eckart streaming within a fluid system, the propagation of the acoustic waves must occur over a dimension that is comparable to or larger than the acoustic attenuation length. As the attenuation length is known to be inversely proportional to the square of actuation frequency [68], significant Eckart streaming is generated within high-frequency acoustofluidic systems only if they have sufficiently large length scales.

(2) *Boundary-driven streaming*: This streaming refers to the flows that arise from viscous dissipations of acoustic waves in the boundary layer region near a solid boundary. Specifically, the no-slip condition near a solid boundary results in steep velocity gradients in a thin boundary layer region of fluid whose thickness is given as $\delta_v = (2\mu/(\rho\omega))^{1/2}$, where μ and ρ are the fluid viscosity and density, respectively, and ω is the (angular) frequency [69]. These velocity gradients cause strong dissipations of acoustic waves and result in the generation of flow vorticity within this region. In case of standing waves, with fixed nodal and antinodal positions, this yields a streaming flow within the boundary layer, referred to as Schlichting streaming or inner boundary layer streaming [70]. At the edge of the boundary layer, this boundary layer streaming flow provides a

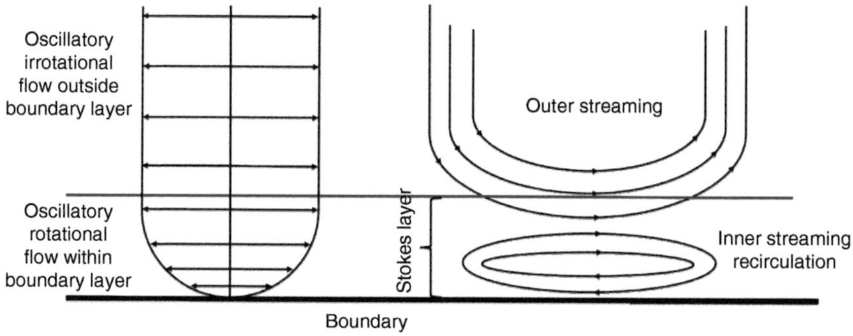

Figure 2.2 Schematic illustration of boundary-driven streaming. The dotted line represents the Stokes layer that separates the (rotational) inner streaming flow and (irrotational) outer streaming flow. Source: Adapted from Sadhal [66].

slip velocity for the outer flow, and results in a counter-rotating streaming flow outside the boundary layer that is referred to as the outer boundary layer streaming or Rayleigh streaming [5]. Figure 2.2 shows a schematic representation of the generation of boundary layer streaming, depicting the unsteady oscillatory flow as well as the resulting inner and outer boundary layer streaming flows. Boundary layer-driven streaming is the primary form of streaming flows observed in microfluidic devices that employ an oscillating boundary (e.g. channel walls, sharp edges or cilia, and bubble interface) [42, 46, 71–82] actuated via acoustic waves with frequencies in the kHz–MHz range.

In addition to these classifications, streaming flows are also classified based on the magnitude of streaming velocity, relative to the fluid's leading-order acoustic response. Specifically, when the velocity of the harmonic acoustic component is larger than the acoustic streaming velocity, the streaming is referred to as slow streaming. Conversely, when the acoustic streaming velocity is larger than the velocity of the harmonic acoustic component, it is referred to as fast streaming.

Typically, streaming flows are analyzed via a perturbation approach based on the identification of an appropriate large or small parameter. One such choice pertains to the definition of a frequency-based Reynolds number, where the velocity is taken to be the product of frequency and a length scale. Correspondingly, the Womersley number, α is defined as $\alpha = (\rho\omega a^2/\mu)^{1/2}$, where ρ is the fluid density, μ is the fluid viscosity, ω is the (angular) frequency, and a is a suitable length scale. Another commonly adopted choice for a small parameter is based on the relative size of the velocity amplitude with a frequency-based velocity, i.e. $U_0/\omega a$, where U_0 is velocity amplitude of fluid oscillations. We note that while the decomposition of the fluid response into an acoustic and streaming component is relatively straightforward, the appropriate small or large parameters for such decompositions to be meaningful are not always identified in the literature. Similarly, derivations of leading-order and streaming governing equations from the full compressible Navier–Stokes equations typically employ inherent assumptions that are not always made mathematically transparent [83]. We remark that from a strict mathematical perspective, the slow streaming theory is based on a rigorous asymptotic perturbation expansion approach; however, the same cannot be said about the fast streaming theory [18]. In what follows, we will provide the derivations of the governing equations for the slow streaming theory and a brief discussion of the typical approach for modeling fast streaming.

2.3.2.1 Modeling Approach for Slow Streaming

In the slow streaming regime, the fluid variables, pressure, density, velocity, and temperature can be decomposed as [84]

$$p = p_0 + \varepsilon\tilde{p}_1 + \varepsilon^2\tilde{p}_2 + O(\varepsilon^3) + \dots \tag{2.12a}$$

$$\rho = \rho_0 + \varepsilon\tilde{\rho}_1 + \varepsilon^2\tilde{\rho}_2 + O(\varepsilon^3) + \dots \tag{2.12b}$$

$$\boldsymbol{v} = \boldsymbol{v}_0 + \varepsilon\tilde{\boldsymbol{v}}_1 + \varepsilon^2\tilde{\boldsymbol{v}}_2 + \boldsymbol{O}(\varepsilon^3) + \dots \tag{2.12c}$$

$$T = T_0 + \varepsilon\tilde{T}_1 + \varepsilon^2\tilde{T}_2 + O(\varepsilon^3) + \dots \tag{2.12d}$$

where $\varepsilon \ll 1$ is a nondimensional smallness parameter that prescribes the order of the acoustic velocity $\boldsymbol{v}_1 = \varepsilon\tilde{\boldsymbol{v}}_1$ and is typically set as the acoustic Mach number, $\varepsilon = |\boldsymbol{v}_1|/c_0$, where c_0 is the speed of sound in the fluid at equilibrium. In slow streaming, the streaming velocity $\boldsymbol{v}_2 = \varepsilon^2\tilde{\boldsymbol{v}}_2$ is smaller than the acoustic velocity and is assumed to be of second order. The zeroth-order fields describe the fluid behavior at equilibrium (i.e. in the absence of acoustic actuation). While the presence of a background flow can be considered by assuming a nonvanishing zeroth-order velocity, many microfluidic applications do not have an underlying net flow in the absence of acoustics. Under this scenario, the zeroth-order velocity can be taken to be zero. Noting this, we define

$$\boldsymbol{v}_1 = \varepsilon\tilde{\boldsymbol{v}}_1, p_1 = \varepsilon\tilde{p}_1, \rho_1 = \varepsilon\tilde{\rho}_1, T_1 = \varepsilon\tilde{T}_1 \tag{2.13a}$$

$$\boldsymbol{v}_2 = \varepsilon^2\tilde{\boldsymbol{v}}_2, p_2 = \varepsilon^2\tilde{p}_2, \rho_2 = \varepsilon^2\tilde{\rho}_2, T_2 = \varepsilon^2\tilde{T}_2 \tag{2.13b}$$

Substituting Eqs. (2.12)(a-d) into the Eqs. (2.7)–(2.9), the first-order balance of mass, momentum, and energy can be derived as

$$\frac{\partial \rho_1}{\partial t} + \rho_0(\boldsymbol{\nabla} \cdot \boldsymbol{v}_1) = 0 \tag{2.14}$$

$$\rho_0\frac{\partial \boldsymbol{v}_1}{\partial t} = -\boldsymbol{\nabla} p_1 + \mu\boldsymbol{\nabla}^2\boldsymbol{v}_1 + \left(\mu_B + \frac{\mu}{3}\right)\boldsymbol{\nabla}(\boldsymbol{\nabla} \cdot \boldsymbol{v}_1) \tag{2.15}$$

$$\rho_0 T_0\frac{\partial s_1}{\partial t} = \kappa\boldsymbol{\nabla}^2 T_1 \tag{2.16}$$

Further, the equation of state (Eq. (2.10)) yields

$$p_1 = c_0^2\rho_1 \tag{2.17}$$

Equations (2.14)–(2.17) constitute the first-order system of equations that, when complemented with boundary conditions, can be solved to obtain the harmonic component of the fluid response. Following a similar approach for the second-order terms, combined with a subsequent time-averaging operation, yields the following second-order system of equations

$$\left\langle \frac{\partial \rho_2}{\partial t} \right\rangle + \langle \nabla \cdot (\rho_0 \boldsymbol{v}_2) \rangle + \langle \nabla \cdot (\rho_1 \boldsymbol{v}_1) \rangle = 0 \tag{2.18}$$

$$\left\langle \rho_0 \frac{\partial \boldsymbol{v}_2}{\partial t} \right\rangle + \left\langle \rho_1 \frac{\partial \boldsymbol{v}_1}{\partial t} \right\rangle + \langle \rho_0 (\nabla \boldsymbol{v}_1) \boldsymbol{v}_1 \rangle = -\langle \nabla p_2 \rangle + \mu \langle \nabla^2 \boldsymbol{v}_2 \rangle + \left(\mu_B + \frac{\mu}{3} \right) \langle \nabla (\nabla \cdot \boldsymbol{v}_2) \rangle \tag{2.19}$$

$$\left\langle (\rho_0 T_1 + \rho_1 T_0) \frac{\partial s_1}{\partial t} \right\rangle + \langle \rho_0 T_0 \nabla s_1 \cdot \boldsymbol{v}_1 \rangle = \mu \left\langle \left(\nabla \boldsymbol{v}_1 + (\nabla \boldsymbol{v}_1)^T - \frac{2}{3} (\nabla \cdot \boldsymbol{v}_1) \boldsymbol{I} \right) : \nabla \boldsymbol{v}_1 \right\rangle + \mu_B \left\langle (\nabla \cdot \boldsymbol{v}_1)^2 \right\rangle + \kappa \langle \nabla^2 T_2 \rangle \tag{2.20}$$

where $\langle N \rangle$ represents the time average of N over an oscillation period. This system of equations, coupled with the appropriate boundary conditions, can be used to obtain the time-averaged fluid response. Note that both the first- and second-order problems are linear. Therefore, the perturbation approach can be viewed as the simplification of the original nonlinear problem into two sets of linear problems that allow successive solutions (i.e. are not two-way coupled). It should also be noted that the validity of this approach is predicated on the assumption that the series in Eq. (2.12) are convergent. In other words, the streaming solution must be of an order higher than the acoustic solution. This assumption provides a mathematically rigorous definition of the slow streaming regime. We note that this perturbation approach for slow streaming has served as the foundation of extensive numerical analyses for acoustofluidic applications and has provided important insights into emerging fluid manipulation systems [63, 85–91].

2.3.2.2 Modeling Approach for Fast Streaming

For cases where the streaming velocity is of similar or larger order than the acoustic velocity, the above series decomposition is non-convergent and becomes invalid. For such cases, it has been proposed to decompose the fluid variables into their harmonic and time-averaged steady components as

$$p = p_1(x, y, z, t) + p_{dc}(x, y, z) \tag{2.21a}$$

$$\rho = \rho_1(x, y, z, t) + \rho_{dc}(x, y, z) \tag{2.21b}$$

$$\boldsymbol{v} = \boldsymbol{v}_1(x, y, z, t) + \boldsymbol{v}_{dc}(x, y, z) \tag{2.21c}$$

Substitution of these expressions into Eqs. (2.7)–(2.10) and subsequent time-averaging yields a set of nonlinear equations that can be solved to obtain the harmonic and streaming solutions. We remark that, in contrast to slow streaming, where the oscillatory and mean systems of equations can be solved sequentially, fast streaming results in a set of fully coupled equations such that the oscillatory and mean components need to be solved simultaneously. We refer the interested reader to Zarembo [92], Friend and Yeo [19] for further details.

2.3.3 Distinction Between Lagrangian and Eulerian Fluid Velocity and Stokes' Drift

The confusion regarding the distinction between Lagrangian and Eulerian fluid velocity can be a significant cause of misunderstanding in acoustic streaming flows. This has been described by Chu and Apfel [93] in relation to a system consisting of a vibrating piston at the open end of a tube. In such a system, the Eulerian mean of the velocity field is nontrivial, but the mean mass flux of the system is zero. This apparent paradox between the mean flow velocity and absence of mass flow can be explained by recognizing the fact that it is the Lagrangian mean flow and not the Eulerian flow that represents the mean velocity of a fluid particle. Therefore, it is the Lagrangian mean velocity that is expected to be zero in the absence of mass flow, and the Lagrangian mean velocity is indeed zero for this system. Specifically, the boundary condition at the oscillating wall can be written as

$$\boldsymbol{v}(\mathbf{r}, t)|_{\mathbf{r}=\mathbf{r}_0+\mathbf{d}(\mathbf{r}_0,t)} = \frac{\partial}{\partial t}\mathbf{d}(\mathbf{r}_0, t) \tag{2.22}$$

where $\mathbf{r}_0$ denotes the position of oscillating surface at rest and $\mathbf{d}(\mathbf{r}_0, t)$ denotes a time-harmonic oscillation of the surface. In other words, the no-slip boundary condition at an oscillating wall demands that the fluid velocity be zero at the displaced position of the interface and not at the mean (undisturbed) position [94] (see Figure 2.3a). Furthermore, a Taylor series expansion of this boundary condition shows that the appropriate mean Eulerian velocity boundary condition at the second order is $\boldsymbol{v}_2 = -\langle(\nabla \boldsymbol{v}_1)\xi_1\rangle$. In acoustic literature, this difference between Lagrangian and Eulerian velocity is frequently described by introducing the notion of Stokes' drift. Specifically, Stokes drift is given as $\langle(\nabla \boldsymbol{v}_1)\xi_1\rangle$, where ξ_1 is the

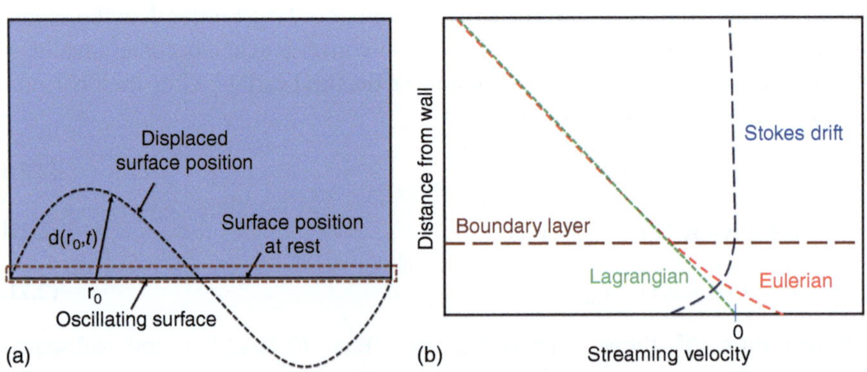

Figure 2.3 Schematic plot of Eulerian velocity, Lagrangian velocity, and Stokes drift. (a) Schematic illustration of the displaced surface position and surface position at rest. The no-slip condition should be satisfied at the displaced position of the surface. (b) shows a zoom-up of the velocity distribution in the region near the oscillating surface (denoted by brown box in panel (a)). The Eulerian velocity is equal to and opposite of Stokes drift at the wall; therefore, Lagrangian velocity is zero at the oscillating wall.

first-order displacement field, obtained by integrating the first-order velocity field $\int \boldsymbol{v}_1 dt$. Using this, the Lagrangian flow velocity is given as

$$\boldsymbol{v}^L = \boldsymbol{v}_2 + \langle (\nabla \boldsymbol{v}_1)\xi_1 \rangle \tag{2.23}$$

where $\boldsymbol{v}_2$ is the mean Eulerian flow velocity, and $\boldsymbol{v}^L$ is the mean Lagrangian velocity. As shown in Figure 2.3b, the Stokes drift and the mean Eulerian $\boldsymbol{v}_2$ at the oscillatory boundary are equal and opposite, resulting in a zero Lagrangian velocity. The above expression also indicates that proper care must be taken to resolve the first-order velocity field gradients near the boundary to accurately obtain the Stokes' drift and Lagrangian velocity in numerical analyses [83, 94]. To partially mitigate this need, the first- and second-order governing equations have recently been formulated directly based on the Lagrangian mean velocity; this makes the decomposition of boundary conditions into their first- and second-order straightforward and offers numerical advantages [83].

2.3.4 Acoustic Streaming Near Solid Particles

In this section, we briefly discuss the generation of streaming around solid particles. As indicated earlier, sound waves in infinite media are typically characterized by a lack of vorticity. However, the presence of solid boundaries (and accompanying no-slip boundary conditions) results in the generation of vorticity within a thin boundary layer near the fluid–solid interface, which causes streaming flow both within and outside of the boundary layer. The understanding of streaming around solid particles is particularly important for acoustic levitation techniques where solid particles are positioned at velocity nodes or antinodes by leveraging the acoustic radiation force on these particles in an acoustic field. Analytical solutions for streaming around solid particles have been obtained for several cases, including for different positions of spherical particles, prolate and oblate spheroids, and traveling or standing acoustic waves [95–98]. These analyses typically build on the perturbation expansion approach by identifying a suitably small or large parameter to obtain streaming solutions both within and outside the Stokes boundary layer. Given the small thickness of the Stokes boundary layer for high-frequency acoustic waves, the outer streaming appears to be governed by a slip velocity at the interface obtained from the inner streaming solution. Therefore, if the details of the solution within the Stokes layer are not of interest, the outer streaming can be analyzed by assigning an appropriate slip boundary condition at the interface [66, 99–101]. While the specific details of these solutions are beyond the scope of this chapter, we remark that several interesting streaming patterns ensue from changes in the shape or position of a solid particle relative to velocity node/antinode of the acoustic wave (see Figure 2.4).

2.3.5 Acoustic Streaming Near Fluid–Fluid Interfaces

The streaming near a fluid–fluid interface such as a droplet or bubble placed in an acoustic field differs from the case of streaming around a solid particle due to the addition of interfacial mobility. In this class of streaming flows, the streaming

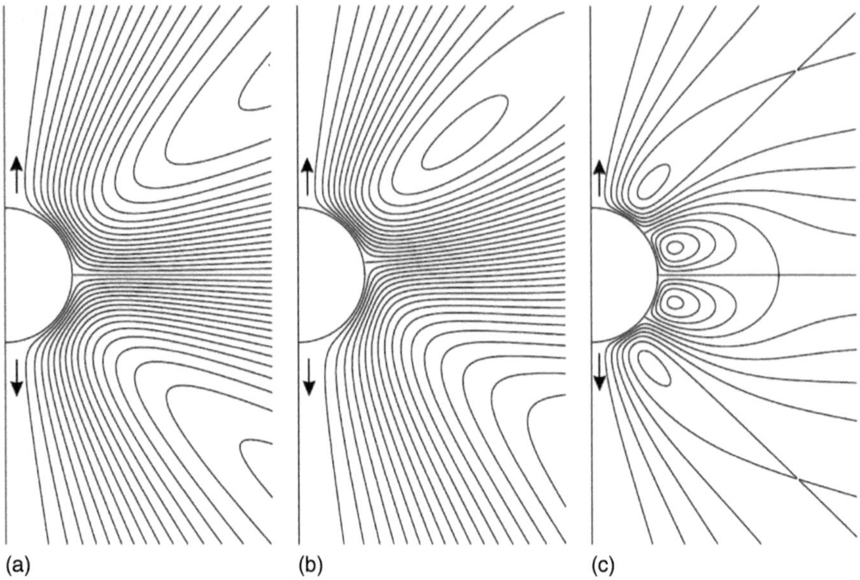

Figure 2.4 Streaming flow around a small rigid sphere placed at (a) velocity antinode, (b) slightly below the antinode, and (c) at the velocity node, in a plane standing acoustic wave field. Source: Rednikov and Sadhal [102]/Reproduced with permission from Cambridge University Press.

motion is observed both within and outside the droplet. Here, the classical results from Riley [103] have been generalized by Zhao et al. [104] to consider a liquid drop positioned at the velocity antinode of the acoustic field. Specifically, these results are obtained by modifying the interface boundary condition from no slip to a combination of zero normal velocity, continuity of velocity, and shear stress. The details of this analytical solution can be found in [104]. We note that this solution predicts the absence of recirculating layer when the droplet viscosity is sufficiently low. Unlike solid particles, where a thin Stokes layer results in significant shear stresses leading to vorticity generation, the added interfacial mobility in the case of a droplet mitigates this effect. Therefore, while droplets with sufficiently high viscosity do have a recirculating region like solid particles, progressive reduction of droplet viscosity results in the eventual cessation of the recirculating layer. Indeed, the limiting case of streaming around bubbles is characterized by the lack of a recirculating layer; this further results in the outer streaming direction being opposite that for a solid particle. The recirculating layer does appear for the case of surface-contaminated bubbles, which behave similar to rigid particles. The theoretical developments for streaming around a bubble have followed a similar approach as the droplet case by generalizing Riley's classical solution to consider a shear-free boundary condition at the bubble interface [105]. Depending on the modes of bubble oscillations (volume pulsation, radial, translational, etc.), different streaming flows can be obtained [106] (See Figure 2.5).

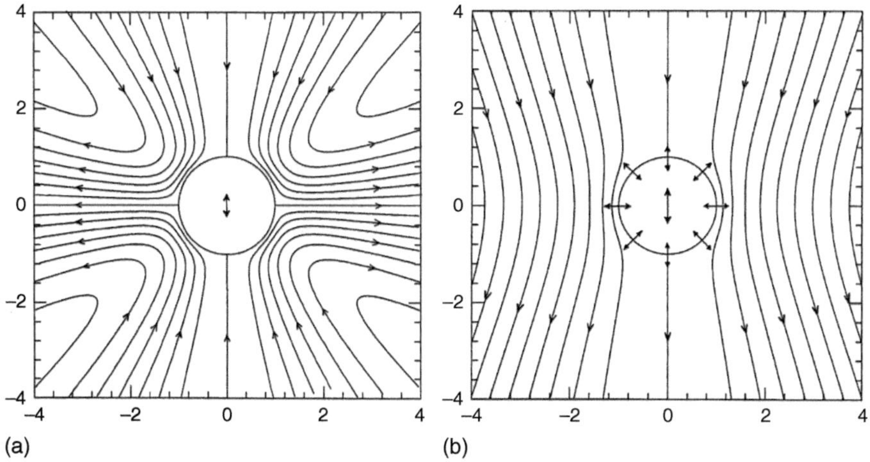

Figure 2.5 Streamlines of the Lagrangian streaming outside the boundary layer of a bubble undergoing: (a) purely translational oscillations, and (b) both translational and radial oscillations. Source: Longuet-Higgins [106]/Reproduced with permission from The Royal Society.

2.4 Forces in Acoustofluidic Systems

In this section, we describe the typical forces experienced by foreign inclusions (e.g. a droplet, bubble, and solid particle) within a fluid system subjected to an acoustic field. The particles in an acoustic field experience a variety of forces, including a primary acoustic radiation force due to direct interaction with acoustic wave, a secondary acoustic radiation force due to multiparticle interaction with acoustics, and a streaming-induced drag force. The combination of these forces results in a net particle motion that is commonly referred to as acoustophoretic motion. In the following, we discuss the physical origins and mathematical descriptions of these forces.

2.4.1 Primary Acoustic Radiation Force

A particle within a fluid system subjected to acoustic actuation experiences a time-averaged primary acoustic radiation force. The interaction of solid particles with acoustic waves and the resulting acoustic radiation forces have been rigorously studied in acoustic literature since the nineteenth century. We note that these forces are sometimes described by the term "acoustic radiation pressure," but the use of this term is a bit of misnomer since it refers to a force, which is a vector quantity. From a continuum mechanics perspective, this force can be understood as the time average of the traction applied by the fluid on the solid particle. Noting this, for a general solid particle with its boundary denoted as $S(t)$ and its corresponding outward unit normal denoted as $\boldsymbol{n}$, the net force is given as $\int_{S(t)} \boldsymbol{\sigma n}\, dS$, where $\boldsymbol{\sigma} = -p\boldsymbol{I} + \mu(\boldsymbol{\nabla v} + (\boldsymbol{\nabla v})^T) + \left(\mu_B - \frac{2\mu}{3}\right)(\boldsymbol{\nabla}.\boldsymbol{v})\boldsymbol{I}$, is the Cauchy stress in the Newtonian fluid. Given the time-scale separation between the acoustic actuation and the mean fluid response, this expression can be time-averaged to obtain

$$\boldsymbol{F} = \left\langle \int_{S(t)} \boldsymbol{\sigma n}\, dS \right\rangle \tag{2.24}$$

Under the assumptions stipulated in the perturbation expansion approach, the Cauchy stress tensor can be decomposed into corresponding first- and second-order components, and acoustic radiation force can be written as

$$\boldsymbol{F} = \int_{S_0} \langle \boldsymbol{\sigma}_2 - \rho_0 \boldsymbol{v}_1 \otimes \boldsymbol{v}_1 \rangle \boldsymbol{n}\, dS \tag{2.25}$$

where $\boldsymbol{\sigma}_2$ is the second-order component of the Cauchy stress tensor, and S_0 is the equilibrium configuration of the boundary $S(t)$. Based on this general expression of the radiation force, several researchers have derived expressions to describe radiation forces for specific types of acoustic waves (e.g. traveling wave, standing wave), fluid media (e.g. inviscid, viscous), particle shapes (e.g. spherical, cylindrical), and particle positions (e.g. velocity node, velocity antinode). For instance, King derived the expression for primary radiation force on a rigid spherical incompressible particle in an inviscid fluid subjected to a standing wave field as

$$\boldsymbol{F}_{\boldsymbol{s}}^{\boldsymbol{rad}} = \pi \rho_0 |A|^2 (kR_0)^3 \frac{5 - 2\lambda_p}{3(2 + \lambda_p)} \sin(2kd) \tag{2.26}$$

where ρ_0 is the fluid density at equilibrium, A is the complex amplitude of the velocity potential of the incident acoustic field, $k = \omega/c$ is the acoustic wavenumber, ω is the driving (angular) frequency, c is the speed of sound in the fluid, R_0 is the radius of sphere, $\lambda_p = \rho_0/\rho_p$ with ρ_p being the particle density, and d is the distance between the equilibrium center position and the nearest velocity node plane of a standing acoustic wave. For heavy particles (with $\rho_p > 0.4\rho_0$), the acoustic radiation force pushes the particles to pressure nodes (or velocity antinodes). In contrast, lighter particles (with $\rho_p < 0.4\rho_0$) are pushed toward the pressure antinodes (or velocity nodes).

This expression was later generalized by Yosioka and Kawasima [107] for a compressible spherical particle in an acoustic wave field. The general expression for a compressible particle was then employed to obtain approximations of the primary acoustic radiation force on a liquid drop ($\boldsymbol{F}_{\boldsymbol{d}}^{\boldsymbol{rad}}$) and a bubble ($\boldsymbol{F}_{\boldsymbol{b}}^{\boldsymbol{rad}}$) in a standing acoustic wave field in the limit of long sound wavelength (i.e. $kR_0 \ll 1$) as

$$\boldsymbol{F}_{\boldsymbol{d}}^{\boldsymbol{rad}} = \frac{\pi}{3} \rho_0 |A|^2 (kR_0)^3 \left(\frac{5 - 2\lambda_p}{2 + \lambda_p} - \frac{\lambda_p c^2}{\tilde{c}^2} \right) \sin(2kd) \tag{2.27a}$$

$$\boldsymbol{F}_{\boldsymbol{b}}^{\boldsymbol{rad}} = \frac{\pi \rho_0 |A|^2 k R_0 \left(1 - \frac{\omega_0^2}{\omega^2}\right) \sin(2kd)}{\left(1 - \frac{\omega_0^2}{\omega^2}\right)^2 + \delta_{rad}^2} \tag{2.27b}$$

where R_0 is now the equilibrium radius of particle (droplet or bubble), $\lambda_p = \rho_0/\rho_p$ with ρ_p being the equilibrium particle density, d is the distance between the equilibrium center position and the nearest velocity node plane of the standing acoustic

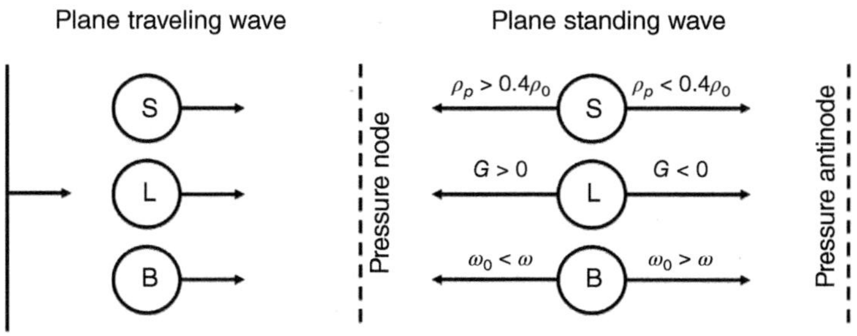

Figure 2.6 Motion of immersed foreign inclusions within a weakly dissipating plane traveling and standing wave field. *S* denotes a rigid sphere, *L* denotes a liquid droplet, and *B* denotes a bubble. $G = \frac{5-2\lambda_p}{2+\lambda_p} - \frac{\lambda_p c^2}{\tilde{c}^2}$ denotes the ratio that appears in Eq. (2.27a). Source: Adapted from Doinikov [108].

wave, $\delta_{rad} = kR_0$ is the radiation damping constant, and ω_0 is the fundamental resonance frequency of the bubble given as

$$\omega_0 = \frac{\tilde{c}}{R_0}\left(\frac{3}{\lambda_p}\right)^{\frac{1}{2}} \tag{2.28}$$

We note that while the behavior of a droplet is distinct from that of a solid particle in general, the above expression for force on a droplet consistently reduces to that for the force on a solid in the limit case of $c/\tilde{c} \to 0$. Further, it can be observed from the radiation force expression for a bubble that large bubbles (with $\omega_0 < \omega$) are forced toward the pressure nodes (or velocity antinodes) while small bubbles (with $\omega_0 > \omega$) are pushed toward the pressure antinodes (or velocity nodes). Figure 2.6 provides a summary of the motion of a solid particle (*S*), liquid drop (*L*), and a gas bubble (*B*) under the action of a plane traveling or standing acoustic wave.

The above expressions represent some of the earliest theoretical developments for behavior of particles in an acoustic field and are strictly valid only for an ideal (i.e. nonviscous and non-heat-conducting) fluid. Since then, the viscous nature of fluids has been accounted for by several authors, including Doinikov [109] and Settnes and Bruus [110]. More recently, Karlsen and Bruus [111] have extended this analysis to thermoviscous fluids. In the following, we consider an immersed particle of radius $a \ll \lambda$ (where λ is the wavelength in the fluid), mass density ρ_P and compressibility κ_P. For this case, the primary radiation force can be written as [111]

$$\boldsymbol{F}^{rad} = -\pi a^3 \left[\frac{2\kappa_0}{3} Re\left[f_1^* p_1^* \nabla p_1\right] - \rho_0 Re\left[f_2^* \boldsymbol{v}_1^* \cdot \nabla \boldsymbol{v}_1\right]\right] \tag{2.29}$$

where $\kappa_0 = 1/\left(\rho_0 c_0^2\right)$ is the compressibility of the liquid, $Re[N]$ denotes the real part of quantity N, the asterisk denotes complex conjugates, and the coefficients f_1 and f_2 are given as

$$f_1 = 1 - \frac{\kappa_P}{\kappa_0} \tag{2.30a}$$

$$f_2 = \frac{2(\rho_P - \rho_0)(1-\gamma)}{2\rho_P + \rho_0(1-3\gamma)}; \quad \gamma = -\frac{3}{2}\left[1 + i(1+\tilde{\delta})\right]\tilde{\delta}; \quad \tilde{\delta} = \frac{\delta}{a}; \quad \delta = \sqrt{\frac{2\mu}{\omega\rho_0}} \tag{2.30b}$$

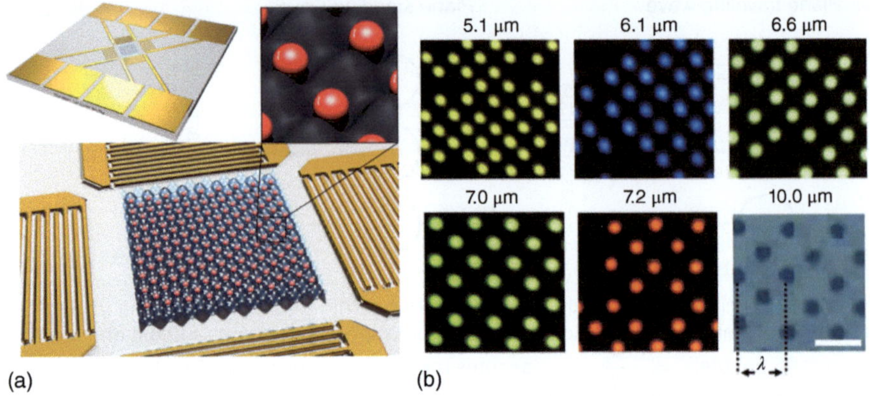

Figure 2.7 Acoustic radiation force-based 2D single-cell patterning. (a) Schematic illustration of a device generating four sets of intersecting surface acoustic waves to realize single-cell patterning per pressure node. (b) Patterning of different sizes of spherical particles in a two-dimensional acoustic grid. Source: Reproduced from Collins et al. [124], Springer Nature/CC BY 4.0.

We note that the above expression for the primary acoustic radiation force neglects the fluid viscosity at the second-order level. In contrast, Doinikov [112] considers fluid viscosity at the second-order level, and therefore, accounts for the presence of a streaming flow that is generated solely due to a particle's interaction with an acoustic field. In fact, Doinikov's analysis predicts an inversion of acoustic radiation force direction such that heavier particles can be pushed toward the pressure antinode (or velocity node), provided that the thickness of the viscous boundary layer is larger than the particle radius. These predictions were confirmed by Baasch et al. [113] via a numerical analysis where streaming around the solid particle was found to contribute significantly to the net acoustic radiation force.

From an application perspective, primary radiation force on particles has been leveraged for many exciting applications, including particle trapping [114, 115], cell separation [31, 116, 117], particle/cell patterning [118–121], and cell washing [122, 123]. Figure 2.7 shows a representative example of an acoustofluidic device comprised a microfluidic chamber bonded to a piezoelectric substrate. An acoustic standing wave field is generated via interdigital transducers patterned onto a piezoelectric substrate, which is used for the trapping of spherical particles and cells of different sizes at the pressure nodes of the acoustic field with one cell per acoustic well. This technology allows the study of isolated phenomena in a small cell population and can realize cell trapping and observation for long durations to enable drug discovery applications.

2.4.2 Secondary Acoustic Radiation Force

In the case of multiparticle systems or systems with particles in the vicinity of walls, the scattering from each particle results in secondary radiation forces that can cause attraction or repulsion between particles. This secondary acoustic radiation force is typically employed in microfluidic applications to form particle aggregates or to trap particles/cell near an oscillating boundary such as sharp edges [81], cilia [125], or bubble surfaces [75, 126]. Different expressions for secondary acoustic radiation force have been derived for various combinations of particles (solid particles, liquid drops, or bubbles) under consideration. We note that the literature in this field is rather extensive, and a complete overview is beyond the scope of this chapter. However, in the following, we discuss expressions of the secondary radiation force for some combinations of these particles.

2.4.2.1 Forces Between Two Rigid Spheres

The interaction force between two rigid spheres was first derived by König [3]. For the case where spheres are free to move due to an acoustic field, the force on the jth rigid sphere, $F_j (j = 1, 2)$ is given as

$$\boldsymbol{F}_j = \frac{3\pi\rho_0 R_1^3 R_2^3}{L^4} \langle \boldsymbol{U}_1(\boldsymbol{U}_2 \cdot \boldsymbol{e}_j) + \boldsymbol{U}_2(\boldsymbol{U}_1 \cdot \boldsymbol{e}_j) + \boldsymbol{e}_j(\boldsymbol{U}_1 \cdot \boldsymbol{U}_2) - 5\boldsymbol{e}_j(\boldsymbol{U}_1 \cdot \boldsymbol{e}_j)(\boldsymbol{U}_2 \cdot \boldsymbol{e}_j)\rangle \tag{2.31}$$

where R_j is the radius of jth sphere, e_j is the unit vector pointing from the equilibrium center of sphere j to the equilibrium center of the other sphere, $\langle\ \rangle$ denotes the time average, and U_j is the velocity of jth sphere relative to the velocity of the host fluid. This velocity is given as $U_j = 2(\rho_0 - \rho_f)v_I(r_j)/(\rho_0 + 2\rho_j)$, where $v_I(r_j)$ is the fluid velocity generated by the incident acoustic field at the location of jth sphere in the absence of both spheres, and ρ_j is the density of jth sphere. This analysis was further extended for a compressible fluid by Doinikov and Zavtrak [127], which resulted in the addition of three long-range terms (proportional to L^3, L^2, and L) to the above expression. Further, it was observed that all the force terms depend on the re-scattering phase, kL. Therefore, the sign of the interaction force oscillates with increases in L.

2.4.2.2 Forces Between Two Bubbles

The interaction force between two pulsating bubbles was first studied experimentally as well as theoretically by C.A. Bjerknes and V.F.K. Bjerknes. In the acoustic literature, it is customary to refer to this interaction force as the secondary Bjerknes force [128]. This force can be expressed as

$$\boldsymbol{F}_B = \frac{2\pi\rho_0 |A|^2 R_{10} R_{20}}{L^2 \left(1 - \frac{\omega_1^2}{\omega^2}\right)\left(1 - \frac{\omega_2^2}{\omega^2}\right)} \tag{2.32}$$

where R_{10} and R_{20} are the bubble radii at equilibrium, L is the distance between their centers at equilibrium, and ω_1 and ω_2 are their fundamental resonant frequencies. A positive value of this force indicates mutual attraction between the bubbles, while a negative value indicates repulsion. It can be clearly observed from the above expression that the interaction force is attractive in all cases except when the driving frequency is between the fundamental frequencies of the bubbles, in which case the interaction force becomes repulsive. The above expression was derived under several simplifying assumptions, including (i) an ideal incompressible host fluid,

(ii) adiabatic law is valid for the gas within the bubbles, (iii) a weak incident field, and (iv) a large separation between bubble centers, $L \gg R_{10}, R_{20}$. However, in cases where these assumptions are not satisfied, several interesting behaviors have been observed in the bubble systems, including the formation of so-called bubble grapes, where bubbles form clusters that are bigger than their resonance size and do not undergo coalescence or breakup as long as the acoustic field is switched on. This phenomenon was later explained by Zabolotskaya [129] who pointed out that it is crucial to consider the influence of bubbles' scattered fields on each other's pulsations to explain the formation of bubble grapes. This analysis revealed that the interaction force can switch from attractive to repulsive as the bubbles come closer to each other, leading to the formation of a bubble pair with stable separation distance. This theory has been further expanded by Doinikov and Zavtrak [130, 131] to account for multiple re-scatterings of acoustic fields between two bubbles as well as their oscillations. Further, Doinikov also reported the changes to interaction forces for cases where fluid viscosity is considered.

2.4.2.3 Forces Between a Solid Particle and a Bubble

The interaction force between a solid particle and a bubble is expressed as

$$\boldsymbol{F}_p = \frac{4\pi\rho_0|A|^2R_{b0}^2R_p^3}{L^5\left[\left(1-\frac{\omega_{bo}^2}{\omega^2}\right)^2+\delta_b^2\right]}\left(\frac{\rho_p-\rho_0}{\rho_0+2\rho_p}\right)\boldsymbol{e}_p \tag{2.33}$$

where R_{bo} is the equilibrium radius of the bubble, R_p is the solid particle radius, ρ_p is the solid particle density, ω_{bo} is the monopole resonant frequency of the bubble, and $\boldsymbol{e}_p$ is the unit vector pointing from the center of the solid particle at equilibrium to that of the bubble. It can be observed from Eq. (2.33) that solid particles with a density larger than the host liquid will experience an attractive force toward the bubble, while solid particles that are lighter than the host liquid are repelled away from the bubble.

This expression assumes an ideal, incompressible host liquid and considers the oscillations of the solid particle to be a result of the scattered acoustic field of the bubble. However, the solid particle oscillations also arise from the incident acoustic field. Taking this into consideration, the interaction force can be expressed as

$$\boldsymbol{F}_p = \left(\frac{2}{3}\right)\pi R_p^3R_{b0}^2(\rho_p-\rho_0)L^{-3}Re\{\dot{R}_b[\boldsymbol{U}_p-3(\boldsymbol{e}_p\cdot\boldsymbol{U}_p)\boldsymbol{e}_p]\} \tag{2.34}$$

where $\dot{R}_b$ is the radial pulsation velocity of the bubble and $\boldsymbol{U}_p$ is the particle's translational velocity.

Similar to the situation for bubble–bubble interaction, the consideration of fluid compressibility would yield additional long-range force terms that are important at large separation distances. On the other hand, for small distances, multiple re-scattering between the bubble and solid particle becomes important. Doinikov and Zavtrak [130, 131] have reported a numerical study demonstrating that at small distances, the interaction force can change significantly both in magnitude and direction.

2.4.2.4 Forces Between a Liquid Drop and a Bubble

The interaction forces between a bubble and fluid droplet have received significant interest from researchers, motivating the development of biomedical ultrasonic tools to study the interaction of bubbles with blood plasma components (e.g. red blood cells). Here, a liquid drop model can be viewed as a better representation of a red blood cell than a rigid sphere model. As a first approximation, if the separation distance between the bubble and liquid drop is sufficiently large to ignore their shape oscillations, then the interaction force can be expressed by the same expression as that for the interaction between a bubble and a solid particle. However, for small separations, the shape oscillations become important. This scenario was first analyzed by Doinikov, who expressed the interaction force as an infinite series and employed recurrence formulas to obtain the solution. While this analysis ignored the drop's compressibility, it allowed for modeling of the drop's shape oscillations. The numerical solution from Doinikov predicted several interesting phenomena for the case when the driving frequency is close to the natural frequencies of the shape modes of the bubble or the droplet. Specifically, as the distance between the bubble and the droplet decreases, the interaction force can change sign. Further, it was found to increase in magnitude faster than the prediction for the case between a bubble and solid particle (Eq. (2.33)). For droplets lighter than the host fluid, this force switches from repulsion to attraction as the separation distance is reduced, allowing a light droplet to be attracted to a bubble if they are brought sufficiently close. Conversely, for droplets heavier than the host fluid, the interaction force changes from attraction to repulsion with a decrease in separation distance, allowing a heavy droplet and a bubble to form a stable pair with a constant separation distance.

Having described some representative expressions for secondary radiation forces in acoustic fields, we remark that these forces have been employed for a wide variety of applications such as particle assembly [132, 133], cluster formation [108, 134], and acoustic imaging [135–137]. Figure 2.8 shows an example of a microfluidic platform that employs microbubbles trapped in microcavities along the channel sidewalls to enable 3D rotation of *Caenorhabditis elegans*. An acoustic field is utilized to induce bubble oscillation and generate acoustic streaming and radiation force fields within the microchannel. As a result, *C. elegans* are trapped within the device by multiple oscillating bubbles and can be rotated by varying degrees to expose *C. elegans* for imaging and observation.

2.4.3 Hydrodynamic Drag Force

In addition to the direct interaction with the acoustic field, an immersed particle can also experience a drag force arising from the streaming generated within the fluid itself. This case is frequently considered to numerically obtain the trajectories of the tracer particles in acoustofluidic microdevices. In the absence of wall effects, the drag force can be expressed as

$$\boldsymbol{F}^{drag} = 6\pi\mu a(\boldsymbol{v}_{str} - \boldsymbol{v}_P) \tag{2.35}$$

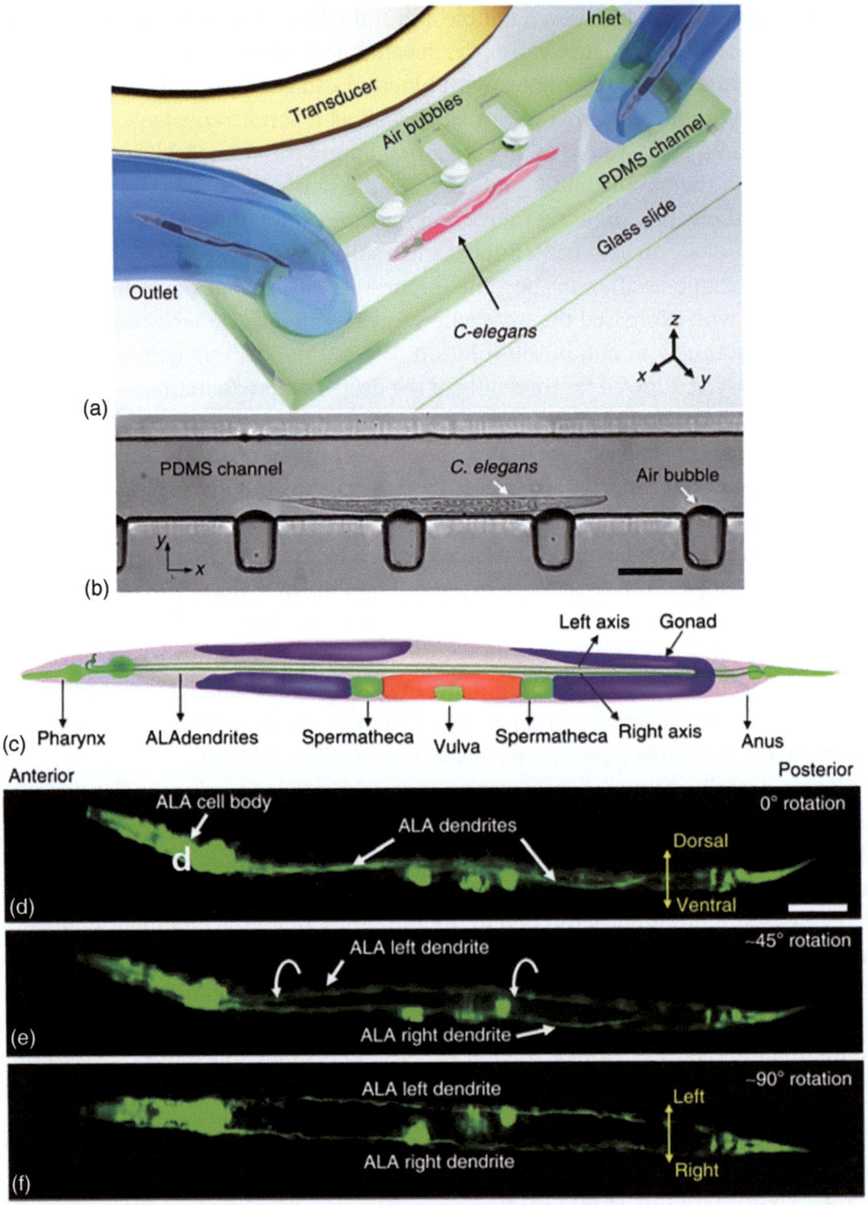

Figure 2.8 (a) Schematic illustration of the experimental setup for an acoustofluidic rotational manipulation device. Air microbubbles are trapped within the sidewall microcavities of the channel. A piezoelectric transducer that generates acoustic waves is placed next to the microfluidic channel. (b) Optical image showing a mid-L4 stage *C. elegans* trapped by multiple oscillating microbubbles. Scale bar: 100 μm. (c) Schematic illustration of *C. elegans* showing the anatomy of ALA cell body and dendrites. (d–f) Fluorescence image of a *C. elegans* at 0°, 45°, and 90° rotations. Source: Reproduced from Ahmed et al. [138], Springer Nature/CC BY 4.0.

where $\boldsymbol{v}_P$ is the particle's velocity and $\boldsymbol{v}_{str}$ is the streaming velocity. We remark that $\boldsymbol{v}_{str}$ is typically taken to be the Eulerian second-order streaming velocity $\boldsymbol{v}_2$; however, this choice is usually not rigorously justified and other choices including the Lagrangian streaming velocity exist [83]. Noting this and neglecting the presence of other particles in the flow, the particle motion is then predicted via Newton's second law

$$m_P \frac{d\boldsymbol{v}_P}{dt} = \boldsymbol{F}^{rad} + \boldsymbol{F}^{drag} \tag{2.36}$$

where $\boldsymbol{F}^{rad}$ represents the radiation force on the particle. Further, the inertia of particles is frequently ignored for microfluidic applications, and the particle's velocity, $\boldsymbol{v}_P$ is simply determined via the balance of acoustic radiation force and streaming drag to yield

$$\boldsymbol{v}_P = \boldsymbol{v}_{str} + \frac{\boldsymbol{F}^{rad}}{6\pi\mu a} \tag{2.37}$$

We note that while particles in acoustofluidic systems experience both acoustic streaming- induced forces as well as radiation forces, often only one of these is leveraged for the specific application at hand. For instance, fluid manipulation applications tend to focus on the generation of acoustic streaming, while particle manipulation applications tend to rely on acoustic radiation forces and aim to suppress acoustic streaming. Equation (2.35) indicates that the motion of larger particles can be expected to be dominated by acoustic radiation forces, while smaller particles are dominated by streaming-induced hydrodynamic forces. Recent investigations have identified a transition range of particle size for which the motion is significantly affected by both the streaming and radiation forces. Nonetheless, we remark that this transition size can be expected to vary depending on the specific experimental settings and should be investigated on a case-by-case basis.

Overall, we remark that both acoustic streaming and acoustic forces have been extensively leveraged in microfluidic applications for a variety of fluid and particle manipulation applications. We refer the readers to excellent references [18–20, 139] and tutorial series [29] for a more detailed discussion of the applications of acoustic technologies in microfluidics.

2.5 Conclusions and Perspectives

Acoustic technologies offer an exciting new direction to address many fundamental and challenging problems in chemistry, biomedicine, and engineering. They are particularly appealing for biomedical applications since they offer excellent biocompatibility, contact-less and label-free manipulation of fluids and particles, and the ability to penetrate deep into biological tissues.

In this chapter, we have described the basic theories of acoustic wave propagation in fluids and solid media. We introduced the concept of acoustic streaming and outlined its common classifications. We further discussed the radiation forces experienced by foreign inclusions (e.g. solid particles, liquid drops, bubbles) in acoustic systems. It should be noted that classical analytical expressions for these forces employ inherent assumptions pertaining to the acoustic field and particle (e.g. field symmetry, particle shape, particle: wavelength ratio). These assumptions are often violated in emerging acoustic applications that employ acoustic wave propagation in complex media, where acoustic wave propagation in enclosed environments leads to multiple scattering from arbitrary particle shapes. Consequently, numerical approaches are typically needed to study acoustic phenomena in such systems. Nonetheless, classical theories continue to provide valuable insights into the physical mechanisms governing the interaction of acoustics with solid/fluid particles and serve as the foundation for the development of more rigorous theoretical and numerical methods for the ever-expanding use of acoustic technologies in novel, emerging applications.

Recent advances in micro/nano-scale acoustics and their integration with other technologies have proven immensely fruitful for a variety of emerging applications in healthcare and bioengineering. Future improvements in our ability to focus acoustic energy with improved spatial and temporal resolution are bound to lead to increased clinical adoption of acoustic technologies. Proof-of-concept demonstrations of acoustic technologies integrated into microfluidic, magnetic, and optical systems have been already reported, leading to considerable excitement concerning their potential use for novel applications such as targeted drug delivery, medical imaging, minimally invasive surgery, and cell cutting/transportation. Continued advancement in our fundamental understanding of acoustics will not only enable optimization and translation of these exciting technologies but will also facilitate further exploration and research in multidisciplinary, integrated approaches to truly revolutionize biomedicine and healthcare.

In summary, acoustics represents an exciting technology to realize novel, paradigm-changing approaches to address important challenges in biology, medicine, and engineering. Continued improvements in these technologies and corresponding theoretical developments will undoubtedly lead to development and translation of acoustic technologies for an even richer range of applications and provide promising solutions for future biomedical and healthcare applications.

References

1 Chladni, E.F.F. (2015). *Treatise on Acoustics: The First Comprehensive English Translation of EFF Chladni's Traité d'Acoustique*. Springer.

2 Faraday, M. (1831). XVII On a peculiar class of acoustical figures; and on certain forms assumed by groups of particles upon vibrating elastic surfaces. *Philosophical Transactions. Royal Society of London* 121: 299–340.

3 Konig, W. (1891). Hydrodynamics acoustical analysis. *Annals of Physical Chemistry* 42 (3): 353–370.

4 Kundt, A. (1866). Ueber eine neue Art akustischer Staubfiguren und über die Anwendung derselben zur Bestimmung der Schallgeschwindigkeit in festen Körpern und Gasen. *Annalen der Physik* 203 (4): 497–523.

5 Strutt, J.W. (1884). On the circulation of air observed in Kundt's tubes, and on some allied acoustical problems. *Philosophical Transactions. Royal Society of London* 175: 1–21.

6 Lurton, X. (2004). *An Introduction to Underwater Acoustics*. Acoustical Society of America.

7 Hartmann, W.M. (2013). *Principles of Musical Acoustics*. Springer.

8 Mehta, M., Johnson, J., and Rocafort, J. (1999). *Architectural Acoustics: Principles and Design*. Prentice-Hall, Incorporated.

9 Lurton, X. (2002). *An Introduction to Underwater Acoustics: Principles and Applications*, vol. 2. Springer.

10 Sutton, J.L. (1979). Underwater acoustic imaging. *Proceedings of the IEEE* 67 (4): 554–566.

11 Blackstock, D.T. (2001). *Fundamentals of Physical Acoustics*. Acoustical Society of America.

12 Mason, W.P. (2012). *Physical Acoustics V14: Principles and Methods*, vol. 14. Elsevier.

13 Merino-Martínez, R., Sijtsma, P., Snellen, M. et al. (2019). A review of acoustic imaging methods using phased microphone arrays. *CEAS Aeronautical Journal* 10 (1): 197–230.

14 Sarvazyan, A.P., Urban, M.W., and Greenleaf, J.F. (2013). Acoustic waves in medical imaging and diagnostics. *Ultrasound in Medicine & Biology* 39 (7): 1133–1146.

15 Cannon, R. and Aminzadeh, F. (2013). Distributed acoustic sensing: state of the art. In: *SPE Digital Energy Conference*. OnePetro.

16 Muanenda, Y. (2018). Recent advances in distributed acoustic sensing based on phase-sensitive optical time domain reflectometry. *Journal of Sensors* 2018: 1–16. https://doi.org/10.1155/2018/3897873.

17 Basaeri, H., Christensen, D.B., and Roundy, S. (2016). A review of acoustic power transfer for bio-medical implants. *Smart Materials and Structures* 25 (12): 123001.

18 Ding, X., Li, P., Lin, S.C.S. et al. (2013). Surface acoustic wave microfluidics. *Lab on a Chip* 13: 3626–3649.

19 Friend, J. and Yeo, L.Y. (2011). Microscale acoustofluidics: microfluidics driven via acoustics and ultrasonics. *Reviews of Modern Physics* 83 (2): 647–704.

20 Laurell, T. and Lenshof, A. (2014). *Microscale Acoustofluidics*. Royal Society of Chemistry.

21 Fink, M., Montaldo, G., and Tanter, M. (2003). Time-reversal acoustics in biomedical engineering. *Annual Review of Biomedical Engineering* 5 (1): 465–497.

22 Lee, A. (2013). The third decade of microfluidics. *Lab on a Chip* 13: 1660–1661.

23 Hajar, R. (2012). The art of listening. *Heart Views* 13 (1): 24.

24 Bishop, P.J. (1980). Evolution of the stethoscope. *Journal of the Royal Society of Medicine* 73 (6): 448–456.

25 Aldrich, J.E. (2007). Basic physics of ultrasound imaging. *Critical Care Medicine* 35 (5): S131–S137.

26 Valluru, K.S., Chinni, B.K., Rao, N.A. et al. (2009). Basics and clinical applications of photoacoustic imaging. *Ultrasound Clinics* 4 (3): 403–429.

27 Liu, W. and Yao, J. (2018). Photoacoustic microscopy: principles and biomedical applications. *Biomedical Engineering Letters* 8 (2): 203–213.

28 Jeon, S., Kim, J., Lee, D. et al. (2019). Review on practical photoacoustic microscopy. *Photoacoustics* 15: 100141.

29 Bruus, H., Dual, J., Hawkes, J. et al. (2011). Forthcoming Lab on a Chip tutorial series on acoustofluidics: acoustofluidics—exploiting ultrasonic standing wave forces and acoustic streaming in microfluidic systems for cell and particle manipulation. *Lab on a Chip* 11 (21): 3579–3580.

30 Wu, M., Ozcelik, A., Rufo, J. et al. (2019). Acoustofluidic separation of cells and particles. *Microsystems & Nanoengineering* 5: 32. https://doi.org/10.1038/s41378-019-0064-3.

31 Ding, X., Peng, Z., Lin, S.-C.S. et al. (2014). Cell separation using tilted-angle standing surface acoustic waves. *Proceedings of the National Academy of Sciences* 111 (36): 12992–12997.

32 Laurell, T., Petersson, F., and Nilsson, A. (2007). Chip integrated strategies for acoustic separation and manipulation of cells and particles. *Chemical Society Reviews* 36 (3): 492–506.

33 Xie, Y., Nama, N., Li, P. et al. (2016). Probing cell deformability via acoustically actuated bubbles. *Small* 12 (7): 902–910.

34 Guo, F., Li, P., French, J.B. et al. (2015). Controlling cell–cell interactions using surface acoustic waves. *Proceedings of the National Academy of Sciences* 112 (1): 43–48.

35 Barani, A., Paktinat, H., Janmaleki, M. et al. (2016). Microfluidic integrated acoustic waving for manipulation of cells and molecules. *Biosensors & Bioelectronics* 85: 714–725.

36 Zhou, Y. (2016). The application of ultrasound in 3D bio-printing. *Molecules* 21 (5): 590.

37 Nawaz, A.A., Chen, Y., Nama, N. et al. (2015). Acoustofluidic fluorescence activated cell sorter. *Analytical Chemistry* 87 (24): 12051–12058.

38 Faridi, M.A., Ramachandraiah, H., Iranmanesh, I. et al. (2017). MicroBubble activated acoustic cell sorting. *Biomedical Microdevices* 19 (2): 1–7.

39 Jakobsson, O., Grenvall, C., Nordin, M. et al. (2014). Acoustic actuated fluorescence activated sorting of microparticles. *Lab on a Chip* 14 (11): 1943–1950.

40 Huang, T.J. (2019). Acoustofluidics: Merging acoustics and microfluidics for biomedical applications. *The Journal of the Acoustical Society of America* 145 (3): 1786–1786.

41 Ozcelik, A., Ahmed, D., Xie, Y. et al. (2014). An acoustofluidic micromixer via bubble inception and cavitation from microchannel sidewalls. *Analytical Chemistry* 86 (10): 5083–5088.

42 Ahmed, D., Chan, C.Y., Lin, S.-C.S. et al. (2013). Tunable, pulsatile chemical gradient generation via acoustically driven oscillating bubbles. *Lab on a Chip* 13 (3): 328–331.

43 Huang, P.-H., Ren, L., Nama, N. et al. (2015). An acoustofluidic sputum liquefier. *Lab on a Chip* 15 (15): 3125–3131.

44 Qiu, W., Joergensen, J.H., Corato, E. et al. (2021). Fast microscale acoustic streaming driven by a temperature-gradient-induced nondissipative acoustic body force. *Physical Review Letters* 127 (6): 64501.

45 Green, R., Ohlin, M., Wiklund, M. et al. (2014). Applications of acoustic streaming. In: *Microscale Acoustofluidics* (ed. T. Laurell and A. Lenshof), 312–336. Royal Society of Chemistry.

46 Tang, S., Ayan, B., Nama, N. et al. (2016). On-chip production of size-controllable liquid metal microdroplets using acoustic waves. *Small* 12 (28): 3861–3869.

47 Bose, N., Zhang, X., Maiti, T.K., and Chakraborty, S. (2015). The role of acoustofluidics in targeted drug delivery. *Biomicrofluidics* 9 (5): 052609.

48 Salari, A., Appak-Baskoy, S., Coe, I.R. et al. (2021). Dosage-controlled intracellular delivery mediated by acoustofluidics for lab on a chip applications. *Lab on a Chip* 21 (9): 1788–1797.

49 Zervantonakis, I.K. and Arvanitis, C.D. (2016). Controlled drug release and chemotherapy response in a novel acoustofluidic 3D tumor platform. *Small* 12 (19): 2616–2626.

50 Qi, A., Yeo, L.Y., and Friend, J.R. (2008). Interfacial destabilization and atomization driven by surface acoustic waves. *Physics of Fluids* 20 (7): 74103.

51 Tan, M.K., Friend, J.R., Matar, O.K., and Yeo, L.Y. (2010). Capillary wave motion excited by high frequency surface acoustic waves. *Physics of Fluids* 22 (11): 112112.

52 Kinsler, L.E., Frey, A.R., Coppens, A.B., and Sanders, J.V. (2000). *Fundamentals of Acoustics*. Wiley.

53 de Hoop, A.T. (2001). *Handbook of Radiation and Scattering of Waves: Acoustic Waves in Fluids, Elastic Waves in Solids, Electromagnetic Waves*. Acoustical Society of America.

54 Filippi, P., Habault, D., Lefebvre, J.P. et al. (2000). *Acoustics: Basic Physics, Theory and Methods*. Acoustical Society of America.

55 Meyer, E. (2012). *Physical and Applied Acoustics: An Introduction*. Elsevier.

56 Pierce, A.D. (2019). *Acoustics: An Introduction to Its Physical Principles and Applications*. Springer.

57 Rossing, T.D. and Rossing, T.D. (2014). *Springer Handbook of Acoustics*. Springer.

58 Jerome, T.S., Ilinskii, Y.A., Zabolotskaya, E.A., and Hamilton, M.F. (2019). Born approximation of acoustic radiation force and torque on soft objects of arbitrary shape. *The Journal of the Acoustical Society of America* 145 (1): 36–44.

59 Zhang, L. and Marston, P.L. (2012). Axial radiation force exerted by general non-diffracting beams. *The Journal of the Acoustical Society of America* 131 (4): EL329–EL335.

60 Hahn, P., Leibacher, I., Baasch, T., and Dual, J. (2015). Numerical simulation of acoustofluidic manipulation by radiation forces and acoustic streaming for complex particles. *Lab on a Chip* 15 (22): 4302–4313.

61 Glynne-Jones, P., Mishra, P.P., Boltryk, R.J., and Hill, M. (2013). Efficient finite element modeling of radiation forces on elastic particles of arbitrary size and geometry. *The Journal of the Acoustical Society of America* 133 (4): 1885–1893.
62 Garbin, A., Leibacher, I., Hahn, P. et al. (2015). Acoustophoresis of disk-shaped microparticles: a numerical and experimental study of acoustic radiation forces and torques. *The Journal of the Acoustical Society of America* 138 (5): 2759–2769.
63 Bruus, H. (2012). Acoustofluidics 2: perturbation theory and ultrasound resonance modes. *Lab on a Chip* 12 (1): 20–28.
64 Bruus, H. (2012). Acoustofluidics 7: the acoustic radiation force on small particles. *Lab on a Chip* 12 (6): 1014–1021.
65 Eckart, C. (1948). Vortices and streams caused by sound waves. *Physics Review* 73 (1): 68–76.
66 Sadhal, S.S. (2012). Acoustofluidics 13: analysis of acoustic streaming by perturbation methods. *Lab on a Chip* 12 (13): 2292–2300.
67 Wiklund, M., Green, R., and Ohlin, M. (2012). Acoustofluidics 14: applications of acoustic streaming in microfluidic devices. *Lab on a Chip* 12: 2438–2451.
68 Squires, T.M. and Quake, S.R. (2005). Microfluidics: fluid physics at the nanoliter scale. *Reviews of Modern Physics* 77 (3): 977–1026.
69 Landau, L.D. and Lifshitz, E.M. (1975). *The Classical Theory of Fields*, 2e. Butterworth-Heinemann.
70 Schlichting, H. (1932). Berechnung ebener periodischer Grenzschichtstromungen. *Physikalishce Zeitschrift* 33: 327–335.
71 Huang, P.-H., Chan, C.Y., Li, P. et al. (2018). A sharp-edge-based acoustofluidic chemical signal generator. *Lab on a Chip* 18 (10): 1411–1421.
72 Bachman, H., Chen, C., Rufo, J. et al. (2020). An acoustofluidic device for efficient mixing over a wide range of flow rates. *Lab on a Chip* 20 (7): 1238–1248.
73 Chindam, C., Nama, N., Ian Lapsley, M. et al. (2013). Theory and experiment on resonant frequencies of liquid-air interfaces trapped in microfluidic devices. *Journal of Applied Physics* 114 (19): 194503.
74 Xie, Y., Chindam, C., Nama, N. et al. (2015). Exploring bubble oscillation and mass transfer enhancement in acoustic-assisted liquid-liquid extraction with a microfluidic device. *Scientific Reports* 5 (1): 1–9.
75 Ren, L., Nama, N., McNeill, J.M. et al. (2019). 3D steerable, acoustically powered microswimmers for single-particle manipulation. *Science Advances* 5 (10): eaax3084.
76 Doinikov, A.A., Gerlt, M.S., and Dual, J. (2020). Acoustic radiation forces produced by sharp-edge structures in microfluidic systems. *Physical Review Letters* 124 (15): 154501.
77 Huang, P.-H., Nama, N., Mao, Z. et al. (2014). A reliable and programmable acoustofluidic pump powered by oscillating sharp-edge structures. *Lab on a Chip* 14 (22): 4319–4323.
78 Huang, P.-H., Chan, C.Y., Li, P. et al. (2015). A spatiotemporally controllable chemical gradient generator via acoustically oscillating sharp-edge structures. *Lab on a Chip* 15 (21): 4166–4176.

79 Kaynak, M., Ozcelik, A., Nama, N. et al. (2016). Acoustofluidic actuation of in situ fabricated microrotors. *Lab on a Chip* 16 (18): 3532–3537.

80 Huang, P.-H., Xie, Y., Ahmed, D. et al. (2013). An acoustofluidic micromixer based on oscillating sidewall sharp-edges. *Lab on a Chip* 13 (19): 3847–3852.

81 Ozcelik, A., Nama, N., Huang, P. et al. (2016). Acoustofluidics: acoustofluidic rotational manipulation of cells and organisms using oscillating solid structures (Small 37/2016). *Small* 12 (37): 5230.

82 Huang, P., Zhao, S., Bachman, H. et al. (2019). Acoustofluidics nanomaterial synthesis: acoustofluidic synthesis of particulate nanomaterials. *Advancement of Science* 6 (19): 1970113.

83 Nama, N., Huang, T.J., and Costanzo, F. (2017). Acoustic streaming: an arbitrary Lagrangian-Eulerian perspective. *Journal of Fluid Mechanics* 825: 600–630.

84 Nyborg, W.L. (1958). Acoustic streaming near a boundary. *The Journal of the Acoustical Society of America* 30 (4): 329–339.

85 Nama, N., Huang, P.H.P.-H., Huang, T.J.T.J., and Costanzo, F. (2014). Investigation of acoustic streaming patterns around oscillating sharp edges. *Lab on a Chip* 14 (15): 2824–2836.

86 Muller, P.B., Barnkob, R., Jensen, M.J.H., and Bruus, H. (2012). A numerical study of microparticle acoustophoresis driven by acoustic radiation forces and streaming-induced drag forces. *Lab on a Chip* 12 (22): 4617–4627.

87 Devendran, C., Albrecht, T., Brenker, J. et al. (2016). The importance of travelling wave components in standing surface acoustic wave (SSAW) systems. *Lab on a Chip* 16 (19): 3756–3766.

88 Nama, N., Barnkob, R., Mao, Z. et al. (2015). Numerical study of acoustophoretic motion of particles in a PDMS microchannel driven by surface acoustic waves. *Lab on a Chip* 15 (12): 2700–2709.

89 Nama, N., Huang, P.-H., Huang, T.J., and Costanzo, F. (2016). Investigation of micromixing by acoustically oscillated sharp-edges. *Biomicrofluidics* 10 (2): 24124.

90 Barnkob, R., Nama, N., Ren, L. et al. (2018). Acoustically driven fluid and particle motion in confined and leaky systems. *Physical Review Applied* 9 (1): 14027.

91 Ding, X., Li, P., Lin, S.-C.S. et al. (2014). Lab-on-a-chip technologies enabled by surface acoustic waves. In: *Microscale Acoustofluidics* (ed. T. Laurell and A. Lenshof), 354–398. Royal Society of Chemistry https://books.rsc.org/books/edited-volume/1076/Microscale-Acoustofluidics.

92 Zarembo, L.K. (1971). Acoustic streaming. In: *High-Intensity Ultrasonic Fields* (ed. L.D. Rozenberg), 135–199. Springer https://link.springer.com/book/10.1007/978-1-4757-5408-7.

93 Chu, B.T. and Apfel, R.E. (1982). Acoustic radiation pressure produced by a beam of sound. *The Journal of the Acoustical Society of America* 72 (6): 1673–1687.

94 Bradley, C.E. (1996). Acoustic streaming field structure: the influence of the radiator. *The Journal of the Acoustical Society of America* 100 (3): 1399–1408.

95 Lee, C.P. and Wang, T.G. (1989). Near-boundary streaming around a small sphere due to two orthogonal standing waves. *The Journal of the Acoustical Society of America* 85 (3): 1081–1088.

96 Rednikov, A. and Riley, N. (2002). A simulation of streaming flows associated with acoustic levitators. *Physics of Fluids* 14 (4): 1502–1510.

97 Rednikov, A.Y., Riley, N., and Sadhal, S.S. (2003). The behaviour of a particle in orthogonal acoustic fields. *Journal of Fluid Mechanics* 486: 1–20.

98 Rednikov, A.Y. and Sadhal, S.S. (2004). Steady streaming from an oblate spheroid due to vibrations along its axis. *Journal of Fluid Mechanics* 499: 345–380.

99 Riley, N. (1998). Acoustic streaming. *Theoretical and Computational Fluid Dynamics* 10 (1): 349–356.

100 Vanneste, J. and Bühler, O. (2011). Streaming by leaky surface acoustic waves. *Proceedings of the Royal Society A Mathematical, Physical and Engineering Sciences* 467 (2130): 1779–1800.

101 Lei, J., Glynne-Jones, P., and Hill, M. (2013). Acoustic streaming in the transducer plane in ultrasonic particle manipulation devices. *Lab on a Chip* 13 (11): 2133–2143.

102 Rednikov, A.Y. and Sadhal, S.S. (2011). Acoustic/steady streaming from a motionless boundary and related phenomena: generalized treatment of the inner streaming and examples. *Journal of Fluid Mechanics* 667: 426–462.

103 Riley, N. (1966). On a sphere oscillating in a viscous fluid. *Quarterly Journal of Mechanics and Applied Mathematics* 19 (4): 461–472.

104 Zhao, H., Sadhal, S.S., and Trinh, E.H. (1999). Internal circulation in a drop in an acoustic field. *The Journal of the Acoustical Society of America* 106 (6): 3289–3295.

105 Doinikov, A.A. and Bouakaz, A. (2010). Acoustic microstreaming around a gas bubble. *The Journal of the Acoustical Society of America* 127 (2): 703–709.

106 Longuet-Higgins, M.S. (1970). Viscous streaming from an oscillating spherical bubble. *Proceedings of the Royal Society A Mathematical, Physical and Engineering Sciences* 1998 (454): 725–742.

107 Yosioka, K. and Kawasima, Y. (1955). Acoustic radiation pressure on a compressible sphere. *Acta Acustica United with Acustica* 5 (3): 167–173.

108 Doinikov, A. (2003). Acoustic radiation forces: classical theory and recent advances. *Recent Research Developments in Acoustics* 661 (2): 39–67.

109 Doinikov, A.A. (1931). Acoustic radiation pressure on a rigid sphere in a viscous fluid. *Proceedings of the Royal Society A Mathematical, Physical and Engineering Sciences* 1994 (447): 447–466.

110 Settnes, M. and Bruus, H. (2012). Forces acting on a small particle in an acoustical field in a viscous fluid. *Physical Review E, Statistical, Nonlinear, and Soft Matter Physics* 85 (1): 1–12.

111 Karlsen, J.T. and Bruus, H. (2015). Forces acting on a small particle in an acoustical field in a thermoviscous fluid. *Physical Review E* 92 (4): 043010.

112 Doinikov, A.A. (1994). Acoustic radiation pressure on a compressible sphere in a viscous fluid. *Journal of Fluid Mechanics* 267: 1–22.

113 Baasch, T., Pavlic, A., and Dual, J. (2019). Acoustic radiation force acting on a heavy particle in a standing wave can be dominated by the acoustic microstreaming. *Physical Review E* 100 (6): 061102.

114 Chen, Y., Fang, Z., Merritt, B. et al. (2016). Onset of particle trapping and release via acoustic bubbles. *Lab on a Chip* 16 (16): 3024–3032.

115 Hammarström, B., Evander, M., Barbeau, H. et al. (2010). Non-contact acoustic cell trapping in disposable glass capillaries. *Lab on a Chip* 10 (17): 2251–2257.

116 Li, P., Mao, Z., Peng, Z. et al. (2015). Acoustic separation of circulating tumor cells. *Proceedings of the National Academy of Sciences* 112 (16): 4970–4975.

117 Bhagat, A.A.S., Bow, H., Hou, H.W. et al. (2010). Microfluidics for cell separation. *Medical and Biological Engineering and Computing* 48: 999–1014.

118 Ozcelik, A., Rufo, J., Guo, F. et al. (2018). Acoustic tweezers for the life sciences. *Nature Methods* 15 (12): 1021–1028.

119 Shi, J., Ahmed, D., Mao, X. et al. (2009). Acoustic tweezers: patterning cells and microparticles using standing surface acoustic waves (SSAW). *Lab on a Chip* 9 (20): 2890–2895.

120 Tran, S.B.Q., Marmottant, P., and Thibault, P. (2012). Fast acoustic tweezers for the two-dimensional manipulation of individual particles in microfluidic channels. *Applied Physics Letters* 101 (11): 114103.

121 Chen, Y., Ding, X., Steven Lin, S.-C. et al. (2013). Tunable nanowire patterning using standing surface acoustic waves. *ACS Nano* 7 (4): 3306–3314.

122 Li, S., Ding, X., Mao, Z. et al. (2015). Standing surface acoustic wave (SSAW)-based cell washing. *Lab on a Chip* 15 (1): 331–338.

123 Li, A., Kusuma, G.D., Driscoll, D. et al. (2021). Advances in automated cell washing and concentration. *Cytotherapy* 23 (9): 774–786.

124 Collins, D.J., Morahan, B., Garcia-Bustos, J. et al. (2015). Two-dimensional single-cell patterning with one cell per well driven by surface acoustic waves. *Nature Communications* 6 (1): 1–11.

125 Dillinger, C., Nama, N., and Ahmed, D. (2021). Ultrasound-activated ciliary bands for microrobotic systems inspired by starfish. *Nature Communications* 12 (1): 1–11.

126 McNeill, J.M., Nama, N., Braxton, J.M., and Mallouk, T.E. (2020). Wafer-scale fabrication of micro-to nanoscale bubble swimmers and their fast autonomous propulsion by ultrasound. *ACS Nano* 14 (6): 7520–7528.

127 Doinikov, A.A. and Zavtrak, S.T. (1991). The koenig force in a compressible fluid. *Journal of Applied Mechanics and Technical Physics* 32 (6): 880–882.

128 Bjerknes, V. (1906). *Fields of Force: Supplementary Lectures, Applications To Meteorology; A Course of Lectures in Mathematical Physics Delivered December 1 to 23, 1905*. Columbia University Press.

129 Zabolotskaya, E.A. (1984). Interaction of gas-bubbles in a sound field. *Soviet Physics Acoustics* 30 (5): 365–368.

130 Doinikov, A.A. and Zavtrak, S.T. (1995). On the mutual interaction of two gas bubbles in a sound field. *Physics of Fluids* 7 (8): 1923.

131 Doinikov, A.A. and Zavtrak, S.T. (1996). On the "bubble grapes" induced by a sound field. *The Journal of the Acoustical Society of America* 99 (6): 3849–3850.

132 Gong, Z. and Baudoin, M. (2019). Particle assembly with synchronized acoustic tweezers. *Physical Review Applied* 12 (2): 24045.

133 Liu, W., Gao, H., Liu, K. et al. (2022). A review on particle assembly in standing wave acoustic field. *Journal of Nanoparticle Research* 24 (4): 1–22.

134 Zavtrak, S.T. (2004). Radiation forces in classical and quantum physics. In: *ÉTudes On Theoretical Physics: Collection of Works Dedicated to 65th Anniversary of the Department of Theoretical Physics of Belarusian State University* (ed. L.M. Barkovsky, I.D. Feranchuk, and Y.M. Shnir), 339–352. https://play.google.com/store/books/details?pcampaignid=books_read_action&id=MPLICgAAQBAJ&pli=1.

135 Dayton, P., Klibanov, A., Brandenburger, G., and Ferrara, K. (1999). Acoustic radiation force in vivo: a mechanism to assist targeting of microbubbles. *Ultrasound in Medicine & Biology* 25 (8): 1195–1201.

136 Zhao, S., Borden, M., Bloch, S.H. et al. (2004). Radiation-force assisted targeting facilitates ultrasonic molecular imaging. *Molecular Imaging* 3 (3): 15353500200404116.

137 Palanchon, P., Tortoli, P., Bouakaz, A. et al. (2005). Optical observations of acoustical radiation force effects on individual air bubbles. *IEEE Transactions on Ultrasonics, Ferroelectrics, and Frequency Control* 52 (1): 104–110.

138 Ahmed, D., Ozcelik, A., Bojanala, N. et al. (2016). Rotational manipulation of single cells and organisms using acoustic waves. *Nature Communications* 7 (1): 1–11.

139 Zhang, P., Bachman, H., Ozcelik, A., and Huang, T.J. (2020). Acoustic microfluidics. Annu Rev. *Analytical Chemistry* 13: 17–43.

3

Materials for Acoustic Wave Generation and Modulation

Noé Jiménez

Universitat Politècnica de València, Instituto de Instrumentación para Imagen Molecular, Camino de Vera s/n, València 46022, Spain

3.1 Introduction

3.1.1 Generation and Detection of Ultrasound

Acoustic waves have been used in medicine since the times of Hippocrates of Kos in fifth century BCE, when the sounds of the body were listened through auscultation during a physical examination. However, since the last century, modern technologies for acoustic wave detection, and also generation, have enabled more reliable medical imaging techniques, which are now common in the clinical practice. These include standard ultrasound imaging and a broad spectrum of acoustic imaging modalities such as Doppler ultrasound to detect flow speed, ultrasound elastography to measure tissue elasticity, quantitative ultrasound to evaluate bones, contrast ultrasound imaging to enhance ultrasound sensitivity and specificity, and, recently, functional ultrasound imaging to measure brain activity, ultrasound localization microscopy to image blood vessels at super resolution, or photo-acoustic techniques to provide molecular imaging, among others.

Beyond imaging, when the intensity of the acoustic waves is increased, intense vibrations can induce biological effects on tissue. In this way, focused ultrasound is emerging as a therapeutic modality with a whole branch of promising modalities. These include now classical treatments for kidney stone fragmentation using extracorporeal shock wave lithotripsy, high-intensity focused ultrasound to locally ablate tissue through thermal effects, emerging local drug delivery applications such as sonodynamic therapy, sonoporation, or blood-brain barrier opening, or promising therapeutic applications such as histotripsy to mechanically destroys tissue through cavitation creating a lesion with very sharp boundaries. In addition, a broad range of ultrasound transducers are used as biomedical tools such as ultrasonic baths for cleaning and sterilization of medical and surgical instruments, sonotrodes for ultrasonic processing and sonochemistry, and a wide range of ultrasonic motors, actuators and sensors in medical instrumentation.

Acoustic Technologies in Biology and Medicine, First Edition.
Edited by Adem Ozcelik, Ryan Becker, and Tony Jun Huang.

3.1.2 Technologies for Ultrasound Transducers

At the root of these biomedical applications is the generation and detection of acoustic waves using a transducer, i.e. a device that converts an electric signal into a mechanical vibration to generate an acoustic wave, and, typically, they can also generate an electric signal when detecting an acoustic or elastic wave. There exists several technologies to convert mechanical to electrical energy, including piezoelectric transducers (based on the piezoelectricity of the materials), magnetostrictive transducers (based on the magnetostriction of ferromagnetic materials), capacitive micromachined ultrasonic transducers (CMUT, based on capacitive microelectromechanical systems or MEMS), piezoelectric micromachined ultrasonic transducers (PMUT, based on miniaturized flexural piezoelectric films), optoacoustic transducers (based on the photoacoustic effect), electrodynamic transducers (based on Faraday's induction law), or electrohydraulic transducers (based on a fast bubble cavitation generated by a spark plug immersed in water). Covering all these ultrasound generation techniques in detail is beyond the scope of this chapter. Here, we will focus on piezoelectric transducers, as they are the most common transducers for medical applications.

3.2 Piezoelectricity

Piezoelectric materials are the most common in medical ultrasound transducers, most ultrasound imaging probes and modern therapeutic ultrasound devices are based on this technology. Piezoelectricity, discovered by the brothers Pierre and Jacques Curie in 1880 [1], is the property of several materials to accumulate electrical charge when the medium is deformed, and, conversely, to deform under the action of an external electric field. The piezoelectric effect occurs in solids with a nonsymmetrical crystalline structure. When the asymmetric crystalline structure is deformed under external stress, a polarization occurs, as seen in Figure 3.1. This causes a flow of electrons to accumulate at one end of the crystal until electrostatic equilibrium is reached [2]. In this way, an electric field appears inside the piezoelectric when it is deformed. It should be noted that the piezoelectric effect also occurs in the reverse direction, i.e. a piezoelectric crystal subjected to an electric potential difference will deform until electrostatic equilibrium is reached inside the crystal. Thus, piezoelectric materials are used both as sensors/receivers (direct piezoelectric effect) and actuators/emitters (reverse piezoelectric effect).

3.2.1 Model Equations

The complete modeling of piezoelectric material dynamics is a complex problem, mainly due to the tensor character of the stresses together with the material anisotropy, both in the elastic and piezoelectric properties, of piezoelectric solids. In addition, the materials exhibit viscoelastic behavior and other phenomena such as nonlinearity, hysteresis or dependence of elasticity and piezoelectricity

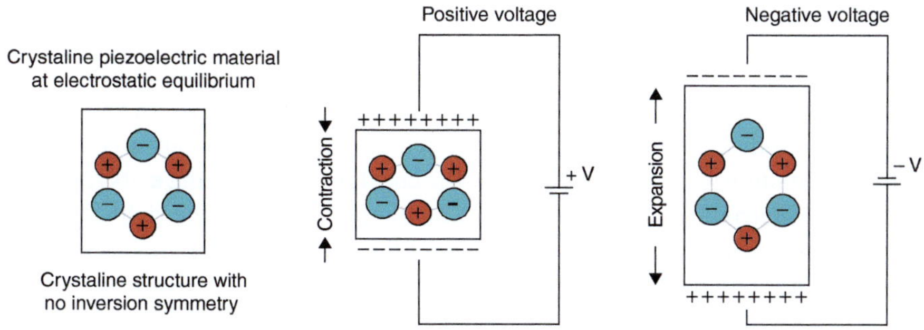

Figure 3.1 Piezoelectric effect in a crystal with non-inversion symmetry. When the material is at electrostatic equilibrium the centers of positive and negative charges are not at the same position. When the material is compressed the crystal structure is deformed and due to asymmetric charge distribution an electrical potential is observed. On the other hand, when the crystal is expanded the opposite effect is observed and a negative electric potential is generated. The inverse piezoelectric effect also appears: the material deforms when an electrical potential is imposed to maintain electrostatic equilibrium.

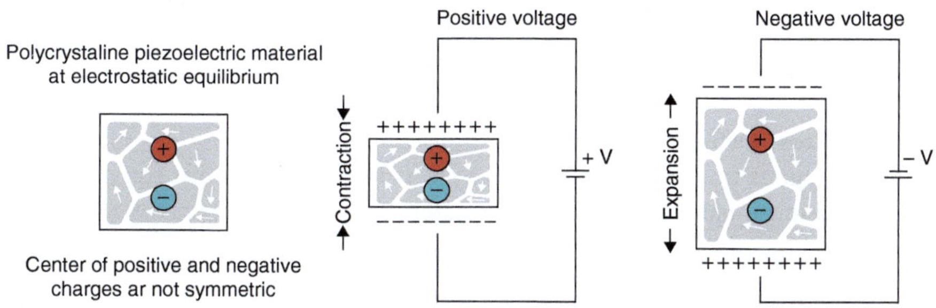

Figure 3.2 Piezoelectric effect in a polycrystaline material. When the material is at electrostatic equilibrium the centers of positive and negative charges are not in the same position. When the material is compressed the polycrystaline structure is deformed and due to asymmetric charge distribution an electrical potential is observed. On the other hand, when the polycrystaline is expanded the opposite effect is observed and a negative electric potential is generated. The inverse piezoelectric effect also appears: the material deforms when an electrical potential is imposed to maintain electrostatic equilibrium.

on temperature. However, there are simplified configurations that allow their description by means of simple models, such as the vibration of a piezoelectric ceramic in thickness model. Indeed, this case is of particular interest in engineering since the coupling with the surrounding medium is very efficient: the deformations produced by the piezoelectric can be transmitted to another medium in the form of acoustic or longitudinal elastic waves, e.g. as those used for ultrasound imaging (Figure 3.2).

3.2.1.1 Stress-Charge Formulation

In the piezoelectric material the ratio between the stress tensor, $\boldsymbol{T}$ (N/m^2), the strain tensor, $\boldsymbol{S}$ (dimensionless), the electric field vector, $\boldsymbol{E}$ (V/m) and electric displacement vector, $\boldsymbol{D}$ (C/m^2), are given by the piezoelectric constitutive equations [3]

$$\boldsymbol{T} = \mathbf{c}^E\mathbf{S} - \mathbf{e}^t\boldsymbol{E} \tag{3.1}$$

$$\boldsymbol{D} = \boldsymbol{\varepsilon}^{S}\boldsymbol{E} + \mathbf{e}\boldsymbol{S} \tag{3.2}$$

where $\boldsymbol{c}^E$ (N/m^2) is the matrix of elastic moduli for constant electric field and $\boldsymbol{\varepsilon}^S$ (F/m) is the permittivity matrix of the material for constant strain. In general, the superscript indicates the magnitude that remains invariant when a constant is defined. Finally, $\mathbf{e}$ (C/m^2) is the piezoelectric constant matrix, being $\mathbf{e}^t$ its respective transposed matrix. In the system (3.1)–(3.2), the first terms on the right-hand side represent the elastic and electrical behavior of the material, while the second terms represent the coupling between the two quantities, i.e. the piezoelectricity.

3.2.1.2 Strain-Charge Formulation

The constitutive piezoelectric Eqs. (3.1)–(3.2) in stress-charge formulation can be formulated strain-charge formulation, resulting in the governing equations

$$\boldsymbol{S} = \boldsymbol{s}^{E}\boldsymbol{T} + \boldsymbol{d}\mathbf{E} \tag{3.3}$$

$$\boldsymbol{D} = \boldsymbol{\varepsilon}^{T}\boldsymbol{E} + \boldsymbol{d}^{t}\boldsymbol{T} \tag{3.4}$$

where $\boldsymbol{s}^E$ (N/m^2) is the elastic compliance matrix for constant electric field and $\boldsymbol{\varepsilon}^T$ (F/m) is the permittivity matrix of the material for constant stress. Finally, $\boldsymbol{d}$ (C/N) is the piezoelectric constant matrix, being $\boldsymbol{d}^t$ its respective transposed matrix.

3.2.1.3 Stress-Field Formulation

In an analogous way, we can also derive piezoelectric constitutive equations for a stress-field formulation. In this way, the system can be written as

$$\boldsymbol{T} = \boldsymbol{c}^{D}\boldsymbol{S} - \boldsymbol{h}^{t}\boldsymbol{D} \tag{3.5}$$

$$\boldsymbol{E} = \boldsymbol{\beta}^{S}\boldsymbol{D} - \boldsymbol{h}\boldsymbol{S} \tag{3.6}$$

where $\boldsymbol{\beta}^S$ is the inverse permittivity at constant strain, and $\boldsymbol{h}$ is the piezoelectric constant matrix for the stress-field formulation.

3.2.1.4 Strain-Field Formulation

Finally, in a strain-field formulation, the piezoelectric equations are

$$\boldsymbol{S} = \boldsymbol{s}^{D}\boldsymbol{T} + \boldsymbol{g}^{t}\boldsymbol{D} \tag{3.7}$$

$$\boldsymbol{E} = \boldsymbol{\beta}^{T}\boldsymbol{D} - \mathbf{g}\boldsymbol{T} \tag{3.8}$$

where $\boldsymbol{\beta}^T$ is the inverse permittivity at constant stress, and $\mathbf{g}$ (Vm/N) is the piezoelectric constant matrix for the strain-field formulation.

3.2.2 The Piezoelectric Constants

Writing the system in either way is interesting to design specific sensors or actuators, and one should keep in mind that piezoelectric material manufacturers and laboratories indifferently measure one or the other in the characterization of the piezoelectric constants. It is worth remembering that

$$\mathbf{c}^E = \frac{1}{\mathbf{s}^E}, \quad \mathbf{c}^D = \frac{1}{\mathbf{s}^D} \quad \varepsilon^S = \frac{1}{\beta^S}, \quad \varepsilon^D = \frac{1}{\beta^D} \tag{3.9}$$

but on the other hand

$$\mathbf{c}^E \neq \mathbf{c}^D, \quad \mathbf{s}^E \neq \mathbf{s}^D \quad \boldsymbol{\varepsilon}^S \neq \boldsymbol{\varepsilon}^T, \quad \boldsymbol{\beta}^S \neq \boldsymbol{\beta}^T \tag{3.10}$$

Therefore, special care must be taken when choosing one set of elastic and electrical constants. In three spatial dimensions, the system of constitutive Eqs. (3.1–3.2) for a piezoelectric material can be written as

$$\begin{bmatrix} S_1 \\ S_2 \\ S_3 \\ S_4 \\ S_5 \\ S_6 \\ D_1 \\ D_2 \\ D_3 \end{bmatrix} = \begin{bmatrix} s_{11}^E & s_{12}^E & s_{13}^E & s_{14}^E & s_{15}^E & s_{16}^E & d_{11} & d_{12} & d_{13} \\ s_{21}^E & s_{22}^E & s_{23}^E & s_{24}^E & s_{25}^E & s_{26}^E & d_{21} & d_{22} & d_{23} \\ s_{31}^E & s_{32}^E & s_{33}^E & s_{34}^E & s_{35}^E & s_{36}^E & d_{31} & d_{32} & d_{33} \\ s_{41}^E & s_{42}^E & s_{43}^E & s_{44}^E & s_{45}^E & s_{46}^E & d_{41} & d_{42} & d_{43} \\ s_{51}^E & s_{52}^E & s_{53}^E & s_{54}^E & s_{55}^E & s_{56}^E & d_{51} & d_{52} & d_{53} \\ s_{61}^E & s_{62}^E & s_{63}^E & s_{64}^E & s_{65}^E & s_{66}^E & d_{61} & d_{62} & d_{63} \\ d_{11} & d_{12} & d_{63} & d_{14} & d_{15} & d_{16} & \varepsilon_{11}^T & \varepsilon_{12}^T & \varepsilon_{13}^T \\ d_{21} & d_{22} & d_{63} & d_{24} & d_{25} & d_{26} & \varepsilon_{21}^T & \varepsilon_{22}^T & \varepsilon_{23}^T \\ d_{31} & d_{32} & d_{63} & d_{34} & d_{35} & d_{36} & \varepsilon_{31}^T & \varepsilon_{32}^T & \varepsilon_{33}^T \end{bmatrix} \begin{bmatrix} T_1 \\ T_2 \\ T_3 \\ T_4 \\ T_5 \\ T_6 \\ E_1 \\ E_2 \\ E_3 \end{bmatrix} \tag{3.11}$$

where we introduced the Voigt notation convention for the components $1 = xx$, $2 = yy$, $3 = zz$, $4 = xy$, $5 = yz$, and $6 = zx$. However, to analyze an ultrasonic transducer, this general set of equations can be simplified because most of the piezoelectric constants are negligible (or zero) due to the specific crystalline structure of the material and poling direction.

3.2.3 Longitudinal Motion in a Piezoelectric Material

Generally, in piezoelectric materials, such as quartz or lead zirconate titanate (PZT), many of the piezoelectric constants are zero (or negligible) due to the symmetry of the crystal, i.e. the piezoelectric effect only occurs efficiently in some particular directions.

3.2.3.1 A Simple Piezoelectric Model

For the present example, we will consider vibrations that occur only in thickness, ignoring what happens in other directions by reducing the system of equations to variations in the x_3 direction, as shown in Figure 3.3. This models the thickness motion of a slab, for example a disc, of piezoelectric material with two electrodes on its faces, polling direction x_3, and a high piezoelectric constant h_33. Thus, the system of equations given by (3.5) and (3.6) is reduced to

$$T_3 = c_{33}^D S_{33} - h_{33} D_3 \tag{3.12}$$

$$E_3 = \beta_{33}^S D_3 - h_{33} S_3 \tag{3.13}$$

These equations describe the system in the x_3 direction assuming that the other piezoelectric and elastic moduli are zero (which is not the case in a real ceramic).

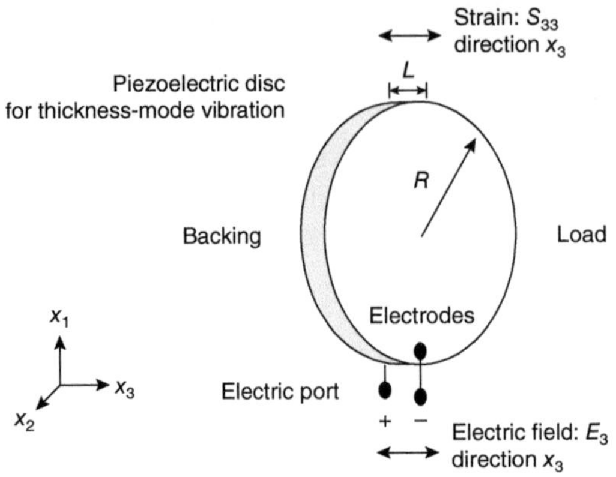

Figure 3.3 Piezoelectric disc for vibration in thickness mode.

3.2.3.2 Waves in the Piezoelectric Material

To close the system, we can use Newton's Second Law to make a balance of forces

$$\rho_p \frac{\partial^2 \xi_3}{\partial t^2} = \frac{\partial T_3}{\partial x_3} \tag{3.14}$$

where ρ_p is the density in the piezoelectric material and ξ_3 is the displacement of the material relative to its equilibrium position in the direction x_3. In this equation the first term expresses the product of the mass and the acceleration of the system, and the second the force produced by the stress field. The deformation is defined as a function of the displacement, ξ_3, as

$$S_{33} = \frac{\partial \xi_3}{\partial x_3} \tag{3.15}$$

and the particle velocity as

$$v_3 = \frac{\partial \xi_3}{\partial t} \tag{3.16}$$

Thus, the deformation is related to the particle velocity as

$$\frac{\partial S_{33}}{\partial t} = \frac{\partial}{\partial t}\left(\frac{\partial \xi_3}{\partial x_3}\right) = \frac{\partial}{\partial x_3}\left(\frac{\partial \xi_3}{\partial t}\right) = \frac{\partial v_3}{\partial x_3} \tag{3.17}$$

Finally, we differentiate with respect to time the constitutive equations to obtain

$$\rho_p \frac{\partial v_3}{\partial t} = \frac{\partial T_3}{\partial x_3}, \tag{3.18}$$

$$\frac{\partial T_3}{\partial t} = c_{33}^D \frac{\partial v_3}{\partial x_3} - h_{33} \frac{\partial D_3}{\partial t}, \tag{3.19}$$

$$\frac{\partial E_3}{\partial t} = \beta_{33}^S \frac{\partial D_3}{\partial t} - h_{33} \frac{\partial v_3}{\partial x_3} \tag{3.20}$$

These equations describe the dynamics of the system in the x_3 direction. Equations (3.18) and (3.19) are equivalent to the constitutive relations of acoustics for longitudinal motion in one dimension. For example, in open circuit these can be combined into a wave equation where the sound speed is

$$c_p = \sqrt{\frac{c_{33}^D}{\rho_p}} \tag{3.21}$$

Like any other solid material, piezoelectric media transmit waves. The main difference with the acoustic equations is the coupling with the partial differential equation (PDE) electric equation through the piezoelectric constant h_{33}. The coupling implies that variations on the electric field will induce particle velocity gradients and, therefore strain, and these perturbations will propagate as a longitudinal motion through the material. Equations (3.18) and (3.19) can be solved directly by applying numerical or analytical methods when the boundary conditions are known. For example, Richard Krimholtz et al., proposed in 1970 a model commonly referred to as the Krimholtz, Leedom, and Matthaei model (KLM) equivalent circuit [4] to describe the motion of a piezoelectric layer in the frequency domain. This kind of model allows a great flexibility to add additional elements to the transducer, such as backing layers, impedance matching layers, and electronic components, as we will show below.

3.3 Piezoelectric Materials

There exist several types of materials that exhibit piezoelectricity. These include crystals, piezoelectric ceramics, piezoelectric polymers, and piezoelectric composites. However, in all these materials, piezoelectricity is only observed at low temperatures, when nonsymmetric microscopic crystalline structure holds under deformation. Above a critical temperature, called the Curie temperature, T_C, ferroelectric materials undergo a phase transition, shifting from a low-temperature ferroelectric phase to a high-temperature paraelectric phase [5]. Therefore, above T_C both ferroelectricity and piezoelectricity disappear and the material vibrates as common (visco)elastic solid. Therefore, for practical applications a higher Curie temperature is recommended to achieve better stability and operational range.

The electromechanical coupling factors, k_{ij}, are usually used to measure the performance of a piezoelectric material. They quantify the conversion of mechanical to electrical energy. Common electromechanical coupling factors provided by manufactures are defined as

$$k_{33} = \frac{d_{33}}{\sqrt{S_{33}\varepsilon_{33}}} \qquad k_{31} = \frac{d_{31}}{\sqrt{S_{11}\varepsilon_{33}}} \qquad k_{15} = \frac{d_{15}}{\sqrt{S_{44}\varepsilon_{11}}} \tag{3.22}$$

where, following a Voigt notation, subscripts 4 and 5 refer to planes normal to the 1 and 2 axis, respectively. Therefore, they indicate how efficiently the material deforms in a given direction under the action of an electric field in other direction. Note these quantities are dimensionless.

3.3.1 Piezoelectric Crystals

Piezoelectric crystals can be found in nature, for example, quartz (SiO_2), berlinite, sucrose, Rochelle salt ($NaKC_4H_4O_64H_2O$), topaz, and a tourmaline group of minerals. However, natural crystals present weak piezoelectric properties as compared with synthetic materials, e.g. piezoelectric ceramics, as we will see below. Crystalline structures can also be synthesized in the laboratory to obtain piezoelectric substances such as gallium orthophosphate ($GaPO_4$) and langasite ($La_3Ga_5SiO_{14}$).

3.3.2 Piezoelectric Ceramics

Piezoelectric ceramics are substances largely used in industry because it can be synthesized with great flexibility and at relatively low cost [6]. These include most of the standard piezoelectric as the common lead zirconate titanate ($Pb[Zr_x\ Ti_{1-x}]O_3$), known as PZT. The basic composition of PZT is doped with several elements to achieve different properties, which are usually proprietary formulas of manufactures. For example, PZT-4 is doped with Fe, PZT-5 with Nb-doping, PZT-6 with Cr-doping, or PZT-7 with La-doping. Some commonly used compositions include lead titanate ($PbTiO_3$), potassium niobate ($KNbO_3$), zinc oxide (ZnO), or sodium tungstate ($NaWO_3$). Piezoelectric ceramics such as PZT are very efficient materials showing a high electromechanical coupling factor. They also show a high stiffness and high dielectric constant, while they present low elastic and dielectric losses [7].

However, the lead content of most piezoceramics, harmful for living organisms during usage, manufacturing and disposal, have risen the attention to use lead-free synthetic piezoceramics [8], such as long time known barium titanate ($BaTiO_3$), and to design more efficient materials such as lithium niobate ($LiNbO_3$), lithium tantalate ($LiTaO_3$), aluminium nitride (AlN), scandium–aluminum nitride, sodium niobate ($NaNbO_3$) or bismuth ferrite ($BiFeO_3$), etc [9]. However, manufacturing some of these materials can result in a large environmental impact due to mining and processing.

3.3.3 Piezoelectric Polymers

Polymerization can be used to obtain piezoelectric substances such as polyvinylidene fluoride (PVDF) films and copolymers of vinylidene fluoride (VDF) with trifluoroethylene (TrFE), or with tetrafluroethylene (TeFE) [10]. Piezopolymers present desirable mechanical and functional properties such as flexibility, durability and biocompatibility, and low-cost manufacturing, while they feature a low dielectric constant and a low acoustic impedance, which provides an excellent matching with water or tissues [11]. However, they are less efficient than piezoceramics and because they are currently manufactured in thin films, their practical use for actuators is limited, and they are mostly used for sensors.

3.3.4 Piezoelectric Composites

Finally, piezocomposites can also be manufactured by mixing two phases, usually a stiff piezoceramic matrix filled by a soft polymer [12]. Piezoelectric composites are efficient, presenting a good electromechanical coupling factor. In addition, they avoid the mechanical fragility of piezoceramics. These materials also present a low acoustic impedance improving the impedance matching with water. Composites are manufactured in several ways, the active phase can be made of piezoceramic rods or fibers, blended by a viscoelastic resin, molded in bulky shapes or thin films. Piezoelectric coating has been proposed by polling a mix of PTZ powder and epoxy resin.

3.4 Ultrasound Transducers

The piezoelectric material is the main element of the transducer, as it transforms electric fields into elastic waves and vice versa. However, it is important to note that, in practice, the piezoelectric element will never vibrate uncoupled from surrounding materials. Transducers are usually composed of more elements.

3.4.1 Elements of a Transducer

In a real transducer we usually have, at least, a load on both sides of the piezoelectric element: the radiation load and backing layer, respectively. Commonly more elastic layers are located between the load and the piezoelectric element to enhance the transmission of acoustic energy. In addition, the electrical port is connected to some electrical network, the signal generator and amplifier, connection wires, etc. The complete set of mechanical and electrical elements that allow us to use the piezoelectric effect to emit or to detect acoustic waves is called a transducer. A basic example for a longitudinal transducer, where the piezoelectric vibrates in thickness mode, is the one shown in Figure 3.4.

3.4.2 The Piezoelectric Slab

The slab of piezoelectric material is the main element of the transducer. As in any vibrating solid, the geometry of the material and its properties govern its dynamics.

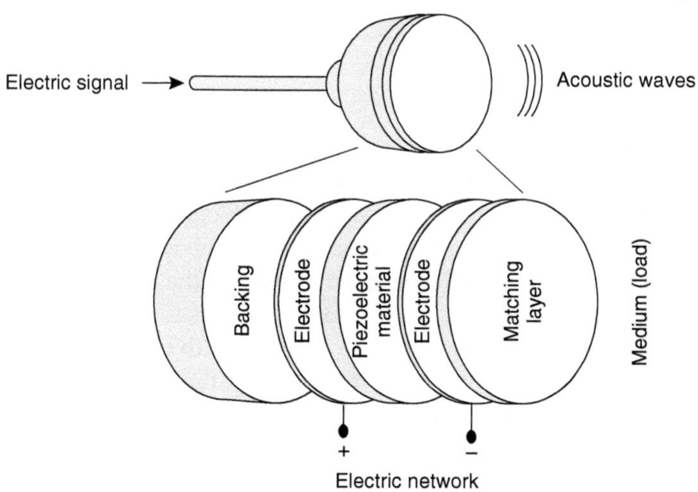

Figure 3.4 Model of a basic ultrasound piezoelectric transducer for vibration in thickness mode.

For example, a disc of piezoelectric material exhibit longitudinal resonance modes. For the thickness mode, as described before, the longitudinal modes of a piezoelectric slab of thickness L appear when $L = n\lambda/2$, where λ is the wavelength in the piezoelectric material, and $n = 1,3,5,\ldots$, i.e. at odd multiples of the first longitudinal resonance frequency. Note that a disc will also show radial modes, that can appear at lower frequencies, but typically the coupling strength is larger for the first axial mode ($L = \lambda/2$). In addition, this mode is efficiently coupled with longitudinal waves in fluids or tissues, because the piezoelectric produces a uniform axial deformation along the disc surface, compressing the fluid and generating acoustic waves. However, note that eigenmodes also depend on boundary conditions, i.e. the surrounding material impedance and its geometry, the clamping strength on the piezoelectric boundaries, but also the electric impedance on its electrodes.

3.4.3 Matching Layers

On the front side of the ceramic is a layer of elastic solid material called an impedance matching layer. The function of this layer is to improve the energy transfer from the piezoelectric layer to the medium to be radiated. This is necessary because of the large difference in acoustic impedance between the piezoelectric material and the radiation load, in many cases water or tissue for biomedical applications. While this layer is mandatory in piezoceramics and piezoelectric crystals designs, while it can be neglected when using soft piezoelectric polymers.

3.4.3.1 Classical Matching Layer Design

The design of matching layers considers that the transmission coefficient of acoustic energy between a layer of material located between the piezoelectric material and the radiation medium should be the maximum at a design frequency. In theory, the transmission can reach the unity when two conditions are fulfilled. The first condition is that the impedance of the matching material should be the geometric mean of the load and the piezoelectric medium, as [9]

$$Z_m = \sqrt{Z_p Z_L} \tag{3.23}$$

where $Z_m = \rho_m c_m$, $Z_p = \rho_p c_p$, and $Z_L = \rho_L c_L$ are the impedance of the matching layer, the piezoelectric material, and the load; and c_m, c_p, and c_L; and ρ_m, ρ_p, and ρ_L are their sound speed and mass density, respectively. For biomedical ultrasound transducers, typical acoustic impedance for piezoceramics is about 15–30 MRayl, tissue or water impedance is about 1.5 MRayl, so the material of the layer should be tuned to have specific impedance values, about 3–15 MRayl. This is achieved using epoxy, polyurethane or polystyrene layers, that can be doped with metal powder to increase its density. The second condition is that the length of the layer, L_m, should be one quarter of the wavelength in the matching material

$$L_m = \frac{\lambda_m}{4} \tag{3.24}$$

where $\lambda_m = c_m/f_0$ is the wavelength in the matching layer material at the design frequency f_0. This last equation implies that transmission will only be the unity at one specific frequency. For therapeutic ultrasound transducers and actuators this can be of interest because they are typically designed for narrow frequency range. For example, therapeutic ultrasound transducers for high-intensity focused ultrasound radiate long sinusoidal bursts, with thousands of cycles at frequencies around 1 MHz, i.e. a narrow matching bandwidth is enough. However, for imaging a short pulse of about one cycle is needed for a good axial resolution, therefore, a large bandwidth is required.

3.4.3.2 Multiple Matching Layer Design

A solution to increase the bandwidth is to include several matching layers, typically two or three. This creates an impedance gradient and acoustic energy can smoothly penetrate for detection into the transducer (or radiate for emission) [13]. If one consider a total of N layers, the material of each layer can be chosen to fulfill the following condition [14]

$$Z_n = \sqrt[N+1]{Z_P^{(N-n+1)} Z_L^n} \tag{3.25}$$

where $Z_n = \rho_n c_n$ is the impedance, ρ_n the density, and c_n the sound speed of the nth layer. Note that for $N = 1$, this equation collapse to (3.23). The length of the nth matching layer is given by

$$L_n = \frac{\lambda_n}{4} \tag{3.26}$$

where $\lambda_n = c_n/f_0$ is the wavelength in the nth matching layer. One should keep in mind that perfect transmission is never achieved due to viscoelastic losses, the more layers are added, the more important the losses are, limiting the total thickness of the set of impedance matching layers. For example, ultrasound imaging transducers usually are manufactures with one or two matching layers.

3.4.3.3 Broadband Matching Layer Design

Other designs include the electromechanical coupling and the broadband considerations, resulting in a modified impedance matching condition of Eq. (3.23) as [15]

$$Z_m = Z_P^{1/3} Z_L^{2/3} \tag{3.27}$$

for a matching single-layer design. For double matching layer we can use [16]

$$Z_{m1} = Z_P^{4/7} Z_L^{3/7}, \quad \text{and} \quad Z_{m2} = Z_P^{1/7} Z_L^{6/7} \tag{3.28}$$

to improve the bandwidth of the transmitted spectrum.

3.4.4 Backing Layer

Another element of the transducer is the backing layer or simply known as *backing*. By adjusting the properties of this layer we can design transducers with a long temporal response and narrow bandwidth or, conversely, with a short impulse response and wide bandwidth [17]. When the impedance difference between the piezoelectric element and the backing is very high, the energy is almost entirely reflected at the rear interface, maximizing the radiation toward the front face. For example, if a very low impedance backing is present, such as air, the quality factor of the resonance of the piezoelectric transducer is high, and its bandwidth is very narrow. This is desirable in actuators such as therapeutic ultrasound transducers.

However, for imaging, a broad frequency range is needed for axial resolution. To damp the resonance of the piezoelectric slab, the backing material should be absorbent and impedance matched with the piezoelectric. In imaging transducers the impedance difference between the active element and the backing is usually small and damping is introduced into the backing material through viscoelastic losses and doping. In this way, the resonance of the transducer is damped and a broad bandwidth response is obtained. However, increasing the operating frequency range in this way leads to a loss of efficiency because part of the energy is not radiated. In addition, absorbing backing materials can heat up under high-intensity ultrasound, therefore, attention should be paid when using long and high voltage pulsed burst signals for imaging transducers such as pulse-coded excitation techniques or elastography.

Materials to design backing layers include conductive epoxy, whose impedance is about 6 MRayl and large attenuation coefficients (120 dB/mm at 30 MHz), or doped epoxy such as tungsten-loaded epoxy to increase its impedance up to 8–20 MRayl, between others.

3.4.5 Electrical Impedance Matching Network

Finally, it is common for the ultrasonic transducer to require an electrical impedance matching network to optimize the energy transfer through the electrical port. This network couples the electrical impedance between the excitation/detection system and the piezoelectric element. For narrow bandwidth systems, this network can be built using a few passive electronic components. However, the complexity of the network increases for broadband ultrasound transducers as those used for imaging.

3.5 Ultrasound Beams

Once the ultrasound transducer vibrates in thickness mode, the vibrating surface generates an acoustic beam that propagates, usually, through water or soft tissue. The specific features of ultrasound beam mainly depend on the source size and frequency and its geometry.

3.5.1 Circular Aperture Transducers

One simple but common case of transducer is the case of a flat circular source vibrating in thickness mode. In this case, we can consider a flat circular surface of radius a, which moves uniformly with velocity $v_0 \exp(i\omega t)$ normal to its surface, and which is fixed on a rigid infinite panel, called a baffle, where the velocity is zero.[1] To obtain the

1 The baffle is introduced to simplify the analysis, similar results are obtained without considering this boundary condition.

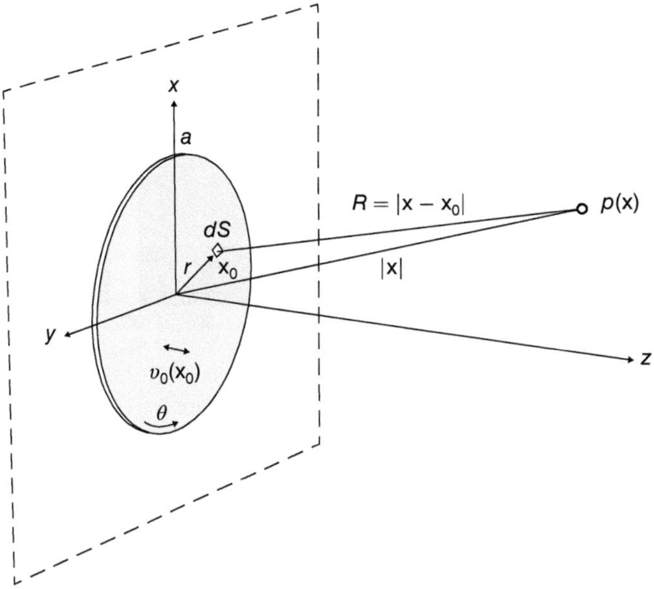

Figure 3.5 Diagram of a flat circular transducer and calculation points.

acoustic field, we will divide the circular surface into infinitesimal surface elements dS, each of which will act as a single source. Thus, using the Rayleigh–Sommerfeld integral, the acoustic field at any point is given by

$$p(\mathbf{x}, t) = \frac{i\omega\rho_0 v_0}{2\pi}\int_S \frac{e^{i(\omega t - kR)}}{R} dS \tag{3.29}$$

Although this integral is difficult to solve in general, we can find a general solution for the acoustic field in series form [18, 19]. Although the complete solution is valid at any point in space, the series converges slowly and the solution is not compact (Figure 3.5). However, if we restrict the validity only along the axis of the transducer, we can have the compact analytical solution along the z coordinate and on the symmetry axis, when $x = 0$ and $y = 0$, and therefore $r = 0$ and $\theta = 0$. In this case, $R = \sqrt{r_0^2 + z^2}$, leading to

$$p(z, t) = p_0 \left[e^{i(\omega t - kz)} - e^{i(\omega t - k\sqrt{a^2+z^2})}\right] \tag{3.30}$$

where $p_0 = c_0\rho_0 v_0$. The solution on the acoustic axis is built by the superposition of two waves, a plane wave, given by the term $\exp(-ikz)$, and a wave traveling from the edge of the piston, given by the term $\exp(-ik\sqrt{a^2 + z^2})$. The plane wave describes the radiation from the flat piston, while the second wave introduces diffraction effects at the piston edge. Taking the absolute value of Eq. (3.30) we obtain the amplitude of the field along the axis

$$|p(z, t)| = 2p_0 \left|\sin\left(\frac{1}{2}kz\left[\sqrt{1 + \frac{a^2}{z^2}} - 1\right]\right)\right| \tag{3.31}$$

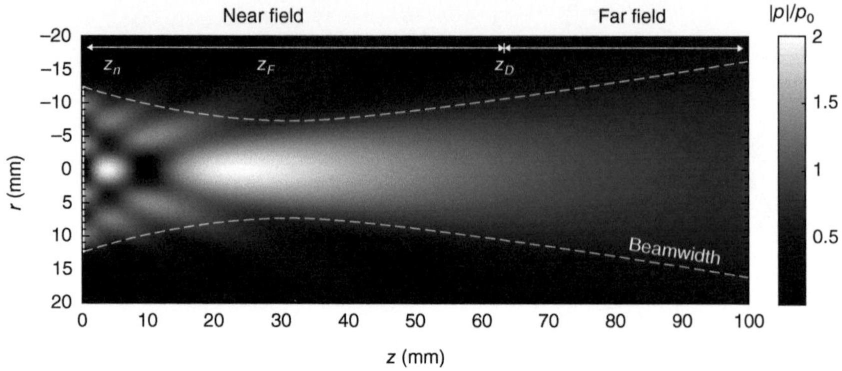

Figure 3.6 Example of the field radiated by a circular piston of aperture $2a = 25$ mm and frequency 244 kHz.

Near the surface, these waves interfere constructively and destructively and oscillations are present, therefore the pressure varies between 0 and $2p_0$ (Figure 3.6).

The maxima in the radiation field of a flat piston will occur at the distances from the axis that satisfy

$$z_n = \frac{(ka)^2 - (n\pi)^2}{2\pi kn} \tag{3.32}$$

where $n = 1,3,5, \ldots$. The oscillating area near the piston is known as the Fresnel region. This function has an ultimate maximum, which appears at a distance from the transducer of $z_1 = ka^2/2\pi - \pi/2k$. Since $ka^2/2\pi \gg \pi/2k$ in most acoustic problems, the Fresnel distance is defined as the distance at which the last maximum approximately appears

$$z_F = \frac{ka^2}{2\pi} \tag{3.33}$$

From the latter maximum, the field starts to decrease progressively. The diffraction distance, also known as Rayleigh distance, is given by

$$z_D = \frac{ka^2}{2} \tag{3.34}$$

and defines the point from which the field amplitude begins to decay inversely with distance at a rate of $1/z$. This length marks the transition between the near field and the far field. It is worth noting there that transducers with other geometries such as rectangular radiating surfaces show a roughly similar pattern, an oscillating near field, and beyond the Rayleigh distance the field decays with the inverse of the distance.

3.5.2 Focused Transducers

For certain applications, surfaces with curved geometry are also used, so that, due to the curvature, acoustic waves converge toward a given point and the energy is

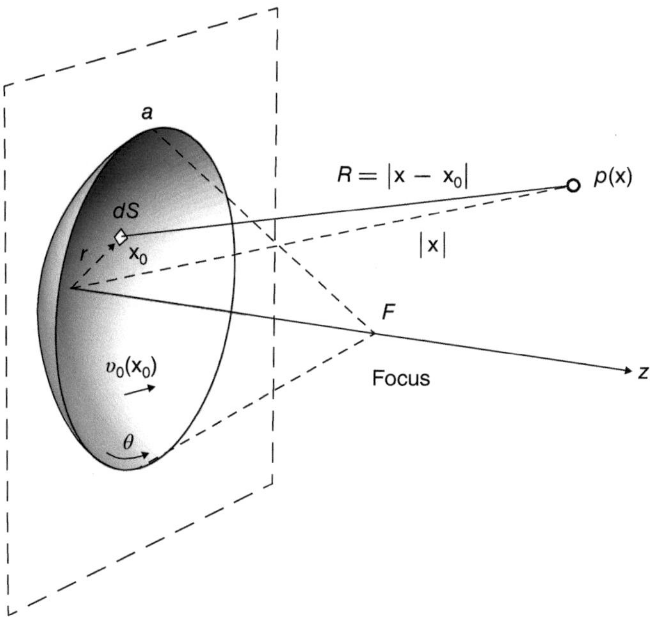

Figure 3.7 Schematic of a focused acoustic transducer and calculation points.

concentrated on a focal zone. In the case of the spherical cap of radius of curvature F and aperture radius a, as shown in Figure 3.7, we can also find analytical solutions of Eq. (3.29), such as those obtained on the symmetry axis by [20], or for the whole field but in paraxial approximation by [21] or [22].

A complete solution for the amplitude of the on-axis field is given by

$$p(z, \omega) \approx 2p_0 \left| \frac{1}{1 - \tilde{z}} \sin\left(k \frac{z - R_{max}}{2}\right) \right| \tag{3.35}$$

where F is the radius of curvature of the piston, $\tilde{z} = z/F$, a is the average aperture and R_{max} is calculated as

$$R_{max} = F\sqrt{1 + (1 - \tilde{z})^2 - 2(1 - \tilde{z})\sqrt{1 - \frac{a^2}{F^2}}} \tag{3.36}$$

This solution is valid for all apertures, but for moderately focused surfaces, when $F \gg a$, using a paraxial approximation of the diffraction we obtain

$$p(z, \omega) \approx 2p_0 \left| \frac{1}{1 - \tilde{z}} \sin\left(G \frac{1 - \tilde{z}}{2\tilde{z}}\right) \right| \tag{3.37}$$

where $G = z_D/F = ka^2/2F$ is the transducer gain, which measures the increase in sound pressure at the geometric focus relative to the pressure at the source surface.

As in the case of the circular surface, we can also distinguish between near-field and far-field zones. Within the near field we find three distinct regions. Near the piston, for $z < z_N$, the near Fresnel zone extends, where there are strong oscillations due to constructive and destructive interference. We define the Fresnel distance in this case as

$$z_N = \frac{F}{1 + F\lambda/a^2} \tag{3.38}$$

The Fraunhofer zone comprises the acoustic focus and the geometric focus of the source. In this region, the beamwidth is minimal, and the acoustic energy is highly concentrated. It is important to note that the acoustic focus, i.e. the area of maximum energy, does not generally coincide with the geometric focus due to diffraction effects. If there are no nonlinear effects and the medium is homogeneous, the acoustic focus appears closer to the transducer than the geometric focus. Beyond the Fraunhofer zone, $z > z_F$, where

$$z_F = \frac{F}{1 - F\lambda/a^2} \tag{3.39}$$

extends the far Fresnel zone, in which there are also oscillations, but they are softer and of smaller amplitude. Progressively, they disappear.

Figure 3.8 shows a schematic of these near-field regions. An important parameter to characterizing a focused transducer is its gain, G, defined as the field amplitude at the geometric focus, $z = F$, compared with the field at the surface of the source, given by

$$G = \left|\frac{p_F}{p_0}\right| = p_0\frac{ka^2}{2F} \tag{3.40}$$

It is critical to note that since the maximum amplitude value of the field occurs before the geometric focus, the maximum amplitude at the acoustic focus is somewhat larger than p_F. The point at which, in a paraxial approximation, a maximum occurs in the field given by Eq. (3.37), is given by [23]

$$z_{max} = F\frac{3G}{2G + \sqrt{G^2 + 72}} \tag{3.41}$$

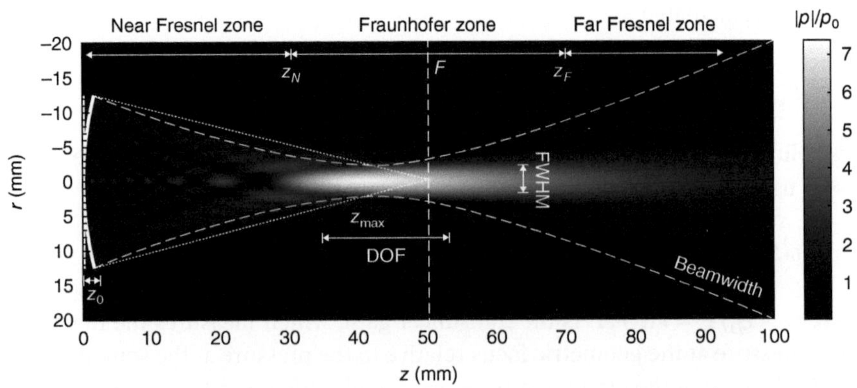

Figure 3.8 Example of the field produced by an aperture focused transducer $2a = 25$ mm, frequency 1 MHz, and focal $F = 50$ mm.

Note that the position of the acoustic focus depends on the initial amplitude [24, 25], but if the vibration amplitude is small we can consider linear propagation and, therefore, this effect is negligible.

On the other hand, it is relevant to characterize the focus in terms of its lateral and axial width (or depth of field), defined at half the maximum pressure in the acoustic focus

$$FWHM = 0.7\frac{F\lambda}{a} \tag{3.42}$$

for the transverse width, and

$$DOF = 1.8\frac{F^2\lambda}{a^2} \tag{3.43}$$

for the depth of field in the axial direction. Other points of interest are the exit plane, the distance of which is given by $z = F - \sqrt{F^2 - a^2}$, or the coverage angle, given by $\beta = \arcsin a/F$.

3.5.3 Phased-Array Transducers

For many applications transducers are composed of many small piezoelectric elements. In these transducers each active element is excited with an individual electric signal, and in analogy with phased-array antennas for electromagnetic RADAR systems, they are named phased-array transducers. These transducers are used for SONAR systems in underwater acoustics, and are also very common in biomedical applications for ultrasound imaging. Phased-array transducers can be manufactured in several configurations such as linear arrays, convex probes, sectorial arrays, two-dimensional (2D) transducer arrays, or small endocavitary ultrasound transducer to perform internal examinations such as endovaginal or endorectal probes. An example of a linear phased-array probe is shown in Figure 3.9.

By setting individual delays for each excitation signal, the transducer can modulate the transmitted wavefront to transmit focusing beams and control the location of the focus, e.g. in a linear array we can control its depth and lateral position. In addition, phased-array transducers offer a great flexibility to conform other wavefronts, such as steered quasi-plane wavefronts as those used for plane-wave compounding imaging or even spherically diverging beams. In many imaging modalities a partial aperture is used for each transmission, and usually this is controlled by using channel apodization. This consists on setting a spatial window to the channel amplitude, usually a Gaussian function, to reduce lateral lobes.

3.6 Acoustic Lenses

As we have presented previously, beam shaping and focusing can be obtained by curved radiating surfaces, for example, using spherically shaped piezoelectric ceramics. However, in most cases using a curved piezoelectric transducer increases the cost and the complexity of the transducer. In addition, the coupling with tissues

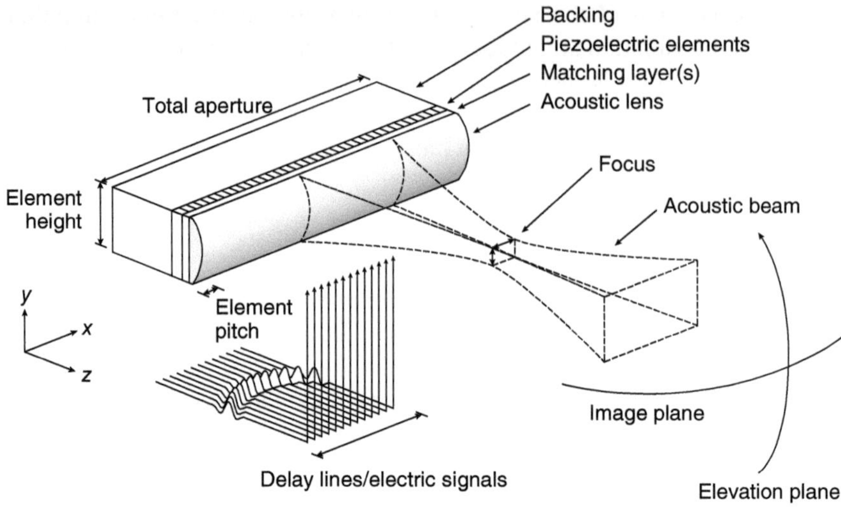

Figure 3.9 Linear phased-array probe using apodization to excite the array with a partial aperture. By controlling the delay of each signal, the system can dynamically move the focus in transmission.

is complicated due to convex shape of the surface. Therefore, most transducers incorporate an acoustic lens. Acoustic lenses are the mechanical analog for optical. They can be fabricated using several techniques, which we have classified into refractive, diffractive, reflective, resonant, and holographic.

3.6.1 Refraction by Bulky Lenses

A wavefront traveling through a flat interface between two materials with different sound speeds will be refracted following Snell's law. When this surface is curved, the wavefront can be collimated to a given point or area, and different curvature strategies can be used to shape the resulting beam profile.

3.6.1.1 Spherical Lenses

For example, a plane wave traveling through a concave lens with *spherical* curvature results in focusing when the sound speed of the lens material, c_m, is higher than the sound speed of the surrounding medium, c_0. When the sound speed of the lens is slower than the medium, $c_m <_c 0$, a convex profile can be used to focus the beam, as shown in Figure 3.10.

The geometrical focus of the beam, F, is given be the refraction index as [26]

$$F = \frac{R}{1 - c_0/c_m} \tag{3.44}$$

where R is the radius of curvature of the spherical lens. However, spherical lenses are not perfect and result in geometric aberrations, which are important for large apertures.

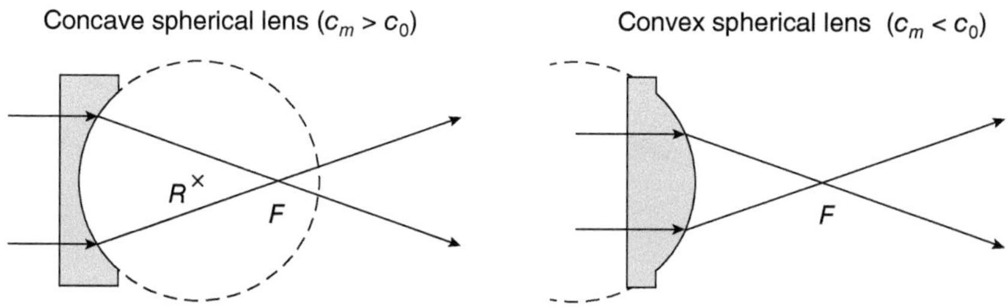

Figure 3.10 Spherical acoustic lenses.

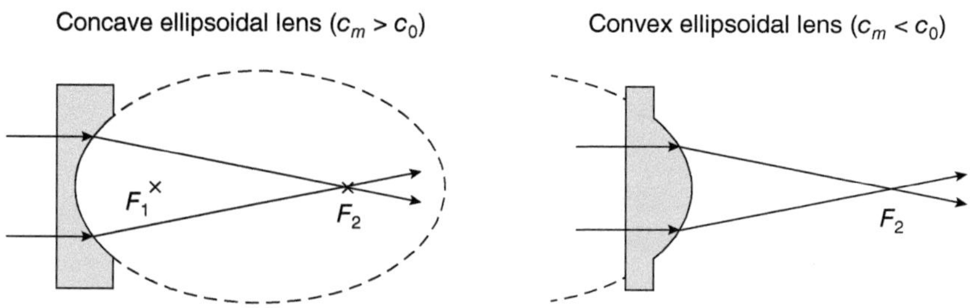

Figure 3.11 Ellipsoidal acoustic lenses.

3.6.1.2 Ellipsoidal Lenses

In this case, a better solution to focus on a point is given by an *ellipsoidal* lens, being the larger focus of the ellipsoid the geometrical focus of the acoustic lens.

In this case, the ellipsoidal lens is designed by setting [27, 28]

$$\epsilon = \frac{c_0}{c_m} \tag{3.45}$$

as the eccentricity of the lens, and the focal spot is generated around the second focus of the ellipse F_2, as shown in Figure 3.11. Aspherical lenses, in analogy with optics, can also be designed. Refractive lenses are usually included in most transducers designs, from medical imaging probes to industrial sensors.

3.6.1.3 Axicon Lenses

Other strategy is to produce a conical wavefront instead of a focused one by using a conical lens, named *axicon* [29, 30]. Axicons can be convex (conical) or concave, as shown in Figure 3.12, as a function of the sound speed ratio c_0/c_m. Instead of a focused beam, axicons generate a (truncated) Bessel beam, i.e. a beam showing a large focal spot with a good lateral resolution.

For moderately focused axicons, the focal length of the Bessel beam F_2, as shown in Figure 3.12, is given by

$$F_2 = \frac{a}{(c_m/c_0 - 1)\alpha} \tag{3.46}$$

where a is the half aperture of the lens, and α is the conical angle of the axicon with respect to the perpendicular plane of the lens. While Bessel beams can be generated

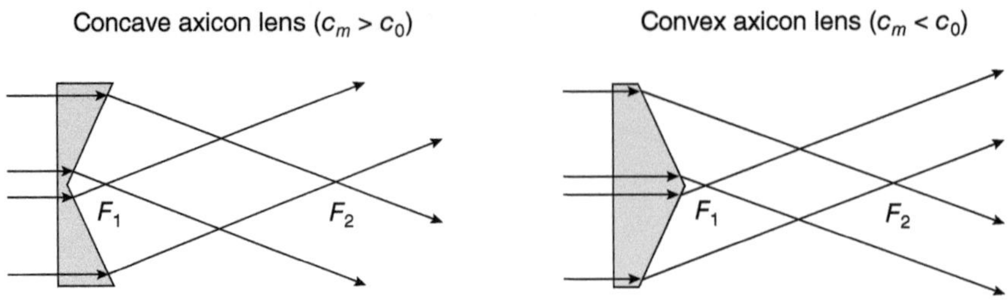

Figure 3.12 Axicon acoustic lenses.

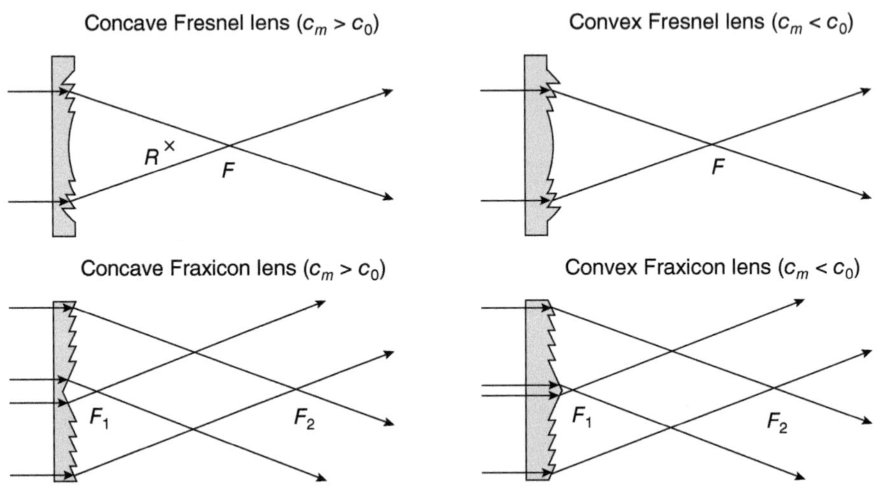

Figure 3.13 Fresnel and fraxicon acoustic lenses.

by annular phased-array sources [31], axicon lenses allow the generation of these limited diffraction beams in a low-cost manner.

3.6.1.4 Frensel and Fraxicon Lenses

Refractive acoustical lenses are bulky when a large aperture and short focal, i.e. large f-number given by $2a/F$, or large gain is required. In this case, the wave should travel through a thick material layer and ultrasound suffers from strong absorption, that can heat and damage the transducer elements. To solve this problem, *Frensel* lenses can be used for focused beams [32], as shown in Figure 3.13. This procedure produces a blazed profile of concave or convex surfaces, resulting in a rough but thin lens. In analogy, *fraxicon* lenses can be designed for Bessel beam generation [33], these lenses presenting a sawtooth radial profile.

3.6.1.5 Lenses for Vortex Generation

In addition, other exotic beams can also be generated using refractive lenses, such as vortex beams. Vortex beams are wavefronts containing a phase dislocation, typically at the beam axis. The phase of the field twists around this phase singularity, and resembling an acoustical whirlpool. At the axis, the field is null due to destructive

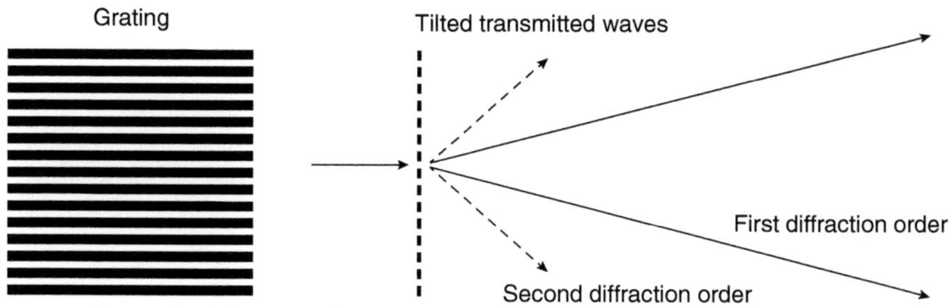

Figure 3.14 Diffraction grating over a period set of apertures, waves are diffracted at angles θ_n.

interference and, therefore, the focal spot of a vortex beam shows a toroidal shape instead of an elliptical one. To generate such a complex beams, one can produce spiral Fresnel lenses or spiral fraxicons [33], e.g. manufacturing them by three-dimensional (3D)-printing techniques. Helicoidal surfaces can also be used [34]. These beams can be used for many biomedical applications such as particle trapping, manipulation, and imaging.

3.6.2 Diffraction by Gratings

Another mechanism to modulate a wavefront is to use the diffraction over an engineered grating. A grating should be highly reflecting to enhance scattering and block the passage of waves in given areas.

3.6.2.1 Cartesian Diffraction Grating

For example, if a set of linear apertures or slits are set on a perfectly reflecting plate, due to the continuity of the transverse wavevector at the interface, ultrasound is transmitted only at particular angles, as shown in Figure 3.14.

These angles, θ_n, are given by

$$\theta_n = \arccos\left(n\frac{\lambda}{d}\right) \tag{3.47}$$

where $n = 0,1,2,\ldots$ is the diffraction grating order, $\lambda = c_0/f$, and a is the separation distance of the slits. We can see that the diffraction grating angles are chromatic, as they depend on frequency. In addition, θ_n only take real values for $|n\lambda/d| \leq 1$, meaning that higher diffraction grating angles only occur in the form of a propagating mode when the slit separation distance d is large compared to the wavelength λ. Using diffraction grating phenomena on axisymmetric geometries one can collimate acoustic beams.

3.6.2.2 Asymmetric Diffraction Grating

When a circular grating is used, as shown in Figure 3.15, by manufacturing a plate with equispaced circular apertures at radial locations [35]. The radial position of each aperture is given by

$$r_m = m \cdot d \tag{3.48}$$

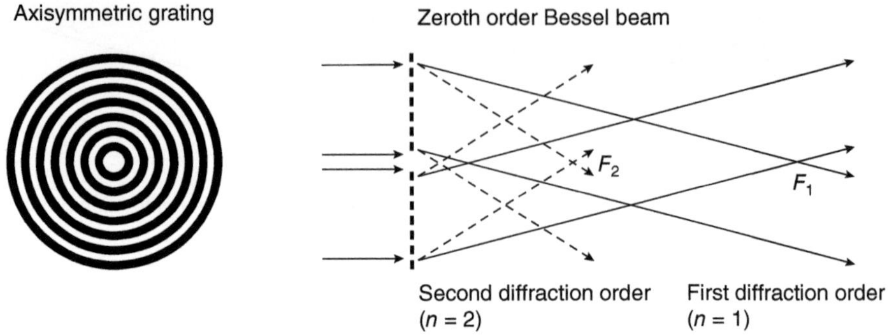

Figure 3.15 Axisymmetric diffraction grating, generates a zeroth-order Bessel beam.

where m is the index of the circular aperture, and d is the separation between them.

Due to asymmetric diffraction grating, a conical wavefront is generated. Its conical angle, α_n, is given for each diffraction grating order n as $\sin \alpha_n = n\lambda/d$. Therefore, and in analogy with axicon lenses, the acoustic field is collimated at the axis in the form of a zeroth-order Bessel beam [35], extending up to a focal distance of

$$F_n = \frac{ad}{n\lambda}\sqrt{1-\left(\frac{n\lambda}{d}\right)^2} \tag{3.49}$$

where a is the half aperture of the lens. We can observe that if higher diffraction orders are excited, they generate secondary elongated focal spots near the source.

3.6.2.3 Fresnel Zone Plates

If a focused beam is desired, then a Fresnel zone plate can be used [36, 37]. The radial location of the mth blocking rings is given by the recursive formula

$$r_{m+1}^2 = r_m^2 + \left(n^2\lambda^2 + n\lambda 2\sqrt{r_m^2 + F^2}\right) \tag{3.50}$$

where F is the focus of the Fresnel zone plate, m is the index of the aperture, and n is the diffraction grating order, as shown in Figure 3.16.

The edge of the mth aperture is given

$$r_m^{2\prime} = \left(\sqrt{r_0^2 + F^2} + m\frac{\lambda}{2}\right)^2 - F^2 \tag{3.51}$$

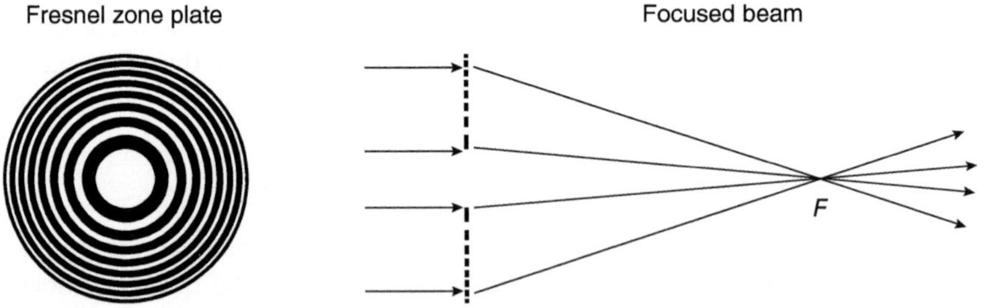

Figure 3.16 Fresnel zone plate produces a focused beam. Only the first diffraction order is shown.

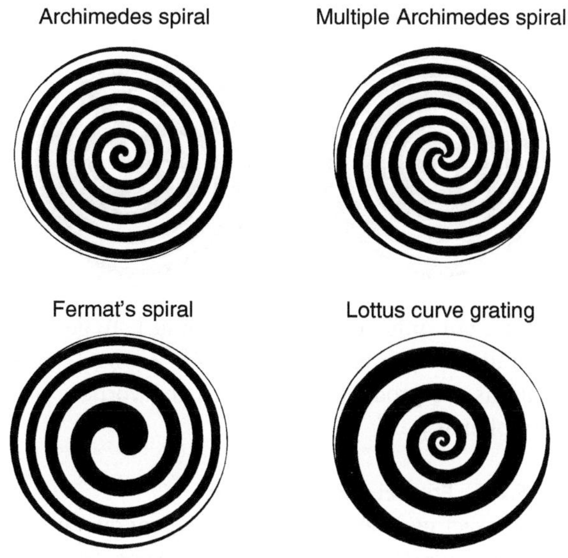

Figure 3.17 Archimedean spiral diffraction gratings. Archimedes spiral generates a high-order vortex beam, multiple arms spirals generate vortices with higher topological charge where the topological charge is proportional to the number of arms. Fermat's spiral produces aberrated focused and the Lottus curve generates unfocused vortex beam.

and the location of the center of each annuls is given by

$$r_m^2 = \left(\sqrt{r_0^2 + F^2} + m\lambda\right)^2 - F^2 \tag{3.52}$$

Diffraction gratings work efficiently only in a limited range of focal distances. For large focal distances, higher diffraction grating orders appear and secondary foci emerge near the source. However, these lenses are easy to manufacture, the designs are simple, and they are stable because they do not depend on the properties of the lens material, only the geometry and the sound speed of the surrounding medium.

3.6.2.4 Archimedean Spiral Gratings

Exotic beams can also be generated using diffraction gratings. For example, the grating can be carved following an Archimedean spiral [38], given by the families of polar curves

$$r(\theta) = \frac{d}{2\pi}\theta^{\gamma} \tag{3.53}$$

The first case is the *Archimedes spiral grating* for $\gamma = 1$. In these spirals the space between winds is constant, given by d, as shown in Figure 3.17.

Therefore, diffraction grating occurs at angles $\sin \alpha_n = n\lambda/d$, like in axisymmetric gratings. However, the phase rotates along the polar coordinate. Therefore, instead of a zeroth-order, a first order Bessel beam is generated [38]. In general, multiple arms spirals generate vortices with higher topological charge where the topological charge is proportional to the number of arms. Higher-order Bessel beams can be generated using Archimedes spiral of multiple arms, as shown in Figure 3.17. Higher diffraction orders also generate high-order Bessel beams.

Vortices can also be generated by Archimedean spirals using $\gamma \neq 1$ see e.g. Ref. 39. Diverging spirals, like the Lottus curve for $\gamma = 2$, can result in a vortex

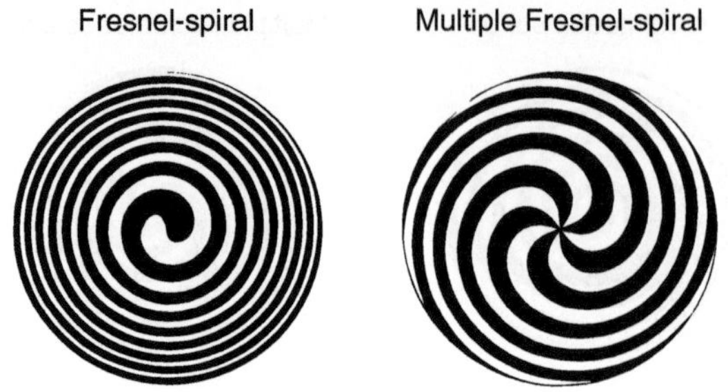

Figure 3.18 Fresnel-spiral zone plate lens, generates a sharply focused vortex beam. The topological charge of the vortex can be controlled using multiple-arm spiral.

beam, but unfocused [40, suppl. mat.]. To generate a focused beam, Archimedean spirals, like the Fermat's spiral for $\gamma = 0.5$, can result in focusing, however, aberrations are produced when focusing vortices using a Fermat's spiral grating [40, suppl. mat.].

3.6.2.5 Fresnel-Spiral Zone Plate

The exact geometrical solution to generate an optimally focused vortex beam the *Fresnel-spiral zone plate* given by the polar curve [40]

$$r_m(\theta)^2 = \left[\sqrt{r_0^2 + F^2} + \lambda\left(\frac{M\theta}{2\pi} - m\right)\right]^2 - F^2 \tag{3.54}$$

where F is the focal of the lens, M is the total number of the arms, m is the index of each arm, and $r_0^2 = (F + \lambda)^2 - F^2$ is the radius of the central opaque area, as shown in Figure 3.18.

In a classical Fresnel zone plate, waves irradiating the zone plate diffract from the opaque zones and constructively interfere at the desired focus. However, in the case of a Fresnel-spiral grating, waves diffracted from opposite angles arrive at the focus in the opposite phase due to the fact that the path difference between both trajectories is $\lambda/2$. Then, in the focal plane, the destructive interference leads to the generation of a hollow focused beam, i.e. a focused vortex beam. One can notice that the Fermat's curve is similar to the Fresnel spiral, but only the Fresnel-spiral curve is the geometrically optimal structure to design binary diffraction gratings producing sharp vortex focusing. Fresnel-spiral using multiple arms allow sharp acoustic focusing, while the topological charge of the vortex beam can be directly related to the number of arms of the Fresnel-spiral. Note that focusing is produced by diffraction grating; therefore, the separation between opaque and transparent areas should be of the order of half wavelength. Thus, the product between the wavenumber and the aperture of the lens should be high enough to allow for various windings.

3.6.3 Reflection by Curved Surfaces

Reflection on curved surfaces can be used to focus an acoustic beam. Shaped by evolution, concave surfaces are present in almost every ear design on living mammals. Spherical acoustic mirrors were used as early warning devices before the World War II, to detect incoming bombers by their engine noise, for example, as those located at the coast of Denge, UK. These concrete mirrors can enhance the sound amplitude at

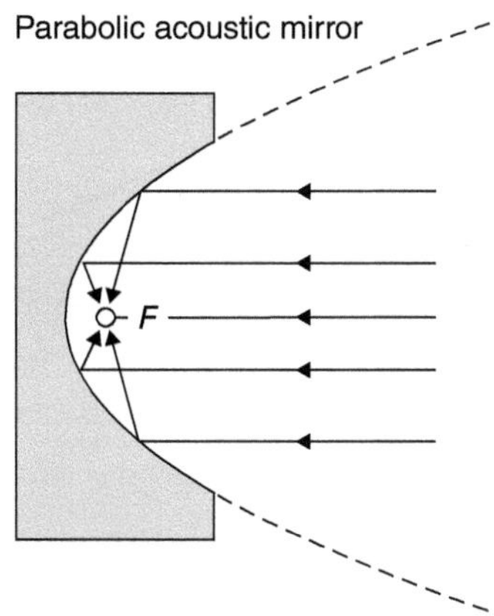

Figure 3.19 Paraboloidal acoustic mirror.

the focal spot. However, spherical mirrors suffer from strong aberrations when they present a large aperture.

3.6.3.1 Parabolic Reflectors

To sharply focus a plane wavefront a better solution than a spherical mirror is to use a parabolic surface, as shown in Figure 3.19. In parabolic mirrors, a plane wave is collimated toward the focus of the paraboloid. Note that, if the wavefront is not plane, then the acoustic focal spot is not sharp and will suffer from aberrations.

These mirrors are used in several acoustic devices. Paraboloids are used in highly directive microphones for acoustics, and also for therapeutic ultrasound applications like extracorporeal shockwave therapy to fragment kidney stones. In this last case, the source is usually a cylindrical electrodynamic transducer.

3.6.3.2 Ellipsoidal Reflectors

Other transducer used for the first designs of lithotriptors for extracorporeal shock-wave therapy was an electrohydraulic source, that consists in a spark plug. When the voltage is high enough to break the dielectric, a spark is generated leading to a sudden rise of temperature, creating a collapsing bubble and an intense spherical acoustic wave. However, this diverging wavefront should be focused again on the desired target, the kidney stone.

To focus a spherical wave an ellipsoidal reflector can be used, as shown in Figure 3.20. In this case, the point source should be located on the first focus of the ellipsoid, and the diverging wavefront, after being reflected on the surface, converges to the second focus of the ellipsoid. Note that, in practice, usually truncated ellipsoids are used due to space constraints.

3.6.4 Holograms

3.6.4.1 Field Projections

By using the Helmholtz–Kirchhoff integral, the field produced by an arbitrary complex source can be projected at any point of the space. For example, if the acoustic

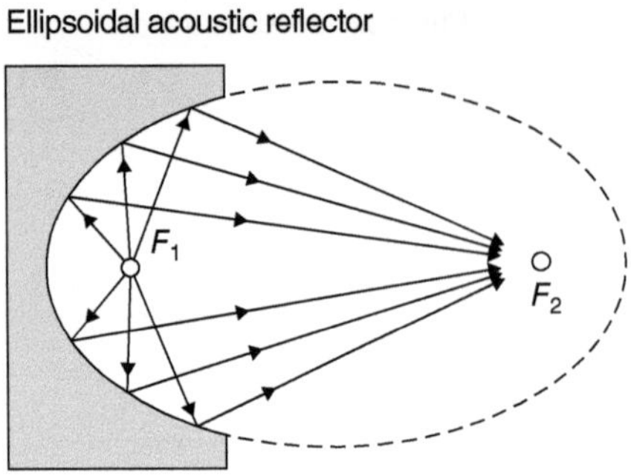

Figure 3.20 Ellipsoidal reflector, focusing a point source located at the first focus of the ellipsoid toward the second one.

field is known in a surface located at a distance $z = z_0$, defined as $P(x, y, z_0)$, we can project that field to another plane located at $z = z_h$, to obtain $P(x, y, z_h)$, as

$$P(x,y,z_h) = -\frac{1}{2\pi}\iint_{-\infty}^{+\infty} \frac{\partial P(x,y,z_0)}{\partial n}\frac{e^{-ikR}}{R}dxdy \tag{3.55}$$

where $R = |\mathbf{r}_h - \mathbf{r}_0|$ is the distance between each point in the original plane, $\mathbf{r}_0 = (x_0, y_0, z_0)$, and each point of the target plane, $\mathbf{r}_h = (x_h, y_h, z_h)$, and $\partial/\partial n = \mathbf{n} \cdot \nabla$ indicates differentiation from the surface normal $\mathbf{n}$. Since $\partial P/\partial n = -ik\rho_0 c_0 v_0$ holds for a plane wave, for a source vibrating at normal speed v_0 Eq. (3.55) results in the Rayleigh–Sommerfeld integral. We can find a more efficient expression for this calculation using the *angular spectrum* of the field, given by as

$$P(k_x,k_y,z) = \iint_{-\infty}^{+\infty} P(x,y,z)e^{-i(k_x x+k_y y)}dxdy \tag{3.56}$$

where k_x and k_y are the transverse components of the wave vector. Note that Eq. (3.56) can be calculated very efficiently using a 2D spatial fast Fourier transform. Thus, we can project the angular spectrum of the source in a z_0 plane onto a z_h plane as

$$P(k_x,k_y,z_h) = P(k_x,k_y,z_0)\cdot H(k_x,k_y,\Delta z) \tag{3.57}$$

where $\Delta z = z_h - z_0$ is the separation distance between planes, and $H(k_x, k_y, \Delta z)$ is the propagator that models the diffraction, given by

$$H(k_x,k_y,\Delta z) = e^{-i\Delta z\sqrt{k^2-k_x^2-k_y^2}} \tag{3.58}$$

Using the inverse spatial Fourier transform

$$P(x,y,z) = \frac{1}{4\pi^2}\iint_{-\infty}^{+\infty} P(k_x,k_y,z)e^{i(k_x x+k_y y)}dk_x dk_y \tag{3.59}$$

we can retrieve the field from the angular spectrum as

$$P(x,y,z_h) = \frac{1}{4\pi^2}\iint_{-\infty}^{+\infty} P(k_x,k_y,z_0)e^{-i(k_x x+k_y y)}e^{-i(z_h-z_0)\sqrt{k^2-k_x^2-k_y^2}}dk_x dk_y \tag{3.60}$$

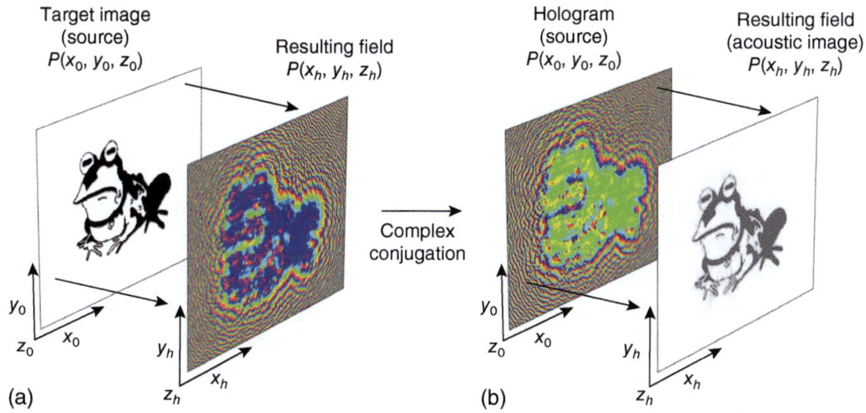

Figure 3.21 (b) Projection process of the field generated by an acoustic source with an arbitrary surface $P(x_0, y_0, z_0)$, in this case an image. At a distance z_h the projected field is given by $P(x_h, y_h, z_h)$. (a) If the source now radiates the complex conjugate of the projected field, at a distance of z_h we reconstruct a diffraction-limited acoustic image, which resembles the original source.

These expressions allow the calculation of a field in one plane from the measurement of the field in another plane. This opens the door to different applications, such as the characterization of the radiation of a source from near-field measurements, known as acoustic holography techniques (Figure 3.21) [41].

3.6.4.2 Synthesis of Acoustic Images

One of the advantages of using these methods is that the field information can be projected both forwards and backwards, which makes it possible to calculate the field produced by a source, but also vice versa, i.e. to know which source would produce a target field. In essence, the time-reversal symmetry of the wave equation allows us the back projection processes. These methods are based on techniques developed in optics by Denis Gabor [42] who discovered that a 3D representation of an object could be obtained from the interference between a reference light beam and the waves reflected by the object. As the pattern of the reflected waves could be printed on a photographic plate, the real object was not needed to reconstruct the virtual image once the wavefront was captured. He called the capturing surface (and not the 3D representation) the *hologram*,[2] and for this he was awarded with the Nobel Prize in Physics in 1971.

Using these optical holograms, light beams can be modulated to generate visible images. In the same way, acoustic holograms can also be used to synthesize acoustic images, i.e. to concentrate sound or vibrations in an arbitrary area of space, and simultaneously allow other areas to remain silent or at rest. Acoustic holograms modify the mechanical wavefront in such a way that they are able to modulate both their phase and, in some cases, their amplitude. For acoustic waves at ultrasonic frequencies, such as those commonly used in biomedical applications around 1 MHz,

2 The word holography comes from the Greek *holos* meaning "complete" and *gram* meaning "drawing."

acoustic holograms can be fabricated by 3D printing elastic polymers, resulting in acoustic lenses with a roughness scale of a few millimeters [43].

3.6.4.3 Biomedical Applications of Holograms

These acoustic holograms have found a recent application in the biomedical field. For example, they have been proposed for targeted ultrasound therapies in neurology [44]. Focused ultrasound is commonly used for multiple neurological therapies, and offers many desirable aspects, as it does not use ionising radiation, resulting in completely noninvasive, safe, and cost-effective treatments. Examples of these treatments are thermal ablation of thalamic nuclei by focused ultrasound for the treatment of essential tremor, opening the blood-brain barrier for localized and noninvasive drug delivery, or focal neuromodulation applications by mechanical action of ultrasound on central nervous system tissues. However, on their way to the brain, ultrasound beams are affected by aberrations introduced by skull bones, which introduce strong attenuation and refraction by irregularly modifying the wavefront. Aberration and refraction results in a poor and uncontrolled focusing. Moreover, in many cases the area of therapeutic interest is not a single point, but a structure with a particular 3D shape, as occurs in therapies on deep brain nuclei, e.g. the hippocampus. Acoustic holographic lenses can overcome these limitations. Since holograms offer very precise manipulation of the wavefront, low-cost 3D printed holographic lenses can be designed to compensate for aberrations introduced by bones and, simultaneously, focus mechanical energy in an arbitrary area, e.g. they can generate several focal points, allowing several structures of the central nervous system to be targeted at the same time. For example, using holograms we can simultaneously open the blood-brain barrier on two spots of the brain, i.e. at the left and right hemispheres [45] (Figure 3.22).

These acoustic holograms go beyond traditional ultrasound beams to synthesize complex acoustic images, allowing the generation of curved or self-bending beams [44], or even focal zones in the shape of the target in the deep-brain using small

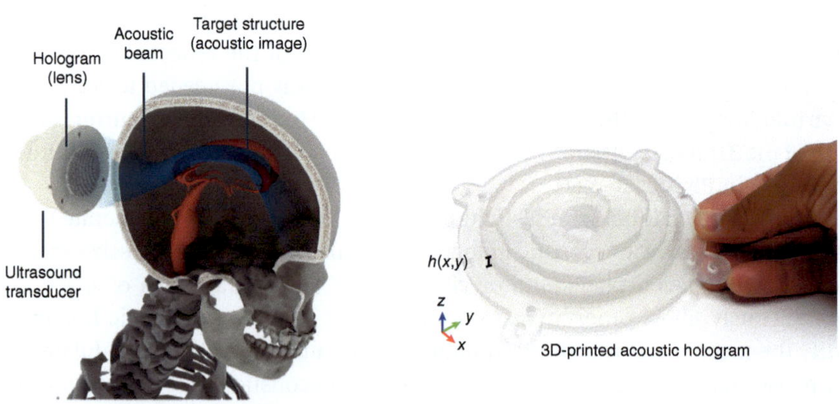

Figure 3.22 Application of acoustic holograms for therapeutic ultrasound in neurology, and example of 3D printed holographic lens. Source: Jimenez-Gamb et al. [46]/American Physical Society.

apertures [47]. Holograms also make it possible to synthesize other complex acoustic fields within the skull, such as vortex beams [46]. These beams have a phase singularity on the axis, which opens the door to noninvasive, noncontact trapping and manipulation of small objects such as drug aggregates or clots within the central nervous system. Beyond neurology, holograms has been also proposed for large volume hyperthermia [48]. Holographic lenses open new avenues to generalize emerging applications of therapeutic ultrasound, such as localized drug delivery or neuromodulation procedures, through noninvasive and very low-cost systems.

References

1 Curie, J. and Curie, P. (1880). Développement par compression de l'électricité polaire dans les cristaux hémièdres à faces inclinées. *Bulletin de minéralogie* 3 (4): 90–93.

2 Kino, G.S. (1987). *Acoustic Waves: Devices, Imaging, and Analog Signal Processing*. Englewood Cliffs, NJ: Prentice-Hall.

3 Arnau, A. (2008). *Piezoelectric Transducers and Applications, Springer Science & Business Media*. Berlin, Heidelberg: Springer-Verlag.

4 Krimholtz, R., Leedom, D.A., and Matthaei, G.L. (1970). New equivalent circuits for elementary piezoelectric transducers. *Electronics Letters* 6 (13): 398–399.

5 Liu, Z., Wu, H., Ren, W., and Ye, Z.-G. (2022). Piezoelectric and ferroelectric materials: fundamentals, recent progress, and applications. In: *Reference Module in Chemistry, Molecular Sciences and Chemical Engineering*. Elsevier. ISBN 978-0-12-409547-2. https://doi.org/10.1016/B978-0-12-823144-9.00069-8. https://www.sciencedirect.com/science/article/pii/B9780128231449000698.

6 Chilibon, I. and Marat-Mendes, J.N. (2012). Ferroelectric ceramics by sol–gel methods and applications: a review. *Journal of Sol-Gel Science and Technology* 64 (3): 571–611.

7 Tressler, J.F., Alkoy, S., and Newnham, R.E. (1998). Piezoelectric sensors and sensor materials. *Journal of Electroceramics* 2 (4): 257–272.

8 Panda, P.K. (2009). Environmental friendly lead-free piezoelectric materials. *Journal of Materials Science* 44 (19): 5049–5062.

9 Rathod, V.T. (2020). A review of acoustic impedance matching techniques for piezoelectric sensors and transducers. *Sensors* 20 (14): 4051.

10 Fukada, E. (2000). History and recent progress in piezoelectric polymers. *IEEE Transactions on Ultrasonics, Ferroelectrics, and Frequency Control* 47 (6): 1277–1290.

11 Xie, L., Wang, G., Jiang, C. et al. (2021). Properties and applications of flexible poly(vinylidene fluoride)-based piezoelectric materials. *Crystals* 11 (6): 644.

12 Akdogan, E.K., Allahverdi, M., and Safari, A. (2005). Piezoelectric composites for sensor and actuator applications. *IEEE Transactions on Ultrasonics, Ferroelectrics, and Frequency Control* 52 (5): 746–775.

13 Inoue, T., Ohta, M., and Takahashi, S. (1987). Design of ultrasonic transducers with multiple acoustic matching layers for medical application. *IEEE Transactions on Ultrasonics, Ferroelectrics, and Frequency Control* 34 (1): 8–16.

14 Alvarez-Arenas, T.E.G. (2004). Acoustic impedance matching of piezoelectric transducers to the air. *IEEE Transactions on Ultrasonics, Ferroelectrics, and Frequency Control* 51 (5): 624–633.

15 Desilets, C.S., Fraser, J.D., and Kino, G.S. (1978). The design of efficient broad-band piezoelectric transducers. *IEEE Transactions on Sonics and Ultrasonics* 25 (3): 115–125.

16 Fang, H.J., Chen, Y., Wong, C.M. et al. (2016). Anodic aluminum oxide–epoxy composite acoustic matching layers for ultrasonic transducer application. *Ultrasonics* 70: 29–33.

17 Kossoff, G. (1966). The effects of backing and matching on the performance of piezoelectric ceramic transducers. *IEEE Transactions on Sonics and Ultrasonics* 13 (1): 20–30.

18 King, L.V. (1934). On the acoustic radiation field of the piezo-electric oscillator and the effect of viscosity on transmission. *Canadian Journal of Research* 11 (2): 135–155. https://doi.org/10.1139/cjr34-080.

19 Douglas Mast, T. and Yu, F. (2005). Simplified expansions for radiation from a baffled circular piston. *The Journal of the Acoustical Society of America* 118 (6): 3457–3464. https://doi.org/10.1121/1.2108997.

20 O'Neil, H.T. (1949). Theory of focusing radiators. *The Journal of the Acoustical Society of America* 21 (5): 516–526. https://doi.org/10.1121/1.1906542.

21 Lucas, B.G. and Muir, T.G. (1982). The field of a focusing source. *The Journal of the Acoustical Society of America* 72 (4): 1289–1296. https://doi.org/10.1121/1.388340.

22 Chen, X., Schwarz, K.Q., and Parker, K.J. (1993). Radiation pattern of a focused transducer: a numerically convergent solution. *The Journal of the Acoustical Society of America* 94 (5): 2979–2991. https://doi.org/10.1121/1.407329.

23 Makov, Y.N., Espinosa, V., Sánchez-Morcillo, V.J. et al. (2006). Strong on-axis focal shift and its nonlinear variation in low-Fresnel-number ultrasound beams. *The Journal of the Acoustical Society of America* 119 (6): 3618–3624. https://doi.org/10.1121/1.2197805.

24 Camarena, F., Adrián-Martínez, S., Jiménez, N., and Sánchez-Morcillo, V. (2013). Nonlinear focal shift beyond the geometrical focus in moderately focused acoustic beams. *The Journal of the Acoustical Society of America* 134 (2): 1463–1472. https://doi.org/10.1121/1.4812865.

25 Jiménez, N., Camarena, F., and González-Salido, N. (2017). Dynamic nonlinear focal shift in amplitude modulated moderately focused acoustic beams. *Ultrasonics* 75: 106–114.

26 Fry, W.J. and Dunn, F. (1962). Ultrasound: analysis and experimental methods in biological research. In: Nastuk, W.L. *Physical Techniques in Biological Research. IV. Special Methods*, Academic Press Inc.: New York City 261–394.

27 Schlussler, L. (1980). The design and test results for an acoustic lens with elliptic surfaces. *The Journal of the Acoustical Society of America* 67 (2): 699–701.

28 Dunn, F. and Fry, F.J. (1980). Acoustic elliptic lenses: a historical note. *The Journal of the Acoustical Society of America* 68 (1): 350.

29 McLeod, J.H. (1954). The axicon: a new type of optical element. *JOSA* 44 (8): 592–597.

30 Burckhardt, C.B., Hoffmann, H., and Grandchamp, P.-A. (1973). Ultrasound axicon: a device for focusing over a large depth. *The Journal of the Acoustical Society of America* 54 (6): 1628–1630.

31 Hsu, D.K., Margetan, F.J., and Thompson, D.O. (1989). Bessel beam ultrasonic transducer: fabrication method and experimental results. *Applied Physics Letters* 55 (20): 2066–2068.

32 Hadimioglu, B., Rawson, E.G., Lujan, R. et al. (1993). High-efficiency Fresnel acoustic lenses. *1993 Proceedings IEEE Ultrasonics Symposium*, 579–582. IEEE.

33 Jiménez-Gambín, S., Jiménez, N., Benlloch, J.M., and Camarena, F. (2019). Generating Bessel beams with broad depth-of-field by using phase-only acoustic holograms. *Scientific Reports* 9 (1): 1–13. https://doi.org/10.1038/s41598-019-56369-z.

34 Jiménez, N., Ealo, J., Muelas-Hurtado, R.D. et al. (2021). Sub-wavelength acoustical vortex beams using self-demodulation. *Physical Review Applied* 15 (5): 054027. https://doi.org/10.1103/PhysRevApplied.15.054027.

35 Jimenez, N., Romero-García, V., Picó, R. et al. (2014). Acoustic Bessel-like beam formation by an axisymmetric grating. *EPL (Europhysics Letters)* 106 (2): 24005.

36 Farnow, S.A. and Auld, B.A. (1974). Acoustic Fresnel zone plate transducers. *Applied Physics Letters* 25 (12): 681–682.

37 Jimenez, N., Romero-García, V., Pico, R. et al. (2015). Nonlinear focusing of ultrasonic waves by an axisymmetric diffraction grating embedded in water. *Applied Physics Letters* 107 (20): 204103.

38 Jiménez, N., Picó, R., Sánchez-Morcillo, V. et al. (2016). Formation of high-order acoustic Bessel beams by spiral diffraction gratings. *Physical Review E* 94 (5): 053004.

39 Wang, T., Ke, M., Li, W. et al. (2016). Particle manipulation with acoustic vortex beam induced by a brass plate with spiral shape structure. *Applied Physics Letters* 109 (12): 123506. https://doi.org/10.1063/1.4963185.

40 Jiménez, N., Romero-García, V., García-Raffi, L. et al. (2018). Sharp acoustic vortex focusing by Fresnel-spiral zone plates. *Applied Physics Letters* 112 (20): 204101.

41 Maynard, J.D., Williams, E.G., and Lee, Y. (1985). Nearfield acoustic holography: I. Theory of generalized holography and the development of NAH. *The Journal of the Acoustical Society of America* 78 (4): 1395–1413.

42 Gabor, D. (1948). A new microscopic principle. *Nature* 161: 777–778.

43 Melde, K., Mark, A.G., Qiu, T., and Fischer, P. (2016). Holograms for acoustics. *Nature* 537 (7621): 518–522.

44 Jiménez-Gambín, S., Jiménez, N., Benlloch, J.M., and Camarena, F. (2019). Holograms to focus arbitrary ultrasonic fields through the skull. *Physical Review Applied* 12 (1): 014016.

45 Jiménez-Gambín, S., Jiménez, N., Pouliopoulos, A.N. et al. (2022). Acoustic holograms for bilateral blood-brain barrier opening in a mouse model. *IEEE Transactions on Biomedical Engineering* 69 (4): 1359–1368. https://doi.org/10.1109/TBME.2021.3115553.

46 Jiménez-Gambín, S., Jiménez, N., and Camarena, F. (2020). Transcranial focusing of ultrasonic vortices by acoustic holograms. *Physical Review Applied* 14 (5): 054070. https://doi.org/10.1103/PhysRevApplied.14.054070.

47 Andrés, D., Jiménez, N., Benlloch, J.M., and Camarena, F. (2022). Numerical study of acoustic holograms for deep-brain targeting through the temporal-bone window. *Ultrasound in Medicine and Biology* 48 (5): 872–886. https://doi.org/10.1016/j.ultrasmedbio.2022.01.010.

48 Andrés, D., Vappou, J., Jiménez, N., and Camarena, F. (2022). Thermal holographic patterns for ultrasound hyperthermia. *Applied Physics Letters* 120 (8): 084102. https://doi.org/10.1063/5.0081565.

4

Ultrasound and Ultrasonic Imaging in Medicine: Recent Advances

Tuğba Ö. Onur

Zonguldak Bülent Ecevit University, Department of Electrical-Electronics Engineering, Zonguldak 67100, Türkiye

4.1 Introduction

In the 1940s, infrasonic technology, which was first used in radars during World War II, started to see use in research as a new method for medical imaging, and by the 1970s, it had been adopted for use in diagnostic medicine. In the last 20 years, ultrasound imaging has become one of the most widely used imaging modalities due to its high biocompatibility, high tissue penetration depth, and portability. Because acoustic waves are nonionizing, ultrasound is an attractive alternative to other methods as there is a lower risk of harm to the patient.

Thanks to the rapid development of ultrasonic imaging technology, new methods have evolved to further improve imaging quality. For example, tissue harmonic imaging (THI) has been adapted along with conventional ultrasound imaging to further improve imaging quality. The purpose of THI is to send sound waves with a lower frequency value into the body to obtain higher-quality images by detecting higher frequency sound waves by making use of intra-body vibrations.

4.2 Ultrasound Waves

Ultrasound is a type of acoustic wave that is generated via mechanical vibrations and whose frequency is above the limit of human hearing (the human ear can hear sounds between 20 Hz and 20 kHz). In general, sounds with a frequency greater than 20 kHz are defined as ultrasound. Although there are various types of waves, "longitudinal compressional waves" are the most common acoustic waves that are used for diagnostic imaging. The particles that vibrate within the medium alternate between states of compression and rarefaction as the wave travels through the medium. In this way, energy is transported to the medium in a direction parallel

Acoustic Technologies in Biology and Medicine, First Edition.
Edited by Adem Ozcelik, Ryan Becker, and Tony Jun Huang.

to the vibrations of its particles. Because ultrasound is the vibration of atoms or molecules around their equilibrium positions, it is considered a form of mechanical energy [1].

4.2.1 Types of Ultrasonic Waves

The vibration direction of atoms and the direction of energy emission should not be confused with each other, as they occur independently from one another. In the wave representation of motion, the amplitude-distance axes clearly show the direction of propagation and the direction of the distance axis. However, the direction of vibration is not clear using this representation. There are three major types of ultrasonic waves, all containing unique vibration and propagation directions. Thus, it is important to differentiate between these wave types in terms of their features.

i. **Longitudinal Waves**
   In longitudinal waves, the vibration and energy emission directions are the same. This is also called a compressional pressure wave. This wave is used in ultrasonic evaluation with imaging probes in general (radiating perpendicular to the surface). It can propagate into any material medium, such as solids, liquids, and gases. Longitudinal waves are the most common type of acoustic waves encountered and are the wave type we know from everyday life. It is important to note that the propagation speed of longitudinal waves is higher than that of other ultrasonic waves.
ii. **Transverse Waves**
   For transverse waves, the vibration and energy emission directions are perpendicular to each other. This is also called a shear wave, which is used in ultrasonic evaluation with angular probes. It can only propagate in solid media, not in liquids or gases. The propagation speed of transverse waves is about half that of a longitudinal wave.
iii. **Surface Acoustic Waves**
   The vibratory motion of surface acoustic waves is an ellipse perpendicular to the direction of emission. As the amplitude changes, the size of the ellipse will change accordingly. Surface acoustic waves have been given their name because they only propagate on the surface of a substrate and do not penetrate into a material. The attenuation of surface acoustic waves is very rapid, with 100% of the energy dissipated at a depth of approximately one wavelength of the material. The speed of surface acoustic waves is slightly lower than transverse waves [2].

4.2.2 Behavior of Ultrasound Waves at Interfaces

Ultrasonic waves experience both refraction and reflection as they propagate through anisotropic media, specifically at the interfacial boundaries between the different materials of which the media is comprised. The laws governing the reflection and refraction of light in optics also apply to acoustic waves. For example, the refraction of ultrasound waves occurs is governed by Snell's Law, similar to optics:

$$\frac{\text{Sin}\theta_1}{\text{Sin}\theta_2} = \frac{V_1}{V_2} \tag{4.1}$$

where,

θ_1: Angle of incidence
θ_2: Angle of refraction
V_1: Propagation speed of ultrasound in medium 1
V_2: Propagation speed of ultrasound in the second medium 2.

Reflection and refraction of ultrasound waves occur when they pass through surfaces or tissues with different acoustic impedances. Biological tissues are inhomogeneous, anisotropic materials that contain many different densities. When an ultrasound beam encounters a boundary between these interfaces, it refracts and changes direction, with some of the waves reflected back to the transducer. Additionally, the ultrasound beam decreases in power as it passes through the material as the acoustic waves are attenuated. There are three main factors that contribute to the attenuation of acoustic waves. The first and most prevalent contributing factor to beam attenuation is energy absorption by the surrounding media. Some of the energy of the ultrasound beam passing through the material is absorbed and converted to heat. The second factor contributing to attenuation is the scattering of acoustic waves as they pass through a material. The third factor contributing to attenuation is the expansion of the ultrasound beam. The energy density of an ultrasound beam is inversely proportional to its cross-sectional area. As the beam travels through the tissue, it expands, the cross-sectional area increases, and the energy density passing through the tissue decreases [2]. The attenuation of the ultrasound beam depends on the properties of the material it is traveling through, the material thickness, and the frequency of the wave. As the material thickness and frequency increase, the attenuation also increases. Because penetration depth is a function of frequency, low-frequency ultrasound waves are commonly used to image deep tissue. However, resolution is also a function of frequency, with lower frequencies yielding lower resolutions. Thus, a careful balance between penetration depth and resolution must be considered.

4.2.3 Ultrasound Power and Intensity

The power of an ultrasound beam can be expressed in various units; however, in practice, this is commonly reported in Watts (W). A power of 1 W corresponds to an energy rate of 1 joule per second. The ultrasound intensity is equal to the amount of energy flowing through a unit surface per unit of time and is expressed in W/cm^2 [1]. Ultrasound equipment should only be used appropriately by people fully trained in its safety. These people should have a full awareness of the ultrasound equipment settings, maintenance, and the effects of machine settings on power levels. In addition, for medical ultrasound imaging applications in medical diagnosis, examination times should be kept short but long enough to obtain a useful diagnostic result.

4.2.4 Ultrasound Applications

Outside of medical imaging, ultrasound has found utility across a wide range of industries. The most common use it has found has been in the adoption of ultrasonication as a cleaning process, which utilizes ultrasonic cavitation in a liquid to remove foreign contaminants from the surface of a material. Ultrasonication has shown to be an extremely effective technique for cleaning contaminated machine parts, medical equipment, electronic components, and surfaces with thin recesses and protrusions that are difficult to clean using conventional methods. More recently, research in ultrasonication has been focused on its use in the textile industry. It was found that ultrasonication can be used in combination with certain enzymes to achieve removal of foreign substances such as dirt, and oil from textile materials and eliminate the disadvantages of using enzymes alone to clean textiles due to their large molecular size and low mobility. The main contribution of ultrasound to cleaning textiles during wet finishing processes is primarily due to cavitation in the liquor. Cavitation occurs as a result of the explosion of small energetic bubbles, which are formed as a result of the movement of the ultrasound waves in the liquor. An increase in mass transfer from liquid to solid is observed as a result of cavitation at the solid/liquid interface [3]. This strong mobility resulting from cavitation improves the transfer of dirt from the fabric surface to the liquor, as well as the transport of large-molecule enzymes to the fiber surface, and as a result, the total reaction rate will increase. Moreover, ultrasonication is good for the environment as an alternative to common detergents since it requires less electricity and water.

Intense competition pushes manufacturers to produce cheaper and cleaner products. Under these conditions, it becomes necessary for a company's survival to adopt new technologies that give their products a leading edge. Ultrasonication, which is environmentally friendly, is also important in this context. For example, dirt removal, which requires mass transfer, is very slow when done with conventional methods and requires a high energy potential. For this reason, ultrasonication can clean products in a shorter time with much less energy or the need for additional chemicals. Ultrasonication is also environmentally friendly and is a useful tool to reduce the carbon footprint for the textile industry.

Ultrasound has also found use in the food industry by using cavitation to change the mechanical, physical, and chemical/biochemical properties of a process, thus shortening the reaction time, increasing the yield, and reducing the costs compared to traditional food processing methods. Foods that contain proteins, vitamins, carbohydrates, enzymes, fats, minerals, water, and other organic components are considered complex materials. Therefore, the processing and preservation of foodstuffs require different practices depending on the type of food [4]. In this context, thermal processes, which are used to increase the shelf life of foods, are one of the most widely used processing methods to provide microbial and enzyme inactivation. However, exposure of foods to high temperatures during this process can cause undesirable changes in the qualities of the food, namely the texture, taste, color, smell, and nutritional (vitamins, proteins, etc.) components [5]. In addition, the increasing consumer demand for food products that have a natural taste and odor, do not contain additives or preservatives, and have a high nutritional value, creates the need to develop nonthermal, innovative approaches in food processing [6]. In response to these demands, the food industry has turned to new technologies such as the use of high pressure, percussive, electric, and magnetic fields to provide high-quality foods with "fresh" properties that are functionally enhanced. Among these emerging technologies, there is an increasing interest in the use of ultrasound [7]. There are two main categories for the use of ultrasound in food processing applications. The first category focuses on the replacement of traditional food processing technologies with ultrasound [8]. Applications such as food cutting [9], emulsification/homogenization [10], sterilization/pasteurization [11], meat tenderizing [12], and degassing [13] belong to this category. The second category is focused on assistive technologies that accelerate traditional techniques so that processing can be completed more efficiently and quickly. In this latter category, the disadvantages of conventional techniques used during the procedure are eliminated or improved through the use of ultrasound. These food processes include extraction [14], freezing [15], thawing [16], brining [17], filtration [18], and drying/dehydration [19].

In addition to its use in the food industry, ultrasound is also one of the most commonly used imaging methods in modern medicine. The most important difference between ultrasound and other imaging methods is that ultrasound uses nonionizing forms of penetrating energy to image patients in the form of sound waves at frequencies that cannot be heard by the human ear. Another significant advantage is that images can be formed in real time, which allows the user to observe dynamic biological processes such as the motion of tissue or blood. The acoustic waves form echoes that are transformed into electrical signals, which are used to create images. The frequency, the strength of the acoustic signals, and the arrival time of the echoes are the basis factors for image formation.

4.3 Ultrasonic Imaging

The purpose of all diagnostic ultrasound imaging applications is to detect and visualize the acoustic energy reflected from organ interfacial boundaries. Ultrasound is capable of producing grayscale, high-resolution images and providing information about the direction and velocity of blood flow in the vessels in real time. Thanks to these features, ultrasound imaging has become an important and versatile medical imaging tool.

However, obtaining clinically relevant images with ultrasound is not a straightforward process. It requires a combination of technical skills to acquire high-quality images, including diagnostic skills to interpret the location of the image they are trying to obtain and knowledge of the underlying physical principles governing ultrasound. The user should have a good understanding of the basis of the interactions of acoustic waves with tissue and the methods and tools used to acquire and optimize medical ultrasound images. With proper training, a sonographer can obtain high-quality images with ultrasound, providing a wealth of information that can be used for diagnostic examination and avoiding errors that may result from omission of information or misinterpretation of the image. Ultrasound imaging and Doppler ultrasound are based on the interactions of acoustic waves, which are scattering, refraction, and reflection of sound energy. The amplitude of the reflected energy is used to create ultrasound images, and the frequency changes due to sound scattering can be used to calculate the position and velocity information of moving objects, such as blood flow in the body [20].

The user needs to control many variables in order to generate, detect, and process ultrasound imaging data. In addition, ultrasound operators should understand the methods used in the generation of ultrasound data and the theory and operation of the devices that detect, display, and store the acoustic signals produced in clinical examinations. Modern ultrasound imaging uses a digital system that carries out signal and image processing, numerical calculations, and system input–output of the data. A medical ultrasound imaging system used for imaging consists of three main components: a scanner, a display, and a user interface (Figure 4.1). To form an image, high-frequency sound waves are first generated using a transducer and

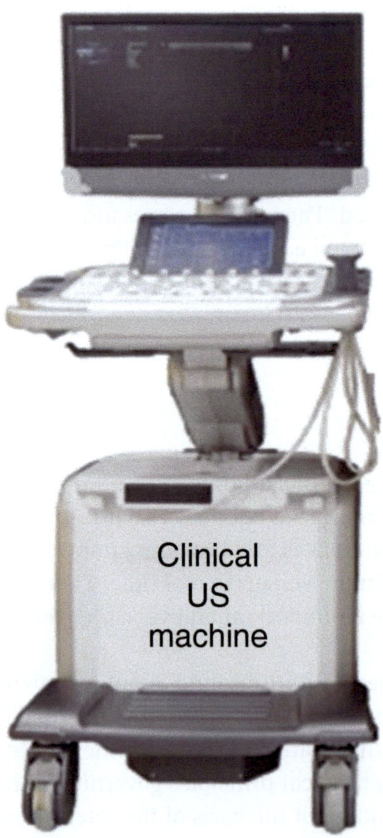

Figure 4.1 Medical ultrasound systems are used for diagnostic purposes. Source: Reproduced from Kim et al. [21], Springer Nature/CC BY 4.0.

delivered to the tissue. The formation of an ultrasound image starts with the return of the signal beam produced and sent from the transmitter arrays in the transducer section of the device to the receiving array. After the ultrasound signal beam is sent to the body, the sound wave echo (echo) signals return (reflect) from the touch interface and then reaches the receiver array elements, where the signal will be processed. The position of the echo signal returned to the transducer is then sorted into a column. These columns are then processed to generate a grayscale image on the screen of the ultrasound device.

Ultrasound images soft tissues by sending a high-frequency acoustic wave, typically between 2 and 15 MHz, into the tissue. Ultrasound does not propagate in the air, so the acoustic waves need to be adequately coupled into the tissue to obtain images. Some of the most common uses for medical ultrasound imaging are for imaging organs in the abdominal cavity (liver organs, pancreas, bile, diaphragm, spleen), for obstetrics (baby development, genital organs, kidneys), breast ultrasonography, and sonography of the thyroid, vascular structures, muscles, joints, and heart. It is also possible to view the blood flow of vessels in the brain using transcranial Doppler. Ocular echography is also used to visualize the eye muscles with ultrasound. To summarize briefly, it is possible to use the ultrasonography technique for imaging many different organs in the body. The organs that cannot be imaged by ultrasound are those that have material properties that do not allow the propagation of sound; although images can be obtained, they may not be suitable for diagnostic use. Because ultrasound cannot travel through air, the lungs cannot be visualized using this technique because they are filled with air sacs. Ultrasound also cannot propagate through bone, so the imaging of important structures such as the bones that make up the ear cannot be imaged either. To optimize the ultrasound parameters for specific organs, different probes are used by the sonographer.

Ultrasound imaging has a very large number of significant applications in medicine, such as monitoring pregnancies, providing guidance for needle biopsies, and providing insight into patient conditions to help physicians decide the best treatment options. Ultrasound has been used in medicine for more than 40 years, and is routine practice in obstetrics it has become an indispensable tool for gynecological examinations and pregnancies.

In pregnancy: Sonograms are one of the most widely utilized ultrasound medical procedures. They are performed to monitor the health of the mother and child, and generate images of the fetus during pregnancy. Between four and five weeks of pregnancy, a due date and many potential health problems can be determined. At six to seven weeks, ultrasound can detect the fetal heartbeat and determine if there is a multiple pregnancy (twins, triplets, etc.). At 8–9 weeks, the arms and legs of the fetus can be seen, with more details observed at 10–18 weeks. Parents can determine the baby's gender within 19–21 weeks. At 31 weeks, part of the baby is clearly visible; for example, the face and all the different expressions of it can be clearly made out, the limbs and their movements can also be seen; and at 35–37 weeks, the baby's position (head down or wedge) can be checked and a nonstress test can be performed.

In diagnostic imaging: Ultrasound is also used to diagnose a wide variety of conditions affecting the organs and soft tissue areas of the body. The images obtained by using ultrasound are in real time, as opposed to a single snapshot like an X-ray image.

In the medical field, aside from imaging, ultrasound is also used for therapeutic purposes. Ultrasound heat therapy is one of the most commonly used treatment methods in physical therapy. The acoustic waves used in ultrasound heat therapy are generally between 800.000 and 3000.000 Hz and are used at higher energy levels than those seen in diagnostic imaging applications to increase the generation of heat within the tissue. The heat generated by ultrasound therapy increases blood flow to the region, decreases pain, and accelerates healing of the tissue. In physical therapy, ultrasound is also used to diagnose various conditions related to the musculoskeletal system, such as herniated disc, neck hernia, muscle tendon compression in the shoulder, frozen shoulder, carpal tunnel syndrome, bursitis, tendinitis, muscle strain, ligament injury, and meniscus damage.

4.3.1 Ultrasonic Imaging System

Ultrasound sends high-frequency sound waves into the tissue on a pulse-reflection basis. Echoes that return to the tissue surface are retrieved by the probe and used as the signal for image formation. It is a frequently used diagnostic method due to its ease of use, portability, reliability, and absence of radiation risk. Ultrasonic imaging systems generate acoustic waves with frequencies greater than 20 kHz, which is the upper limit of human hearing. Although ultrasonic waves are limited between 20 and 100 kHz, this can sometimes be as high as 30 MHz, depending on the application. The ultrasound frequency used for diagnosis varies between 2 and 10 MHz. To generate the acoustic waves, the piezoelectric (pressure electricity) phenomenon is used to obtain high frequencies. Crystals such as quartz, lithium sulfate, cadmium sulfate, zinc oxide, tourmaline, barium, titanate, and lead titanate have piezoelectric properties. When mechanical deformation is applied, the piezoelectric crystal generates a voltage. When the opposite is done, the crystals will create mechanical energy by expanding and compressing. The piezoelectric effect enables the conversion between mechanical and electrical energy. Ultrasound transducers utilize the piezoelectric effect to generate sound waves and to convert received sound waves into an electrical signal for image generation [22]. When the transducer is excited by an electrical signal, it generates a high-frequency ultrasonic pulse that travels into the human body by the inverse piezoelectric effect. Reflected echoes are retrieved by the same probe and displayed on the screen after processing. Thanks to ultrasound imaging, we can easily obtain images of vital organs and visualize many important processes in the body in a safe manner [23].

4.3.1.1 Transducer

The transducer converts electrical energy into sound energy and vice versa. A voltage is applied to the crystal consisting of lead zirconate titanate (PZT) placed on

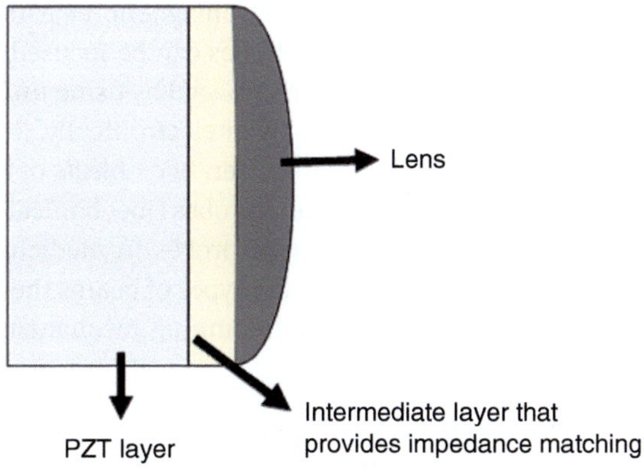

Figure 4.2 The front surface of the probe.

the front of the transducer. As a result, the crystals contract or expand transversely, longitudinally, or radially to create vibrations, which in turn create the compressional longitudinal waves that are used in ultrasound. When a returning echo reaches these crystals, they produce a corresponding electric voltage. Thanks to these crystals, it is possible to both produce and perceive sound [20].

Since the crystal has very high impedance compared to the tissue, ultrasound coupling gel is used to match the impedance between the probe and the tissue. There is also an intermediate layer to further provide impedance matching, a backing layer to absorb spurious acoustic waves, and a lens to focus the acoustic waves [24]. Figure 4.2 shows the front surface of the probe to help reader for visualization.

The frequency of the transducer the sonographer uses will be different according to the thickness of the tissue to be examined. High frequency increases resolution, but there is a corresponding decrease in penetration depth. Because of this, high frequencies are normally only used in the examination of superficial tissues. Low frequencies are used to examine deeper tissues [22].

The transducer is used both to send ultrasonic waves and to receive returning waves. Ultrasonic waves are sent into the tissue 0.1% of the time, while reflected waves are received by the probe the remaining 99.9% of the time.

4.3.1.2 Probes

The head that carries the transducer is called the "probe" and it is a contact element, providing transmission between the ultrasound and the patient. They are produced in different types according to the body area to which they are applied, the beam shape they will generate, and the frequency range they use. Every probe houses piezoelectric crystals, which expand or contract longitudinally, transversely, or radially toward their surroundings, depending on the polarity of the applied voltage, the geometry of the crystal, and its initial polarization. PZT and other common polarized ceramic crystals are used as the piezoelectric crystals in most probes. The appropriate crystal thickness is selected to produce the ultrasound wave at the desired frequency.

Ultrasound waves emanating from an unfocused probe expand and scatter after a certain distance, so it is important that the focusing of the waves is properly controlled. The classical electromagnetic near-field and far-field concepts can be used to find this distance. Probes can be focused or unfocused. The objects within the focal range can be clearly imaged by using unfocused probes. In addition, focusing can be done mechanically or electronically. Resolution determines the power of the device to distinguish between two objects of a specific size. Although there are two main types of ultrasound probes (mechanical and electronic), electronic probes are the most commonly used probes in medicine. Probes can be further divided into three groups based on the types of beams they form: linear, convex, and phased array. The arrangement and scanning mechanisms of the elements differ between each type of probe [20, 23].

1) *Linear probes:* Linear probes use rectangular scans of elements arranged along a line. These probes are used to examine superficial structures and organs (e.g. breast, thyroid); however, they can occasionally be used to measure deeper structures or babies if their radius grows. Typical structures for linear probe heads usually vary between 64 and 192 elements. Frequency ranges for linear probes are usually between 3.5 MHz for low-frequency probes and 7.5 MHz for high-frequency probes.
2) *Convex probes:* Convex or curvilinear probes scan in a wide arc and produce a fan beam. Because they have a wider aperture than linear probes, they are more commonly used for transabdominal imaging, typically operating between 3.5 and 5 MHz, although higher-frequency microconvex probes (~7.5 MHz) are also used. The main application areas for larger convex probes are abdominal (radiology) and obstetric studies. Microconvex probes are more commonly used for gynecological or urological purposes. Scanning angles can vary between 40° and 150°.
3) *Sector probes:* Sector probes scan in the form of a fan. It is used to display small areas.

Phased array transducers are usually used for beam-steering applications. Electronic control over beam geometry and direction can be provided with the use of phased array transducers, and they can be adapted to provide optimal energy deposition patterns for a given therapeutic application. Their use in conjunction with modern medical imaging for therapy guidance allows precise targeting, online monitoring, and posttreatment evaluation of ultrasound-mediated bioeffects. Phased array transducers are functionally categorized according to their types, frequencies, number of elements, and size of elements. Phased array transducers are mostly the angle beam type; however, they can also be direct contact or immersion transducers. The number of elements of the phased array transducers is commonly 16 to 128; for some, it can be as many as 256. The ability to focus and direct phased array transducers can improve with an increase in the number of elements. In addition, beam steering capability gets better with a smaller element width. They are mainly used in echocardiography since they have a minimal footprint that allows them to image between the ribs and have ability to show the whole heart.

Linear and convex probes are mostly used for imaging the deeper tissues/organs. The resolution of the image obtained with convex probes decreases with depth. In addition to the above scanning formats for B-mode scanning, which provides simultaneous scans of the body that can be viewed as a two-dimensional (2D) image on the screen, there are also radial, compound, and trapezoidal scanning formats [25].

4.3.1.3 Central Processing Unit

The ultrasonic waves reflected from the tissues are detected by the transducer, which generates a low-amplitude voltage. The generated voltage requires amplification, which is performed in the central processing unit. It is the source that also provides the electrical energy required by the ultrasound device and transducer. This unit also increases the quality of the obtained image and transmits the image to the output display [20].

4.3.1.4 Output Display

The data obtained by processing ultrasonographic waves and transforming them into images is displayed on the output display, usually through the monitor. It can also be displayed in color or grayscale. The pixel intensity of the image is used to indicate the amount of sound waves that have been reflected/received by the transducer and absorbed by the tissue [20].

4.3.2 Focus

There are two primary methods for achieving the focus of the probe. First, mechanical focusing is provided by acoustic lenses placed in front of the piezoelectric crystals. The second method utilizes electronic focusing, where the focal length and beam direction can be changed by triggering the crystals at different times, providing electronic delay. In electronic focusing, crystals are formed into groups of seven or eight crystals to obtain a focused ultrasound beam. The focal length and beam direction can be changed by triggering each crystal with a certain delay and timing.

4.3.3 Resolution

In ultrasonographic imaging systems, resolution is directly proportional to frequency. As the frequency of ultrasound increases, the wavelength decreases, thus increasing the resolution. Resolution in ultrasound indicates the ability to distinguish two adjacent points. There are two types of resolution [24].

(1) Axial resolution discriminates between different tissues in the same direction as the ultrasound beam. Half of the transmitted pulse time determines the axial resolution.
(2) Lateral resolution distinguishes between two points perpendicular to the direction of propagation of the ultrasound beam. Lateral resolution is highest in the focal region. However, in order to examine deep tissues, it is necessary to use lower frequencies.

4.3.4 Beamforming

The process of directing and focusing acoustic pulses is known as beamforming or bundling. Beamforming is a general signal processing method used to control the direction of reception or transmission of a signal in a transducer array. The transducer's array elements can be fired at different times to control focus and direction, and the signals sent are sent with a time delay and are focused at a certain point. In the same way, the system time delay is applied and aligned with the returning signals reflected from the tissue. The data is then collected to obtain a strong signal [25, 26]. After focusing, radio frequency (RF) data is generated.

4.4 Sound-Tissue Interactions in Ultrasonography

Sound waves propagate at different speeds in different environments. The product of the frequency of the wave and the wavelength gives the speed of the sound in the medium.

$$V = \lambda f \tag{4.2}$$

At high frequencies, the resolution will increase; however, since the attenuation of the sound wave in tissue will increase, the penetration depth will decrease. The propagation speed of sound varies in tissues, so sound waves should be used at a frequency suitable for the tissue to be examined. Ultrasonic sound waves traveling through tissue undergo interactions such as absorption, reflection, refraction, and scattering.

4.4.1 Reflection

In ultrasonic imaging, the reflected part of the sound wave creates the image. As shown in Figure 4.3, when a wave passes from a medium with different acoustic impedances to a second medium, some of the wave is reflected and the rest passes into the other medium. Since the incident wave and the reflected wave are in the same medium, the wavelength is the same. In this case, $\theta_i = \theta_r$ will be equal to each other. However, the length and angle of the passing wave will be different [27].

The reflection rate depends on the acoustic impedance of the tissue and the angle of incidence of the sound.

1) Acoustic impedance refers to the ability of the tissue to reflect the sound wave. In simple terms, it can be considered as the resistance of the tissue to the sound wave [22].

$$Z = \rho V \tag{4.3}$$

In Eq. (4.3), Z is the acoustic impedance, ρ is the density of the tissue, and V is the speed of sound. Ultrasonic echoes are made up of reflections and backscatters of the acoustic waves as they interact with the tissue boundaries.

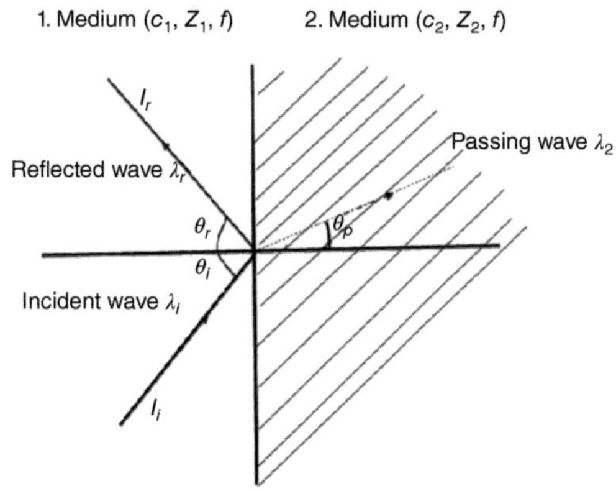

Figure 4.3 The incident wave is reflected from the interface, creating reflected, and passing waves.

Back reflections from tissue interfaces occur because the tissue layers have different acoustic impedances. The acoustic impedance can be expressed as the ratio of the pressure on an imaginary surface in a sound wave to the velocity of particle flow across the surface. The change in impedance causes ultrasonic energy to be reflected back. Larger differences in impedance between tissues cause more energy to be reflected back, and thus a higher amplitude echo is obtained. The remaining acoustic energy continues to move deeper into the tissue. The acoustic impedance of water closely matches the impedance of soft tissue, so very few reflections are obtained at that interface, which can in fact be used as a coupling interface between ultrasound and soft tissue. Soft tissue interfaces allow the propagation of the ultrasonic signal. However, since there is a large difference in impedance between soft tissues and bone, most of the acoustic energy is reflected back by the bone, and the remaining energy is absorbed by the bone. The difference in acoustic impedance between tissue and air is so great that ultrasonic waves cannot enter air-containing organs and an image cannot be obtained.

As ultrasonic sound waves pass through the tissues, the difference between the impedances of the tissues is determined by the amount of reflection. As this difference increases, the sound reflection also increases [22] (Table 4.1).

2) *Sound arrival angle*: The reflection, refraction, and scattering rates change according to the arrival angle of the ultrasonic sound wave. The sound wave should be sent at certain angles so that the reflection rate does not increase [22].

4.4.2 Refraction

When a sound wave passes from one medium to another, its frequency does not change. Only its speed and wavelength adapt to the environment. The change of direction of sound waves in a changing medium is expressed as refraction. Refraction is an undesirable situation in the image, because it causes resolution reduction, image distortions, and artifacts [22].

Table 4.1 Acoustic parameters of some tissues.

Tissue type	Intensity P_0 (g/cm^3)	Velocity c (m/s)	Characteristic impedance Z(rayl)$\cdot 10^{-6}$	Attenuation α (1/cm)
Blood	1.055	1580	1.667	0.034
Skull	1.738	2770	4.814	1.500
Brain	1.030	1460	1.504	0.060
Fat	0.937	1479	1.386	0.080
Heart muscle	1.048	1546	1.620	0.815
Kidney	1.040	1572	1.635	0.090
Liver	1.064	1570	1.670	0.149
Lung	0.400	658	0.263	4.300
Muscle	1.070	1566	1.675	0.15
Water	1.000	1500	1.500	0.0002

4.4.3 Absorption

Some of the energy of the sound wave is absorbed by the tissues and converted into heat, with higher frequencies increasing the thermal absorption. Depending on the thickness of the tissue to which the sound wave is transmitted, reflection, refraction, or scattering occurs. If the wavelength of the sound is larger than the medium, reflection occurs; if it is small, scattering occurs; and in case of equality, refraction occurs. The amount of ultrasonic waves absorbed is directly proportional to the depth of the tissue and the frequency [22].

4.4.4 Attenuation

Attenuation of an acoustic wave occurs due to a decrease in ultrasonic intensity as it passes through a medium. Absorption, scattering, beam deflection, reflection, and refraction are the primary contributing factors that cause attenuation.

4.4.4.1 Attenuation by Reflection, Refraction, and Deflection

Reflections on both smooth and rough surfaces contribute to ultrasound attenuation. In addition, refractions attenuate the ultrasonic intensity because they affect the direction of the beam traveling on the surface of the interface. In the case of attenuation via deflection, although the total power does not change, as the cross-sectioned beam area increases, the intensity decreases inversely. This type of weakening is also called geometric attenuation.

In general, the two biggest causes of attenuation that can be encountered during ultrasound imaging are absorption and scattering. As an ultrasonic wave propagates through the tissue, it is weakened by the sum of the absorption and scattering it has undergone. Attenuation is also expressed as the variation of the decrease in the pressure and intensity of the ultrasonic wave depending on the distance.

$$I(z) = I(z = 0)\exp(-\mu z) \tag{4.4}$$

$$p(z) = p(z = 0)\exp(-\alpha z)$$

In Eq. (4.4), I is the ultrasound intensity, μ is the intensity attenuation coefficient, and α represents the pressure attenuation coefficient. Both symbols are expressed in cm^{-1}. In practice, μ and α are generally denoted in dB cm^{-1}.

$$\mu\left(\text{dB cm}^{-1}\right) = \frac{-1}{z} 10 \log \frac{I(z)}{I(z = 0)} \tag{4.5}$$

$$\alpha\left(\text{dB cm}^{-1}\right) = \frac{-1}{z} 20 \log \frac{p(z)}{p(z = 0)} \tag{4.6}$$

Since absorption and scattering vary with frequency, attenuation also varies with frequency. In most tissues, there is an almost linear relationship between the attenuation coefficient (dB cm^{-1}) and frequency. The typical attenuation coefficient for soft tissues is $1\ \text{dB cm}^{-1}/\text{MHz}$, meaning that the attenuation coefficient for a 3 MHz ultrasonic beam is $3\ \text{dB cm}^{-1}$. Attenuation is commonly defined by many experts as the distance at which the wave loses half of its energy or amplitude. In addition, attenuation is less in muscles than it is in other soft tissues. This is because muscle contains a lot of water, which does not attenuate the acoustic energy very much. The half-power distance of the bone is very small compared to the muscles, which explains why bones act as a barrier for ultrasound. The half-distances of air and lung tissue are also very low, which is why ultrasound cannot image lungs.

4.4.4.2 Attenuation by Scattering

If, on rough surfaces, the ultrasonic beam encounters structures that are the same size or smaller than the wavelength of the ultrasonic wave, the wave will scatter in all directions. This phenomenon also causes the sound wave to attenuate in the direction of propagation.

The angular variation of the scattering and the magnitude of the scattering vary according to the shape, size, physical, and acoustic properties of the reflector. The total scattered ultrasonic power depends on the relationship between the size (d) of the structure encountered and the wavelength of the ultrasonic wave. For structures with dimensions much smaller than the infrasonic wavelength, the scattering power is expressed by the equation given in Eq. (4.7), where λ denotes the wavelength of the ultrasonic wave and W_s corresponds to the total scattered ultrasonic power.

$$W_s \propto \frac{d^6}{\lambda^4} \tag{4.7}$$

4.4.4.3 Attenuation by Absorption

One of the causes of attenuation of the ultrasonic wave is the absorption that occurs as a result of the conversion of acoustic energy into heat. There are two important mechanisms that contribute to this absorption: classical absorption and relaxation. Classical absorption occurs as a result of the friction of moving particles during the passage of the ultrasonic wave through the tissue. This attenuation is expressed by the absorption coefficient β_{class}, and is proportional to the square of the frequency (f) of the ultrasonic wave.

$$\beta_{class} = Af^2 \tag{4.8}$$

In Eq. (4.8), A is a constant, including the coefficients of stickiness and heat.

The second absorption mechanism is relaxation. The tissue undergoes a relaxation time (τ), and the relaxation time is expressed as the time it takes for the molecule to return to its original position after the ultrasound wave has passed. The relaxation process is expressed by the relaxation absorption coefficient β_r.

$$\beta_r = \frac{\beta_0 f^2}{1 + \left(\frac{f}{f_R}\right)^2} \tag{4.9}$$

f_R stands for relaxation frequency. Moreover, the maximum value of f_R expression is obtained when $f_R = (2\pi\tau)^{-1}$. The total absorption coefficient β is expressed as the sum of the classical absorption coefficient and the relaxation absorption coefficient.

$$\beta = \beta_{class} + \beta_r = Af^2 + \sum_n \frac{\beta_0 f^2}{1 + \left(\frac{f}{f_R}\right)^2} \tag{4.10}$$

4.4.4.4 Time Gain Reduction (TGR) and Depth Gain Reduction (DGR)

Ultrasound is attenuated by soft tissues, and an electronic correction is required to compensate for this decrease in echo amplitude. This correction process is done by integrating time gain reduction (TGR) and depth gain reduction (DGR) units into the ultrasonic devices. TGR amplifies echo amplitudes from deep within the body. This process needs to be manually controlled for every exam, so six to eight sliding keys are placed onto the control unit for the user to modify and optimize the image quality. The image depth displayed on the screen is divided according to the number of these keys, and the user manipulates the slider in that region to increase the amplitude of the image at the desired depth. Therefore, to compensate for the effects of absorption and focusing, the depth dimension of the image should also be divided into strips, which are provided by DGR. Thanks to the combined effect of these gains, a gradual increase is achieved to compensate for absorption effects.

4.5 Ultrasonic Imaging Methods

In ultrasonography, two basic diagnostic methods are used: real-time imaging and Doppler ultrasonography.

4.5.1 Real-Time Imaging

In real-time imaging, while the probe is moved over the body, the number of images appearing on the screen within the time unit is increased, allowing the movement to be observed visually. Between 500 and 3000 ultrasound pulses per second are sent to the tissue with the probe. The echoes returning from the tissues are recorded rapidly, resulting in a high framerate. To further optimize the framerate, different methods are used to display these returning echoes [22].

4.5.1.1 A-Mode

A-mode, which is not used frequently today, is one of the oldest and easiest modes developed for imaging. This mode creates a 1D image dependent on the depth, according to the intensities of the signals reflected from the tissue layers of different densities. This operating mode is also called pulse echo. The name A-mode originated from the word amplitude, and as such, it is also known as amplitude mode [28].

To generate an image in A-mode, a short pulse at the transmitter output is applied to the crystal of the transducer, as shown in Figures 4.4 and 4.5. Part of the energy packet produced by the ultrasound wave propagating in the medium is reflected at a tissue interfacial boundary where there is a discontinuity in characteristic impedance of the tissues. The reflected pulses are amplified by the receiver and displayed on a screen, with the horizontal axis being time and the vertical axis being the amplitudes of the echoes [27].

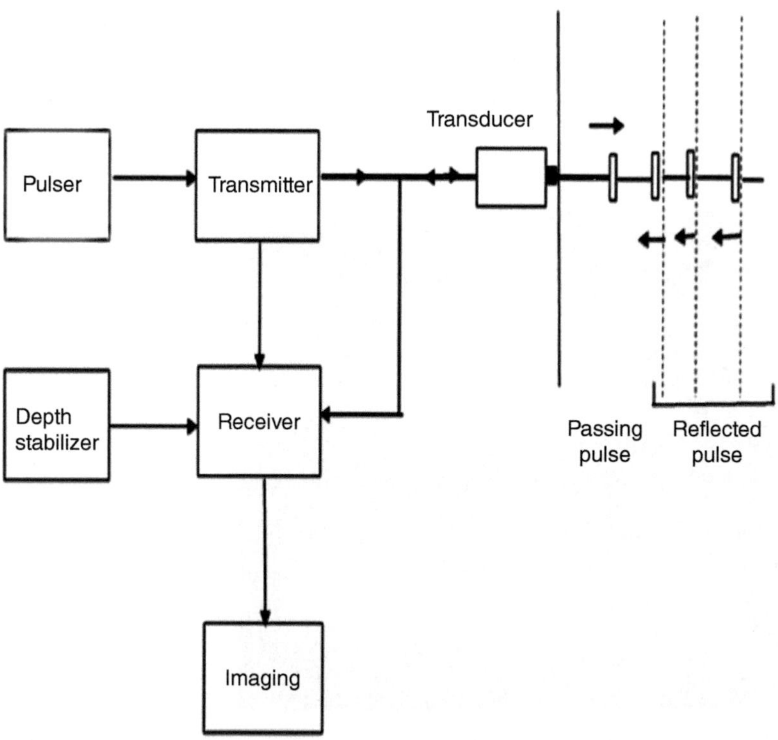

Figure 4.4 Block diagram of an A-mode imaging system.

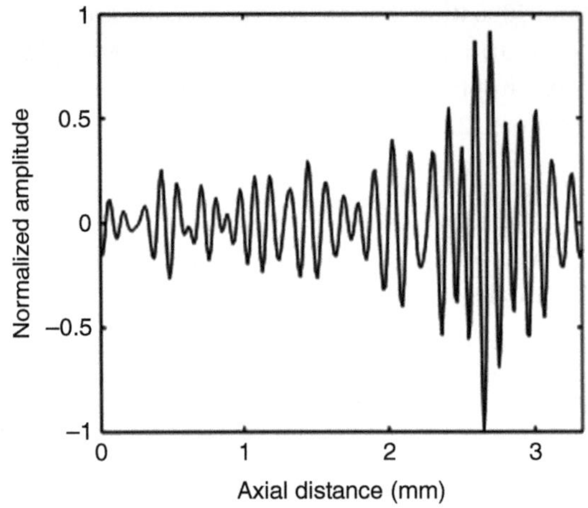

Figure 4.5 An example of A-mode ultrasound imaging. Source: Quinsac et al. [29]/Hindawi/Licensed under CC BY 4.0.

4.5.1.2 M-Mode

M-mode is an enhanced version of A-mode. This mode is also known as motion mode because it is used to dynamically image the movement of tissues and organs. This type of mode is useful in many different medical applications, such as investigations into arterial elasticity, left ventricular function, or presence of pneumothorax. If the surfaces are being imaged in M-mode move, the points generated in the image show a corresponding movement. Immobile surfaces, on the other hand, form straight lines. Time information is shown on the horizontal axis and depth information on the vertical axis. It can be used for imaging moving organs such as the heart [20, 27, 29–31] (Figure 4.6).

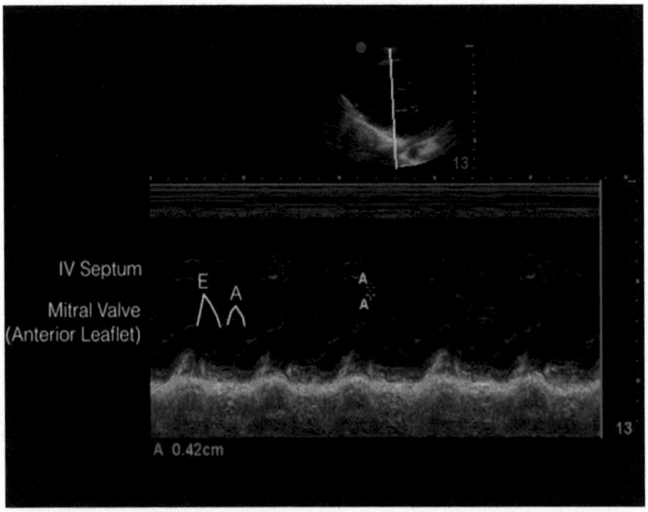

Figure 4.6 An example of M-mode imaging. Source: Reproduced with permission from Turandot et al. [32], ELSEVIER.

4.5.1.3 B-Mode

B-mode is the most commonly used imaging method, also known as brightness mode. This mode is used to generate 2D images. Since B-mode imaging can be created very quickly, it can be used to examine simple and moving structures in real time. The degree of brightness of the pixels indicates the intensity of echoes reflected from the tissues and is coded via grayscale according to the brightness levels [20] (Figures 4.7 and 4.8).

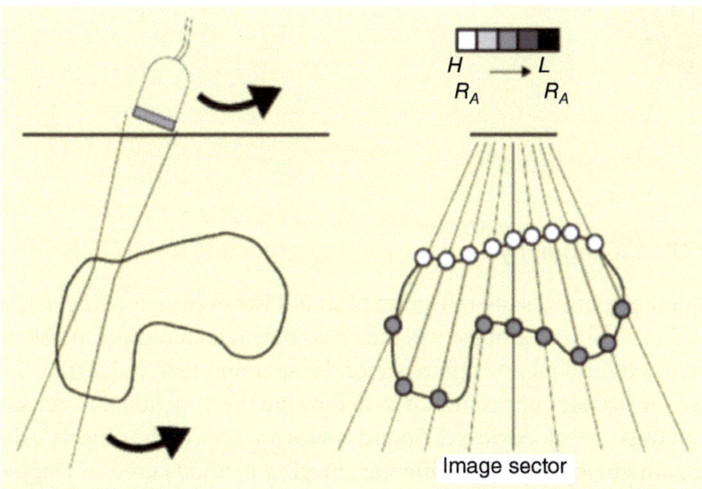

Figure 4.7 An example of a B-mode imaging system. Source: Williams [33]/Reprinted with permission from Elsevier.

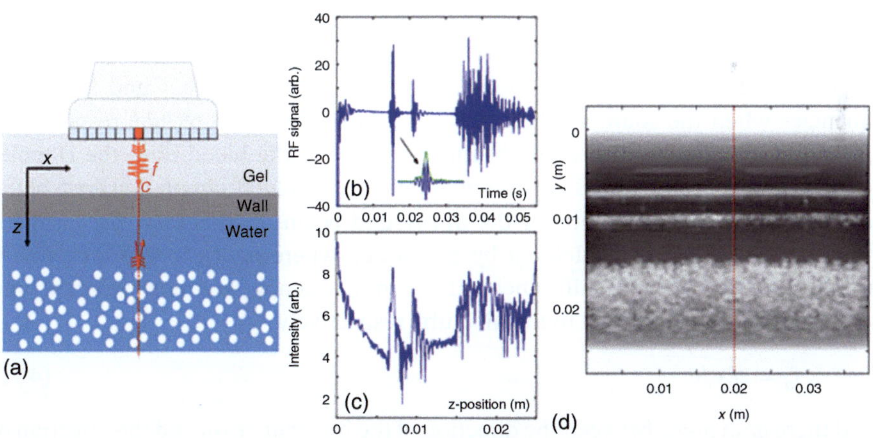

Figure 4.8 An example of B-mode imaging: (a) Illustration of the imaging setup. Acoustic waves generated by the probe travel through the tissue, and reflected waves that return to the probe are captured and used to generate an image. (b) Raw signal obtained from the transducer. (c) Intensity signal obtained from the transducer (d) 2D Image generated using the processed signal (b). Source: Reproduced from Poelma [34], Springer Nature/CC BY 4.0.

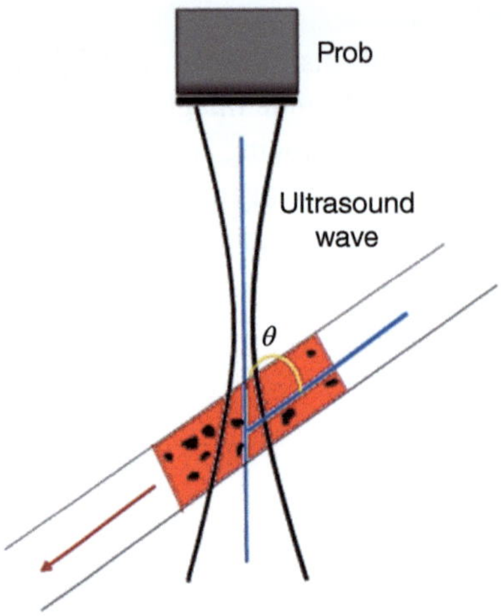

Figure 4.9 Measurement of blood flow velocity with Doppler ultrasound.

4.5.2 Doppler Ultrasonography

Doppler ultrasonography utilizes the reflection of sound waves from moving objects to obtain information about those objects as they pass through an ultrasound beam, which is expressed as the frequency difference of the sent and received signal. The most common use for Doppler ultrasound is to determine the direction and velocity of blood flow in an organ or blood vessel. Sound waves are sent to the objects being measured at a certain angle. There are different imaging methods used in Doppler ultrasonography depending on the specific application. For example, continuous wave Doppler is used for high-velocity blood flows, while in pulsed wave Doppler, the transmitted waves are not continuous but in the form of pulse sequences. In color flow imaging, blood flow velocity and direction are encoded with colors of varying shades to better visualize the dynamics of the movements being imaged [24] (Figure 4.9).

For sound transmission, the Doppler shift occurs with higher and lower frequencies when the source and the receiver move closer and recede, respectively. In addition, when an ultrasound beam is exposed to the blood flow, the Doppler effect can also be observed. In this manner, in the blood circulation, red blood cells act as mobile scatterers for the ultrasound beam. In this case, the basis for Doppler equation can be defined by Eq. (4.11), where f_d, f_t, v, and c represent the Doppler frequency shift, incident beam frequency, scatterer velocity, and propagation speed of sound in the medium, respectively.

$$f_d = 2f_t v/c \tag{4.11}$$

If there is an angle between the direction of the incident beam and the direction of blood flow, the scatterer velocity term v in Eq. (4.11) can be replaced with the velocity in the direction of flow that is obtained by cosine of the angle. The mathematical definition can be found in Eq. (4.12).

$$f_d = 2f_t \cos\theta v/c \tag{4.12}$$

The above equations form the basis for clinical application of the Doppler principle [24].

4.5.2.1 Continuous Wave Doppler

Continuous wave Doppler uses a specialized probe comprised of a pair of piezoelectric crystals, with one crystal sending a continuous sound wave to the tissues and a second one that detects returning echoes. The start signal is subtracted from the returning signals to obtain the Doppler shift, which is converted into an audio signal after being amplified. Because the sound wave is continuous, the tissue depth from which the waveform was obtained cannot be determined. It is used to detect the movement of blood flowing in the vessel rather than the movement of the vessel wall.

4.5.2.2 Duplex Doppler

Duplex Doppler systems combine Doppler with real-time image acquisition. With B-mode imaging, real-time images of the region of interest are obtained to find the vessel that is being examined, and the Doppler shift obtained from the reflected echoes is displayed graphically. Using this method, arterial flow velocities can be measured and displayed on a graph, allowing hemodynamic characterization and detection of vascular pathologies.

4.5.2.3 Color Doppler

In color Doppler, the blood flow velocity is displayed visually as an overlay in the real-time image. By applying this method, it is possible to evaluate the visualization of blood vessels and their hemodynamic characteristics. With real-time B-mode, the Doppler shift in the echoes coming from the region determined on the screen is calculated. Only the vascular structures in the selected area are colored according to the flow direction and velocity, while the structures outside the selected area are displayed in grayscale. According to the calculated Doppler shift, the blood flow direction and velocity are determined, and a colored intensity scale is assigned based on the direction of the blood flow. The color output is displayed in different intensities (light and dark) according to the blood flow rate. Clinical applications of the method include visualization of carotid, cardiac, and peripheral vascular blood flow, abdominal deep veins imaging, fetal imaging, organ and tumor perfusion studies. Displaying the flow in color enables the visualization of small vascular structures that cannot be seen using duplex Doppler. The distinction between vascular structures and other structures can be made well, and thus, more anatomical details about the examined structure are obtained. Since it provides visualization of the arterial lumen, local abnormalities can be easily recognized such as thrombi and atherosclerotic plaques. In addition, ulcerated plaques can be visualized, and stenoses can be detected and measured.

4.5.3 Real-Time Artifacts in Imaging

There are many sources of artifacts that affect image quality and texture definition under examination. If the causes of these artifacts are known, it will be easier to recognize them and know how to prevent them. Major artifacts include acoustic shadows, back wall enrichment, reverberation, side wings, and refraction.

Acoustic shadows: Acoustic shadows refer to the black band-shaped shadow formed behind hyperechoic structures that reflect sound, such as kidney stones, calcifications, and bones. While these structures reflect most of the sound, they absorb the rest, so the formations behind them cannot be seen.

Back wall enrichment: Back wall enrichment is also referred to as acoustic reverberation. It is the brightness that occurs distal to cystic formations and is important in the identification of the cyst. Since cystic formations absorb sound less than other structures, the sound beam will be stronger when it reaches the distal side of the cystic formation. Because the echoes reflected from here will be stronger, the soft tissue distal to the cyst will be brighter than the surrounding soft tissues.

Reverberation: Reverberation generally occurs if a very strong reflector is close to the transducer surface. Many reflections occur between the surface of the transducer and the reflector, and these reflections appear as a series of repeating lines in the image. Artifacts of this type are quite common at the gas-soft tissue or gas–liquid interfaces. It is an artifact that is generally observed at the soft tissue-bone interfaces due to the absorption of sound waves by the bone.

Side lobes: In addition to the strong main acoustic beam, the transducer produces two or more side lobes of lower power at the edges of the main lobe. These side lobes are caused by the deflection and refraction of the beam and interference. It increases with the increase in transducer size, resulting in artifacts in the images generated by the returned echo.

Refraction: Another cause of artifacts is the refraction phenomenon mentioned previously. Refraction occurs at two tissue boundaries with different acoustic properties. Depending on the difference in acoustic properties, the incoming signal deviates at various levels.

4.5.4 Factors Affecting Image Quality

Axial (depth) resolution: Axial resolution is parallel to the sound beam sent to the tissue to be examined and is used to distinguish two different structures in different directions. Axial resolution is determined by the length of the transmitted audio pulse. The resolution is inversely proportional to the pulse length, and the pulse length is shortened by shortening the wavelength of the sound beam or decreased by increasing the frequency.

Lateral (horizontal) resolution: Lateral resolution is perpendicular to the sound beam sent to the tissue to be examined and is used to distinguish features of two different structures in the same direction. The width of the sound beam determines the lateral resolution. To separate two adjacent objects, the acoustic beam should be narrower than the distance separating them. In the case of using a thin acoustic beam, the formations perpendicular to the acoustic beam can be viewed separately. Reducing the transducer size and increasing the frequency causes the acoustic beam width to be reduced. The thickness of the acoustic beam is not constant and starts to decrease after exiting the probe. By considering this, adjusting the transducers to form the thinnest region of the acoustic beam at the depth of the tissue to be examined changes the focal zone, which increases lateral resolution. The focal zone is the point that has the highest lateral resolution. After the focal zone region, the sound beam begins to expand again.

Ultrasonographic zone: The thickness of the acoustic beam is not constant. After it exits the probe, it creates a linear extension defined as the near zone and then transitions to the remote zone, which begins to expand in a fan-like manner. Axial and lateral resolution in the near zone is high. Increasing the frequency and transducer diameter causes the near zone to shorten and the far zone to form closer to the transducer.

Q phenomenon: The Q phenomenon refers to the duration and purity of the sound wave. If the frequency spectrum is large, the Q-factor is low. The Q-factor is the ratio of the center frequency to the bandwidth. A high Q-factor transducer indicates a narrow bandwidth.

4.6 Tissue Harmonic Imaging (THI)

In order to understand the benefits of THI imaging, first of all, it is necessary to know how traditional ultrasonic imaging is carried out. In traditional imaging, the ultrasonic system transmits and receives a sound pulse at a specific frequency. The difference between transmitted and received signals occurs due to the attenuation in the received signal while it is passing through the tissues. In THI, the returning signal includes not only the main signal but also a harmonic frequency that is twice as high as the operating frequency. Once the main and harmonic signals are received together by the system, it distinguishes between these two elements and only processes the harmonic signal [35].

4.6.1 The Occurrence of Harmonic Signals

As an example, when an acoustic wave encounters a bubble, it transmits two types of reflections back. The first of these is a reflection of the returning main signal, which is what is used in conventional imaging. The second signal is one that is formed by the vibration of the bubble due to its interaction with the acoustic wave. The vibration generates harmonics at a frequency twice that of the original wave. The generation of harmonic signals is not something that is formed by the ultrasonic system itself. These signals are produced by the body due to tissue interactions. To exemplify this, sending an acoustic wave through the body can be considered. The wave attenuates and changes due to tissue properties as it travels through the body. The sound wave behaves like a pressure wave when it comes into contact with the tissue. When the wave reaches the tissue, first the tissue is compressed, and the speed of sound increases in the compressed region. The speed of sound decreases when the tissue loosens. Since sound speeds vary in different parts of the waveform, the wave is distorted. This distortion in the waveform creates the harmonic wave component. The final waveform contains the main returning signal as well as the harmonics resulting from the distortion [32].

4.6.2 The Separation of Harmonic Signals from the Main Signal

There have been two methods used to generate harmonic images. In systems operating according to the second-harmonic imaging frequency, only the transmitted signals are received, which are twice the frequency of these. To separate the harmonic signal from the main signal, a narrow frequency band gap must be chosen, at the cost of lower spatial resolution. It is necessary to use a high-pass or narrow-band pass filter to separate the main signal and the harmonic signal. This causes a decrease in contrast resolution due to the overlapping of the harmonic signal from main signal. As a result, this method causes decrease in spatial and contrast resolution [36].

In addition to the traditional THI, a newly developed technique uses a broadband harmonic imaging method that utilizes a phase inversion technique. Using this method, the entire bandwidth of the converter is used, and a higher-quality image is obtained at every depth. The phase inversion technique depends on the digital signal processing capabilities of the ultrasonic imaging system.

Two pulses are sent to the body sequentially, and the phase of the second of the pulses is inverted relative to the first. While these pulse waves are reflected from the linear tissue, they are added to and annihilate each other. However in reflections from nonlinear regions, this operation does not occur. Thus, as a result of the 180° phase difference, the main and odd harmonic signal components are suppressed, but the even harmonic components, specifically the amplitude of the second-harmonic signal, are greatly increased. The result is a superior image in both contrast and spatial resolution [36].

4.6.3 The Advantages of Harmonic Signals

Infrasonic waves produced by harmonic signals have very important advantages. One of the first and most important advantages is the reduction of ambient artifacts (reverberation due to intra-body gases), which are encountered especially in the examination of difficult patients and significantly reduce the image quality, and the reduction of artifacts (shadowing due to stones, echo enrichment due to liquid-filled cystic structures) that may occur due to the structures examined. Another advantage is better lateral resolution in harmonic imaging due to thinner beam widths. Also, in harmonic imaging, images are sharper, less noisy, and have higher contrast.

4.7 Recent Advances in Ultrasound Imaging for Medicine

Ultrasound has been widely used in medicine for the last 20 years, and has become a gold standard in medical imaging techniques. Remarkable work has been done in elastography, development of ultrasound contrast materials for improved imaging [37], and super-resolution imaging fields. In particular, it is very important in terms of determining the elastic properties of tissues, cancer diagnosis, evaluation of fibrous tumors, detection of plaque formations in the artery, detection of COVID-19 in the recent pandemic, etc. [38]. However, elastic properties may not be clearly determined using conventional ultrasound imaging. Therefore, new ultrasound imaging technologies have been developed with high-resolution quantitative elastography capabilities [39, 40]. Super-resolution imaging has been achieved by tracking microbubbles as they travel through the vasculature [41]. In order to realize these new technologies, micro-machined transducers have been developed to create new ultrasound imaging probes for 3D and 2D array converters [42]. 2D phased array transducers with high-intensity focused ultrasound (HIFU) enable rapid diagnosis and abdominal cancers as well as treatment of brain diseases without any craniotomy or burr hole trephination [43]. In addition, noninvasive neuromodulation tools have been developed in combination with MRI guidance, enabling the treatment of brain diseases [44, 45].

The delivery of drugs via ultrasound and the examination of microbubbles are other topics of interest in the field of medicine. Since ultrasound is a time-varying acoustic pressure wave, acoustic cavitation occurs by the oscillation and collapse of microbubbles subjected to an acoustic wave. Therefore, drug delivery using focused ultrasound and spatially targeted ultrasound-mediated microbubbles has been rapidly growing area of research in recent years [46].

Another exciting application for ultrasound therapy in recent years is histotripsy, which uses high-intensity ultrasound to ablate and destroy tissue noninvasively [47]. In histotripsy, repetitive ultrasound pulses of high amplitude without cavitation are applied to the target soft tissue. Unlike HIFU, this technique can also be used in cancer treatment as it provides a clean margin line between the treated and surrounding soft tissue.

Besides the methods mentioned above, other ultrasound-related research areas, such as photoacoustic and ultrasound neurostimulation, are developing rapidly. Photoacoustic imaging (PAI) uses pulsed laser excitation to produce sound in the body, and the signal is received by an ultrasound transducer. However, since the focal point size of the laser is several times smaller than the ultrasound beam, higher image resolution can be achieved compared to general ultrasound. Many of the algorithms developed in the ultrasound imaging field, such as ultrasound elastography, can be applied to PAI.

References

1 Bergmann, L. (1954). *The Ultrasound*. Stutgart: S.Hirzel Verlag.

2 Blitz, J. (1971). *Ultrasonic Methods and Application*. London: Butterworths.

3 Moholkar, V.S., Nierstrasz, V.A., and Warmoeskersen, M.M.C.G. (2003). Intensification of mass transfer in wet textile process by power ultrasound. *Autex Research Journal* 3: 129–138. Holland.

4 Ercan, S.S. and Soysal, C. (2013). Use of ultrasound in food preservation. *Natural Science* 5: 5–13.

5 Demirdöven, A. and Baysal, T. (2008). The use of ultrasound and combined technologies in food preservation. *Food Reviews International* 25 (1): 1–11.

6 Rastogi, N.K. (2011). Opportunities and challenges in application of ultrasound in food processing. *Critical Reviews in Food Science and Nutrition* 51 (8): 705–722.
7 Soria, A.C. and Villamiel, M. (2010). Effect of ultrasound on the technological properties and bioactivity of food: a review. *Trends in Food Science & Technology 2010* 21 (7): 323–331.
8 Tao, Y. and Sun, D.W. (2015). Enhancement of food processes by ultrasound: a review. *Critical Reviews in Food Science and Nutrition* 55 (4): 570–594.
9 Arnold, G., Zahn, S., Legler, A., and Rohm, H. (2011). Ultrasonic cutting of foods with inclined moving blades. *Journal of Food Engineering* 103 (4): 394–400.
10 Shanmugam, A. and Ashokkumar, M. (2017). Ultrasonic preparation of food emulsions. In: *Ultrasound in Food Processing: Recent Advances* (ed. M. Villamiel, A. Montilla, J.V. García-Pérez, et al.), 287–311. United Kingdom: John Wiley & Sons Ltd.
11 Khandpur, P. and Gogate, P.R. (2016). Evaluation of ultrasound based sterilization approaches in terms of shelf life and quality parameters of fruit and vegetable juices. *Ultrasonics Sonochemistry* 29: 337–353.
12 Hu, J., Ge, S., Huang, C. et al. (2018). Tenderization effect of whelk meat using ultrasonic treatment. *Food Science & Nutrition* 6 (7): 1848–1857.
13 Villamiel, M., Verdurmen, R., and Jong, P. (2000). Degassing of milk by high-intensity ultrasound. *Milchwissenschaft-Milk Science International* 55 (3): 123–125.
14 Lianfu, Z. and Zelong, L. (2008). Optimization and comparison of ultrasound/microwave assisted extraction (UMAE) and ultrasonic assisted extraction (UAE) of lycopene from tomatoes. *Ultrasonics Sonochemistry* 15 (5): 731–737.
15 Comandini, P., Blanda, G., Soto-Caballero, M.C. et al. (2013). Effects of power ultrasound on immersion freezing parameters of potatoes. *Innovative Food Science & Emerging Technologies* 18: 120–125.
16 Nowak, K.W., Zielinska, M., and Waszkielis, K.M. (2019). The effect of ultrasound and freezing/thawing treatment on the physical properties of blueberries. *Food Science and Biotechnology* 28 (3): 741–749.
17 Cárcel, J.A., Benedito, J., Bon, J., and Mulet, A. (2007). High intensity ultrasound effects on meat brining. *Meat Science* 76 (4): 611–619.
18 Muthukumaran, S., Kentish, S.E., Stevens, G.W. et al. (2007). The application of ultrasound to dairy ultrafiltration: the influence of operating conditions. *Journal of Food Engineering* 81 (2): 364–373.
19 Bozkir, H., Ergün, A.R., Serdar, E. et al. (2019). Influence of ultrasound and osmotic dehydration pretreatments on drying and quality properties of persimmon fruit. *Ultrasonics Sonochemistry* 54: 135–141.
20 Ünal, D. (2008). Imaging techniques used in medicine. Master Thesis, Gazi University Division of Physics Education, Ankara.
21 Kim, J., Park, S., Jung, Y. et al. (2016). Programmable real-time clinical photoacoustic and ultrasound imaging system. *Science Reports* 6: 35137.
22 Shijie, W., Xiaodong, C., Jing, B., and Daoyin, Y. (2008). Ultrasonic transmitter for coded excitation in medical endoscope imaging system. In: International

Symposium on Photoelectronic Detection and Imaging 2007: Laser, Ultraviolet, and Terahertz Technology. Edited by Zhou, Liwei. Proceedings of the SPIE, Volume 6622 (2008), article id. 662217, 1–7, Beijing, China.
23 Ulu, M.O. (2008). Medical applications of particle detectors. Institute of Natural and Applied Sciences, University of Çukurova, Adana.
24 Güngör, M.A (2008). The investigation of temperature rise formed by diagnostic ultrasound on tissue due to nonlinear propagation and making a simulation programme. Master Thesis, Institute of Natural and Applied Sciences, Gazi University, Ankara.
25 Whittingham, T.A. (2007). Medical diagnostic applications and sources. *Progress in Biophysics and Molecular Biology* 93 (1–3): 84–110.
26 Shu, Y. (2011). Real-time ultrasound biomicroscopy with optoacoustic arrays. Master Thesis, Delaware University Electrical and Computer Engineering.
27 Kurnaz, M.N (2006). Segmentation of ultrasound images by using incremental neural networks. PhD Thesis, Graduate School of İstanbul Teknik University.
28 Kalender, O. and Kavalcı, R. (2001). *Ultrasonography*, 96–115. The Journal of Turkish Military Academy.
29 Quinsac, C., Basarab, A., and Kouamé, D. (2012). Frequency domain compressive sampling for ultrasound imaging. *Advances in Acoustics and Vibration* 2012: 1–16.
30 Thomenius, K.E. (2004). Instrumentation design for ultrasonic imaging. In: *Standard Handbook of Biomedical Engineering and Design, Digital Engineering Library*, 25.1–25.16. New York: McGraw-Hill.
31 Panagiota, K. (2011). Simulation of ultrasound brain cancer imaging. Master's Thesis, National Technical University of Athens, Faculty of Electrical and Computer Engineering.
32 Turandot, S., Sebastian, D., Siadecki, R. et al. (2015). M-mode ultrasound applications for the emergency medicine physician. *The Journal of Emergency Medicine* 49 (5): 686–692.
33 Williams, D. (2012). The physics of ultrasound. *Anaesthesia And Intensive Care Medicine* 13 (6): 264–268.
34 Poelma, C. (2016). Ultrasound imaging velocimetry: a review. *Experiments in Fluids* 58 (3): 1–28.
35 Lee, J. and Chang, J.H. (2019). Dual-element intravascular ultrasound transducer for tissue harmonic imaging and frequency compounding: development and imaging performance assessment. *IEEE Transactions on Biomedical Engineering* 66 (11): 3146–3155.
36 Haerten, R., Lowery, C., Becker, G. et al. (1999). *Ensemble Tissue Harmonic Imaging*. Erlangen, Germany: Tha Technology and Clinical Utility.
37 Lee, H., Kimm, H., Han, H. et al. (2017). Microbubbles used for contrast enhanced ultrasound and theragnosis: a review of principles to applications. *Biomedical Engineering Letters* 7: 59–69.
38 Kwon, S.J. and Jeong, M.K. (2017). Advances in ultrasound elasticity imaging. *Biomedical Engineering Letters* 7 (2): 71–79.

39 Sandrin, L., Tanter, M., Catheline, S., and Fink, M. (2002). Shear modulus imaging with 2D transient elastography. *IEEE Transactions on Ultrasonics, Ferroelectrics, and Frequency Control* 49: 426–435.
40 Bruce, M., Kolokythas, O., Ferraioli, G. et al. (2017). Limitations and artifacts in shear-wave elastography of the liver. *Biomedical Engineering Letters* 7 (2): 81–89.
41 Errico, C., Pierre, J., Pezet, S. et al. (2015). Ultrafast ultrasound localization microscopy for deep super-resolution vascular imaging. *Nature* 527: 499–502.
42 Lee, W. and Roh, Y. (2017). Ultrasonic transducers for medical diagnostic imaging. *Biomedical Engineering Letters* 7 (2): 91–97.
43 Kim, Y. (2017). Clinical application of high-intensity focused ultrasound ablation for uterine fibroids. *Biomedical Engineering Letters* 7 (2): 99–105.
44 Elias, W.J., Lipsman, N., Ondo, W.G. et al. (2016). A randomized trial of focused ultrasound thalamotomy for essential tremor. *New England Journal of Medicine* 375: 730–739.
45 Chang, W.S. and Chang, J.W. (2017). Focused ultrasound treatment for central nervous system diseases: neurosurgeon's perspectives. *Biomedical Engineering Letters* 7 (2): 107–114.
46 Han, M., Hur, Y., Hwang, J., and Park, J. (2017). Biological effects of blood-brain barrier disruption using a focused ultrasound. *Biomedical Engineering Letters* 7 (2): 115–120.
47 Xu, Z., Raghavan, M., Hall, T.L. et al. (2007). High speed imaging of bubble clouds generated in pulsed ultrasound cavitational therapy—histotripsy. *IEEE Transactions on Ultrasonics, Ferroelectrics, and Frequency Control* 54: 2091–2101.

5

Photoacoustic Imaging and Sensing for Biomedical Applications

Amalina B. E. Attia, Ruochong Zhang, Mohesh Moothanchery, and Malini Olivo

*Institute of Materials Research and Engineering (IMRE), Agency for Science, Technology and Research (A*STAR), Singapore 138634, Singapore*

5.1 Introduction

Photoacoustic imaging (PAI) is a promising biomedical imaging technique combining light illumination and acoustic detection [1–3]. It has taken off in the past decades and has been investigated for several biomedical imaging applications. Unlike pure optical techniques, including optical coherence tomography [4, 5] and fluorescence imaging [6], PAI does not rely on backscattered photons occurring in biological tissues, which limits the use of pure optical imaging techniques for deep tissue imaging while maintaining high resolution. PAI depends on the detection of acoustic signals due to pulsed optical illumination-generated thermal expansion of tissue. Hence, it can achieve imaging depths up to centimeters with high resolution. Since acoustic scattering is much less than that of light, acoustic signal can propagate much longer in biological tissue without attenuation.

Nanosecond-pulsed (<10 ns) lasers are commonly used for illumination in PAI. The thermal expansion due to temperature increase from the absorption of optical energy will generate acoustic waves, which can be detected using high-frequency ultrasonic transducers. PAI depends mainly on the endogenous chromophores (hemoglobin, melanin, lipid, water, etc.) in human tissue. By illuminating biological tissue at multiple wavelengths depending on the individual characteristic absorption spectra of the chromophores, PAI can image and quantify these chromophores to understand the hemodynamic changes in disorders for effective disease diagnosis and management. The ability to measure oxygen saturation in vasculatures by differentiating oxy-hemoglobin (HbO_2) and deoxy-hemoglobin (Hb) in a label-free manner enables PAI to be a future candidate for diagnostic imaging tool in clinic. PAI can be used for longitudinal monitoring of blood perfusion, vasodilation, and oxygenation status, which are the characteristics of ischemia, hypoxia, etc. [7]. Depending on the melanin absorber in skin, PAI can also be used for melanoma, pigmented lesions, and hair follicle imaging applications. For the sake of improving the contrast or in the case of similar absorption spectra for different chromophores,

Acoustic Technologies in Biology and Medicine, First Edition.
Edited by Adem Ozcelik, Ryan Becker, and Tony Jun Huang.

exogenous contrast agents like methylene blue [8] and indocyanine green (ICG) [9] can also be used [10–12].

PAI systems are classified into three main configurations, tomography, mesoscopy, and microscopy, depending on the light illumination and ultrasound detection geometry (Figure 5.1) [13]. In photoacoustic tomography (PAT), illumination at near-infrared (NIR) wavelengths and detection of acoustic signals using array transducers [14, 15] are commonly used. Use of PAT systems based on single-element transducers [16, 17] can be time consuming, making them difficult for clinical imaging applications. Arranging multiple transducer elements in an array allows acoustic signals to be detected at different angles, to achieve real-time imaging of the region of interest with a larger field of view (FOV). Depending on the orientation of the transducer elements, cross-sectional two-dimensional (2D) or volumetric (3D) imaging can be done. This transducer element array can be mounted stationary, attached to translation stages [18, 19], or in hand-held mode [20, 21], depending on the region of interest and application. The 2D transducer element array can either be configured in linear, curved, or a combination of linear and curved orientations. Linear transducers arrays are common in commercial ultrasound systems, and hence PAI imaging can be incorporated as an add-on to those systems by incorporating lasers [22, 23]. The limited angle for the detection of acoustic signals compared to curved transducer array systems in the region of interest is a limitation of these systems [19].

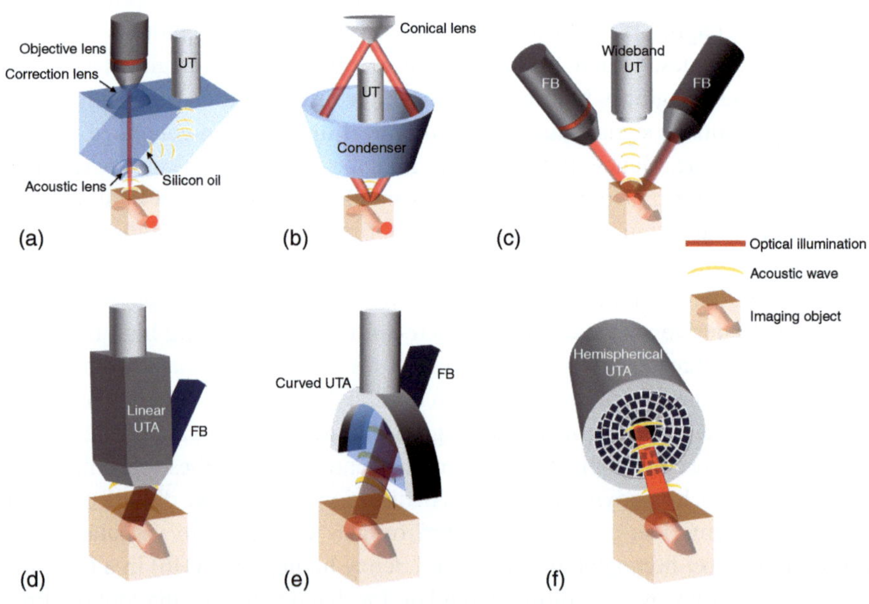

Figure 5.1 Different schematics of photoacoustic imaging (PAI) systems. (a) Optical resolution-photoacoustic microscopy (OR-PAM). (b) Dark field acoustic resolution-photoacoustic microscopy (AR- PAM). (c) Raster scanning optoacoustic mesoscopy (RSOM); photoacoustic tomography (PAT) with (d) linear (e) curved (f) hemispherical ultrasound transducer array. FB, fiber bundle; UT, ultrasound transducer; UTA, ultrasound transducer array.

Arranging transducer elements in 3D enables detection of acoustic waves from all directions to provide volumetric images from a single laser pulse. The FOV for those systems can be larger [24, 25] or smaller [26], depending on the applications. There are several PAT system manufacturers for applications to image vasculature and intrinsic chromophores in deep tissues up to centimeters, including FUJIFILM VisualSonics, iThera Medical, ENDRA Life Sciences, TomoWave Laboratories, and Seno Medical. These PAT systems are not suitable for high-resolution imaging.

Photoacoustic microscopy (PAM), a high-resolution PA modality, can further be classified into two categories: acoustic-resolution PAM (AR-PAM) and optical-resolution PAM (OR-PAM), depending on the type of focusing [27]. An imaging depth greater than 1 mm and an optical diffusion limit of <50 μm resolution in biological tissues can be achieved using AR-PAM exploiting acoustic focusing. The frequency and aperture (NA) of the transducer will determine the lateral resolution in AR-PAM. A combination of 50 MHz transducer having 0.44 NA can achieve <50 μm lateral resolution and up to 3 mm imaging depth [28, 29]. In dark field AR-PAM setup, the illuminating beam is in a ring shape after passing through a conical lens, focusing on the sample via an optical condenser with a hole in the center. An ultrasonic transducer with a matching focal length as the optical condenser is placed in the center of the optical condenser. The implemented dark field excitation is advantageous to avoid strong PA signals from the surface of skin, which in turn helps to achieve high-contrast images from deep tissue.

To resolve a single capillary (~5 μm) using AR-PAM, higher transducer frequencies >400 MHz should be used, which reduce the penetration depth to microns (<100 μm). When the illuminating light beam is tightly focused, further improvements in the lateral resolution can be achieved, giving way to OR-PAM [30]. There are several configurations used in OR-PAM, and the most prominent one is by using an optoacoustic combiner, a combination of right angle and rhomboid prisms separated by a thin layer of silicon oil. The silicon oil behaves as a transmission–reflection filter, which transmits the optical beam and reflects the acoustic beam. The NA of the transducer attached on top of the rhomboid prism is dependent on the acoustic lens connected to the underside of the rhomboid prism. Image acquisition in PAM is done by raster scanning of the optoacoustic combiner using 2D motorized stages. The PA signals obtained by the transducer in biological tissue will form an A-line. During 1D movement of the stage, multiple A-lines were acquired to form a B-scan, which can provide 2D images. Multiple B-scans are acquired during the entire raster scanning of the 2D stages to provide 3D volumetric PA images. Imaging speed limitation due to motorized stages can be improved using a microelectromechanical systems (MEMS) scanner [31–34]. Use of MEMS scanners can be used to develop miniaturized systems for endoscopic devices and clinical translation.

When the initial AR-PAM design was modified using a combination of wide detection band transducer and fiber bundle laser excitation, this specific configuration was defined as raster-scanning optoacoustic mesoscopy (RSOM) [35]. A single high-frequency (normally ≥50 MHz) transducer with a wide detection bandwidth (11–99 MHz) was used to achieve an axial resolution of 10 μm and lateral resolution of 40 μm [36]. The acquired data sets can be separated by the frequency band (e.g.

11–33 MHz [low] and 33–99 MHz [high]) followed by combining them together, to differentiate superficial high-resolution vasculatures as well as deep bigger vascular structures. This single- wavelength (532 nm) configuration has been used for investigating inflammatory skin conditions, such as psoriasis [36] and eczema [37]. The system with a multiple-wavelength laser was used to resolve melanin and monitor hemodynamic response in the superficial vasculature of skin [38].

All different PAI systems and configurations described here have their own unique advantages for clinical applications, including deep tissue imaging, high-resolution in vivo vascular imaging, cellular-level imaging, noninvasive imaging, and label-free imaging.

5.2 Photoacoustic Imaging Applications

5.2.1 PAI of Breast Cancer

Breast cancer is one of the most common cancers in women, with 2.3 million women diagnosed in 2020 alone [39, 40]. Mammography, the most common screening method used to diagnose breast cancer is painful, involves ionizing radiation and exhibits low accuracy for dense breast tissues [41]. Ultrasonography is the next common method, mainly used to differentiate malignant and benign cysts [42]. Magnetic resonance imaging (MRI) and other nuclear imaging modalities [43] are less common for breast cancer diagnosis. The ability to differentiate between the oxygenation of the vasculature in the tissue makes optical imaging a reliable imaging modality over others or in combination with others [44], since hypoxia and angiogenesis are the hallmark markers of cancer. However, the use of optical imaging for breast cancer screening is impeded by the limited penetration depth of light in biological tissues. Combining the advantages of optical illumination and ultrasound detection, PAI could achieve deeper imaging depth at higher spatial resolution. The capability to achieve imaging depths up to several centimeters could benefit several clinical applications, and one main application particularly attractive is in the field of breast cancer. There are many PA systems developed by different groups to push the boundaries of the maximum imaging depth in breast tissue while maintaining high spatial and temporal resolution discussed herein on both in vivo breast tissue and ex vivo excised tissues.

5.2.1.1 In Vivo Imaging

It was Oraevsky et al. [45] in 2001 who pioneered the use of PA for breast imaging. Several other studies followed, showing the feasibility of different PA systems to detect malignancies in breasts in vivo [46–51]. One such initial clinical study was reported by Manohar et al. using the Twente PA mammoscope, a clinical prototype for in vivo breast imaging. The system used a single wavelength laser at 1064 nm for illumination and an array of 588 ultrasound detector elements for acoustic detection. The system has undergone several iterations to achieve a resolution of ~3.5 mm and FOV of $9 \times 8\,\text{cm}^2$ with an image acquisition speed of 10 minutes [52]. This PA system successfully detected 32 of the 33 lesions [25]. Rather than lipids or collagen, the presence of vasculature on the high-intensity lesions was noticed in malignant breast with the contrast independent of breast tissue density [25, 52]. The diagnosis of tumors in dense breast tissue is not possible on X-Ray mammograms [46]. A 3D handheld probe having 256 transducer elements (central frequency – 4 MHz) was used to demonstrate the capability of imaging denser breast tissue using PA [53]. Using a tunable optical parametric oscillator (OPO) for multiple wavelength illumination, volumetric imaging of the breast with spatial maps of HbO_2, Hb, and total hemoglobin (HbT) chromophores was achieved at high resolution [54]. The system was also capable of detecting high microvascular density in the tissues, an indicator of breast cancer [55, 56].

The ability to detect oxygenation status of blood vessels helps greatly in the assessment and monitoring of hemodynamic changes in the blood vessels in the course of cancer treatment [57]. Post chemotherapy, low oxygen saturation (indicating hypoxia) was observed in the peritumoral blood vessels (Figure 5.2) via hemispherical transducer array and by using the breast misshapement algorithm. Similarly, intratumoral blood vessels were apparent, due to the increase in blood perfusion from chemotherapy. The tumor size measured by ultrasound was observed while all these finer changes occurred, showing the advantages of PAI for clinical imaging applications. In the same study, differences in the blood

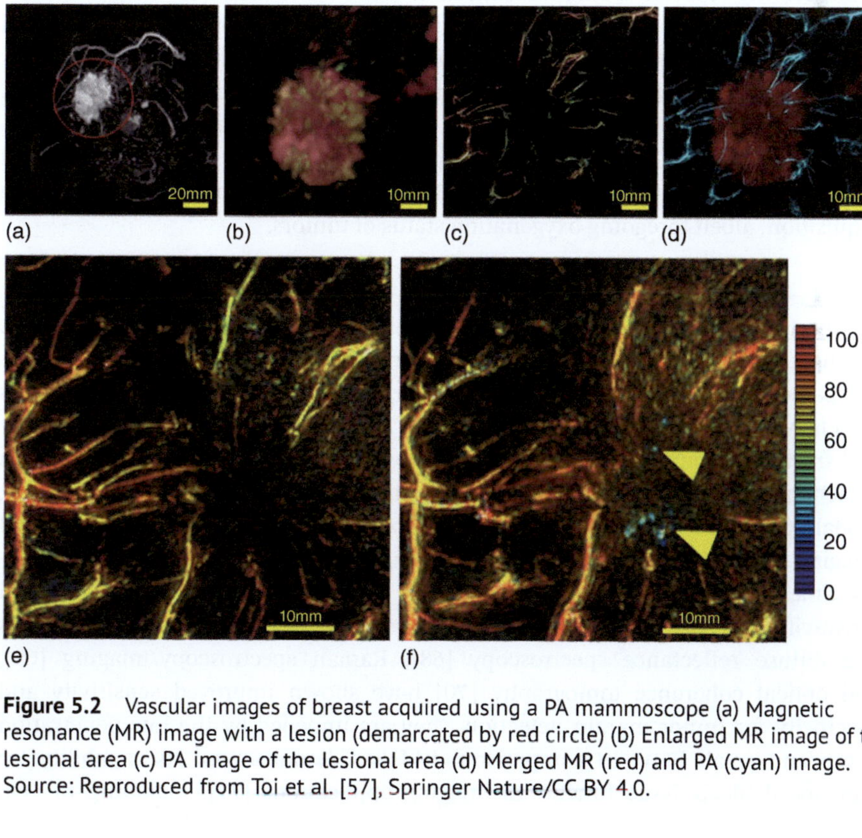

Figure 5.2 Vascular images of breast acquired using a PA mammoscope (a) Magnetic resonance (MR) image with a lesion (demarcated by red circle) (b) Enlarged MR image of the lesional area (c) PA image of the lesional area (d) Merged MR (red) and PA (cyan) image. Source: Reproduced from Toi et al. [57], Springer Nature/CC BY 4.0.

vessels were observed between diseased and contralateral breasts, which introduced the need for analyzing fine blood vessels for the detection of early breast cancer. Imagio from Seno Medical Instruments, a CE-marked PAI system, demonstrated the capability of ultrasound combined with PAI for determining the malignancy of the breast cancer tissue based on different hemoglobin (HbO_2 and Hb) measurements [58]. Differentiating between benign and malignant tissues allows for the downgrading of the benign tumor classification, hence reducing potential benign tumor biopsies [59].

Diot et al. imaged the distribution of fats and water in normal and affected breast tissue in addition to Hb, HbO_2, and total blood volume (TBV) using a handheld PAI system with 28 discrete illumination wavelengths from 700 to 970 nm [60]. Increased vasculature, which can be detected as Hb, HbO_2, and TBV in the periphery, could be related to angiogenesis in cancer. Tumor regions have a 30-fold increase in TBV compared to the normal neighboring regions. Disruption of fat and water signals was observed in the region of tumor, as the tumor will replace the existing fat and water layers. Understanding these cancer features in breast tissues using PAI could potentially help in the guidance of surgery as well as treatment planning.

Since breast is closer to heart and lungs, motion artifacts could affect the imaging; hence, advancement in correction for those motion artifacts needs to be done through post-processing algorithms along with faster imaging acquisition [57]. Lin et al. recently reported a single breath-hold PAI system which can largely reduce motion artifacts [24]. The capability of the system was demonstrated on a 27-year-old healthy volunteer, and the system could provide images from the nipple to the chest wall in 15 seconds. The ring detectors used in the system can help to switch between 3D and 2D imaging modes to address the limited view drawback [13]. Almost 90% of the breast cancer lesions were precisely detected with high spatial and temporal resolution. Single-wavelength imaging at 1064 nm was recommended for fast image acquisition, albeit foregoing oxygenation status of tumors.

5.2.1.2 Ex Vivo Imaging

Lumpectomy or breast-conserving surgery (BCS) combining radiation therapy is among the most efficient and popular alternative in early-stage breast cancers compared to total mastectomy [61] However, the re-surgery rate in BCS patients to achieve tumor-free margin remains high (∼20–70%) [62, 63]. Histopathology and frozen sections are the current gold standard procedures for tumor margin detections [64, 65], which are limited by their long processing times. Imaging modalities, such as ultrasound imaging, are operator dependent and lack functional imaging capabilities [66]. Although X-ray mammography has a shorter processing time [67], it has limitations in imaging dense breast tissue due to reduced sensitivity and specificity. While optical spectroscopy and imaging techniques like diffuse reflectance spectroscopy [68], Raman spectroscopy/imaging [69], and optical coherence tomography [70] have shown improved sensitivity and specificity for tumor margin detection, they are impeded by the low penetration depth and small FOV. In comparison, PAI has demonstrated its capability for high-speed, deep-tissue functional imaging applications [71], including in vivo breast cancer imaging. In addition, PAI has demonstrated its usability mainly for tumor margin detection in ex vivo breast specimen as well. The ability to combine PAI along with ultrasound makes a synergistic system for clinical studies.

Li et al. studied the distribution of fat and Hb in excised breast tissue using multiple wavelengths in the range of 1100–1250 nm [72]. PA imaging was done on 12 fixed specimens for assessing tumor margins, in which normal breast tissue was considered composed of fatty tissue with scattered connective tissue with the rest as cancerous. This technique could provide a sensitivity of 100% and specificity of 75%. Specificity can be improved by better classification of dense connective tissue since it is being considered cancerous. A multimodal ultrasound and PAT system has been made for intraoperative applications, using absorption contrast of fat at 1197 nm, with significant reduction of the scan time to 2 minutes on a $10 \times 10\,cm^2$ imaging area [73]. This system could provide 85.5% sensitivity and 90% specificity, from a 66-patient clinical study, through machine learning algorithms (Figure 5.3).

Goh et al. differentiated the distribution of fat and Hb in excised tissues to classify between negative and positive margins using wavelengths in the range 700–1100 nm [19]. A continuous lipid layer was observed on the edges of the tumor samples with negative margins, whereas discontinuous lipid layers were found on samples with positive margins [60]. A total of 15 excised specimens were imaged soon after surgery for the presence of such lipid layers. Six margins per sample were analyzed and correlated with histology, which provides 100% sensitivity and 97.6% specificity. The two negative margins that were incorrectly identified as positive margins while affecting the sensitivity, interestingly, were small margins below 0.5 mm. In fact, 2 mm is the limit for acceptable margins for ductal carcinoma in situ (DCIS) by the Society of Surgical Oncology, which is well above the margins incorrectly identified [75, 76].

An OR-PAM system using 266 nm ultraviolet (UV) laser at 266 nm provided histology-like images of unprocessed, fixed breast samples [77]. Even though it required 7 hours to image $1\,cm \times 1\,cm$ area, the capability of detecting cancer cell clusters in the tissue specimen makes UV-PAM a promising candidate. However, another version of UV-PAM using multiple focus illuminations and one-dimensional ultrasound array detection improved the imaging time by 40 times [78].

5.2.2 PAI for Skin Imaging

Skin, being the largest organ in the body, functions as a barrier to the environment and first line of defense against pathogens. Composed of multiple layers, superficial blood vessels reside in the connective tissue of the inner dermis layer of the skin, while the outermost epidermis layer consists of multiple layers of flattened cells. Any abnormalities in the skin's structural architecture and blood vessels, commonly due to intrinsic aging and exposure to extrinsic environmental factors, usually result in cutaneous disorders. The standard clinical practice to diagnose these disorders is through visual inspection, which is corroborated by skin

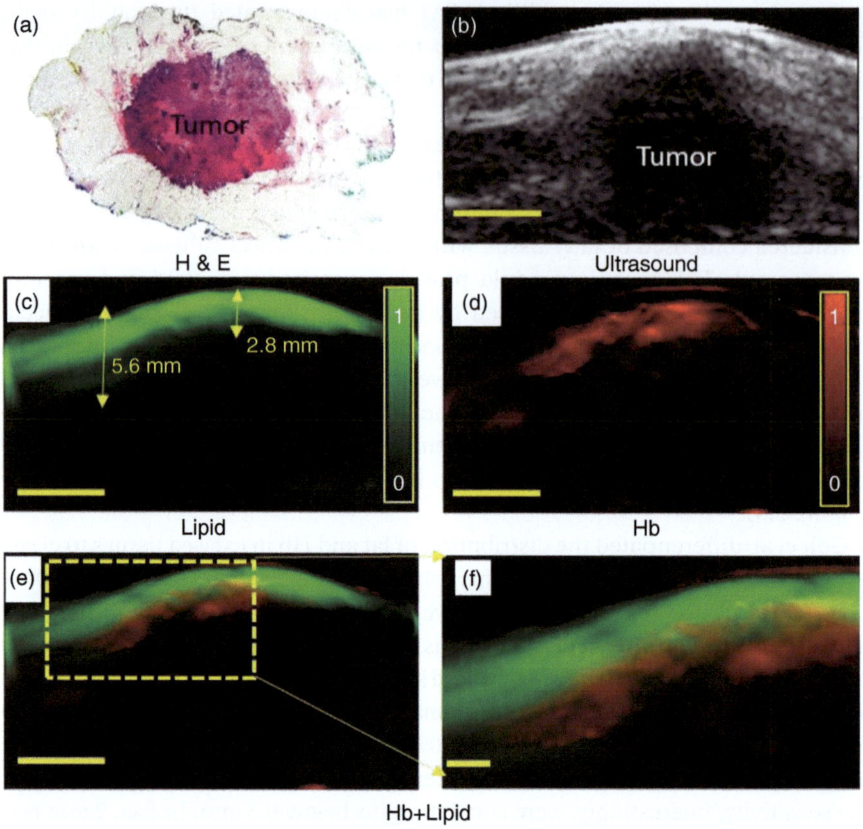

Figure 5.3 Two-dimensional (2D) US and PA imaging of breast tumor having a negative margin. (a) Hematoxylin and Eosin (H & E) stained image. (b) US image. (c) Lipid distribution map, showing a thin lipid layer immediately above the tumor and broader lipid layer adjacent to the tumor. (d) Hemoglobin (Hb) distribution map: showing intense signals on the tumor rim. (e) Merged images of lipid (c) and Hb (d) maps. (f) Zoomed image of marked square region in E representing continuous lipid layers as the negative tumor margin characteristic. Scale bars, 5 mm. Source: Reproduced with permission from Balasundaram et al. [74], ELSEVIER.

biopsies and histopathology. The biggest drawback in this is that skin biopsies are unaesthetically pleasing as they leave a scar and do not reveal the physiology and vascular dysfunctions of the skin. Noninvasive skin imaging can potentially replace skin biopsies as it gives insight into the skin's structural features and physiological function. Clinical skin imaging has been widely reported using optical imaging up to superficial depths due to the pronounced light scattering in the biological tissue. Thus, PAI is an ideal candidate for noninvasive skin imaging. PAI in mesoscopic configuration provides visualization of the multiple layers of skin and the cutaneous vasculature in the vascular plexus [79]. Tomographic PAI, on the other hand, is suitable to image skin structures that are bigger and infiltrated deeper, such as skin cancer.

5.2.2.1 PAI of Skin Cancer

There are three categories of skin cancer: basal cell carcinoma (BCC), squamous cell carcinoma (SCC), and melanoma, of which the first two types constitute nonmelanoma skin cancers. Of the three types, melanomas are aggressive and originate from pigmented lesions. Skin cancers are usually treated surgically to remove them with an allowance of noncancerous tissue around the skin cancer. To reduce this allowance, the skin cancer can be removed layer by layer for a more precise surgical removal in Mohs micrographic surgery (MMS). Surgical intervention methods have proven to be time-consuming and imprecise, limitations that can be addressed with preoperative imaging of the skin cancer lesion to visualize its size and depth to reduce surgery time and allow for tissue sparing.

PAI presents itself as a potential imaging tool for preoperative imaging of skin cancers, and its use has been reported to determine the dimensions of in situ skin cancers from their endogenous contrast. There are many studies exploiting multispectral PAI with handheld transducer scanner configurations to obtain the depth of in situ skin tumors with the idea to guide their surgical removal. As skin cancers can occur anywhere in the body and especially in areas exposed to the sun, for example, the face and neck, the flexible handheld scanner is suitable for imaging these hard-to-reach areas. The use of tomographic PAI to spatially map nonmelanoma skin cancers by using their melanin and Hb contrast, showing the delineation of the skin cancer in contrast to surrounding normal tissue, has been reported by Chua et al. [80]. Notably, nonmelanoma skin cancers look pigmented in those of higher Fitzpatrick skin types, whereas they seem reddish or erythematous in those of lower Fitzpatrick skin types (Figure 5.4). Contingent on the appearance, skin cancer can be differentiated by its melanin or Hb PA signals (Figure 5.4). When the extent and depth of these PA signals were quantified to represent the cutaneous lesion size and correlated with the dimensions obtained from histological resections of the same lesions, a high correlation between the two was observed with a linear fit of $r = 0.9$. The multispectral ability of PAI allows the visualization of skin cancer lesions, regardless of presence of melanin in the lesions, for the determination of their dimensions.

Other than obtaining the tumor dimensions, PAI has been utilized to enhance the diagnostic accuracy of skin tumors. Park et al. reported the use of a 3D wide-field multispectral ultrasound-guided PAI system to obtain images of in vivo melanoma of sizes ranging from 1.3 to 30 mm in diameter and 0.6 to 9.1 mm in depth [81]. The melanoma depth obtained from PAI analysis was highly correlated with that obtained from histopathology with an average magnitude of 0.36 mm error. Notably, the system was able to visualize metastasized melanoma beyond the tumor's boundary on the palm. PAI was also used in conjunction with optical coherence tomography and reflectance confocal microscopy to discriminate malignant from benign pigmented skin lesions [82]. In 41 pigmented cutaneous skin lesions, imaging features from each of the three modalities were extracted, and these features exhibited a high correlation of more than 71% in predicting malignancy accuracy.

PAI has also been shown to distinguish between the different types of ex vivo periocular skin cancers lesions from their PA spectral signatures based on their

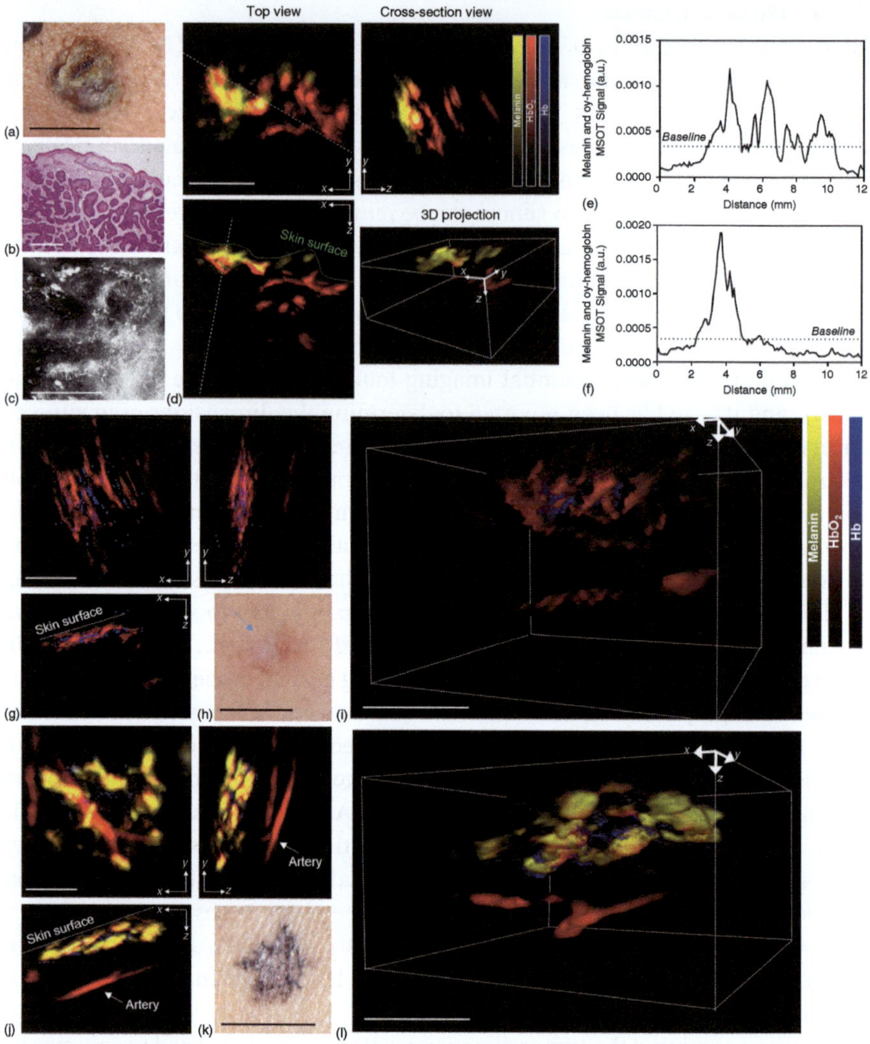

Figure 5.4 Images (clinical and PA) of non-melanoma skin cancer on representative subjects. (a–c) Representative images of a basal cell carcinoma (BCC) on a cheek using hematoxylin and eosin staining and reflectance confocal microscopy, scale bar = 500 µm. (d) PAI maximum intensity projection images from different views and 3D projection of the BCC with the white lines showing where PA signals quantified to give tumor (e) length and (f) depth. PAI images of a BCC (g–i) a patient with Fitzpatrick skin type II and (j–l) a patient with Fitzpatrick skin type IV. Source: Reproduced with permission from Chuah et al. [80], ELSEVIER.

chromophore composition [83]. On ex vivo skin cancer tissues, PA spectrum was acquired from 680 to 970 nm for the distinction between tumor and healthy tissue. Malignant melanomas comprised melanin chromophore, while SCCs contain melanin, deoxygenated hemoglobin, and collagen. On the other hand, melanin, collagen, and hemoglobin contribute to the spectra of BCC, the latter of which was predominantly present.

5.2.2.2 PAI of Inflammatory Skin Diseases

Inflammatory skin diseases such as atopic dermatitis and psoriasis result in vascular remodeling of the skin from immune system changes, giving the skin its characteristic red, irritated, and inflamed look. Currently, disease severity of inflammatory skin disorders is scored based on clinical observations and questionnaires and is subjected to intra-observer and inter-observer variations, lacking similar objective scoring measures. With easy access to and abundant vasculature in the skin, PAI techniques assessing the skin surface and vascular structures could potentially serve as objective inflammatory disease indicators to characterize severity and understand how "inflammatory" vasculature plays a role in the skin disorders.

Olivo's team exploited the RSOM configuration for imaging of skin inflammatory conditions, such as eczema, in 91 subjects including healthy subjects [37]. Imaging-based metrics, such as TBV and thickness of the epidermis, were suggested for assessing vascular inflammation, which displayed unambiguous contrast between varying degrees of atopic dermatitis and healthy conditions. TBV and epidermis thickness were elevated in lesional eczema regions as compared to non-lesional sites. An objective measurement index was formulated from these imaging metrics and correlated with the clinical atopic dermatitis scoring index. This study reveals that the RSOM technology has the potential in investigating the pathogenies of atopic dermatitis and objectively classify the disease's severity based on inflammatory vasculature markers. RSOM can also potentially be exploited to evaluate the progress and effectiveness of treatment longitudinally over a prolonged period. It is worthwhile to note that studies on Asian skin are underrepresented, and this study, in investigating RSOM's performance in darker skin phenotypes, will elucidate the possible ramifications of its clinical use in populations with varied skin phenotypes. Compared to atopic dermatitis, Aguirre et al. observed dilated, elongated, and dense capillary loops in psoriatic skin with RSOM, together with other inflammation markers such as increased diameter and density of vessels in the dermis [84] (Figure 5.5).

Recently, Olivo's team also used a multispectral RSOM system to study the changes of oxygen saturation in superficial skin when stimulated with occlusion challenges [38]. Multispectral RSOM was able to detect the hemodynamic changes upon venous and arterial occlusion and release with time. Thus, multispectral RSOM has the potential to study the oxygenation of inflamed skin to investigate the hemodynamics during vascular remodeling and how it factors during treatment. Indeed, Ishida et al. reported the recovery of oxygenation by more than 10% in psoriatic lesions from PAI upon ixekizumab treatment due to reduced oxygen consumption [85].

5.2.2.3 PAI of Wounds

Chronic wounds are a major burden to the healthcare as there are no current tools to evaluate their response to treatment or their healing status. The clinical practice is to visually inspect the wounds while the naked eyes cannot evaluate the inner dermis layers of the skin. Revascularization surgery is touted as an effective intervention for chronic wounds whereby arterial blood flow to the wounds is restored. Hence, it is imperative to determine whether perfusion has been reinstated post-surgery. Clinical assessment methods such as ankle-brachial pressure index and toe blood

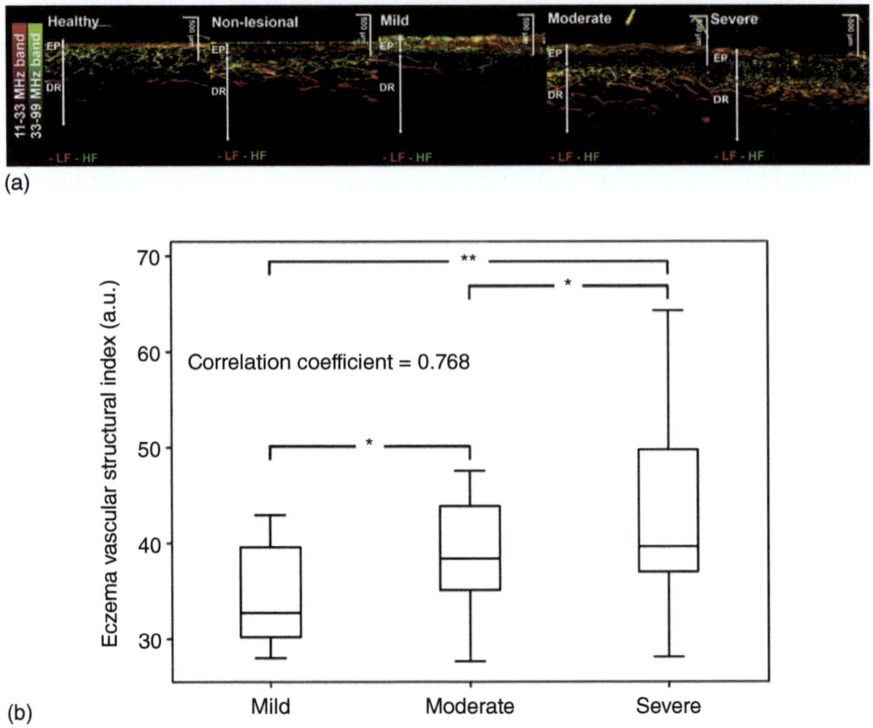

Figure 5.5 RSOM images of healthy and atopic dermatitis skin and classifying their severity using RSOM-derived eczema vascular and structural index (EVSI). (a) Cross-sectional RSOM images of varying atopic dermatitis severities showing demarcated epidermis and dermis regions. All scale bars = 500 μm. (b) Eczema severity classified by EVSI corresponds with patients' clinical severity scores with a correlation coefficient of 0.768. The median values of EVSI are significantly different for the different severities. Source: Reproduced with permission from Yew et al. [37], ELSEVIER.

pressure readings give whole-body perfusion results but not specifically at the site of wounds. Therefore, there is a need for a 3D imaging tool that can visualize the vasculature in deeper depths of the skin to aid in clinical decision-making in terms of wound risk and treatment response. As angiogenesis is vital for wound healing, an imaging tool to detect angiogenesis in the vasculature will help in the management of the wounds and improve the outcomes.

Mantri et al. reported the use of a LED-based PAI tomographic system to detect angiogenesis and classify 22 wounds in 19 patients based on their healing status upon treatment [86]. The wounds were imaged once a week for at least three weeks. Wounds that were healing displayed angiogenesis with new blood vessels forming when visualized by PAI. The PAI signals in the blood vessels intensified significantly with time as the wound healed. On the other hand, no signs of blood vessel formation was visualized by PAI in nonhealing wounds. This study demonstrated that PAI has the capacity in aiding clinical decision-making and how the therapeutic interventions can be effectively managed if the wounds are not showing signs of healing early on.

Wang et al. innovated a portable PA tomography system tailor-made for assessment of chronic ulcers in the foot and validated it with phantom and clinical studies [87]. On a foot of a healthy volunteer, distinct vasculatures can be visualized, signifying good blood circulation. In two subjects with chronic ulcers, the PAI system was capable of visualizing the blood vessels in the wound, and a clear contrast between a well-perfused foot and poor perfusion was seen in the two patients, with the latter giving weaker vascular and background PA signals. While the authors acknowledged that other PA features, such as blood vessel density or oxygen saturation, can be explored in the future, PAI has the potential to be a good clinical assessment of tissue perfusion.

5.3 Photoacoustic Sensing for Biomedical Applications

5.3.1 Noninvasive Temperature Monitoring in Deep Tissue

Temperature measurement in deep tissue is critical for many biomedical applications including, but not limited to, sleep and exercise monitoring, fever treatment, body response to ambience, and temperature control of photothermal therapy [88–92]. However, it is challenging to read out core temperature in real time without invasive catheters or an esophageal thermometer [93], not to mention other local temperature in deep tissue. It has been reported that core temperature can be deduced from surface temperature distribution [94]. A product "Dräger" Tcore using dual-sensor based on heat flux was developed, which can be pasted on patients' forehead in the ICU to continuously calculate core temperature by knowing surface temperature. Although accuracy has been verified with rectal and esophagus temperatures [95, 96], it does not have direct access to local temperature in deep tissue. A dynamic and accurate temperature map of the treated region in deep tissue during thermal therapy is not yet available. PA has been seen as a potential candidate to solve these challenges considering its noninvasiveness, deep penetration, and real-time imaging capabilities. The rationale for temperature monitoring by the PA effect is based on the following equation [97]:

$$p = \mu_a \Gamma(T) F \tag{5.1}$$

where p is the initial pressure generated by the laser-absorbing sample, μ_a represents absorption coefficient, and F is fluence of illumination source. Γ is Grüneisen coefficient correlated to temperature and defined as:

$$\Gamma = \frac{\beta c^2}{C_p} \tag{5.2}$$

where β is thermal expansion coefficient, c is speed of sound, and C_p is heat capacity at constant pressure [98]. Since β and c are temperature dependent [99, 100], Γ and thus p's values can directly reflect temperature, which can be expressed as:

$$\Delta T \propto \frac{\Delta P}{P} \propto \frac{\Delta \Gamma}{\Gamma} \tag{5.3}$$

A PA-guided photothermal therapy for cancer treatment has been reported [97]. In the *ex vivo* experiment, nanoparticles were injected into the target tissue as contrast agent, which has much higher photon absorption than surrounding tissue. The setup is similar to conventional PA/US imaging, except that an additional 800 nm 1 W continuous laser is incorporated to heat cancerous tissue selectively, as shown in Figure 5.6A.

PA signal intensity increased by ~42% in porcine tissue as temperature increased by 9 °C, from 25 to 34 °C with laser radiation. The temperature resolution is 0.16 °C. On the other hand, ultrasound strain changed by less than 1% under the same condition with a temperature resolution of 0.05 °C. Moreover, the signal-to-noise ratio (SNR) of PA and US thermal images is 59.97 and 45.56 dB, respectively. The PA and US thermal images during the four-minute laser therapy are shown in Figure 5.6B. Although progressive temperature elevation can be seen by both methods, PA has a higher contrast-to-noise ratio (CNR) indicating that PA has higher sensitivity in temperature sensing.

Landa et al. [101] reported a 3D real-time temperature mapping method which can provide temperature reading in 40-ms intervals. Data were acquired by a 512-element spherical transducer with a 140° covering angle and a 4 cm radius of curvature. The temperature was estimated based on

$$\frac{\Delta p_0}{p_{0,0}} = \frac{00053\Delta T}{0.0043 + 00053T_0} \tag{5.4}$$

where $p_{0,0}$ and T_0 are initial PA pressure and temperature at starting point. The amplitude of the PA signals will increase by ~2.7% per degree around the typical temperature range in living organisms (~36 °C). In vivo mouse experiments provided a 3D heat map with 150 μm spatial resolution over 120 seconds to monitor the dynamic change during 36 seconds laser heating and cooling.

Note that in the above-mentioned study, only relative temperature changes can be obtained by calculating the PA signal difference between two successive frames. Prior calibration of the baseline is still needed to obtain absolute temperature, which hinders its applications. To overcome this challenge, a thermal memory-based absolute temperature monitoring method was proposed in 2019 [102]. Thermal-energy-memory-based photoacoustic thermometry (TEMPT) is achieved by shining a series of pulses and measuring the ratiometric value of first and Nth PA signals to obtain the baseline Grüneisen parameter $\Gamma_0(T_0)$.

$$\Gamma_0(T_0) = \frac{p_1^2 b(N-1)}{p_N - p_1}, N \geq 2 \tag{5.5}$$

where b is the first-order derivative of Γ, which can be written as:

$$b = \frac{k_T}{\rho C_V} \tag{5.6}$$

Using muscle as an example, linear dependence of Grüneisen coefficient on temperature $k_T = 0.0086$, tissue density $\rho = 1\,\text{g/cm}^3$, and specific heat capacity at constant volume $C_v = 3.8\,\text{J/g/K}$ is known. It can be calculated that

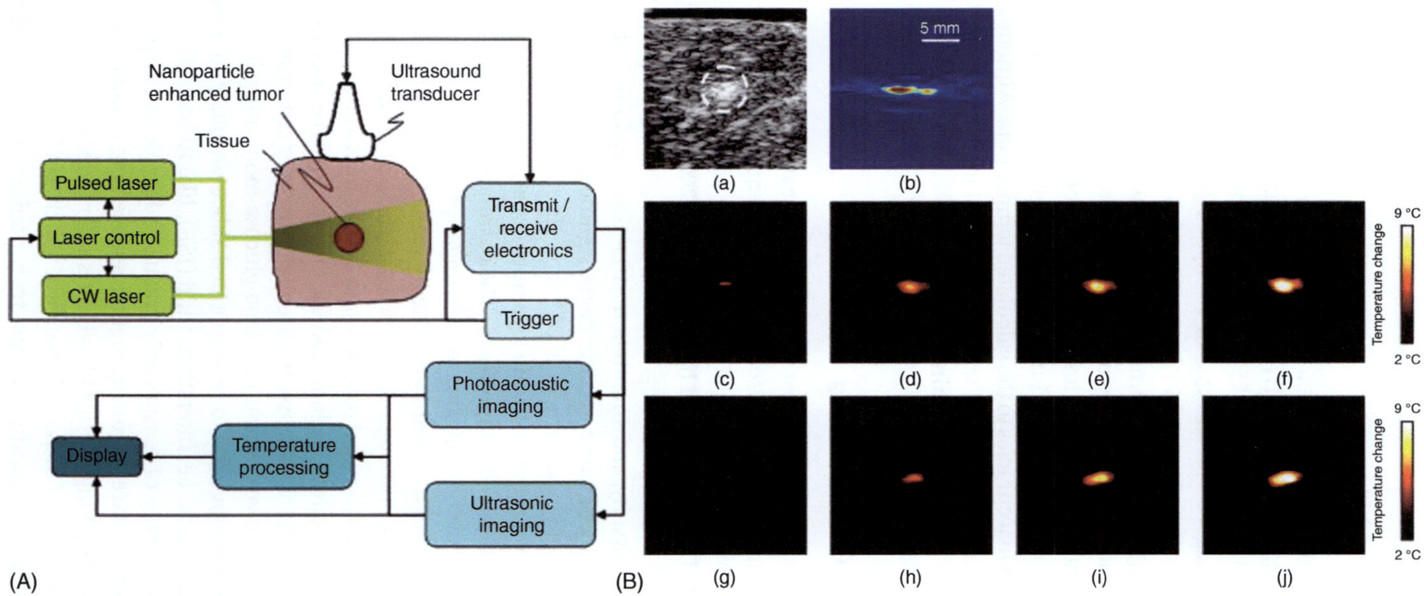

Figure 5.6 (A) photoacoustic-guided photothermal therapy setup. (B) Original (a) US, (b) PA images of porcine tissue with injected nanoparticles. (c–f) PA and (g–j) US thermal images after 60, 120, 180, and 240 seconds of thermal therapy. Field of view is 20 mm × 20 mm. Source: Reproduced from Shah et al. [97], SPIE / CC BY 4.0.

$b \approx 2.3 \times 10^{-3}$ cm^3/J. The baseline temperature T_0 can then be estimated as [103] without prior calibration:

$$T_0 = 115.89\Gamma_0 - 13.14 \tag{5.7}$$

In vivo experiments were carried out to verify TEMPT, together with high-intensity focused ultrasound (HIFU) on a mouse hindlimb. An average error of 0.9 °C was noticed within 20–55 °C at 1.5 cm depth of tissue.

Although researchers have been devoted to explore the possibility of real-time deep tissue temperature monitoring by PA technique, it has not been widely adopted due to several challenges. For example, assumption of linearity of Grüneisen parameter and thus PA signal amplitude may not be valid when temperature goes beyond 53 °C [104]. In addition, tissue properties variations like optical absorption, acoustic speed, and thermal elastic expansion during heating cannot be ignored as they may introduce unexpected inaccuracies to the calculated temperature. Although penetration depth of PA is much deeper than traditional optical methods, it is still limited to several centimeters using NIR illumination.

5.3.2 Noninvasive Glucose Sensing

Blood glucose level is one of the most significant indicators of diabetes mellitus (DM) which is a chronic metabolic disease due to inadequate insulin or cell resistance to insulin. DM was projected to affect 552 million population by 2030, based on 80 countries' data [105]. Long-term hyperglycemia >200 mg/dl or hypoglycemia (<70 mg/dl) will lead to complications, such as stroke, neuropathy, atherosclerosis, nephropathy, retinopathy, and amputation. [106–108]. In addition, diabetes also intensifies the possibility of cause-specific death significantly [109, 110]. Diabetes cannot be cured [111], and management of DM depends on daily blood glucose measures with active control. Frequent insulin injections or bariatric surgery is necessary to alleviate diabetic patients' life expectancy. The current method for daily glucose monitoring is using the well-established in vitro finger-prick glucose meters and disposable testing strips. Patients need to repeatedly extract a small amount of blood every day from fingertips. Although the accuracy of these glucose meters is widely accepted (±20%), it causes physical pain and may lead to infection [112, 113]. Moreover, it only provides a discrete readout, and the testing strips themselves are costly [114]. One compromising way is to use minimally invasive or semi-invasive devices to mitigate pain and achieve continuous blood glucose monitoring (CBGM). They usually incorporate microneedles to be implanted subcutaneously to measure the glucose level in body fluids, i.e. interstitial fluid (ISF), like Abbott FreeStyle Libre. Nevertheless, the sensitivity of sensors may gradually degrade as the protein accumulates on the needle surface, requiring regular replacement to maintain measurement accuracy [115, 116]. Finger-prick blood test is still needed during rapid changes in glucose levels. The cost of disposable patches, which can last for up to 14 days, is high. Therefore, low-cost, continuous, and truly noninvasive devices that can be repeatedly used are highly in demand, aiming to replace those finger-prick devices.

Accuracy of glucose measurement should be the first consideration to assess the performance of all the monitoring devices. The measured data points need to be calibrated and compared with known references. Several indicators can be used to quantify the measurement accuracy from statistical and clinical aspects. For example, coefficient of correlation (r) whose value is within ± 1 shows the degree of correlation between measured data and reference. By calculating R-squared ($R(2)$) value, we can know how well the measured data can represent the reference data. To have an idea of measurement error, mean absolute error (MAE), mean absolute percentage error (MAPE), and root-mean-square error (RMSE) can be calculated. In addition, the degree of agreement can be assessed via the Bland–Altman plot [117]. Apart from these statistical parameters, clinical acceptance is also critical for DM treatment. In 1987, W. L. Clarke proposed a scatterplot named Clarke error grid (CEG), which became the "gold standard" to evaluate the accuracy of glucose measurement from clinical aspect [118]. CEG is differentiated into five zones, where zone A contains values within $\pm 20\%$ error range compared to the reference. Data in zone B have >20% error but still lead to an appropriate treatment. On the contrary, predictions in regions C, D, and E will result in an inaccurate treatment. The target of glucose measuring methods and devices is to make as many data points as possible fall into regions A and B, which are acceptable for clinical use. CEG tells the accuracy of a new glucose monitoring method by consideration of the clinical significance of the prediction error.

During the past decades, electromagnetic (EM) wave, especially light sensing, is an emerging domain of noninvasive glucose sensing due to its rich interactions with essential body compounds. Ishimaru and Zhang et al. [119, 120] summarize different types of EM waves, including radio waves, microwaves, terahertz (THz) waves, visible/IR light, ultraviolet, X-rays, and gamma rays contingent on different wavelengths. Molecules' rotational and vibrational motions lead to light absorption at certain frequencies due to energy resonance with the corresponding quantum energy bands being in the IR region [121]. Thus, IR spectroscopy, including NIR spectroscopy [122] and mid-infrared (MIR) spectroscopy, has become popular in analytical chemistry to characterize complex samples in different states. Table 5.1 summarizes the interactions between IR light and glucose, whose molecular formula is written as $C_6H_{12}O_6$ arranged in pyranose.

As introduced earlier, PA is able to detect the generated ultrasonic wave induced by the light absorption and thermal expansion. Based on specific IR absorption of glucose, PA technique can be a potential candidate for noninvasive glucose monitoring. In conventional PA spectroscopy, we hypothesize that total absorption at a certain wavelength of the sample is directly proportional to received PA signal amplitude, which can be expressed as:

$$[PA(\lambda_1) \quad PA(\lambda_2) \quad \dots \quad PA(\lambda_M)] = k \begin{bmatrix} \varepsilon_1(\lambda_1) & \varepsilon_1(\lambda_2) & \dots & \varepsilon_1(\lambda_M) \\ \varepsilon_2(\lambda_1) & \varepsilon_2(\lambda_2) & \dots & \varepsilon_2(\lambda_M) \\ \dots & \dots & \dots & \dots \\ \varepsilon_N(\lambda_1) & \varepsilon_N(\lambda_2) & \dots & \varepsilon_N(\lambda_M) \end{bmatrix} \times [C_1 \quad C_2 \quad \dots \quad C_N] \tag{5.8}$$

Table 5.1 Glucose absorption peaks and the origin functional groups.

No	Wavelength (nm)	Corresponding functional group
1	9746	C–O–H bending vibration [123–125]
2	9680	ν(C–O–H) or ν(C–O–C) vibration [124–128]
3	9551	C–H bending vibrations [123–125, 129]
4	9290	C–H bending vibrations [123–127, 129]
5	8680	[127]
6	8658	Pyranose ring [130, 131]
7	8244	[126]
8	8000	C–H bending vibrations [123–125, 129]
9	2273	Combination of O–H/C–O stretching [132]
10	2261	νCH + νCCH [133]
11	1688	2νCH [133]
12	1638	First overtone [134]
13	1536	νOH + νCH [133]
14	1408	2νOH [133]
15	1126	3νCH [133]
16	1042	Combination of νCH [134, 135]
17	1018	Combination of νCH [136]
18	939	3νOH [133]
19	930	3νCH_2 [137]
20	910	4νCH [135, 137]

where k is a system coefficient, $\varepsilon(\lambda)$ is the molar extinction coefficient at wavelength λ, and C is molar concentration of glucose. Then, the concentration can be obtained by solving:

$$\mathbf{C} = (\boldsymbol{\varepsilon}^T\boldsymbol{\varepsilon})^{-1}\boldsymbol{\varepsilon}^T(\mathbf{PA}/k) \tag{5.9}$$

which is similar to IR spectroscopy.

Miguel et al. used an external cavity quantum cascade laser (EC-QCL) with MIR wavelengths as an excitation source for a glucose fingerprint region (1000–1220 cm^{-1}) [126]. Detection of ultrasound was performed by a PA cell, which matches the microphone's detection frequency at 50 kHz. They demonstrated noninvasive and continuous glucose monitoring using the setup on human volunteers using oral glucose tolerance test (OGTT) over two to three hours. The noninvasive measurement results were correlated with finger-prick measurements. The mean prediction error is less than 15 mg/dl and all data points fall within CEG zones A and B. Sim et al. combined a raster scanner with PA sensors to add the microscopic imaging capability of skin and mitigate skin condition variation during glucose measurement [138]. The same laser source was used and the index finger was scanned. The obtained imaging resolution was 90 μm. They suggested that better prediction results could be achieved by looking at the relatively darker area between friction ridges on the finger, owing to less secretion of sebum and sweat, as shown in Figure 5.7.

The above-mentioned method relies on the PA amplitude only, yet phase difference information of the collected PA signal is usually overlooked. Zhang et al. proposed a novel method which utilizes inherent time-of-flight information in time-domain PA signals [139]. In this study, both time delay and amplitude information extracted from single PA signal were used separately to predict glucose levels. Then, prediction results were fused to obtain the final prediction result. Measurement accuracy was improved without incorporating additional and complicated instrumentation. According to literature, glucose concentration is directly proportional to PA amplitude [140–143], as shown below

$$c_{glucose} \propto p_0 \propto \eta_{th} \Gamma \mu_a F \tag{5.10}$$

Besides PA amplitude, ultrasound propagation velocity is also linearly proportional to glucose concentration [144, 145]. Thus, if the transducer is fixed, PA peak arrival time is expected to vary with glucose concentration. Velocity can be deduced from the measurement distance and the peak arrival time. Based on linear approximation (Eq. (5.11)), we can derive Eq. (5.12) by substituting $f(x) = 1/x$.

$$f(x) \approx f(a) + f'(a)(x - a) \tag{5.11}$$

$$\frac{1}{x}(x = a) \approx \frac{1}{a} - \frac{1}{a^2}(x - a) \propto -x \tag{5.12}$$

By replacing x with t_{peak}, we can derive Eq. (5.13). Then, it can be seen that concentration is proportional to velocity, as expected.

$$\frac{d}{t_{peak}}(t_{peak} = b) \approx \frac{d}{b} - \frac{d}{b^2}(t_{peak} - b) = m - nt_{peak} = p - qt_{delay} \tag{5.13}$$

$$c_{glucose} \propto v = \frac{d}{t_{peak}} \approx p - qt_{delay} \propto -t_{delay} \tag{5.14}$$

Then, glucose level can be predicted by PA amplitude and time-of-flight information separately and then fused by assigning different weighting factors, which is known as competitive data fusion governed by central limit theorem. The two weighting factors ω_1, ω_2 of the separate predictions are calculated by ${\sigma_1}^2$ and ${\sigma_2}^2$, which are their corresponding variances obtained by calibration (Eq. (5.15)). Theoretically, final result by data fusion is more accurate based on Eq. (5.17) because of the uncertainty reduction [145] and instability compensation.

$$w_i = \frac{{\sigma_1}^2{\sigma_2}^2}{\left({\sigma_1}^2 + {\sigma_2}^2\right){\sigma_i}^2}, \quad i = 1,2 \tag{5.15}$$

$$c_{glucose} = \sum w_i c_{glucosei}, i = 1,2 \tag{5.16}$$

$$\sigma^2 = \frac{{\sigma_1}^2{\sigma_2}^2}{\left({\sigma_1}^2 + {\sigma_2}^2\right)} < {\sigma_i}^2, \quad i = 1,2 \tag{5.17}$$

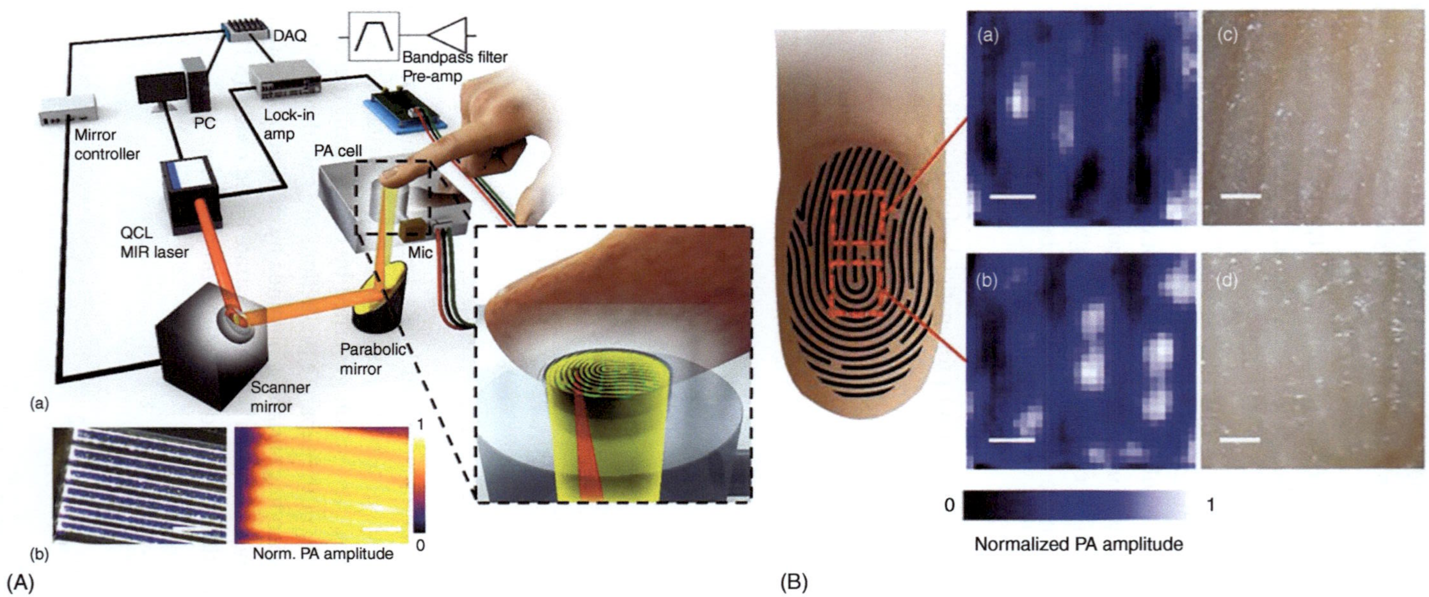

Figure 5.7 (A) (a) Setup of the microscopic PA glucose sensing system. (b) Photoresist structure for imaging resolution evaluation. (B) PA images of two fingertip regions (a and b) and the corresponding micrographs (c and d). Source: Reproduced with permission from Sim et al. [138], Springer Nature/CC BY 4.0.

In vitro experiments using a glucose solution at different concentrations from 0 to 350 mg/dl demonstrated that the data fusion method improved accuracy by 33.63% and 55.80%, respectively, compared to single parameter predictions. CEG analysis results for different methods are shown in [139]. Using the data fusion method, 94.29% and 5.71% of data are in zones A and B. In total, 100% of the data points fall within clinically acceptable regions, indicating excellent glucose prediction accuracy.

They further proposed time-domain photoacoustic waveform spectroscopy (tPAWS) in 2020, by considering the entire time-domain PA waveform [146]. Thermal properties, viscosity, acoustic propagation velocity, and acoustic impedance that contribute to PA waves are all affected by tissue composition [147]. The single wavelength generated PA waveform can be analyzed as a whole to dramatically improve measurement sensitivity and accuracy since data dimensions are enriched. tPAWS deals with the entire time-domain PA waveform using multivariate analysis instead of using single-dimension amplitude to calculate glucose concentration, which is low in complex blood matrix. The multivariate analysis can be expressed as:

$$\boldsymbol{p(t)} = \begin{bmatrix} p_1(t_1) & \cdots & p_1(t_n) \\ \vdots & \ddots & \vdots \\ p_m(t_1) & \cdots & p_m(t_n) \end{bmatrix} = \boldsymbol{SL}^{\mathrm{T}} + \boldsymbol{E} \tag{5.18}$$

where $\boldsymbol{p(t)}$ is a $\boldsymbol{m} \times \boldsymbol{n}$ training data matrix. $\boldsymbol{m}$ rows are observations at different concentrations and $\boldsymbol{n}$ columns are PA signals at different time points. $\boldsymbol{S}$ is an $\boldsymbol{m} \times \boldsymbol{q}$ score matrix and $\boldsymbol{L}$ is $\boldsymbol{n} \times \boldsymbol{q}$ loading matrix with $\boldsymbol{q}$ columns representing principal components. $\boldsymbol{E}$ is residual matrix. The loading and score matrices can be calibrated. Then, multilinear regression (MLR), principal component regression (PCR), and partial least squares regression (PLSR) were conducted to regress $\boldsymbol{S}$ to $\boldsymbol{m} \times 1$ concentration vector $\boldsymbol{C}_{\mathrm{g}}$ and obtain regression coefficients $\hat{\boldsymbol{\beta}}$ $(\boldsymbol{q} \times 1)$:

$$\hat{\boldsymbol{\beta}} = \underset{\boldsymbol{\beta} \in \mathbb{R}^q}{\arg\min} \left\{ (\boldsymbol{C}_{\mathrm{g}} - \boldsymbol{S\beta})^{\mathrm{T}} (\boldsymbol{C}_{\mathrm{g}} - \boldsymbol{S\beta}) \right\} = (\boldsymbol{S}^{\mathrm{T}}\boldsymbol{S})^{-1}\boldsymbol{S}^{\mathrm{T}}\boldsymbol{C}_{\mathrm{g}} \tag{5.19}$$

such that $\boldsymbol{C}_{\mathrm{g}} = \boldsymbol{S}\hat{\boldsymbol{\beta}} + \boldsymbol{E}$. After that, glucose concentration can be predicted as follows when the new PA signal for unknown glucose solution is measured:

$$C_{\mathrm{g},m+1} = p_{m+1}(\boldsymbol{t})\boldsymbol{L}\hat{\boldsymbol{\beta}} \tag{5.20}$$

In vitro experimental results demonstrated that tPAWS is able to recover glucose-related information from noisy PA signals. Using traditional amplitude-based single-dimension prediction, only 40% were in zones A and B of CEG (A: 20%, B: 20%). Applying tPAWS, MLR improved prediction significantly (A: 70%, B: 10%), while PCR and PLSR showed better performance with 100% data in zone A. The accuracy improvements in terms of RMSE are >10 times those of TPAWS compared to traditional method, showing its outstanding prediction performance.

Clinicians and scientists worldwide never stop pursuing truly noninvasive glucose monitoring methods. Despite the promising progress in noninvasive glucose monitoring technique development, there are still several global challenges to be fully addressed. First of all, glucose concentration is relatively low in the complicated blood matrix. To differentiate it from other components, selectivity for optical methods is based on sharp and non-overlapped absorption peaks of a certain analyte. However, this requires tunable or multiple light sources and advanced unmixing algorithms, which may increase the overall system cost and complexity, preventing it from daily usage. Secondly, to achieve continuous monitoring, long-term application of glucose sensors may encounter movement artifacts and body temperature drifting issues during measurement. This could be mitigated by adopting conformable and wearable integrated sensors in the future. Last but not least, the calibration of noninvasive glucose monitoring still requires blood measurements as a reference. Hence, it is necessary to make the noninvasive measurement techniques more stable and robust to reduce the calibration frequency.

References

1 Ntziachristos, V. and Razansky, D. (2010). Molecular imaging by means of multispectral optoacoustic tomography (MSOT). *Chemical Reviews* 110 (5): 2783–2794.

2 Xu, M. and Wang, L.V. (2006). Photoacoustic imaging in biomedicine. *Review of Scientific Instruments* 77 (4): 41101.

3 Attia, A.B.E., Balasundaram, G., Moothanchery, M. et al. (2019). A review of clinical photoacoustic imaging: current and future trends. *Photoacoustics* 16: 100144.

4 Huang, D., Swanson, E.A., Lin, C.P. et al. (1991). Optical coherence tomography. *Science* 254 (5035): 1178–1181.

5 Bhende, M., Shetty, S., Parthasarathy, M.K., and Ramya, S. (2018). Optical coherence tomography: a guide to interpretation of common macular diseases. *Indian Journal of Ophthalmology* 66 (1): 20–35.

6 Sanderson, M.J., Smith, I., Parker, I., and Bootman, M.D. (2014). Fluorescence microscopy. *Cold Spring Harbor Protocols* 2014 (10): pdb.top071795-pdb.top071795.

7 Li, M., Tang, Y., and Yao, J. (2018). Photoacoustic tomography of blood oxygenation: a mini review. *Photoacoustics* 10: 65–73.

8 Ho, C.J.H., Balasundaram, G., Driessen, W. et al. (2014). Multifunctional photosensitizer-based contrast agents for photoacoustic imaging. *Scientific Reports* 4: 5342.

9 Stoffels, I., Morscher, S., Helfrich, I. et al. (2015). Metastatic status of sentinel lymph nodes in melanoma determined noninvasively with multispectral optoacoustic imaging. *Science Translational Medicine* 7 (317): 317ra199.

10 Luke, G.P., Yeager, D., and Emelianov, S.Y. (2012). Biomedical applications of photoacoustic imaging with exogenous contrast agents. *Annals of Biomedical Engineering* 40 (2): 422–437.

11 Wu, D., Huang, L., Jiang, M., and Jiang, H. (2014). Contrast agents for photoacoustic and thermoacoustic imaging: a review. *International Journal of Molecular Sciences* 15 (12): 23616–23639.

12 Weber, J., Beard, P.C., and Bohndiek, S.E. (2016). Contrast agents for molecular photoacoustic imaging. *Nature Methods* 13 (8): 639.

13 Beard, P. (2011). Biomedical photoacoustic imaging. *Interface Focus* 1 (4): 602–631.

14 Dima, A., Burton, N.C., and Ntziachristos, V. (2014). Multispectral optoacoustic tomography at 64, 128, and 256 channels. *Journal of Biomedical Optics* 19 (3): 036021–036021.

15 Needles, A., Heinmiller, A., Sun, J. et al. (2013). Development and initial application of a fully integrated photoacoustic micro-ultrasound system. *IEEE Transactions on Ultrasonics, Ferroelectrics, and Frequency Control* 60 (5): 888–897.

16 Song, K.H., Stoica, G., and Wang, L.V. (2006). In vivo three-dimensional photoacoustic tomography of a whole mouse head. *Optics Letters* 31 (16): 2453–2455.

17 Wang, X., Pang, Y., Ku, G. et al. (2003). Three-dimensional laser-induced photoacoustic tomography of mouse brain with the skin and skull intact. *Optics Letters* 28 (19): 1739–1741.

18 Burton, N.C., Patel, M., Morscher, S. et al. (2013). Multispectral opto-acoustic tomography (MSOT) of the brain and glioblastoma characterization. *NeuroImage* 65: 522–528.

19 Goh, Y., Balasundaram, G., Moothanchery, M. et al. (2018). Multispectral optoacoustic tomography in assessment of breast tumor margins during breast-conserving surgery: a first-in-human case study. *Clinical Breast Cancer* 18 (6): e1247–e1250.

20 Neuschmelting, V., Burton, N.C., Lockau, H. et al. (2016). Performance of a multispectral optoacoustic tomography (MSOT) system equipped with 2D vs. 3D handheld probes for potential clinical translation. *Photoacoustics* 4 (1): 1–10.

21 Schellenberg, M.W. and Hunt, H.K. (2018). Hand-held optoacoustic imaging: a review. *Photoacoustics* 11: 14–27.

22 Sivasubramanian, K., Periyasamy, V., and Pramanik, M. (2018). Non-invasive sentinel lymph node mapping and needle guidance using clinical handheld photoacoustic imaging system in small animal. *Journal of Biophotonics* 11 (1): e201700061.

23 Kim, J., Park, S., Jung, Y. et al. (2016). Programmable real-time clinical photoacoustic and ultrasound imaging system. *Scientific Reports* 6: 35137.

24 Lin, L., Hu, P., Shi, J. et al. (2018). Single-breath-hold photoacoustic computed tomography of the breast. *Nature Communications* 9 (1): 2352.

25 Heijblom, M., Piras, D., van den Engh, F.M. et al. (2016). The state of the art in breast imaging using the Twente Photoacoustic Mammoscope: results from 31 measurements on malignancies. *European Radiology* 26 (11): 3874–3887.

26 Attia, A.B.E., Chuah, S.Y., Razansky, D. et al. (2017). Noninvasive real-time characterization of non-melanoma skin cancers with handheld optoacoustic probes. *Photoacoustics* 7: 20–26.

27 Wang, L.V. (2009). Multiscale photoacoustic microscopy and computed tomography. *Nature Photonics* 3 (9): 503.

28 Maslov, K., Stoica, G., and Wang, L.V. (2005). In vivo dark-field reflection-mode photoacoustic microscopy. *Optics Letters* 30 (6): 625–627.

29 Chuangsuwanich, T., Moothanchery, M., Yan, A.T.C. et al. (2018). Photoacoustic imaging of lamina cribrosa microcapillaries in porcine eyes. *Applied Optics* 57 (17): 4865–4871.

30 Huggenberger, R. and Detmar, M. (2011). The cutaneous vascular system in chronic skin inflammation. In: *Journal of Investigative Dermatology Symposium Proceedings*, 24–32. Elsevier.

31 Moothanchery, M., Bi, R., Kim, J.Y. et al. (2018). Optical resolution photoacoustic microscopy based on multimode fibers. *Biomedical Optics Express* 9 (3): 1190–1197.

32 Kim, J.Y., Lee, C., Park, K. et al. (2015). Fast optical-resolution photoacoustic microscopy using a 2-axis water-proofing MEMS scanner. *Scientific Reports* 5: 7932.

33 Moothanchery, M., Bi, R., Kim, J.Y. et al. (2019). High-speed simultaneous multiscale photoacoustic microscopy. *Journal of Biomedical Optics* 24 (8): 1–7.

34 Moothanchery, M., Dev, K., Balasundaram, G. et al. (2020). Acoustic resolution photoacoustic microscopy based on microelectromechanical systems scanner. *Journal of Biophotonics* 13 (2): e201960127.

35 Omar, M., Gateau, J., and Ntziachristos, V. (2013). Raster-scan optoacoustic mesoscopy in the 25–125 MHz range. *Optics Letters* 38 (14): 2472–2474.

36 Omar, M., Soliman, D., Gateau, J., and Ntziachristos, V. (2014). Ultrawideband reflection-mode optoacoustic mesoscopy. *Optics Letters* 39 (13): 3911–3914.

37 Yew, Y.W., Dinish, U.S., Kuan, A.H.Y. et al. (2020). Raster-scanning optoacoustic mesoscopy (RSOM) imaging as an objective disease severity tool in atopic dermatitis patients. *Journal of the American Academy of Dermatology* 84 (4): 1121–1123.

38 Attia, A.B.E., Moothanchery, M., Li, X. et al. (2021). Microvascular imaging and monitoring of hemodynamic changes in the skin during arterial-venous occlusion using multispectral raster-scanning optoacoustic mesoscopy. *Photoacoustics* 22: 100268.

39 Johnson, R.H., Anders, C.K., Litton, J.K. et al. (2018). Breast cancer in adolescents and young adults. *Pediatric Blood & Cancer* 65 (12): e27397–e27397.

40 Sung, H., Ferlay, J., Siegel, R.L. et al. (2021). Global cancer statistics 2020: GLOBOCAN estimates of incidence and mortality worldwide for 36 cancers in 185 countries. *CA: A Cancer Journal for Clinicians* 71 (3): 209–249.

41 Heywang-Köbrunner, S.H., Hacker, A., and Sedlacek, S. (2011). Advantages and disadvantages of mammography screening. *Breast Care* 6 (3): 199–207.

42 Geisel, J., Raghu, M., and Hooley, R. (2018). The role of ultrasound in breast cancer screening: the case for and against ultrasound. In: *Seminars in Ultrasound CT MRI*, vol. 39(1), 25–34.

43 Quality, O.H. (2016). Magnetic resonance imaging as an adjunct to mammography for breast cancer screening in women at less than high risk for breast cancer: a health technology assessment. *Ontario Health Technology Assessment Series* 16 (20): 1–30.

44 Di Leo, G., Trimboli, R.M., Sella, T., and Sardanelli, F. (2017). Optical imaging of the breast: basic principles and clinical applications. *American Journal of Roentgenology* 209 (1): 230–238.

45 Oraevsky, A.A., Karabutov, A.A., Solomatin, S.V. et al. (2001). Laser optoacoustic imaging of breast cancer in vivo. In: *BiOS 2001 The International Symposium on Biomedical Optics*, vol. 4256. SPIE.

46 Heijblom, M., Piras, D., Xia, W. et al. (2012). Visualizing breast cancer using the Twente Photoacoustic Mammoscope: What do we learn from twelve new patient measurements? *Optics Express* 20 (11): 11582–11597.

47 Ermilov, S.A., Khamapirad, T., Conjusteau, A. et al. (2009). Laser optoacoustic imaging system for detection of breast cancer. *Journal of Biomedical Optics* 14 (2): 024007–024007.

48 Manohar, S., Vaartjes, S.E., van Hespen, J.C.G. et al. (2007). Initial results of in vivo non-invasive cancer imaging in the human breast using near-infrared photoacoustics. *Optics Express* 15 (19): 12277–12285.

49 Heijblom, M., Piras, D., Maartens, E. et al. (2013). Appearance of breast cysts in planar geometry photoacoustic mammography using 1064-nm excitation. *Journal of Biomedical Optics* 18 (12): 126009.

50 Kruger, R.A., Kuzmiak, C.M., Lam, R.B. et al. (2013). Dedicated 3D photoacoustic breast imaging. *Medical Physics* 40 (11): 113301.

51 Kitai, T., Torii, M., Sugie, T. et al. (2014). Photoacoustic mammography: initial clinical results. *Breast Cancer* 21 (2): 146–153.

52 Heijblom, M., Piras, D., Brinkhuis, M. et al. (2015). Photoacoustic image patterns of breast carcinoma and comparisons with magnetic resonance imaging and vascular stained histopathology. *Scientific Reports* 5: 11778.

53 Deán-Ben, X.L. and Razansky, D. (2013). Portable spherical array probe for volumetric real-time optoacoustic imaging at centimeter-scale depths. *Optics Express* 21 (23): 28062–28071.

54 Dean-Ben, X.L., Fehm, T.F., Gostic, M., and Razansky, D. (2016). Volumetric hand-held optoacoustic angiography as a tool for real-time screening of dense breast. *Journal of Biophotonics* 9 (3): 253–259.

55 Filho, A.L., Lopes, J.M., and Schmitt, F.C. (2010). Angiogenesis and breast cancer. *Journal of Oncology* 2010: 576384.

56 Rykala, J., Przybylowska, K., Majsterek, I. et al. (2011). Angiogenesis markers quantification in breast cancer and their correlation with clinicopathological prognostic variables. *Pathology and Oncology Research* 17 (4): 809–817.

57 Toi, M., Asao, Y., Matsumoto, Y. et al. (2017). Visualization of tumor-related blood vessels in human breast by photoacoustic imaging system with a hemispherical detector array. *Scientific Reports* 7: 41970.

58 Neuschler, E.I., Butler, R., Young, C.A. et al. (2018). A pivotal study of optoacoustic imaging to diagnose benign and malignant breast masses: a new evaluation tool for radiologists. *Radiology* 287 (2): 398–412.

59 Menezes, G.L.G., Pijnappel, R.M., Meeuwis, C. et al. (2018). Downgrading of breast masses suspicious for cancer by using optoacoustic breast imaging. *Radiology* 288 (2): 355–365.

60 Diot, G., Metz, S., Noske, A. et al. (2017). Multispectral optoacoustic tomography (MSOT) of human breast cancer. *Clinical Cancer Research* 23 (22): 6912–6922.

61 Fisher, B., Anderson, S., Bryant, J. et al. (2002). Twenty-year follow-up of a randomized trial comparing total mastectomy, lumpectomy, and lumpectomy plus irradiation for the treatment of invasive breast cancer. *New England Journal of Medicine* 347 (16): 1233–1241.

62 Jacobs, L. (2008). Positive margins: the challenge continues for breast surgeons. *Annals of Surgical Oncology* 15: 1271–1272.

63 Atkins, J., Al, M.F., Appleton, C.M. et al. (2012). Positive margin rates following breast-conserving surgery for stage I-III breast cancer: palpable versus nonpalpable tumors. *Journal of Surgical Research* 177 (1): 109–115.

64 Olson, T.P., Harter, J., Munoz, A. et al. (2007). Frozen section analysis for intraoperative margin assessment during breast-conserving surgery results in low rates of re-excision and local recurrence. *Annals of Surgical Oncology* 14 (10): 2953–2960.

65 Creager, A.J., Shaw, J.A., Young, P.R., and Geisinger, K.R. (2002). Intraoperative evaluation of lumpectomy margins by imprint cytology with histologic correlation: a community hospital experience. *Archives of Pathology & Laboratory Medicine* 126 (7): 846–848.

66 Eggemann, H., Ignatov, T., Costa, S.D., and Ignatov, A. (2014). Accuracy of ultrasound-guided breast-conserving surgery in the determination of adequate surgical margins. *Breast Cancer Research and Treatment* 145 (1): 129–136.

67 Thill, M. (2013). MarginProbe: intraoperative margin assessment during breast conserving surgery by using radiofrequency spectroscopy. *Expert Review of Medical Devices* 10 (3): 301–315.

68 Keller, M.D., Majumder, S.K., Kelley, M.C. et al. (2010). Autofluorescence and diffuse reflectance spectroscopy and spectral imaging for breast surgical margin analysis. *Lasers in Surgery and Medicine* 42 (1): 15–23.

69 Keller, M., Vargis, E., Mahadevan-Jansen, A. et al. (2011). Development of a spatially offset Raman spectroscopy probe for breast tumor surgical margin evaluation. *Journal of Biomedical Optics* 16 (7): 77006.

70 Nguyen, F.T., Zysk, A.M., Chaney, E.J. et al. (2009). Intraoperative evaluation of breast tumor margins with optical coherence tomography. *Cancer Research* 69 (22): 8790–8796.

71 Taruttis, A. and Ntziachristos, V. (2015). Advances in real-time multispectral optoacoustic imaging and its applications. *Nature Photonics* 9: 219.

72 Li, R., Wang, P., Lan, L. et al. (2015). Assessing breast tumor margin by multispectral photoacoustic tomography. *Biomedical Optics Express* 6 (4): 1273.

73 Li, R., Lan, L., Xia, Y. et al. (2018). High-speed intra-operative assessment of breast tumour margins by multimodal ultrasound and photoacoustic tomography. *Medical Devices & Sensors* 1 (3): e10018.

74 Balasundaram, G., Goh, Y., Moothanchery, M. et al. (2020). Optoacoustic characterization of breast conserving surgery specimens – a pilot study. *Photoacoustics* 19: 100164.

75 Morrow, M., Van Zee, K.J., Solin, L.J. et al. (2016). Society of Surgical Oncology–American Society for Radiation Oncology–American Society of Clinical Oncology consensus guideline on margins for breast-conserving surgery with whole-breast irradiation in ductal carcinoma in situ. *Practical Radiation Oncology* 6 (5): 287–295.

76 Moran, M.S., Schnitt, S.J., Giuliano, A.E. et al. (2014). Society of surgical oncology-American society for radiation oncology consensus guideline on margins for breast-conserving surgery with whole-breast irradiation in stages I and II invasive breast cancer. *International Journal of Radiation Oncology Biology Physics* 88 (3): 553–564.

77 Wong, T.T.W., Zhang, R., Hai, P. et al. (2017). Fast label-free multilayered histology-like imaging of human breast cancer by photoacoustic microscopy. *Science Advances* 3 (5): e1602168.

78 Imai, T., Shi, J., Wong, T.T. et al. (2018). High-throughput ultraviolet photoacoustic microscopy with multifocal excitation. *Journal of Biomedical Optics* 23 (3): 36007.

79 Hindelang, B., Aguirre, J., Schwarz, M. et al. (2019). Non-invasive imaging in dermatology and the unique potential of raster-scan optoacoustic mesoscopy. *Journal of the European Academy of Dermatology and Venereology* 33 (6): 1051–1061.

80 Chuah, S.Y., Attia, A.B.E., Ho, C.J.H. et al. (2019). Volumetric multispectral optoacoustic tomography for 3-dimensional reconstruction of skin tumors: a further evaluation with histopathologic correlation. *The Journal of Investigative Dermatology* 139 (2): 481–485.

81 Park, B., Bang, C.H., Lee, C. et al. (2021). 3D wide-field multispectral photoacoustic imaging of human melanomas in vivo: a pilot study. *Journal of the European Academy of Dermatology and Venereology* 35 (3): 669–676.

82 von Knorring, T., Israelsen, N.M., Ung, V. et al. (2022). Differentiation between benign and malignant pigmented skin tumours using bedside diagnostic imaging technologies: a pilot study. *Acta Dermato-Venereologica* 102: adv00634–adv00634.

83 Stridh, M.T., Hult, J., Merdasa, A. et al. (2022). Photoacoustic imaging of periorbital skin cancer ex vivo: unique spectral signatures of malignant melanoma, basal, and squamous cell carcinoma. *Biomedical Optics Express* 13 (1): 410. /pmc/articles/PMC8803040/.

84 Aguirre, J., Schwarz, M., Garzorz, N. et al. (2017). Precision assessment of label-free psoriasis biomarkers with ultra-broadband optoacoustic mesoscopy. *Nature Biomedical Engineering* 1 (5): 0068. https://www.nature.com/articles/s41551-017-0068.

85 Ishida, Y., Otsuka, A., Honda, T. et al. (2018). Photoacoustic imaging system visualizes restoration of peripheral oxygenation in psoriatic lesions. *Journal of the European Academy of Dermatology and Venereology* 32 (12): e449–e451. https://onlinelibrary-wiley-com.ejproxy.a-star.edu.sg/doi/full/10.1111/jdv.15032.

86 Mantri, Y., Tsujimoto, J., Donovan, B. et al. (2022). Photoacoustic monitoring of angiogenesis predicts response to therapy in healing wounds. *Wound Repair* *and Regeneration* https://onlinelibrary-wiley-com.ejproxy.a-star.edu.sg/doi/full/10.1111/wrr.12992.

87 Wang, Y., Zhan, Y., Harris, L.M. et al. (2019). A portable three-dimensional photoacoustic tomography system for imaging of chronic foot ulcers. *Quantitative Imaging in Medicine and Surgery* 9 (5): 799.

88 Rossi, R. (2014). Clothing for protection against heat and flames. In: *Protective Clothing*, 70–89. Elsevier.

89 Cagnacci, A., Elliott, J.A., and Yen, S.S. (1992). Melatonin: a major regulator of the circadian rhythm of core temperature in humans. *The Journal of Clinical Endocrinology & Metabolism* 75 (2): 447–452.

90 Zhu, X., Feng, W., Chang, J. et al. (2016). Temperature-feedback upconversion nanocomposite for accurate photothermal therapy at facile temperature. *Nature Communications* 7 (1): 1–10.

91 Waterhouse, J., Drust, B., Weinert, D. et al. (2005). The circadian rhythm of core temperature: origin and some implications for exercise performance. *Chronobiology International* 22 (2): 207–225.

92 Oka, T., Oka, K., and Hori, T. (2001). Mechanisms and mediators of psychological stress-induced rise in core temperature. *Psychosomatic Medicine* 63 (3): 476–486.

93 Nierman, D.M. (1991). Core temperature measurement in the intensive care unit. *Critical Care Medicine* 19 (6): 818–823.

94 Anthony, D., Sarkar, D., and Jain, A. (2016). Non-invasive, transient determination of the core temperature of a heat-generating solid body. *Scientific Reports* 6 (1): 1–10.

95 Mazgaoker, S., Ketko, I., Yanovich, R. et al. (2017). Measuring core body temperature with a non-invasive sensor. *Journal of Thermal Biology* 66: 17–20.

96 Janke, D., Kagelmann, N., Storm, C. et al. (2021). Measuring core body temperature using a non-invasive, disposable double-sensor during targeted temperature management in post-cardiac arrest patients. *Frontiers in Medicine* 8: 666908.

97 Shah, J., Park, S., Aglyamov, S.R. et al. (2008). Photoacoustic imaging and temperature measurement for photothermal cancer therapy. *Journal of Biomedical Optics* 13 (3): 34024.

98 Oraevsky, A.A. and Karabutov, A.A. (2003). Biomedical photonics handbook. In: *Optoacoustic Tomography*, vol. 34, 31–34. Boca Raton-London-New York-Washington, DC: CRC Press.

99 Duck, F.A. (2013). *Physical Properties of Tissues: A Comprehensive Reference Book*. Academic Press.

100 Bamber, J.C. and Hill, C.R. (1979). Ultrasonic attenuation and propagation speed in mammalian tissues as a function of temperature. *Ultrasound in Medicine & Biology* 5 (2): 149–157.

101 Landa, F.J.O., Deán-Ben, X.L., Sroka, R., and Razansky, D. (2017). Volumetric optoacoustic temperature mapping in photothermal therapy. *Scientific Reports* 7 (1): 1–8.

102 Zhou, Y., Li, M., Liu, W. et al. (2019). Thermal memory based photoacoustic imaging of temperature. *Optica* 6 (2): 198–205.

103 Alaeian, M. and Orlande, H.R.B. (2017). Inverse photoacoustic technique for parameter and temperature estimation in tissues. *Heat Transfer Engineering* 38 (18): 1573–1594.

104 Larina, I.V., Larin, K.V., and Esenaliev, R.O. (2005). Real-time optoacoustic monitoring of temperature in tissues. *Journal of Physics D: Applied Physics* 38 (15): 2633.

105 Cho, N., Shaw, J.E., Karuranga, S. et al. (2018). IDF Diabetes Atlas: global estimates of diabetes prevalence for 2017 and projections for 2045. *Diabetes Research and Clinical Practice* 138: 271–281.

106 Harding, J.L., Pavkov, M.E., Magliano, D.J. et al. (2019). Global trends in diabetes complications: a review of current evidence. *Diabetologia* 62 (1): 3–16.

107 Bonora, E., Formentini, G., Calcaterra, F. et al. (2002). HOMA-estimated insulin resistance is an independent predictor of cardiovascular disease in type 2 diabetic subjects: prospective data from the Verona Diabetes Complications Study. *Diabetes Care* 25 (7): 1135–1141.

108 Nather, A., Cao, S., Chen, J.L.W., and Low, A.Y. (2018). Prevention of diabetic foot complications. *Singapore Medical Journal* 59 (6): 291–294.

109 Morrish, N.J., Wang, S.-L., Stevens, L.K. et al., Group WHOMS(2001). Mortality and causes of death in the WHO Multinational Study of Vascular Disease in Diabetes. *Diabetologia* 44 (2): S14.

110 Baena-Díez, J.M., Peñafiel, J., Subirana, I. et al. (2016). Risk of cause-specific death in individuals with diabetes: a competing risks analysis. *Diabetes Care* 39 (11): 1987–1995.

111 Buse, J.B., Caprio, S., Cefalu, W.T. et al. (2009). How do we define cure of diabetes? *Diabetes Care* 32 (11): 2133–2135.

112 Sharon, M.G. (2014). Infection transmission associated with point of care testing and the laboratory's role in risk reduction. *EJIFCC* 25 (2): 188.

113 Inayat, F. and Rai, A.B. (2016). Acute hepatitis C virus infection related to capillary blood glucose meter. *Saudi Medical Journal* 37 (1): 93.

114 Yeaw, J., Lee, W.C., Aagren, M., and Christensen, T. (2012). Cost of self-monitoring of blood glucose in the United States among patients on an insulin regimen for diabetes. *Journal of Managed Care Pharmacy* 18 (1): 21–32.

115 Smith JL. The pursuit of noninvasive glucose: Hunting the deceitful turkey. Revised and Expanded, copyright. 2015.

116 Vashist, S.K. (2012). Non-invasive glucose monitoring technology in diabetes management: a review. *Analytica Chimica Acta* 750: 16–27.

117 Bland, J.M. and Altman, D. (1986). Statistical methods for assessing agreement between two methods of clinical measurement. *The Lancet* 327 (8476): 307–310.

118 Clarke, W.L., Cox, D., Gonder-Frederick, L.A. et al. (1987). Evaluating clinical accuracy of systems for self-monitoring of blood glucose. *Diabetes Care* 10 (5): 622–628.

119 Ishimaru, A. (2017). *Electromagnetic Wave Propagation, Radiation, and Scattering: From Fundamentals to Applications*. John Wiley & Sons.

120 Zhang, R., Liu, S., Jin, H. et al. (2019). Noninvasive electromagnetic wave sensing of glucose. *Sensors* 19 (5): 1151.

121 Wilson, E.B., Decius, J.C., and Cross, P.C. (1955). *Molecular Vibrations: the Theory of Infrared and Raman Vibrational Spectra*. Courier Corporation.

122 Siesler, H.W., Ozaki, Y., Kawata, S., and Heise, H.M. (2008). *Near-Infrared Spectroscopy: Principles, Instruments, Applications*. John Wiley & Sons.

123 Khalil, O.S. (2004). Non-invasive glucose measurement technologies: an update from 1999 to the dawn of the new millennium. *Diabetes Technology & Therapeutics* 6 (5): 660–697.

124 Vasko, P.D., Blackwell, J., and Koenig, J.L. (1971). Infrared and Raman spectroscopy of carbohydrates: Part I: identification of O—H and C—H-related vibrational modes for D-glucose, maltose, cellobiose, and dextran by deuterium-substitution methods. *Carbohydrate Research* 19 (3): 297–310.

125 Tuchin, V. (2008). *Handbook of Optical Sensing of Glucose in Biological Fluids and Tissues*. CRC Press.

126 Pleitez, M.A., Lieblein, T., Bauer, A. et al. (2012). In vivo noninvasive monitoring of glucose concentration in human epidermis by mid-infrared pulsed photoacoustic spectroscopy. *Analytical Chemistry* 85 (2): 1013–1020.

127 Pleitez, M., von Lilienfeld-Toal, H., and Mäntele, W. (2012). Infrared spectroscopic analysis of human interstitial fluid in vitro and in vivo using FT-IR spectroscopy and pulsed quantum cascade lasers (QCL): establishing a new approach to non invasive glucose measurement. *Spectrochimica Acta Part A: Molecular and Biomolecular Spectroscopy* 85 (1): 61–65.

128 Petibois, C., Rigalleau, V., Melin, A.-M. et al. (1999). Determination of glucose in dried serum samples by Fourier-transform infrared spectroscopy. *Clinical Chemistry* 45 (9): 1530–1535.

129 Vasko, P.D., Blackwell, J., and Koenig, J.L. (1972). Infrared and Raman spectroscopy of carbohydrates.: Part II: normal coordinate analysis of α-D-glucose. *Carbohydrate Research* 23 (3): 407–416.

130 Ibrahim, M., Alaam, M., El-Haes, H. et al. (2006). Analysis of the structure and vibrational spectra of glucose and fructose. *Ecletica Quimica* 31 (3): 15–21.

131 Kino, S., Omori, S., Katagiri, T., and Matsuura, Y. (2016). Hollow optical-fiber based infrared spectroscopy for measurement of blood glucose level by using multi-reflection prism. *Biomedical Optics Express* 7 (2): 701–708.

132 Weyer, L. and Workman, J. Jr., (2007). *Practical Guide to Interpretive Near-Infrared Spectroscopy*. CRC Press.

133 Khalil, O.S. (1999). Spectroscopic and clinical aspects of noninvasive glucose measurements. *Clinical Chemistry* 45 (2): 165–177.

134 Yang, W., Liao, N., Cheng, H. et al. (2018). Determination of NIR informative wavebands for transmission non-invasive blood glucose measurement using a Fourier transform spectrometer. *AIP Advances* 8 (3): 35216.

135 Osborne BG, Fearn T, Hindle PH. *Practical NIR Spectroscopy with Applications in Food and Beverage Analysis*. Longman Scientific and Technical; 1993.

136 Šašić, S. and Ozaki, Y. (2001). Short-wave near-infrared spectroscopy of biological fluids. 1. Quantitative analysis of fat, protein, and lactose in raw milk by partial least-squares regression and band assignment. *Analytical Chemistry* 73 (1): 64–71.

137 Golic, M., Walsh, K., and Lawson, P. (2003). Short-wavelength near-infrared spectra of sucrose, glucose, and fructose with respect to sugar concentration and temperature. *Applied Spectroscopy* 57 (2): 139–145.

138 Sim, J.Y., Ahn, C.-G., Jeong, E.-J., and Kim, B.K. (2018). In vivo microscopic photoacoustic spectroscopy for non-invasive glucose monitoring invulnerable to skin secretion products. *Scientific Reports* 8 (1): 1059.

139 Zhang, R., Gao, F., Feng, X. et al. (2017). Noninvasive photoacoustic measurement of glucose by data fusion. *Analyst* 142 (16): 2892–2896.

140 Pai, P.P., Sanki, P.K., Sarangi, S., and Banerjee, S. (2015). Modelling, verification, and calibration of a photoacoustics based continuous non-invasive blood glucose monitoring system. *Review of Scientific Instruments* 86 (6): 64901.

141 Camou, S., Haga, T., Tajima, T., and Tamechika, E. (2012). Detection of aqueous glucose based on a cavity size-and optical-wavelength-independent continuous-wave photoacoustic technique. *Analytical Chemistry* 84 (11): 4718–4724.

142 MacKenzie, H.A., Ashton, H.S., Shen, Y.C. et al. (1998). Blood glucose measurements by photoacoustics. In: *Biomedical Optical Spectroscopy and Diagnostics*, BTuC1. Optical Society of America.

143 Quan, K.M., Christison, G.B., MacKenzie, H.A., and Hodgson, P. (1993). Glucose determination by a pulsed photoacoustic technique: an experimental study using a gelatin-based tissue phantom. *Physics in Medicine and Biology* 38 (12): 1911.

144 Zhao, Z. (2002). *Pulsed Photoacoustic Techniques and Glucose Determination in Human Blood and Tissue*. University of Oulu Finland.

145 Harman-Boehm, I., Gal, A., Raykhman, A.M. et al. (2009). Noninvasive glucose monitoring: a novel approach. *Journal of Diabetes Science and Technology* 3 (2): 253–260.

146 Zhang, R., Luo, Y., Jin, H. et al. (2020). Time-domain photoacoustic waveform analysis for glucose measurement. *Analyst* 145 (24): 7964–7972.

147 Lou, C. and Xing, D. (2010). Photoacoustic measurement of liquid viscosity. *Applied Physics Letters* 96 (21): 211102.

6

Therapeutic Ultrasound

Bar Glickstein, Hila Shinar, and Tali Ilovitsh

Department of Biomedical Engineering, Tel Aviv University, Tel Aviv 6997801, Israel

6.1 Introduction

The use of ultrasound in the field of imaging and diagnostics is well-established and widespread, primarily due to its many advantages in medical imaging: It is noninvasive, safe, easy to operate, cost-effective, and capable of high penetration depths, providing real-time images of deep-seated organs. More recently, ultrasound has started to receive significant attention for its therapeutic potential.

Unlike ultrasonic imaging that aims to probe the tissues that it interacts with, while leaving them unchanged, therapeutic ultrasound aims to generate bioeffects and modify the tissues through various biological mechanisms. Consequently, therapeutic ultrasound often uses higher energy levels compared to imaging applications, which increases the risk of undesired side effects.

Currently, there is an increasing number of ultrasound applications for the treatment of large numbers of diseases and conditions. This was made possible due to a better understanding of the way ultrasound interacts with biological tissues, and due to technological innovations, which allow accurate and localized treatment of the desired areas. Ultrasound treatments are widely used for promoting bone healing, removing kidney stones, and reducing chronic pain. Additional applications, such as the treatment of essential tremor, Parkinson's disease, and prostate tissue ablation, were also recently approved by the United States Food and Drug Administration (FDA). Other exciting applications are constantly emerging.

This chapter explores ultrasound from a therapeutic point of view. First, we describe the potential effects of ultrasound on the human body. The effects are divided into two categories based on their physical nature: mechanical or thermal. Then, we review the current regulations regarding the use of ultrasound for medical treatments. We continue by presenting applications for ultrasound therapy, including FDA-approved applications, as well as applications that are currently being developed in preclinical or clinical trials. Special attention is given to novel technologies and recent improvements to existing treatment techniques. Finally, we discuss future directions and challenges in the field of ultrasound therapy.

Acoustic Technologies in Biology and Medicine, First Edition.
Edited by Adem Ozcelik, Ryan Becker, and Tony Jun Huang.

6.2 Ultrasound-Induced Bioeffects

6.2.1 Introduction

Biological effects induced by ultrasound can be classified into two major physical mechanisms: thermal-based effects and mechanical-based effects. As acoustic waves pass through tissue, they cause an increase in temperature rise through the absorption and conversion of acoustic energy into heat, which can produce a broad range of biological effects, including tissue destruction (necrosis and apoptosis), altered gene expression, and membrane disruption. In addition, deep tissue heating can increase blood circulation in soft tissues, which helps accelerate the body's healing process and decreases pain. Ultrasound-induced mechanical effects cause biological changes that are primarily due to acoustic streaming and cavitation. These mechanical bioeffects result in the production of shear forces, changes in pressure, and the release of various reactive biomolecules. In addition, mechanical bioeffects have been reported to modulate cell membrane properties, alter cellular proliferation, and increase the synthesis of proteins associated with inflammation and injury repair. In general, ultrasound bioeffects are dependent on several factors such as ultrasound exposure parameters, material properties of the tissue, and beam scanning configurations. These factors vary depending on the therapeutic application and the desired outcome.

6.2.2 Thermal Effects

Ultrasound can induce an increase in temperature within biological tissues as its energy is lost to absorption. The thermal effects and changes in cellular activity are strongly dependent on the temperature change. Table 6.1 describes the stages of thermal-induced bioeffects for different temperature ranges.

As the exposure time is increased, there is a corresponding rise in temperature. However, most of the absolute temperature increase occurs during the initial exposure. Different tissues have different thermal absorption coefficients, which indicate the potential change in temperature within the specific tissue.

Table 6.1 Temperature effects.

Temperature range (°C)	Effects
37–39	Extended exposure has no harmful impact.
39–43	Over time, adverse effects may arise.
>41	Fetal issues may arise after extended periods of exposure.
44–46	Protein coagulates.
>45	Enzymes become denatured.
>41.8	Cancer cells fail to reproduce and die.
	This is generally considered a threshold for damage (except eye).

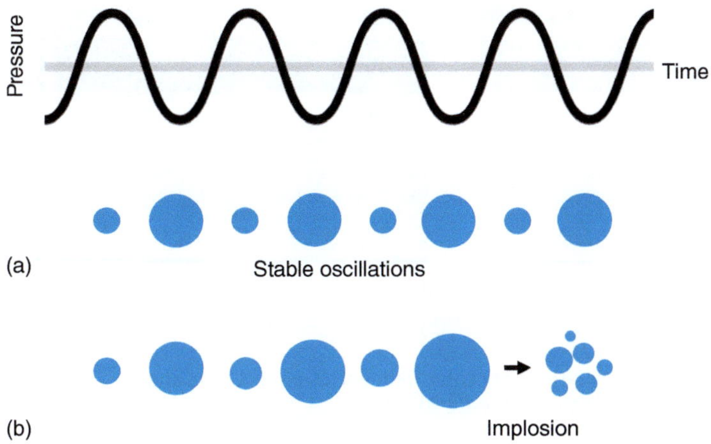

Figure 6.1 Schematic image of acoustic-induced (a) stable and (b) inertial cavitation processes.

6.2.3 Mechanical Effects

6.2.3.1 Cavitation

Cavitation relates the generation of shockwaves to the expansion, contraction, and collapse of bubbles due to the acoustic pressure of the ultrasound beam. There are two major types of cavitation: stable cavitation and inertial cavitation (Figure 6.1). Stable cavitation occurs when the energy of the acoustic wave is absorbed within the bubbles, causing them to oscillate without collapsing, avoiding the production of a shockwave. Inertial cavitation occurs at higher acoustic pressures, where the newly formed bubbles rapidly collapse, which produces a shock wave that can trigger potent mechanical bioeffects such as tissue damage.

6.2.4 Contrast-Enhanced Effects

6.2.4.1 Microbubbles

Microbubbles (MBs) are ultrasound contrast agents that are composed of a gas core and encapsulated by a lipid shell. MBs have an average diameter of 1–7 μm; therefore, they can be safely injected into the bloodstream. The resonance frequency of MBs lies within the 2–15 MHz range, which is within the frequency range that is used for clinical diagnosis. This, combined with the high impedance difference between the MBs and surrounding tissue, causes MB oscillations to produce a strong backscattered echo from an incident ultrasound pulse. In addition to their use as a contrast agent, MB oscillations can also facilitate therapeutic ultrasound applications such as gene-based therapies, blood–brain barrier (BBB) opening, and sonoporation-mediated drug delivery. In the field of high-intensity focused ultrasound (HIFU), studies have combined the use of focused ultrasound together with MBs in order to enhance treatment efficacy and reduce the pressure levels that are required for these processes. More recently, low-frequency ultrasound excitation (below 250 kHz) was shown to cause high amplitude MB oscillation

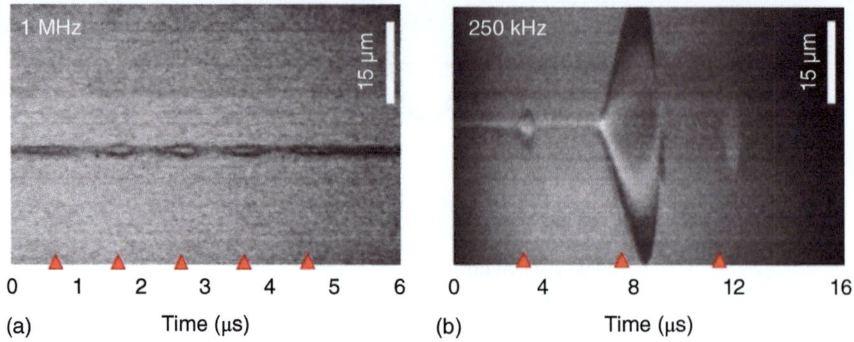

Figure 6.2 Microbubble oscillations at center frequencies of (a) 1 MHz and (b) 250 kHz. Source: Ilovitsh et al. [1], Springer Nature/CC BY 4.0.

at substantially lower pressures compared to higher MHz ultrasonic frequencies (Figure 6.2).

In addition to the enhancement of the MB oscillations, the penetration depth increases at low transmission frequencies due to a reduction in the tissue absorbance at this frequency range, which minimizes attenuation compared to higher frequencies. The focal zone also enlarges under the use of low frequency which helps to treat larger volumes of tissue. By applying low-frequency ultrasound to MBs, it is also possible to induce low-pressure mechanical ablation.

6.2.4.2 Nanobubbles

Nanobubbles (NBs) are submicron bubbles. Their average diameter varies between 100 and 700 nm and is affected by different shells and formulations. The smaller size of the NBs can facilitate their extravasation and accumulation in tumors via an enhanced permeability and retention effect. The use of NBs for ultrasound-mediated applications has been shown in high-contrast ultrasound imaging, molecular imaging, gene delivery, BBB opening, and synergistic thermal HIFU ablation.

6.2.4.3 Nanodroplets

Nanodroplets (NDs) are liquid bubbles that can vaporize into MBs when acoustically activated using ultrasound insonation. When the acoustic pressure increases, the NDs vaporize and undergo a phase change from liquid droplets to gaseous bubbles (ADV, Figure 6.3). Standard NDs have a liquid core stabilized by a shell that can be composed of lipids, polymers, or proteins. Their average diameter ranges from 100 to 750 nm. Like NBs, NDs also have a sufficiently small size for efficient tumor uptake. However, their core is liquid, unlike NBs that have a gaseous core. For this reason, ND applications can be employed for targeted molecular imaging, drug delivery to tumors, gene delivery, and thermal or mechanical ablation. NDs also offer a greater circulatory lifespan compared to standard MBs, thereby increasing the potential applications in which they can be utilized.

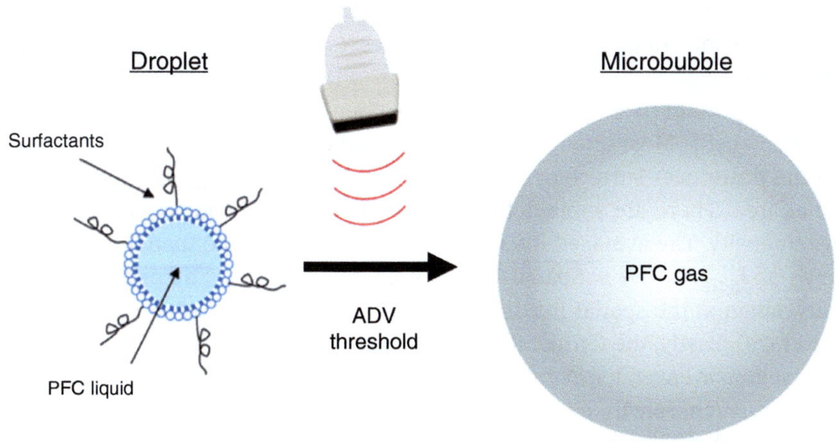

Figure 6.3 Illustration of ultrasound-induced nanodroplet vaporization.

6.2.5 Safety and Regulations

To establish a measure of safety when using ultrasound, two indices have been derived from the two main types of bioeffects described earlier: the thermal index (TI) and the mechanical index (MI). Both indices are unitless parameters that are used for assessing the clinical safety of ultrasound and to give a rough estimate of the safety risk, from either the mechanical or thermal effects that are associated with the insonation at a specific set of parameters.

TI describes the potential temperature rise in the tissue due to ultrasound exposure. The basic definition for *TI* is:

$$TI = \frac{W_0}{W_{DEG}}$$

where W_0 is the source power of the diagnostic ultrasound system, and W_{DEG} is the source power required to increase the tissue temperature by 1 °C. Three different thermal indices have been further developed, aiming to address three specific tissue models: the soft-tissue thermal index (TIS), the bone thermal index (TIB), and the cranial-bone thermal index (TIC). The FDA, which regulates the use of ultrasound devices in the United States of America, defines the maximal allowed value for each TI, based on the heat absorption properties of the relevant tissue. In general, any TI value less than 2 is usually considered safe for extended scanning time durations. TI values greater than 2 are limited in scanning times depending on the value. Additional limitations on the TI values have been added for prenatal examinations.

MI indicates the potential of mechanical effects to occur in the tissue following ultrasound insonation. The MI is defined as the peak negative pressure (PNP) divided by the square root of the center frequency:

$$MI = \frac{PNP}{\sqrt{f}}$$

where the *PNP* is given in MPa, and the frequency *f* is given in MHz.

The general FDA MI limit is 1.9 for diagnostic ultrasound, with some exceptions and specific guidelines for unique cases (such as fetal imaging or eye examination).

Another important parameter that is probed when evaluating ultrasound safety is the beam intensity. Most often, the intensity is indicated either by the total intensity on the transducer surface or by an averaged value over a specific time course or spatial area. There are some definitions commonly used when calculating the average intensity. The averaged intensity over time, taken at the spatial maximum, is known as I_{SPTA} (SPTA = Spatial Peak, Temporal Average). The pulsed-average intensity, taken at the spatial maximum, is defined as I_{SPPA} (SPPA = spatial peak, pulse average). Lastly, the temporal-averaged spatial-averaged intensity is the I_{SATA} (SATA = spatial average, temporal average). Generally speaking, the FDA considers $I_{SPTA} \leq 720\ \text{mW/cm}$ safe for clinical use.

The discussed parameters (the MI, TI, and intensity) introduce universal measurements of ultrasonic output and serve as a valuable tool to estimate the risk for unwanted bioeffects on the body. However, these parameters are not an accurate predictor of the safety of a specific procedure. Therefore, the FDA guidelines require the implementation of the "ALARA" principle. "ALARA" stands for "As Low As Reasonably Achievable": in other words – the acoustic output used in every procedure should be as small as possible. In that manner, the resulting bioeffects are kept minimal.

6.3 Therapeutic Ultrasound Applications

Broadly speaking, therapeutic ultrasound applications can be classified into two major categories: high-intensity and low-intensity ultrasound applications. HIFU treatments utilize mechanical or thermal effects to induce controlled, precise tissue damage for performing noninvasive surgeries using ultrasound. Low-intensity ultrasound, on the other hand, is usually used to stimulate physiological processes [2], such as gene regulation, immunomodulation, and wound healing.

Since the first clinical use of ultrasound in the 1950s for physical therapy [3], a wide variety of therapeutic ultrasound applications have emerged. In this section, we will describe some of the most important, innovative, and exciting therapeutic applications of ultrasound. Due to the great amount of applications in this field, it will not be possible to fully cover every topic; however, there are many great resources that further elaborate on therapeutic ultrasound applications in specific medical areas available in the literature [4–14].

6.3.1 High-Intensity Focused Ultrasound

HIFU is a noninvasive technique in which an ultrasound beam is focused within the body to thermally ablate a local region of interest without damaging any surrounding tissues. HIFU can cause selective tissue necrosis in a well-defined volume, mainly through a combination of heating and cavitation.

To thermally ablate tissue, HIFU treatment parameters are intended to raise the temperature of the selected tissue volume above 55 °C and to maintain this temperature for one second or longer. The heating is very selective and confined to a small region on the order of 1–10 mm. By creating a series of adjacent ablation volumes, a larger area can also be treated. HIFU thermal ablation treatment is usually done via real-time visualization of the treated area, using either magnetic resonance Imaging (MRI) or ultrasound imaging. Each imaging modality has its advantages and disadvantages depending on the area being treated. MRI thermometry can visualize the heated target volume. Usually, a small temperature increase is generated using HIFU for MRI to identify the detection of the affected area inside the body. Then, the transducer position is adjusted until the ultrasound focal plane is at the desired location. High spatial resolution and sensitivity are the main advantages when using MRI guidance. However, HIFU guidance via ultrasound imaging is a low-cost and simple alternative. It provides real-time monitoring, can detect physical parameters of tissues, and obtains feedback about the efficacy of the ablation process. However, the thermally ablated region is not visible using standard imaging unless gas bubbles are utilized with the therapy, which is not always the case. Currently, advanced high-speed ultrasound imaging techniques are being developed to allow ultrasound monitoring of thermal ablation.

Unlike other thermal ablation modalities, such as radiofrequency or laser ablation, HIFU is completely noninvasive. For this reason, HIFU is an attractive modality and has been explored for the treatment of several conditions. For example, HIFU has been FDA-approved for the treatment of uterine leiomyomas and currently in the clinical trial stage for treatment of uterine adenomyosis [15–17]. HIFU is also used for the treatment of benign and malignant solid tumors. Transrectal HIFU treatment of prostate tumors is one of the pilot investigations [18], with both benign prostate hyperplasia and prostate carcinoma being targeted for treatment [19–21]. HIFU for the treatment of liver cancer (mainly for hepatocellular carcinoma) has also been widely investigated [22–24]. In addition, HIFU has been examined as a method of ablation for breast cancer [25]. Further, HIFU has been suggested to treat neurological disorders. MRI-guided HIFU is FDA-approved treatment for essential tremors. Apart from thermal ablation applications, HIFU is also being examined as a therapeutic tool for neuromodulation and drug delivery enhancement.

Although HIFU has shown great potential for noninvasive therapies, there are still several major obstacles that need to be overcome. Firstly, the high-energy acoustic beams that are produced can sometimes burn the tissues that lie between the transducer and the target site instead of the target site alone [26]. Secondly, HIFU treatments can last for a long period of time, with treatment times reaching upward of 6 hours depending on the area being treated [27]. Also, treatments involving the brain are limited by the absorbance of the acoustic waves by the skull, causing heating. In addition, these methods are highly sensitive to patient movement and breathing that might cause undesired damage to healthy tissues that surround the treated area. Still, with improvements in energy delivery schemes and imaging techniques, the safety of HIFU is likely to increase, and its range of applicability will expand.

6.3.2 Histotripsy

In recent years, there has been great interest in developing noninvasive ultrasound tissue ablation techniques via mechanical effects rather than thermal effects. This approach has been defined as histotripsy. The term "histotripsy" is derived from the Greek words "histos" (in Greek: "tissue") and "tripsis" (in Greek: "to crush"), and refers to the disintegration of tissue by short, high-intensity ultrasound pulses. When the acoustic waves are focused within bulk tissue dense, bubble clouds are produced that fractionate and liquefy the tissue. Two different methods are being used in histotripsy to create a pure mechanical damage of soft tissue: cavitation-cloud histotripsy and boiling histotripsy. The first approach, which was pursued by researchers at the University of Michigan, is a highly non-linear and threshold-based phenomenon that uses very short (μs length) ultrasound pulses to generate a dense bubble cloud which collapses and fractionates the tissue. The second technique, developed at the University of Washington, uses longer pulses (ms length) to rapidly generate a mm-sized boiling bubble cloud to mechanically ablate the tissue (Figure 6.4).

The two histotripsy techniques use different mechanisms to achieve tissue ablation. In cavitation-cloud histotripsy the tissue outgases, forming seed bubbles which create a growing cavitation cloud that violently explodes and reforms as it creates more bubbles. Shock waves reflecting from the bubble clouds enhance this chain-reaction effect of bubble creation and inertial cavitation events. As a result of the cavitation cloud cycle effects, the tissue is broken up into sub-micron fragments and becomes liquefied. In boiling histotripsy, heated shock waves are generated, resulting in rapid a temperature increase to over 100 °C and initiation of localized bubbles at the focus which generates a vapor cavity. Shockwave fronts from the rest of the pulse interact with the tissue–vapor cavity interface causing the emission of liquid droplets and tissue micro-fragments that fill the cavity.

Histotripsy has significant potential to be used in many fields as well as in applications where thermal-based ablation is limited, as histotripsy can more effectively damage tissues. Histotripsy has shown great potential for the treatment of tumors

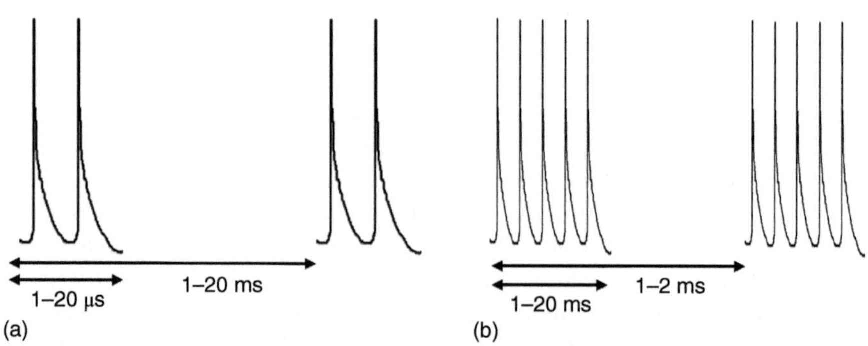

Figure 6.4 Different pulsing schemes used in (a) cavitation cloud histotripsy and (b) boiling histotripsy approaches.

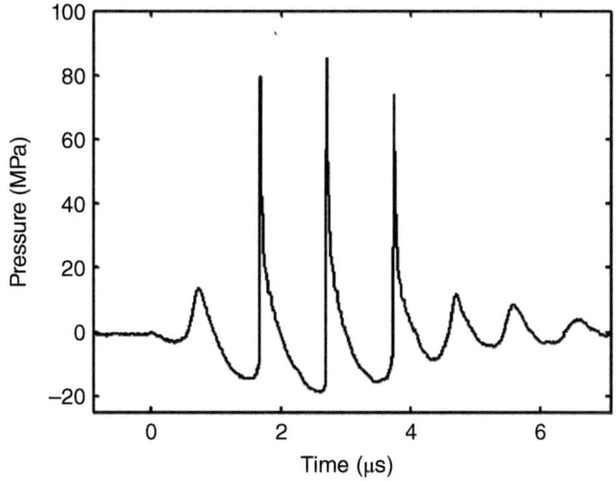

Figure 6.5 Typical histotripsy waveform. Source: Maxwell et al. [42]/With permission of Acoustical Society of America.

in the prostate [28–30], liver [31, 32], kidney [33–35], thrombolysis [36–38], kidney stone erosion [39], and neurological diseases [40].

Histotripsy methods require the use of high pressures for the generation of cavitation nuclei (around 20 MPa [41], Figure 6.5) which raises some safety concerns. One major concern is the possibility of off-target effects. The US pulses need to be precisely focused on the target area in order to avoid cavitation in healthy neighboring cells. Off-target damage following histotripsy of hepatocellular carcinoma was reported in vivo [43]. Another technological challenge is associated with the fabrication of high-intensity focused transducers [44]. For reducing the high pressures that are used for the process, histotripsy can be combined with MBs, NBs, or NDs [32, 45–48]. The use of ultrasound contrast agents both reduces the required pressure and limits the treatment area only to the location where the MBs, NBs, or NDs reside. For further reducing the pressures, the combination of low-frequency ultrasound and contrast agents was suggested, in order to take advantage of the high amplitude oscillations of contrast agents.

The use of targeted MBs coupled with low-frequency excitation to achieve low-energy tumor cell fractionation has been demonstrated in vitro and in vivo [49]. In cell cultures, cell viability was reduced to 16% at a PNP of 800 kPa at the 250 kHz frequency (MI of 1.6) and to 10% at a PNP of 250 kPa at a frequency of 80 kHz (MI of 0.9). In vivo, following intratumoral injection of targeted MBs into tumor-bearing mice and coupled with low-frequency ultrasound, significant tumor debulking and cancer cell death were observed (Figure 6.6).

However, MBs can only be injected locally into the tumor tissue. Scaling down the size of MB contrast agents to the nanometer level shows significant potential for noninvasive cancer therapy.

The usage of NBs as a noninvasive alternative to MBs was recently reported [45]. This study shows that coupled with low-frequency ultrasound insonation, NBs can be used to trigger mechanical effects in tumors following a systemic injection. In vitro, using breast cancer tumor cells the cell viability was reduced to ~17% of live

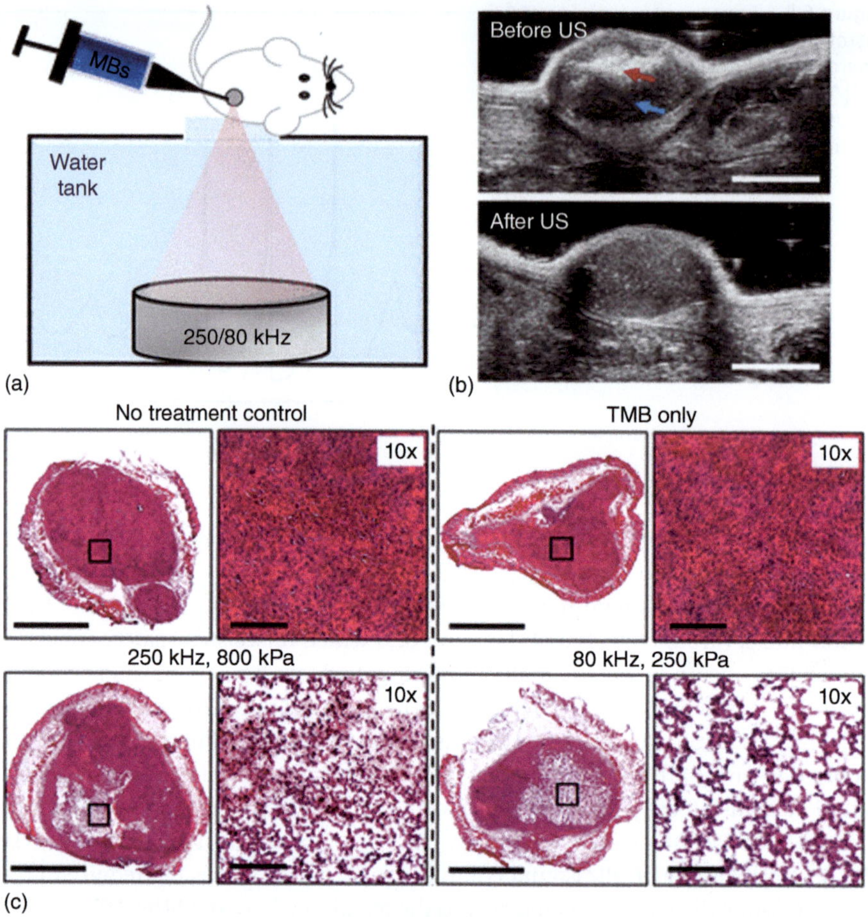

Figure 6.6 Tumor-targeted microbubbles combined with low-frequency ultrasound generate mechanical damage in vivo. (a) Targeted microbubbles (TMB) are intratumorally injected, followed by low-frequency ultrasound application in a custom-made setup. (b) Representative ultrasound images before and after therapeutic ultrasound application. TMBs location and dark shade generated by the TMBs are marked with red and blue arrows, respectively. Scale bar is 4 mm. (c) Histological photomicrographs for no treatment control (NTC), only TMB, and directly treated ultrasound + TMB tumors treated with 250 or 80 kHz US. Source: Bismuth et al. [49], American Chemical Society/CC BY 4.0.

cells. In vivo effective mechanical tumor ablation and tumor tissue fractionation were observed.

Another alternative utilizes NDs. A two-step approach to achieve ND-mediated histotripsy was recently developed [50]. This technique involves the vaporization of the NDs into gaseous MBs, followed by the application of low-frequency ultrasound to implode the MBs and trigger potent mechanical effects in the surrounding tissues. This method generated mechanical lesions in ex vivo chicken liver samples (Figure 6.7).

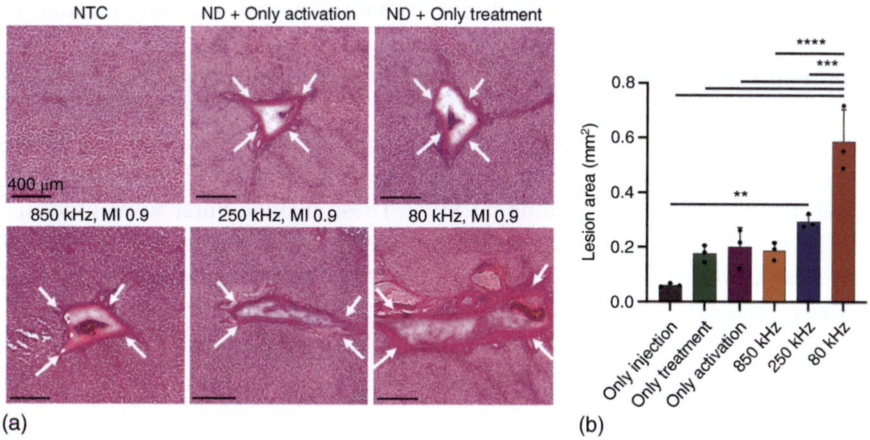

Figure 6.7 Nanodroplet-mediated low-energy histotripsy generates mechanical lesions in ex vivo samples. (a) Histological photomicrographs of no treatment control (NTC), ND + only activation control, ND + only treatment control, ND-mediated histotripsy for ND vaporization followed by low-frequency treatment at an MI = 0.9 and center frequencies of 850, 250, and 80 kHz. (b) Quantification of the lesion area for each group. All experiments were performed in triplicate. Adjusted p values were $^{**}p < 0.01$, $^{***}p < 0.001$, and $^{****}p < 0.0001$. All data are plotted as the mean ± SD. Source: Glickstein et al. [50], Springer Nature/CC BY 4.0.

The development of an ultrasound-guided focused ultrasound system for 3D volumetric ND-mediated histotripsy has been presented in order to enhance the effectiveness of the two-step approach and enable the creation of extensive mechanical damage in tissues [51]. This hybrid system consisted of a rotating imaging transducer, controlled by a motorized rotary, located within a low-frequency therapeutic transducer. By utilizing the rotating imaging transducer, volumetric activation of the NDs can be achieved, leading to the formation of a circular-shaped lesion that makes it possible to treat larger volumes.

These studies suggest that reducing the center frequency enhances the resulting mechanical damage when combining histotripsy with ultrasound contrast agents, thus providing a unique theranostic platform for safe low-energy tumor fractionation, while reducing off-target effects.

6.3.3 Shock Wave Lithotripsy

Shock wave lithotripsy (SWL) is a technique for the fragmentation of kidney stones in the urinary tract. In SWL treatment, stones are disintegrated by the application of short pulses, high-amplitude pressure waves, and acoustic shock waves. The procedure is usually noninvasive, using a lithotripter, which is a device that produces and focuses the acoustic waves. The device is positioned outside the patient's body and is acoustically coupled to them using a water bath. Repeated shock waves hit the stones inside the kidney until they break into small fragments which can be more easily removed or even naturally leave the body through the urinary system.

Current research on SWL focuses on improving the success rate of lithotripsy treatment while reducing the potential risks. Few studies have investigated the influence of different treatment parameters, such as focal zone size, shock wave delivery rate, and power amplitude, on the patient outcomes. Other studies suggest novel alternatives for the traditional SWL. These include burst wave lithotripsy (BWL), which uses continuous ultrasound waves with lower pressure amplitude, and MBs lithotripsy, which is based on the attachment of gas-cored MBs to the stones to enhance the efficiency of the procedure [52].

6.3.4 Drug Delivery and Gene Therapy

By design, cell membranes often prevent large molecules from entering the cells. However, ultrasound waves can modify the permeability of cell membranes through cavitation to allow the delivery of non-permeable extracellular molecules, including drugs, proteins, genes, and a variety of therapeutic agents. This effect, known as sonoporation, provides a targeted delivery platform since the change in the membrane permeability occurs only at the focus area of the transducer. Most sonoporation methods utilize MB contrast agents as they can reduce the cavitation threshold. The use of cell-targeted MBs further enhances the sonoporation effect since the proximity between the adherent MB and the cell surface improves the efficacy of pore creation. Sonoporation efficacy depends mostly on the cell type, the size of the generated pore, and the delivered particle size. The generated pore size is highly correlated with the acoustic parameters and can vary using different acoustic pressures, durations, and pulse repetition frequencies.

Both inertial cavitation and stable cavitation mechanisms can generate pores in the membrane. In inertial cavitation, the MBs collapse, generating a strong jetting flow that can rupture the membrane and cause an intense fluid streaming that results in the perforation of the cell membrane. These changes may be non-recoverable and lead to cell death. Alternatively, stable cavitation can be used, which relies on MB oscillations to form pores. Other mechanisms such as radiation forces and streaming can add to the transfer of process.

The PNP that is used during the process depends on the mechanism that is required and the type of cells that are used. MB inertial cavitation occurs at different PNPs for different frequencies. As mentioned previously, MB oscillations were assumed to be enhanced at their resonance frequency which is typically above 1 MHz. Hence, most MB-mediated sonoporation studies apply insonation at frequencies around 1 MHz [53–55]. However, lower frequencies were shown to enhance MBs oscillations which can increase sonoporation efficacy [1, 56] and allow the transfer of large molecules.

A recent study suggests the use of low-frequency MB-mediated sonoporation to promote the delivery of large molecules [57]. An optimization of the delivery of four fluorescent molecules (ranging from 1.2 to 70 kDa) to breast cancer cells was performed. An optimal PNP of 500 kPa (center frequency of 250 kHz) was most effective

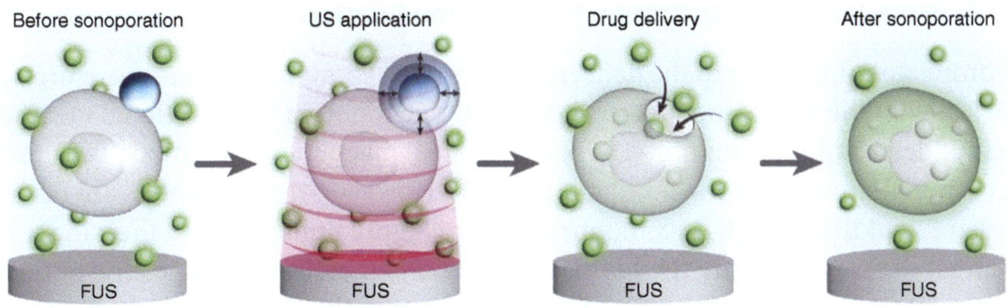

Figure 6.8 Targeted microbubble-mediated sonoporation of cancer cells illustration. Source: Eck et al. [57]/Elsevier/Licensed under CC BY 4.0.

for promoting cell uptake of the molecules. For smaller molecules (1.2 and 4 kDa), ultrasound in combination with MBs resulted in a higher fraction of fluorescent cells (58% and 29%, respectively) compared to larger molecules (20 and 70 kDa) which had lower fractions (10% and 5%, respectively). This study shows that low-frequency insonation of MBs induced high amplitude oscillation in vitro, promoting uptake of large molecules (Figure 6.8).

Ultrasound-mediated drug and gene delivery has shown significant advantages in increasing treatment specificity and efficacy for treatments of a wide range of conditions. However, sonoporation still faces many challenges that need to be addressed in order to ensure proper safety for clinical translation and therefore requires further research and optimization.

6.3.5 Blood–Brain Barrier Opening

The BBB is a unique formation that selectively restricts the exchange of substances between the circulatory system and the brain parenchyma. The BBB plays a critical role in preventing many pathogens and toxic molecules from entering the brain. However, it also poses serious limitations for neurological treatments as it prevents most drugs from reaching the brain, excluding drugs that measure 0.4 kDa or more. In recent years, focused ultrasound has emerged as a noninvasive therapeutic approach that enables the treatment of the brain at low frequencies (below 650 kHz) that can penetrate an intact human skull [1]. When low-energy ultrasound is combined with intravenously injected MBs, they act as efficient theranostic probes that concurrently serve as contrast agents and therapeutic agents, opening the BBB in a noninvasive and localized manner and enabling drug delivery via a variety of mechanisms, including transport through cells and gap junctions to cross from the bloodstream to the brain [58]. Currently, FUS-mediated BBB clinical trials are ongoing for a variety of brain diseases. Several studies have shown the potential of this method for the treatment of Parkinson's disease [59], Alzheimer's disease [60], brain tumors [61, 62], and Huntington's disease [63, 64]. Most methods use MRI-guided FUS systems that allow physicians to localize, monitor, and target the BBB in real time. After treatment, the BBB can heal the microtears that were formed after a certain period of time, the length of which depends on the insonation parameters.

6.3.6 Low-Intensity Ultrasound for Neuromodulation

Neuromodulation refers to the delivery of a targeted stimulus to a specific neurological site with the aim of modulating nervous tissue function. Conventional methods include deep brain stimulation (DBS), transcranial magnetic stimulation (TMS), and transcranial direct current stimulation (tDCS). However, these methods are mostly invasive, and the ones that are noninvasive are highly limited due to their inability to target deep areas inside the brain. Ultrasound is therefore an appealing alternative. It is noninvasive, and yet allows deep penetration through the human skull and inside the brain. Moreover, studies show that the use of MRI guidance for ultrasound therapy results in highly accurate outcomes.

Low-intensity focused ultrasound (LIFU) has been shown to be capable of reversibly modulating region-specific brain function and can potentially be used to enhance or suppress neuronal activity without undesired side effects. Several studies have shown increased and decreased neuronal firing rates [65, 66], the suppression of neural activity [67–69], and variable effects on the peripheral nervous system [70] when LIFU is applied. Although LIFU is a promising tool for neuromodulation, the underlying mechanisms that govern acoustic-mediated neuromodulation are poorly understood and require further research before this technology is developed to the point of feasible clinical use.

6.3.7 Bone Healing

Low-intensity pulsed ultrasound (LIPUS) has been used clinically to promote bone healing for over 20 years. The procedure uses a 1.5 MHz ultrasound probe at intensity of $I_{SATA} = 30\ \text{mW/cm}^2$ to stimulate bone formation.

Unlike the surgical alternative, LIPUS is a noninvasive, low-cost procedure, and it has been shown to have no posttreatment complications. Still, the efficiency of LIPUS for bone repair remains controversial. While many studies have established that the treatment increases healing rates and accelerates healing times for several types of bone fractures (fresh fractures, delayed fractures, and nonunions), the data regarding other types of fractures (such as acute fractures and deep-bone fractures) are inconclusive or inadequate. The mechanisms underlying the basis of LIPUS treatment also need to be further investigated. Studies associate the beneficial effects of LIPUS on bone and tissue healing with several biological processes. For instance, ultrasound has been shown to stimulate cell differentiation, promote cell proliferation, increase local blood flow, and inhibit inflammatory responses [2, 71, 72].

6.3.8 Sonothrombolysis

Blood clots are still a major cause of concern in modern medicine. The available treatment options, usually surgery or medications, may help some patients, yet the risk for complications remains high. Recently, there has been increasing evidence that suggests ultrasound can be used to improve the prognosis of patients with blood clots, using a minimally invasive technique known as sonothrombolysis.

Sonothrombolysis refers to the use of ultrasound for the destruction of blood clots. While the mechanisms underlying this technique are not fully understood, studies have shown that ultrasound-induced mechanical effects are mainly responsible for the lysis of blood clots. An ultrasound beam of high intensity is thereby used to generate acoustic radiation forces, which physically break down the thrombus. When contrast agents (such as MBs) are used alongside the ultrasound during a sonothrombolysis procedure, acoustic cavitation is also produced, and it increases the forces applied to the clot surface, thus enhancing the destruction effect. Additionally, studies show that ultrasound treatment of blood clots allows better penetration of thrombolytic enzymes into the clots [73–75]. Recent work has been dedicated to the optimization of sonothrombolysis treatment parameters, including ultrasound power, frequency, and duty cycle, and the type and concentration of the contrast agents used [74].

6.3.9 Other Applications

Ultrasound has also been found to be beneficial for some cosmetic applications, due to its ability to penetrate through the skin and focus on a specific site with minimal damage to the surrounding tissues. Common cosmetic applications of ultrasound include fat reduction, skin tightening, and scar reduction.

Ultrasound is also used for several physical therapy applications. For instance, low-intensity ultrasound is used for chronic pain management. It is assumed that the heat produced by ultrasound is mainly responsible for this therapeutic effect. Additional applications in this field include treatments for arthritis due to the ability of LIPUS to inhibit inflammatory responses, and for osteoporosis, as LIPUS can be used to increase bone mineral density. Many additional applications for therapeutic ultrasound are currently being explored, such as cataract removal, acceleration of wound healing, and tissue regeneration.

6.4 Conclusions

Therapeutic ultrasound is a rapidly advancing field with a vast range of potential medical applications. Depending on the energy level used, it can induce various bioeffects, including cell death, ablation, and physiological and cellular effects. This chapter explores the physical properties of ultrasound and its induced bioeffects, as well as innovative applications such as tumor ablation, brain therapy, and drug and gene delivery. The use of contrast agents to enhance ablation therapies and potentially lower the pressure threshold is highlighted, along with new directions that combine low-frequency ultrasound to further maximize the bioeffects generated by these agents. In mice, this concept was shown to reduce the pressure threshold by two orders of magnitude. Furthermore, the chapter presents emerging applications and FDA-approved treatments, with promising results from recent clinical trials. The dual functionality of ultrasound as both an imaging and therapeutic tool makes it an attractive technique for developing new treatment methods. Future progress will be further accelerated as breakthrough discoveries provide a greater understanding of the underlying mechanisms and new biological effects are produced, advancing the potential of ultrasound as an effective therapeutic tool.

References

1 Ilovitsh, T., Ilovitsh, A., Foiret, J. et al. (2018). Enhanced microbubble contrast agent oscillation following 250 kHz insonation. *Scientific Reports* 8 (1): 1–15. https://doi.org/10.1038/s41598-018-34494-5.

2 de Lucas, B., Pérez, L.M., Bernal, A., and Gálvez, B.G. (2020). Ultrasound therapy: experiences and perspectives for regenerative medicine. *Genes* 11 (9): 1086. https://doi.org/10.3390/GENES11091086.

3 Miller, D.L., Smith, N.B., Bailey, M.R. et al. (2012). Overview of therapeutic ultrasound applications and safety considerations. *Journal of Ultrasound in Medicine: Official Journal of the American Institute of Ultrasound in Medicine* 31 (4): 623. https://doi.org/10.7863/JUM.2012.31.4.623.

4 Awad, N.S., Paul, V., Alsawaftah, N.M. et al. (2021). Ultrasound-responsive nanocarriers in cancer treatment: a review. *ACS Pharmacology & Translational Science* 2021: 612. https://doi.org/10.1021/acsptsci.0c00212.

5 Bader, K.B., Makin, I.R.S., and Abramowicz, J.S. (2022). Ultrasound for aesthetic applications. *Journal of Ultrasound in Medicine* 41 (7): 1597–1607. https://doi.org/10.1002/JUM.15856.

6 di Biase, L., Falato, E., Caminiti, M.L. et al. (2021). *Focused ultrasound (FUS) for chronic pain management: approved and potential applications. Neurology Research International* 2021: https://doi.org/10.1155/2021/8438498.

7 ter Haar, G. and Mouratidis, P. (2017). The immunological effects of therapeutic ultrasound. *Ultrasound in Medicine & Biology* 43: S233. https://doi.org/10.1016/J.ULTRASMEDBIO.2017.08.1813.

8 Kim, C., Lim, M., Woodworth, G.F., and Arvanitis, C.D. (2022). The roles of thermal and mechanical stress in focused ultrasound-mediated immunomodulation and immunotherapy for central nervous system tumors. *Journal of Neuro-Oncology* 157: 221–236. https://doi.org/10.1007/s11060-022-03973-1.

9 Kotronis, G. and Vas, P.R.J. (2020). Ultrasound devices to treat chronic wounds: the current level of evidence. *The International Journal of Lower Extremity Wounds (IJLEW)* 19 (4): 341–349. https://doi.org/10.1177/1534734620946660.

10 Leinenga, G., Langton, C., Nisbet, R., and Götz, J. (2016). Ultrasound treatment of neurological diseases — current and emerging applications. *Nature Reviews Neurology* 12 (3): 161–174. https://doi.org/10.1038/nrneurol.2016.13.

11 Mei, L. and Zhang, Z. (2021). Advances in biological application of and research on low-frequency ultrasound. *Ultrasound in Medicine & Biology* 47 (10): 2839–2852. https://doi.org/10.1016/J.ULTRASMEDBIO.2021.06.005.

12 Palanisamy, P., Alam, M., Li, S. et al. (2022). Low-intensity pulsed ultrasound stimulation for bone fractures healing: a review. *Journal of Ultrasound in Medicine* 41 (3): 547–563. https://doi.org/10.1002/JUM.15738.

13 Peng, C., Chen, M., Spicer, J.B., and Jiang, X. (2021). Acoustics at the nanoscale (nanoacoustics): a comprehensive literature review. Part II: nanoacoustics for biomedical imaging and therapy. *Sensors and Actuators A: Physical* 332: 112925. https://doi.org/10.1016/J.SNA.2021.112925.

14 Tachibana, K. (2013). Emerging technologies in therapeutic ultrasound: thermal ablation to gene delivery. *Human Cell* 17 (1): 7–15. https://doi.org/10.1111/J.1749-0774.2004.TB00015.X.

15 Li, W., Jiang, Z., Deng, X., and Xu, D. (2020). Long-term follow-up outcome and reintervention analysis of ultrasound-guided high intensity focused ultrasound treatment for uterine fibroids. *International Journal of Hyperthermia : The Official Journal of European Society for Hyperthermic Oncology, North American Hyperthermia Group* 37 (1): 1046–1051. https://doi.org/10.1080/02656736.2020.1807617.

16 Marques, A.L.S., Andres, M.P., Kho, R.M., and Abrão, M.S. (2020). Is high-intensity focused ultrasound effective for the treatment of adenomyosis? A systematic review and meta-analysis. *Journal of Minimally Invasive Gynecology* 27 (2): 332–343. https://doi.org/https://doi.org/10.1016/j.jmig.2019.07.029.

17 Shui, L., Mao, S., Wu, Q. et al. (2015). High-intensity focused ultrasound (HIFU) for adenomyosis: two-year follow-up results. *Ultrasonics Sonochemistry* 27: 677–681. https://doi.org/10.1016/j.ultsonch.2015.05.024.

18 Zhou, Y.-F. (2011). High intensity focused ultrasound in clinical tumor ablation. *World Journal of Clinical Oncology* 2 (1): 8. https://doi.org/10.5306/wjco.v2.i1.8.

19 Blana, A., Walter, B., Rogenhofer, S., and Wieland, W.F. (2004). High-intensity focused ultrasound for the treatment of localized prostate cancer: 5-year experience. *Urology* 63 (2): 297–300. https://doi.org/https://doi.org/10.1016/j.urology.2003.09.020.

20 Luis, A.A., Samuel, P., Atsuko, I. et al. (2020). High intensity focused ultrasound hemigland ablation for prostate cancer: initial outcomes of a United States series. *Journal of Urology* 204 (4): 741–747. https://doi.org/10.1097/JU.0000000000001126.

21 Napoli, A., Alfieri, G., Scipione, R. et al. (2020). High-intensity focused ultrasound for prostate cancer. *Expert Review of Medical Devices* 17 (5): 427–433. https://doi.org/10.1080/17434440.2020.1755258.

22 Cheung, T.T., Fan, S.T., Chan, S.C. et al. (2013). High-intensity focused ultrasound ablation: an effective bridging therapy for hepatocellular carcinoma patients. *World Journal of Gastroenterology* 19 (20): 3083–3089. https://doi.org/10.3748/wjg.v19.i20.3083.

23 Wu, F., Wang, Z.-B., Chen, W.-Z. et al. (2004). Extracorporeal high intensity focused ultrasound ablation in the treatment of patients with large hepatocellular carcinoma. *Annals of Surgical Oncology* 11 (12): 1061–1069. https://doi.org/10.1245/ASO.2004.02.026.

24 Zhang, L., Zhu, H., Jin, C. et al. (2009). High-intensity focused ultrasound (HIFU): effective and safe therapy for hepatocellular carcinoma adjacent to major hepatic veins. *European Radiology* 19 (2): 437–445. https://doi.org/10.1007/s00330-008-1137-0.

25 Peek, M.C.L. and Douek, M. (2017). Ablative techniques for the treatment of benign and malignant breast tumours. *Journal of Therapeutic Ultrasound* 5 (1): 18. https://doi.org/10.1186/s40349-017-0097-8.

26 Kennedy, J.E., Ter Haar, G.R., and Cranston, D. (2003). High intensity focused ultrasound: surgery of the future? *The British Journal of Radiology* 76 (909): 590–599. https://doi.org/10.1259/bjr/17150274.

27 Kennedy, J.E. (2005). High-intensity focused ultrasound in the treatment of solid tumours. *Nature Reviews Cancer* 5 (4): 321–327. https://doi.org/10.1038/nrc1591.

28 Bax, J., Cool, D., Gardi, L. et al. (2008). Mechanically assisted 3D ultrasound guided prostate biopsy system. *Medical Physics* 35 (12): 5397–5410. https://doi.org/10.1118/1.3002415.

29 Dubinsky, T.J., Khokhlova, T.D., Khokhlova, V., and Schade, G.R. (2020). Histotripsy: the next generation of high-intensity focused ultrasound for focal prostate cancer therapy. *Journal of Ultrasound in Medicine* 39 (6): 1057–1067. https://doi.org/https://doi.org/10.1002/jum.15191.

30 Schade, G.R., Hall, T.L., and Roberts, W.W. (2012). Urethral-sparing histotripsy of the prostate in a canine model. *Urology* 80 (3): 730–735. https://doi.org/10.1016/j.urology.2012.05.027.

31 Khokhlova, T.D., Wang, Y.-N., Simon, J.C. et al. (2014). Ultrasound-guided tissue fractionation by high intensity focused ultrasound in an in vivo porcine liver model. *Proceedings of the National Academy of Sciences of the United States of America* 111 (22): 8161–8166. https://doi.org/10.1073/pnas.1318355111.

32 Vlaisavljevich, E., Durmaz, Y.Y., Maxwell, A. et al. (2013). Nanodroplet-mediated histotripsy for image-guided targeted ultrasound cell ablation. *Theranostics* 3 (11): 851–864. https://doi.org/10.7150/thno.6717.

33 Khokhlova, T.D., Schade, G.R., Wang, Y.-N. et al. (2019). Pilot in vivo studies on transcutaneous boiling histotripsy in porcine liver and kidney. *Scientific Reports* 9 (1): 20176. https://doi.org/10.1038/s41598-019-56658-7.

34 Schade, G.R., Wang, Y.-N., D'Andrea, S. et al. (2019). Boiling histotripsy ablation of renal cell carcinoma in the Eker rat promotes a systemic inflammatory response. *Ultrasound in Medicine & Biology* 45 (1): 137–147. https://doi.org/https://doi.org/10.1016/j.ultrasmedbio.2018.09.006.

35 Styn, N.R., Hall, T.L., Fowlkes, J.B. et al. (2012). Histotripsy of renal implanted VX-2 tumor in a rabbit model: investigation of metastases. *Urology* 80 (3): 724–729. https://doi.org/https://doi.org/10.1016/j.urology.2012.06.020.

36 Bollen, V., Hendley, S.A., Paul, J.D. et al. (2020). In vitro thrombolytic efficacy of single- and five-cycle histotripsy pulses and RT-PA. *Ultrasound in Medicine & Biology* 46 (2): 336–349. https://doi.org/https://doi.org/10.1016/j.ultrasmedbio.2019.10.009.

37 Maxwell, A.D., Cain, C.A., Duryea, A.P. et al. (2009). Noninvasive thrombolysis using pulsed ultrasound cavitation therapy - histotripsy. *Ultrasound in Medicine and Biology* 35 (12): 1982–1994. https://doi.org/10.1016/j.ultrasmedbio.2009.07.001.

38 Zhang, X., Owens, G.E., Gurm, H.S. et al. (2015). Noninvasive thrombolysis using histotripsy beyond the intrinsic threshold (microtripsy). *IEEE Transactions on Ultrasonics, Ferroelectrics, and Frequency Control* 62 (7): 1342–1355. https://doi.org/10.1109/TUFFC.2015.007016.

39 Duryea, A.P., Hall, T.L., Maxwell, A.D. et al. (2010). Histotripsy erosion of model urinary calculi. *Journal of Endourology* 25 (2): 341–344. https://doi.org/10.1089/end.2010.0407.

40 Sukovich, J.R., Cain, C.A., Pandey, A.S. et al. (2019). In vivo histotripsy brain treatment. *Journal of Neurosurgery JNS* 131 (4): 1331–1338. https://doi.org/10.3171/2018.4.JNS172652.

41 Vlaisavljevich, E., Maxwell, A., Warnez, M. et al. (2014). Histotripsy-induced cavitation cloud initiation thresholds in tissues of different mechanical properties. *IEEE Transactions on Ultrasonics, Ferroelectrics, and Frequency Control* 61 (2): 341–352. https://doi.org/10.1109/TUFFC.2014.6722618.

42 Maxwell, A.D., Wang, T.-Y., Cain, C.A. et al. (2011). Cavitation clouds created by shock scattering from bubbles during histotripsy. *The Journal of the Acoustical Society of America* 130 (4): 1888–1898. https://doi.org/10.1121/1.3625239.

43 Khokhlova, V.A., Fowlkes, J.B., Roberts, W.W. et al. (2015). Histotripsy methods in mechanical disintegration of tissue: towards clinical applications. *International Journal of Hyperthermia* 31 (2): 145–162. Informa Healthcare. https://doi.org/10.3109/02656736.2015.1007538.

44 Bawiec, C.R., Khokhlova, T.D., Sapozhnikov, O.A. et al. (2021). A prototype therapy system for boiling histotripsy in abdominal targets based on a 256-element spiral array. *IEEE Transactions on Ultrasonics, Ferroelectrics, and Frequency Control* 68 (5): 1496–1510. https://doi.org/10.1109/TUFFC.2020.3036580.

45 Bismuth, M., Katz, S., Mano, T. et al. (2022). Low frequency nanobubble-enhanced ultrasound mechanotherapy for noninvasive cancer surgery. *Nanoscale* 14: 13614–13627. https://doi.org/10.1039/d2nr01367c.

46 Burke, C., Klibanov, A., Sheehan, J., and Price, R. (2011). Inhibition of glioma growth by microbubble activation in a subcutaneous model using low duty cycle ultrasound without significant heating: laboratory investigation. *Journal of Neurosurgery* 114: 1654–1661. https://doi.org/10.3171/2010.11.JNS101201.

47 Tran, B.C., Seo, J., Hall, T.L. et al. (2003). Microbubble-enhanced cavitation for noninvasive ultrasound surgery. *IEEE Transactions on Ultrasonics, Ferroelectrics, and Frequency Control* 50 (10): 1296–1304. https://doi.org/10.1109/TUFFC.2003.1244746.

48 Vykhodtseva, N., McDannold, N., and Hynynen, K. (2004). The use of Optison to reduce the power requirements for focused ultrasound lesion production in the brain - an MRI/histology study in rabbits. *IEEE Ultrasonics Symposium* 2004 (2): 1009–1012. https://doi.org/10.1109/ULTSYM.2004.1417938.

49 Bismuth, M., Katz, S., Rosenblatt, H. et al. (2021). Acoustically detonated microbubbles coupled with low frequency insonation: multiparameter evaluation of low energy mechanical ablation. *Bioconjugate Chemistry* 33: 1069–1071. https://doi.org/10.1021/acs.bioconjchem.1c00203.

50 Glickstein, B., Levron, M., Shitrit, S. et al. (2022). Nanodroplet-mediated low-energy mechanical ultrasound surgery. *Ultrasound in Medicine & Biology* 48 (7): 1229–1239. https://doi.org/https://doi.org/10.1016/j.ultrasmedbio.2022.02.018.

51 Glickstein, B., Aronovich, R., Feng, Y., and Ilovitsh, T. (2022). Development of an ultrasound guided focused ultrasound system for 3D volumetric low energy nanodroplet-mediated histotripsy. *Scientific Reports* 12 (1): 20664. https://doi.org/10.1038/s41598-022-25129-x.

52 Large, T. and Krambeck, A.E. (2019). Emerging technologies in lithotripsy. *Urologic Clinics of North America* 46 (2): 215–223. https://doi.org/10.1016/j.ucl.2018.12.012.

53 Telichko, A.V., Wang, H., Bachawal, S. et al. (2021). Therapeutic ultrasound parameter optimization for drug delivery applied to a murine model of hepatocellular carcinoma. *Ultrasound in Medicine & Biology* 47 (2): 309–322. https://doi.org/10.1016/j.ultrasmedbio.2020.09.009.

54 Wang, M., Zhang, Y., Cai, C. et al. (2018). Sonoporation-induced cell membrane permeabilization and cytoskeleton disassembly at varied acoustic and microbubble-cell parameters. *Scientific Reports* 8 (1): 3885. https://doi.org/10.1038/s41598-018-22056-8.

55 Yang, Y., Li, Q., Guo, X. et al. (2020). Mechanisms underlying sonoporation: Interaction between microbubbles and cells. *Ultrasonics Sonochemistry* 67: 105096. https://doi.org/https://doi.org/10.1016/j.ultsonch.2020.105096.

56 Ilovitsh, T., Feng, Y., Foiret, J. et al. (2020). Low-frequency ultrasound-mediated cytokine transfection enhances T cell recruitment at local and distant tumor sites. *Proceedings of the National Academy of Sciences* 117 (23): 12674–12685. https://doi.org/10.1073/pnas.1914906117.

57 Eck, M., Aronovich, R., and Ilovitsh, T. (2022). Efficacy optimization of low frequency microbubble-mediated sonoporation as a drug delivery platform to cancer cells. *International Journal of Pharmaceutics: X* 4: 100132. https://doi.org/https://doi.org/10.1016/j.ijpx.2022.100132.

58 Tung, Y.-S., Vlachos, F., Feshitan, J.A. et al. (2011). The mechanism of interaction between focused ultrasound and microbubbles in blood-brain barrier opening in mice. *The Journal of the Acoustical Society of America* 130 (5): 3059–3067. https://doi.org/10.1121/1.3646905.

59 Gasca-Salas, C., Fernández-Rodríguez, B., Pineda-Pardo, J.A. et al. (2021). Blood-brain barrier opening with focused ultrasound in Parkinson's disease dementia. *Nature Communications* 12 (1): 779. https://doi.org/10.1038/s41467-021-21022-9.

60 Lipsman, N., Meng, Y., Bethune, A.J. et al. (2018). Blood–brain barrier opening in Alzheimer's disease using MR-guided focused ultrasound. *Nature Communications* 9 (1): 2336. https://doi.org/10.1038/s41467-018-04529-6.

61 Chen, K.-T., Lin, Y.-J., Chai, W.-Y. et al. (2020). Neuronavigation-guided focused ultrasound (NaviFUS) for transcranial blood-brain barrier opening in recurrent glioblastoma patients: clinical trial protocol. *Annals of Translational Medicine* 8 (11): 673. https://doi.org/10.21037/atm-20-344.

62 Mainprize, T., Lipsman, N., Huang, Y. et al. (2019). Blood-brain barrier opening in primary brain tumors with non-invasive MR-guided focused ultrasound: a clinical safety and feasibility study. *Scientific Reports* 9 (1): 321. https://doi.org/10.1038/s41598-018-36340-0.

63 Burgess, A., Huang, Y., Querbes, W. et al. (2012). Focused ultrasound for targeted delivery of siRNA and efficient knockdown of Htt expression. *Journal of Controlled Release: Official Journal of the Controlled Release Society* 163 (2): 125–129. https://doi.org/10.1016/j.jconrel.2012.08.012.

64 Lin, C.-Y., Tsai, C.-H., Feng, L.-Y. et al. (2019). Focused ultrasound-induced blood brain-barrier opening enhanced vascular permeability for GDNF delivery in Huntington's disease mouse model. *Brain Stimulation* 12 (5): 1143–1150. https://doi.org/10.1016/j.brs.2019.04.011.

65 Tsui, P.-H., Wang, S.-H., and Huang, C.-C. (2005). In vitro effects of ultrasound with different energies on the conduction properties of neural tissue. *Ultrasonics* 43 (7): 560–565. https://doi.org/10.1016/j.ultras.2004.12.003.

66 Tyler, W.J., Tufail, Y., Finsterwald, M. et al. (2008). Remote excitation of neuronal circuits using low-intensity, low-frequency ultrasound. *PLoS One* 3 (10): e3511. https://doi.org/10.1371/journal.pone.0003511.

67 Blackmore, J., Shrivastava, S., Sallet, J. et al. (2019). Ultrasound neuromodulation: a review of results, mechanisms and safety. *Ultrasound in Medicine & Biology* 45 (7): 1509–1536. https://doi.org/https://doi.org/10.1016/j.ultrasmedbio.2018.12.015.

68 Dallapiazza, R.F., Timbie, K.F., Holmberg, S. et al. (2018). Noninvasive neuromodulation and thalamic mapping with low-intensity focused ultrasound. *Journal of Neurosurgery* 128 (3): 875–884. https://doi.org/10.3171/2016.11.JNS16976.

69 Kim, H., Park, M.Y., Lee, S.D. et al. (2015). Suppression of EEG visual-evoked potentials in rats through neuromodulatory focused ultrasound. *Neuroreport* 26 (4): 211–215. https://doi.org/10.1097/WNR.0000000000000330.

70 Naor, O., Krupa, S., and Shoham, S. (2016). Ultrasonic neuromodulation. *Journal of Neural Engineering* 13 (3): 31003. https://doi.org/10.1088/1741-2560/13/3/031003.

71 Uddin, S.M.Z., Komatsu, D.E., Motyka, T., and Petterson, S. (2021). Low-intensity continuous ultrasound therapies—a systematic review of current state-of-the-art and future perspectives. *Journal of Clinical Medicine* 10 (12): https://doi.org/10.3390/JCM10122698.

72 Xin, Z., Lin, G., Lei, H. et al. (2016). Clinical applications of low-intensity pulsed ultrasound and its potential role in urology. *Translational Andrology and Urology* 5 (2): 25566–25266. https://doi.org/10.21037/TAU.2016.02.04.

73 Bader, K.B., Bouchoux, G., and Holland, C.K. (2016). Sonothrombolysis. *Advances in Experimental Medicine and Biology* 880: 339. https://doi.org/10.1007/978-3-319-22536-4_19.

74 el Kadi, S., Porter, T.R., Verouden, N.J.W. et al. (2022). Contrast ultrasound, sonothrombolysis and sonoperfusion in cardiovascular disease: shifting to theragnostic clinical trials. *JACC: Cardiovascular Imaging* 15 (2): 345–360. https://doi.org/10.1016/J.JCMG.2021.07.028.

75 Li, J., Xi, A., Qiao, H., and Liu, Z. (2020). Ultrasound-mediated diagnostic imaging and advanced treatment with multifunctional micro/nanobubbles. *Cancer Letters* 475: 92–98. https://doi.org/10.1016/j.canlet.2020.01.028.

7

Application of Ultrasound-Responsive Reagents for Drug Delivery Systems

Hiroshi Kida and Katsuro Tachibana

Department of Anatomy, Fukuoka University School of Medicine, 7-45-1 Nanakuma, Jonan-ku, Fukuoka, 814-0180, Japan

7.1 Historical Background of Research on Bubble Reagents for Medicine

Bubble reagents have long been used in medical applications, most notably as contrast agents in ultrasound imaging. Ultrasound imaging is a diagnostic technique in which information obtained from reflected ultrasound waves directed at scatterers in a living body is reconstructed into an image to depict biological structures and lesions. Ultrasounds propagate through biological tissue as a medium, but when bubbles exist in water-filled blood vessels, the acoustic impedance changes at the gas–liquid interface and is captured as a reflected echo [1]. Bubbles with flexible surfaces resonate with ultrasound and exhibit strong echogenicity [2]. Using this principle, microbubbles (MBs) can be used as ultrasound contrast agents.

In 1969, it was reported that indocyanine green injected through the left ventricle during cardiac catheterization enhanced the echo intensity in M-mode at the aortic root (valve annulus), and research on ultrasound contrast agents began [3]. Subsequent research proved that this phenomenon is due to MBs present in the liquid [4]. Furthermore, it was confirmed that other liquids containing indocyanine green [5] and whole blood [6] could produce the same phenomenon.

In the early days, the contrast effect of MBs was limited to the heart and large blood vessels; thus, they were used exclusively in the cardiovascular field for anatomical structural evaluation of the heart [7], measurement of ejection fractions [8], identification of cardiac shunts [9], and evaluation of valvular disease [10].

Carbon dioxide (CO_2)-filled MBs have been used for liver tumor diagnostic purposes since 1980 [11]. These were effective in depicting hepatic neoplastic lesions but had several limitations; for example, the administration method was limited to intra-arterial administration, and the CO_2 fine bubbles were relatively large, with an average particle size of about 30 μm, so they could only provide contrast in the arterial phase.

Therefore, with the availability of smaller-sized MB ultrasound contrast agents that can be administered intravenously and superior imaging techniques, the CO_2

Acoustic Technologies in Biology and Medicine, First Edition.
Edited by Adem Ozcelik, Ryan Becker, and Tony Jun Huang.

Table 7.1 Clinically applied ultrasound contrast agents.

Trademark name	Launched year	Manufacturer	Adaptive organs	Shell material	Filling gas
Albunex	1993	Mallinckrodt	Heart	Denatured human serum albumin	Air
Levovist	1995	Schering	Heart, kidney, breast, brain	Galactose, palmitic acid	Air
Optison	1998	Amersham	Heart	Cross-linked serum albumin	Perfluoropropane (C3F8) + air
Definity	2001	Bristol-Myers	Heart, liver, kidney	Phospholipid	Perfluoropropane (C3F8) + air
Imagent	2002	Imcor	Heart	Phospholipid	Perfluorohexane (C6F14) + air
SonoVue	2014	Bracco	Heart, liver, breast, urinary tract	Phospholipid	Sulfur hexafluoride (SF6)
Sonazoid	2006	GE Healthcare	Liver, breast	Lipids	Perfluorobutane (C4F10) + air

Source: Adapted from Frinking et al. [13].

imaging method is no longer used today. In addition, early ultrasound contrast agents had a short bubble life and a very short duration of contrast effect. The cross-linked serum albumin was placed at the gas–liquid interface to suppress gas diffusion while decreasing surface tension and extending bubble life, resulting in the product Albunex®, the world's first commercially available ultrasound contrast agent [12]. Since the development of Albunex, various ultrasound contrast agents that have a longer bubble life have been developed (Table 7.1).

Since the gas contained in Albunex was air, its solubility in water was high, and the duration of the contrast effect was limited. In the second and later generations of ultrasound contrast agents, the gases were replaced by high molecular weight and low solubility gases, such as perfluoropropane (C_3F_8) [14], perfluorobutane (C_4F_{10}) [15, 16], and sulfur hexafluoride (SF_6) [17].

In addition, Definity®, Imagent®, SonoVue®, and Sonazoid® were developed with shells composed of phospholipids that provide high echogenicity and stability. This greatly improved the stability of the bubbles and prolonged the duration of the contrast effect. These advances in ultrasound contrast agents provided the foundation for subsequent ultrasound-mediated drug delivery systems (DDS).

7.2 Use of Bubble Reagents as Drug Delivery Systems

7.2.1 Acoustic Cavitation

The effects of ultrasound on living organisms are broadly classified into thermal effects and mechanical effects, with mechanical effects further divided into cavitation and non-cavitation effects [18]. Minnaert resonance is a phenomenon

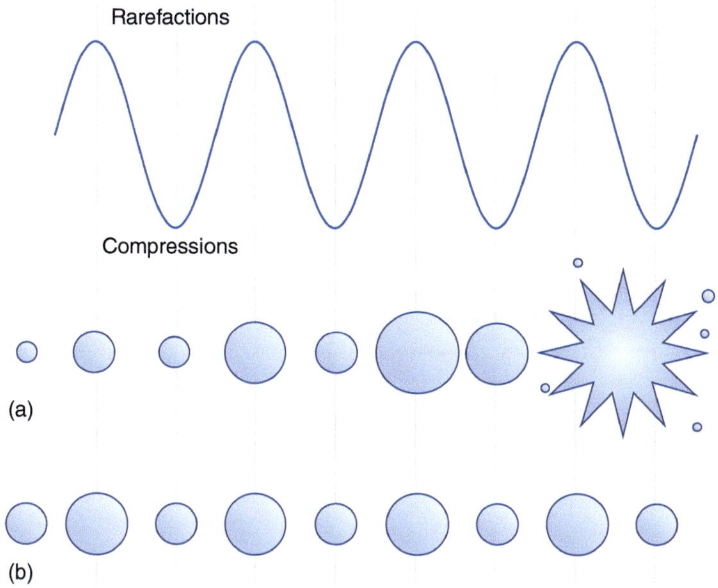

Figure 7.1 Schematic of how ultrasound produces inertial cavitation (a) and non-inertial cavitation (b).

associated with bubbles pulsating at their natural frequency within a liquid in relation to the frequency of sound [19].

$$f = \frac{1}{2\pi\alpha}\left(\frac{3\gamma p_A}{\rho}\right)^{1/2}$$

This equation, expressed as f: frequency, α: bubble radius, γ: polytrope coefficient of the enclosed gas, p_A: ambient pressure, and ρ: water density, shows that frequency and bubble radius are inversely proportional to each other. Ultrasonic irradiation causes a frequency-synchronized repetition of low- and high-pressure states in the acoustic field. Bubbles grow by intermittently oscillating (repeated expansion and contraction), taking in surrounding gas molecules, and eventually burst and collapse [20, 21]. The growth and collapse phenomenon of bubbles under an acoustic field is known as "acoustic cavitation" (Figure 7.1). This phenomenon has been described in various studies and is difficult to strictly define. Nevertheless, Wu et al. stated that "acoustic cavitation refers to activities associated with air or gas bubbles, pockets, and cavities under excitation of acoustic waves" [22]. Paliwal and Mitragotri stated that acoustic cavitation is "the process of formation, pulsation, and collapse of gas-filled cavities in ultrasonicated materials" [23].

Acoustic cavitation can be divided into inertial cavitation and non-inertial cavitation. The phenomenon in which bubbles resonate with ultrasound, grow, and eventually collapse is called inertial cavitation [24]. In contrast, non-inertial cavitation is the process of sustained oscillation of bubbles in a liquid in an acoustic field when the acoustic intensity is insufficient to cause complete collapse of the bubbles [25].

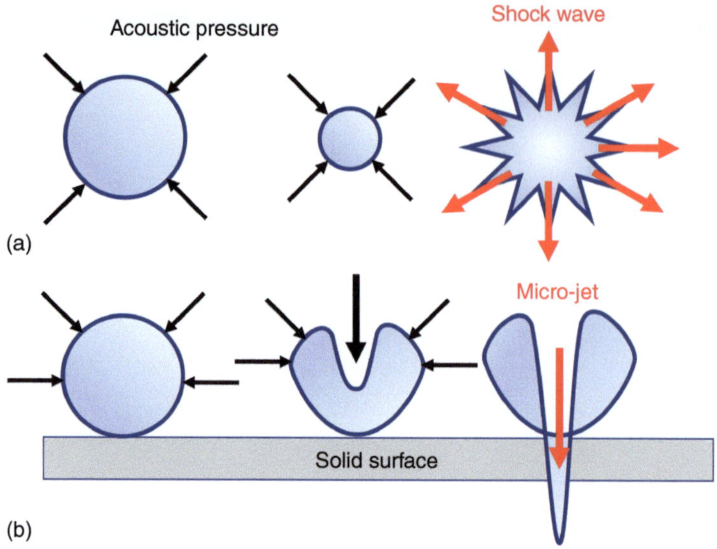

Figure 7.2 Schematic of the formation of radial collapses (a) and asymmetric collapses (b) of microbubbles under ultrasound exposure.

It became clear in the 1990s that ultrasound contrast agent MBs act as cavitation nuclei and lower their threshold during ultrasound irradiation above certain intensity [26]. In an acoustic field, various patterns of microstreaming fluid flow are generated around an oscillating MB [27]. The MB collapses radially [28] or asymmetrically [29] due to inertial cavitation (Figure 7.2).

The first demonstration that the micro-liquid jet flow generated during the disintegration of MBs can be used to enhance drug efficacy was reported by Tachibana and Tachibana in 1995 [30]. Ultrasound irradiation of thrombus in vitro in the presence of an ultrasound contrast agent (MBs) together with a thrombolytic agent (urokinase) was shown to enhance thrombolysis. Following this report, many DDS studies using ultrasound and MBs were conducted and continue to this day.

7.2.2 Importance of Inertial and Non-inertial Cavitation in Improving Drug Permeability

The collapse of ultrasound-irradiated bubbles is thought to be the underlying mechanism that causes transient pores in the cell membrane through which high-velocity micro-jets flow [31]. This allows extracellular drugs or genes to penetrate living cells [31–35]. Unger et al. named their remarkable DDS methodology that utilizes inertial cavitation of MBs “sonoporation” [36]. This clear designation is now widely used. For sonoporation of cell membrane, the distance between the MB and the cell membrane, which can significantly affect the cell membrane, should be less than 5.5 μm and should not exceed the diameter of MBs at most [37].

Both mechanical stress and chemical effects of ultrasound irradiation seem to be involved in the transient perforation of cell membranes [38]. Sonoporation, a drug or gene delivery system that utilizes physical acoustic energy, can deliver various therapeutic modalities, such as small molecule drugs, antibody drugs, proteins, genes (DNA, mRNA), and nucleic acid drugs (siRNA, shRNA, miRNA) into cells.

Non-inertial cavitation also plays an important role in increasing the permeability of biological membranes, which is one of the key roles of DDS using ultrasound and MBs. Oscillation of MBs in non-inertial cavitation in vessels reversibly cleaves tight junctions between endothelial cells, resulting in drug transfer from the vessel into the tissue [39].

One of the histo-anatomical features of the central nervous system is the presence of the blood-brain barrier (BBB) [40]. In brain capillaries, endothelial cells are connected via tight junctions. In addition, the BBB also actively functions as a substance uptake and excretion mechanism. The BBB tightly controls the flow of material in and out of the brain parenchyma and vascular space. A major application for increasing the permeability of endothelial cells is drug delivery through the BBB, often referred to as BBB opening (BBBO) [39] (Figure 7.3a). MBs used in BBBO cannot perforate cell membranes by themselves. In order to deliver genes and nucleic acid drugs to the cytoplasm or nucleus, they must be loaded onto a virus or other carrier, such as liposomes, that can penetrate the cell membrane and be used together [41–43]. Loading MBs used in BBBO with genes and tumor-targeting molecules have been reported to increase the efficiency of gene delivery to brain tumors [44].

Alternatively, non-inertial cavitation promotes endocytosis for drug delivery into cells without membrane perforation [45]. These endocytoses are promoted via the caveolae pathway in vascular endothelial cells of BBB (Figure 7.3b) [46].

The focused ultrasound-mediated intranasal brain drug delivery technique (FUSIN), which combines MBs and high-intensity focused ultrasound (HIFU) as an advanced form of DDS for the CNS, is also being investigated [47]. This method

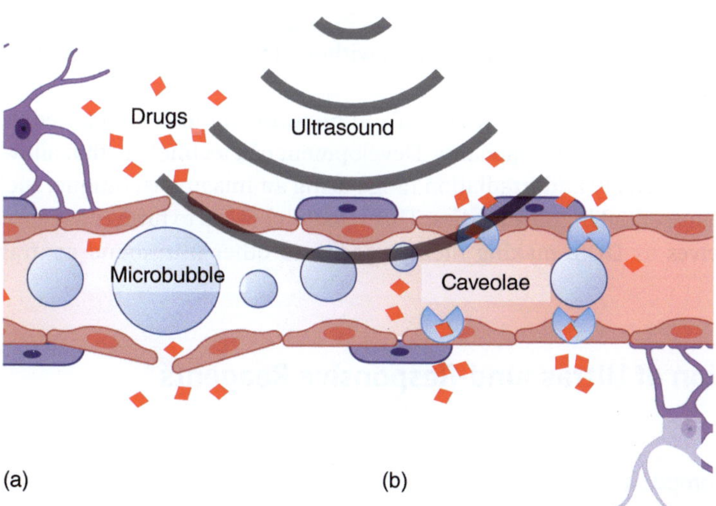

Figure 7.3 Schematic of drug delivery to the brain by ultrasound and microbubbles through the BBB. Non-inertial cavitation causes both BBB opening (a) and caveolae-mediated endocytosis (b).

involves the injection of MBs and drug solutions into the nasal cavity, rather than into the blood vessels, allowing drug delivery to the CNS via the olfactory epithelium by ultrasound irradiation.

Several studies have shown that the combination of ultrasound and MBs can transiently cleave the vascular endothelial cells and deliver gene drugs [48–50] or mesenchymal stem cells [51–55] to treat cardiovascular diseases. Inertial and non-inertial cavitation of MBs by ultrasound seems to be deeply involved in these phenomena.

7.2.3 Targeting and Focusing Using Acoustic Means

To maximize the efficacy of DDS using ultrasound, the range and intensity of ultrasound exposure must be strictly controlled. This is achieved by a method called HIFU or focused ultrasound surgery (FUS) guided on MR imaging [56]. In FUS, as many as 100–1000 ultrasonic oscillators are placed on the concave surface [57].The low-powered, linear ultrasonic beams emitted from the individual elements are focused on a single focal point. As a result, powerful ultrasound is produced at the focal point, producing a therapeutic effect. Recent magnetic resonance imaging-guided FUS (MRgFUS) allows tight control of the focal spot and real-time observation of the focal temperature variation [58]. FUS has been applied to thermocoagulation therapy for essential tremors, Parkinson's disease, and epilepsy by taking advantage of its thermal effects [59–61]. MRgFUS, which allows tight control of irradiation sites and acoustic intensity, is being studied for DDS using acoustic cavitation at low intensities. FUS is also used effectively, especially for BBBO.

The acoustic radiation force (ARF) of the ultrasound acts as a driving force for the MBs, guiding and accumulating them to the affected lesion [54, 62, 63]. Studies using intravascular ultrasound (IVUS) catheters to target vascular endothelial cells show that ARF can accumulate MBs in lesions without any special modifications to the MB surface [54].

Elucidation of the various properties of MBs in response to ultrasound paved the way for their application as DDS modality. Development of a technology that allows precise control of intensity and irradiation range using an imaging technique, such as MRgFUS, was essential in these studies. This research has led to further advances in MBs themselves as DDS, making them superior to other ultrasound contrast agents.

7.3 Variation of Ultrasound-Responsive Reagents for DDS

7.3.1 Shell Composition

Serum albumin is the shell material used in previously described Albunex. Other molecules including biopolymers, such as lactic acid/glycolic acid copolymers (PLGA) [64] and palmitic acid [65], have also been used for developing several bubble shell materials. Among these, phospholipids are the most widely used [14]. Ultrasound irradiation induces the collapse and disappearance of MBs. The collapse is more likely to occur in bubbles with less flexible shells. Phospholipids are highly flexible and biocompatible and thus exhibit high bubble stability and echogenicity [66]. The positive, negative, or neutral charge of MBs can be adjusted depending on the composition of the phospholipids that make up the shell [67–70]. Bubble reagents with a phospholipid shell are called "lipid bubbles" [71]. Bubble formulations calibrated with a phospholipid bilayer are also sometimes called "bubble liposomes" [72], because they are a type of liposome in which the inner space is replaced by gas. The inertial cavitation threshold of MBs varies widely by viscosity and elasticity among different shell compositions. The phospholipid shell, composed of thin, flexible layers, can vibrate at low acoustic pressure. It has been reported that varying the hydrophobic chain lengths in the lipid bubble shell alters the performance of DDS for BBBO [73].

In contrast, higher acoustic pressure is required to reach cavitation for MBs with thicker, stiffer shell layers. For higher sound pressure, hard shell MB fragmentation can propel core gas several microns away, resulting in more aggressive fragmentation, known as sonic cracking [74, 75]. MB contrast agents with polymer shells have a varying qualitative and quantitative response to acoustic pulses than those with lipid shells, according to high-speed optical experiments. The polymer shell does not vibrate significantly in response to ultrasound. In inertial cavitation, bubbles are extruded without shell fracture. A review of 26 studies on drug delivery using various ultrasound parameters and shell compositions of MBs indicated that, compared Optison, Definity ultrasound contrast agent and elevated sonication temperature of 37 °C are associated with efficient uptake of molecules [76].

7.3.2 Improved Stability by Polyethylene Glycol (PEG) Modification

It is possible to change bubble behavior in vivo by applying various chemical modifications to the surface of the bubble shell. For example, it is possible to modulate phagocytosis ability in mononuclear phagocytic cell systems with or without polyethylene glycol (PEG) modification of the fine bubble surface. Mesenchymal cells, such as histiocytes, liver Kupffer cells, and bone marrow sinusoidal endothelium, are collectively called the mononuclear phagocyte system (MPS), and are involved in defense by phagocytosis of invading foreign substances [77]. Generally, MPS supplements and degrades particles larger than 30 nm in diameter that are administered in vivo (in the blood). Ultrasound contrast media with a diameter of 1–5 μm are no exception. Sonazoid, a second-generation ultrasound contrast agent, is designed so that its shell, composed of phosphatidylserine, a type of phospholipid, is recognized and actively phagocytosed by hepatic Kupffer cells, which are part of the MPS. As a result, it acts as a "negative contrast agent" that increases the echoluminance of normal liver tissue and delineates tumor lesions lacking Kupffer cells as low-echo luminance areas [78, 79]. Conversely, when bubble reagents are used as contrast agents or DDS targets other organs, supplementation with MPS has the disadvantage of decreasing retention in circulation. Modification of the bubble shell with PEG prevents trapping by MPS and enhances its retention and the persistence of the contrast effect [80]. However, according to theoretical evaluations, the density, molecular weight, and conformation of PEG chemically bound to phospholipid shells affect inertial cavitation [81]. Specifically, PEG mole fraction increases inertial cavitation, while PEG mole weight decreases it to the inverse 3/5 power in the "brush" regime.

7.3.3 Modification with Targeting Ligands

The surface of the bubble shell can also be modified with various molecules with an affinity for specific body tissues to provide targeted therapy [82]. Targeted MBs bind to specific receptors in the target tissue, delivering the drug in a concentrated manner while minimizing side effects associated with systemic distribution. It is possible to design the particle to target specific tissues, for example, by attaching antibodies to MBs via avidin–biotin interaction, sulfhydryl linkage, or carbodiimide linkage [83, 84], or by binding the ligand to the shell prior to particle formation [85]. Biotinylated antibodies with streptavidin incorporated into the phospholipid shell of MBs are capable of binding in a molar ratio of up to 1 : 4 [86]. While modification by avidin–biotin interactions is easy and robust, the high immunogenicity of avidin hinders its clinical application [86]. Covalent bonding, such as the maleimide–thiol reaction and carboxylic acid-amine coupling, is complicated, while immunogenicity is unlikely to occur.

As a new and different approach, ultrasound contrast agents that can be modified with antibodies using Fc-region-binding polypeptides derived from protein A/G have also been reported [87]. Targeting can also be performed with small-molecule compounds, such as anionic folic acid bound by electrostatic interactions [44].

7.3.4 Drug and Gene Loading

It is possible to load the MB with drugs or genes and deliver them to the lesion at the same time. There are several methods for loading phospholipid shell MBs with drugs and genes. If the MB shell is composed of cationic phospholipids, anionic drugs, and genes, such as DNA or siRNA, they can be adsorbed on its surface by electrostatic interactions (Figure 7.4a) [88–93]. Loading these drugs or genes on the shell reduces the fluidity of the lipids and the elasticity of the MB shell, thereby affecting the acoustic properties of the MB [92]. DNA adsorbed on the surface of MBs enhances resistance to nucleases in the blood [91]. The loaded drug or gene is pulled off as the MB collapses via ultrasound irradiation and is delivered to the cytoplasm through the perforated cell membrane [94]. The loading of genes into MBs improves their transduction efficiency [95, 96]. Hydrophilic drugs can be loaded through avidin–biotin interactions, or they can be covalently connected directly to hydrophilic groups of the phospholipids that make up the MB shell by sulfhydryl linkage, carbodiimide linkage, and targeting ligands (Figure 7.4b) [97].

Hydrophobic or amphiphilic drugs can be retained in the hydrophobic layer of phospholipids that form MB shells (Figure 7.4c) [98, 99]. This method can be used to

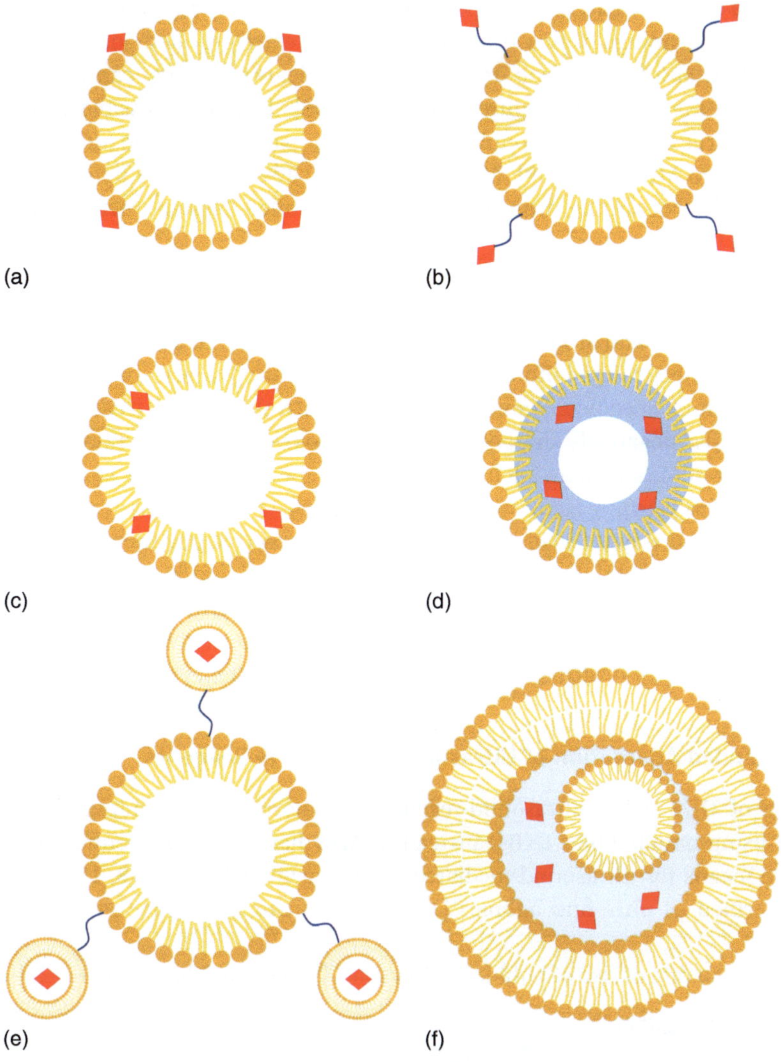

Figure 7.4 Various patterns of the design of drug-loaded microbubbles stabilized with a lipid monolayer. (a) Adsorption of drugs and genes by electrostatic interaction on the surface of positively or negatively charged microbubbles. (b) Direct conjugation of drugs and antibodies to the surface of microbubbles. (c) Addition of hydrophobic drugs to the lipid shell composition of microbubbles. (d) Forming an oil layer dissolved with a hydrophobic drug in the phospholipid shell of the microbubble. (e) Attaching drug-filled liposomes to the surface of the microbubble. (f) Encapsulating the microbubble within a liposome along with the drug.

incorporate siRNA-bound cholesterol into the MB shell [100]. MBs can also be developed by sealing the gas in the inner space of the micelle with oil pre-dissolved with a hydrophobic drug and covered with phospholipid (Figure 7.4d) [101]. Another strategy is to modify the surface of MBs with drug-loaded liposomes and deliver these to the lesion [102–104]. Materials in which MBs are modified with drug-loaded liposomes are called "microbubble–liposome conjugate" or "liposome-microbubble complex." The advantage of this strategy is that even extremely degradable genes, such as mRNA, and highly toxic drugs, such as anticancer agents, can be safely delivered deep into the cell (Figure 7.4e) [105, 106]. It is also possible to load hydrophobic drugs inside phospholipid monolayer particles instead of bilayers and bind them to MBs [107, 108]. MBs can also be loaded into large liposomes or polymer micelles containing drugs [109]. These are called "nested microbubbles" (Figure 7.4f) [110].

Loading of small-molecule drugs into albumin-shell MBs can be accomplished by incubating the drug before or after its formation [111, 112]. The use of nanoparticles consisting of albumin and drugs conjugated to lipid-shell MBs has also been reported [113]. However, this strategy is limited to cases where the drug to be loaded has a high affinity for albumin. Specifically, since albumin is negatively charged, it cannot carry genes that are also negatively charged. One method to overcome this limitation is to incubate albumin MBs with an adenovirus in which the loaded gene is surrounded by capsid proteins [114].

There are several ways to load drugs into MBs composed of polymer shells. The polymer shell is much thicker than the phospholipid shell and has superior drug loading capacity. MBs can be created after the material polymer and the loaded hydrophobic drug are mixed in a certain ratio [115, 116]. Alternatively, hydrophobic drugs can be loaded by infiltrating the MBs in the polymer shell [117].

7.3.5 Extended Adaptation of Ultrasound-Responsive Reagents

The scope of DDS using ultrasound-responsive materials is expanding. A new approach using nanobubbles (NBs) instead of MBs has attracted attention as an alternative means for ultrasound-mediated drug or gene delivery [118, 119]. NBs, officially termed as ultra-fine bubbles [120], are defined as submicron diameter bubbles.

In general, the rate of rise of bubbles follows the "Stokes law" below [121, 122].

$$\boldsymbol{V} = \mathbf{1/18} \times \boldsymbol{gd^2/v}$$

where $\boldsymbol{V}$ (m/s) is the rising velocity of the bubble, $\mathbf{g}$ (m/s^2) is the gravitational acceleration, $\boldsymbol{d}$ (m) is the diameter of the bubble, and $\boldsymbol{v}$ (m^2/s) is the kinematic viscosity of water. This formula shows that the rate of rise of a bubble is proportional to the square of its diameter. Much smaller in particle size than MBs, the rate of rise of NBs will be negligibly small in the calculations. NBs remain stable in water for several months [123].

Similar to MBs, NBs can function as acoustic cavitation nuclei and can be used for sonoporation [124–126]. There are NBs composed of shells, such as phospholipids [127–136], polymers [137, 138], or proteins, such as albumin [139, 140]. NBs, which are smaller than MBs, also exist in nature. In addition to completely artificial NBs, biosynthetic nanoscale bubbles (BNBs), which are biosynthesized and extracted by Halobacterium NRC-1 from protein shells, have been tested for use in drug delivery [140].

Applying the Minnaert resonance [19] described above, it was revealed that a higher frequency of ultrasound is required to resonate NBs, which have a smaller particle size, than that for MBs. Lafond et al. revealed that albumin shell NBs have a broader size distribution range and exhibit more sensitivity to various ultrasound intensities and frequencies than lipid shell MBs [126]. Albumin shell NBs are visualized using high-frequency ultrasound at 40 MHz, and they can be easily collapsed for enhanced cancer cell disruption using low-intensity ultrasound 0.8–1.0 W/cm^2 [125]. It has also been reported that a combination of an echo-imaging device, which has a higher frequency and lower power than those of the sonoporators generally used in DDS studies, and albumin shell NBs can deliver genes to organs in live mice [124].

NBs can also carry small-molecule drugs, such as pDNA or siRNA. Loading methods exist, including adsorption of NBs to surfaces via electrostatic interactions [128, 129, 133, 137–139, 141], incorporation into shells via hydrophobic interactions [131, 132, 134], and formation of NB-liposome complexes [142] or nested NBs [127]. Anionic albumin-shell NBs cannot act as direct carriers of anionic genes, such as pDNA and mRNA. However, even when those genes are mixed with albumin-shell NBs for administration, ultrasound irradiation achieves carrier-free gene delivery. In fact, sonoporation in combination with albumin shell NBs is one of the few methods that can transfect naked mRNA and pDNA into cells in a carrier-free manner [124, 143].

NBs can also be modified with targeting molecules, such as antibodies [128, 132], proteins [140], peptides [85, 135], and small molecules [134, 138] in a manner similar to MBs. There is also a report of NBs modified by nanobodies that are smaller than normal antibodies [131]. Furthermore, NBs reportedly show BBBO effects as well as MBs. pDNA loaded on carriers composed of neuron-binding peptides RVG-R9 and PEG-PEI achieves neuron-selective delivery via BBBO [136].

DDS adapted to ultrasound acoustics extends from ultrasound-responsive bubbles to droplets [144]. Ultrasound-responsive nanodroplets (NDs) are "phase-change emulsions" with a low-boiling liquid core, such as perfluorocarbons, stabilized with artificial shells, such as surfactants. The liquid core is vaporized and inflated by the acoustic energy of the ultrasound above the threshold [145]. It has been reported that both the cavitation [146, 147] and thermal effect [148, 149] or super harmonic focusing [150] of HIFU are involved in the vaporization of the liquid core of NDs. Hydrophobic or amphiphilic drugs can be encapsulated inside NDs [151]. Anionic genes and small-molecule anticancer drugs can be loaded into droplet shells similar to bubbles [152–156]. Anionic genes can be loaded in the shell of a block copolymer of ND cationic poly-amino acids and long-chain fluorocarbons, consisting of a core of liquid perfluorocarbons [157, 158]. Furthermore, the ND shell can also be modified with antibodies [159], peptides [158], or small molecules [160] to target specific cells. Eventually, the vaporized droplets serve as cavitation nuclei in sonoporation [161] and BBBO [160, 162, 163].

The development of nanocarriers similar to NDs, combining the thermal and cavitation actions of ultrasound, has also been attempted. The echogenic low temperature sensitive liposomes (E-LTSLs) developed by Maples et al. are composed of amphiphilic phospholipids and perfluoropentane (PFP) emulsified with 1,3-PD, which, like NDs, are vaporized by the thermal effect of HIFU and release loaded DOX into the tumor lesion [149]. Along with DOX, ammonium bicarbonate and gold nanorods were introduced into folate-bound liposomes, targeting the tumor with folate [164]. The thermal effects of HIFU generate CO_2 gas from the ammonium bicarbonate, and its cavitation effect further releases DOX into the tumor.

Nanoscale metal-organic framework (MOF) of UiO-66-NH2 accurately releases loaded temozolomide (TMZ) into tumors upon ultrasound irradiation through low-frequency oscillations and cavitation effects [165].

7.4 Research on Treatment of Diseases Using Ultrasonic Drug Delivery

7.4.1 Cancer

Of the various diseases, cancer is the most studied target disease for ultrasound-based DDS. Many DDSs have been studied that successfully combine the characteristics of cancer and ultrasound, which differ from normal tissue, and the characteristics of the bubble formulation.

One of the advantages of applying ultrasound to cancer-targeted DDS is the lesion selectivity of ultrasound. Unlike diseases that diffusely damage tissues throughout the body, locally confined tumors are suitable targets for ultrasound irradiation.

Nearly all solid tumors have a hypoxic environment resulting from an imbalance between cell proliferation and angiogenesis [166]. Some cancer cells adapt to the hypoxic environment and become resistant to anticancer drugs and radiation, as well as to invasion, metastasis, and recurrence. This feature is promising for "environmentally targeted therapy by passive targeting" [167]. Release of antitumor drugs loaded in MBs and oxygen to the affected area by ultrasound may improve antitumor effects by eliminating the hypoxic environment [168, 169].

However, in a phenomenon called the Warburg effect, tumor cells produce adenosine triphosphate (ATP) in the glycolytic system rather than through mitochondrial oxidative phosphorylation, even under aerobic conditions [170]. Therefore, the interior of the tumor tissue has a low pH due to increased lactate metabolism. This environment can be used to enhance the antitumor effect of drugs in combination with ultrasound, either by loading them onto pH-responsive nanoparticles and binding them to MBs or by direct loading onto pH-sensitive MBs [171–173].

Excessive angiogenesis in tumors hinders delivery of antitumor drugs due to high interstitial pressure and low perfusion while promoting tumor growth [174]. It was reported that delivery of antiangiogenic drugs by MBs and ultrasound inhibited tumor growth [175, 176].

The surface of the MBs can also be modified with ligands, such as antibodies or peptides with an affinity for surface antigens specific to tumor cell surfaces or tumor microenvironments, to allow MBs to accumulate on the tumor by active targeting.

Epidermal growth factor receptor (EGFR) and human epidermal growth factor receptor 2 (HER2), which are members of the ErbB receptor family of receptor-type tyrosine kinases (RTKs), are overexpressed or mutated in many solid tumors and are deeply involved in tumor growth and metastasis [177]. Ultrasound-responsive agents modified with antibodies that target these receptors actively accumulate in the cancer and deliver the loaded drug or gene. Alternatively, as antibody drugs, they regress tumors via receptor inhibition [178–180].

In addition, cancer cells induce the expression of the immune checkpoint molecule programmed cell death ligand 1 (PD-L1) on their own cells or on immune cells, such as macrophages and dendritic cells [181]. They bind to programmed cell death 1 (PD-1) on T cells and participate in the immune escape of cancer. MBs modified with PD-L1 antibodies accumulate in tumors and enhance the antitumor effect when combined with FUS [182]. It is hypothesized that acoustic cavitation of MBs may have caused mechanical fractionation of the tumor environment, leading to blockade of tumor surface PD-L1 by antibody drugs and activation of cytotoxic T cells.

Alternatively, folate receptors are overexpressed on the surface of many types of tumors [183]. Modification of ultrasound responders by folic acid promotes their accumulation in tumors [44, 164]. In addition, antibodies to surface antigens specific to certain cancer cells have been modified on the surface of ultrasound-responsive reagents and are used for targeting [131]. Targeting of molecules associated with angiogenesis, one of the microenvironmental changes in cancer, will also be performed. There is a significant impact on vascular endothelial growth factor (VEGF) and its receptor (VEGFR2) in angiogenesis. VEGFR2 is weakly expressed in normal tissues, with reported overexpression in bladder, breast, cervical, colon, esophageal, ovarian, pancreatic, oral, and prostate cancers, brain gliomas, renal clear cell carcinoma, NSCLC, and skin melanoma [184]. Anti-VEGFR2 antibodies allow modified ultrasound responders to accumulate in tumor foci. Acoustic cavitation, with ultrasound irradiation following accumulation of the reagent in the lesion, delivers drugs, and genes with antitumor effects into the tumor tissue or even into the cells [107, 185].

Integrin $\alpha_v\beta_3$ on angiogenic tumor vasculature and certain tumor cells is an important family of molecules involved in tumor angiogenesis, consisting of two non-covalently bound transmembrane alpha and beta subunits [186]. Antiintegrin $\alpha_v\beta_3$ antibodies have also been used to surface-modify ultrasound responders for targeting [187, 188]. It has been reported that MBs modified with antibodies against integrin αvβ3 are internalized by tumor vascular endothelial cells and, upon ultrasound irradiation, perforate the cell membrane from the inside, forming a luminal pathway from the vessel cavity to the tumor, contributing to drug delivery [189].

The term "tumor homing peptide" refers to small peptides that have been shown to specialize in homing to the tumor microenvironment, that is, the tumor vasculature, after systemic administration [190]. Peptide sequences with arginine–glycine–aspartic acid (RGD) motifs as tumor homing peptides show a strong affinity for integrins, particularly integrin αvβ3 [191]. Modification of the ultrasound-responsive agent with RGD motif peptides also aids in its accumulation and drug delivery to the tumor [188]. The tumor-homing peptide cysteine–arginine–glutamic acid–lysine–alanine (CREKA) motif, which also binds to integrins, is also used as a modifier to target MBs to tumors [85]. Furthermore, dual or triple targeting combining multiple antibodies or peptides targeting VEGFR2, integrin αvβ3, and other cancer-specific antigens has been reported to enhance the accumulation of these modified MBs in tumors as ultrasound contrast agents [192–194].

One feature of the in vivo behavior of NBs and NDs that is not present in MBs is their ability to penetrate deeper into living tissue. One of them is the enhanced permeability and retention (EPR) effect [195]. In normal tissues, low-molecular-weight compounds can pass through the endothelial cell spaces of capillaries, but high-molecular-weight compounds cannot. However, in solid cancer tissue, this gap is wide, and high-molecular-weight diagnostic and therapeutic agents of several hundred nm can pass through and accumulate specifically in the tumor tissue without being ejected. This effect is called the EPR effect, and many DDS studies are underway today that take advantage of it. While the EPR effect cannot be expected for MBs larger than 1 μm in diameter, NBs and NDs of a few hundred nm in diameter are exactly the particle sizes that can produce the effect [141, 196, 197].

Various anticancer drugs, such as doxorubicin (DOX), cisplatin (CDDP), and paclitaxel (PTX) can be loaded into MBs or ultrasound-responsive droplets and delivered to the affected lesion by inertial cavitation. Indications for DDS by MBs and ultrasound go beyond small-molecule antitumor drugs. Sonoporation by cavitation action perforates cell membranes, allowing delivery of genes and nucleic acid drugs that are only effective intracellularly.

The strategy is to transfect DNA or mRNA encoding proteins with antitumor effects using MBs and ultrasound. Mutations in *p53*, a tumor suppressor gene, play an important role in tumor development, growth, and metastasis [198]. Transfection of the wild-type *p53* gene loaded in MBs alone or in combination with radiation results in improved antitumor effects [199, 200]. The suicide gene therapy strategy involves introducing suicide-inducing transgenes into cancer cells. There are two major suicide gene therapy strategies currently under investigation: CD/5-fluorocytosine (CD/5-FC) and thymine kinase gene/ganciclovir (TK/GCV) [201]. VEGFR2-targeted cationic NBs loaded with a plasmid encoding the double gene for CD-TK accumulated in tumors, while transfection with ultrasound induced apoptosis [128].

Alternatively, RNA interference (RNAi) with miRNAs, siRNAs, and their precursors, shRNAs, can also be applied to treat tumors. Targeting mRNAs of several factors overexpressed in tumors, miRNAs are introduced into the cytoplasm by MBs to downregulate their proteins and exert antitumor effects [88, 202, 203]. Various candidate molecules act as targets for combinations of ultrasound responders and nucleic acid drugs that can knock down specific factors related to the cancer microenvironment.

In many solid tumors, VEGF and VEGFR, both overexpressed by autocrine loops, are closely involved in tumor development and growth [204]. MBs with siRNA targeting *VEGF* mRNA loaded to a cationic polymer carrier and ultrasound irradiation show high antitumor effects [205, 206].

The transcription factor hypoxia-inducible factor 1 (HIF1) is a tightly regulated master regulator of oxygen homeostasis, responding to hypoxia. In growing hypoxic tumors, HIF1 can respond to hypoxia by being deregulated and either increasing oxygen supply or adapting to decreased oxygen availability. It is known that under hypoxic conditions, HIF-1α accumulates due to inhibition of prolyl-4-hydroxylase2 (PHD2), which activates many genes as they adapt to oxygen deprivation, promoting angiogenesis, metastasis, and invasion [207]. The function of PHD2 in carcinogenesis is poorly understood, although there is ample evidence of a positive correlation between HIF and tumorigenesis, metastasis, and poor prognosis [208]. Several studies have shown that modifying MBs with siRNA or shRNA inhibiting HIF1 and delivering them to tumors via ultrasound inhibited tumor growth or even regressed tumors and prolonged survival [209, 210]. However, transfection of *PHD2*-shRNA-loaded PEI with a combination of albumin-shell MBs and ultrasound resulted in significant downregulation of PHD2 expression and increased HIF-1α expression and downstream angiogenesis factors VEGF, TGF-β, and bFGF [211].

Survivin is an inhibitor of the apoptosis protein family. While survivin is virtually absent in normal tissues, it is overexpressed in malignant tumors. Survivin is involved in the development and progression of malignant tumors [212]. The siRNA or shRNA on MBs or NBs is transferred into tumor cells by ultrasound, knocking down the expression of survivin and showing antitumor effects [141, 213].

X-linked inhibitor of apoptosis (XIAP) is a protein that stops cell death by apoptosis via inhibition of caspase production. XIAP is often overexpressed in malignant tumors [214]. MBs loaded with polymeric micelles of siRNA targeting *XIAP* mRNA and ultrasound irradiation restored caspase-3 activity via silencing of *XIAP*, resulting in superior therapeutic efficacy against a human cervical cancer xenograft model in nude mice [215].

Cancer treatments in combination with ultrasound-responsive agents and photodynamic therapy (PDT) have also been proposed. As a relatively noninvasive therapeutic method, PDT involves irradiating a wavelength that is absorbed by a photosensitizing agent and oxygen [216]. By damaging the microvasculature and inducing local inflammation, this procedure directs the death of malignant cells. Xu et al. reported that oxygen delivery to tumors using perfluoropentane-doped oxygen MBs (OPMBs) and ultrasound enhanced the effects of PDT using protoporphyrin IX as a sensitizer and also led to a significant decrease in the expression of HIF1 and related gene expression and prevention of angiogenesis [169]. The multifunctional cationic porphyrin-grafted lipids (CPGLs) that compose the shell of MBs developed by Zhao et al. act as a sensitizer for PDT while carrying siRNA.

MBs carrying HIF1 [210] or FOXA1 [217] siRNA, which are nuclear receptors regulating gene overexpression in breast cancer, were found to exert a synergistic antitumor effect on the lesion when combined with ultrasound and light irradiation.

7.4.2 Central Nervous System Diseases

DDS using MBs and ultrasound is also effective for diseases of the central nervous system (CNS). BBBO or FUSIN combined with MBs and HIFU play a role as an effective DDS for the CNS. In the field of CNS research, the most advanced studies are DDS studies using ultrasound for treatment of malignant tumors.

Delivery of various small-molecule compounds, such as TMZ [218], carboplatin (CBDCA) [219], albumin-bound PTX [220], and etoposide (ETP) [221], has been reported for glioblastoma (GBM) using BBBO or FUSIN techniques in basic studies. There have been reports of delivery of antibody drugs, such as bevacizumab, a humanized monoclonal anti-VEGF antibody [222], ^{89}Zr-cetuximab, a RI-labeled humanized monoclonal anti-EGFR antibody [223], ^{89}Zr-labeled mouse monoclonal anti-CD47 antibody [224], and anti-PD-L1 antibody [47]. In addition, there have been reports of the administration of trastuzumab, an anti-HER2 antibody [225], interleukin-12 (IL-12), a protein drug [226], and NK-92, a human natural killer (NK) cell line [227], for the treatment of metastatic brain tumors. Most of these have been shown to regress tumors and prolong survival.

Not only basic research but also clinical trials of DDS using BBBO are being conducted. The world's first BBBO plus CBDCA trial (NCT02253212) for recurrent GBM is the only BBBO clinical trial to date to demonstrate a survival benefit [228]. In a similar clinical trial (NCT03744026) of BBBO combined with CBDCA, a significant enhancement of the Gd contrast effect of the lesion was observed after irradiation. Effective BBB destruction and CBDCA delivery were suggested [229]. A clinical trial (NCT02343991) of BBBO in combination with liposomal DOX in patients with malignant gliomas also showed enhanced drug delivery by ultrasound irradiation [230].

As with brain tumors, drug delivery studies with ultrasound for Alzheimer's disease are underway. BBBO has been studied in Alzheimer's disease, targeting the hippocampus, which is the focal point of the disease [41, 231]. Basic research has reported Aβ plaque reduction by GSK inhibitors [232], neuroprotection by administration of the *BDNF* gene carried in retroviral vectors [41], reduction of amyloid pathology and improvement of cognitive function by administration of antibodies targeting Aβ, such as aducanumab [233], and reduction of neurogenesis and amyloid pathology by IVIg [231]. Clinical trials have been conducted to confirm the safety of BBBO (NCT02986932) [234] and BBBO (NCT03671889) [235] targeting the hippocampus and olfactory endocortex to induce neurogenesis.

Basic research with BBBO for Parkinson's disease has shown that it can transfer various neurotrophic factors or transcription factors as proteins or genes and reduce disease symptoms. These factors include neutrin (NTN) [236], glial cell line-derived neurotrophic factor (GDNF) [42], BDNF [42], and nuclear factor-erythroid 2-related factor 2 (Nrf2) [237]. It has also been shown that the introduction of anti-α-synuclein antibodies can inhibit their aggregation [238]. The BBBO clinical trial (NCT03608553) for Parkinson's disease with dementia (PDD) confirmed its safety [239].

A different approach to DDS for the CNS than BBBO is sonothrombolysis, or accelerated thrombolysis and lysis, using a thrombolytic agent in combination with ultrasound irradiation for embolic cerebral infarction. Ekosonic endovascular system (EKOS, Boston Scientific), an endovascular treatment device based on this principle, was developed and demonstrated efficacy in clinical trials IMS I and IMS II, but failed to demonstrate superiority over intravenous rtPA alone in the subsequent IMS III (NCT00359424) [240]. Fortunately, the device was eventually commercialized for thrombolytic therapy for pulmonary embolism and was extremely successful [241]. Combined lysis of the thrombus with ultrasound and systemic tissue plasminogen activator for emergent revascularization (CLOTBUST-ER) was expected to have early revascularization effects. Unfortunately, it failed to demonstrate superiority over the control group in its phase III trial [242]. Thus, clinical trials for sonothrombolysis for cerebral infarction have been difficult.

However, basic research is steadily advancing this therapy as DDS by adding targeting peptides against thrombus to MBs or by loading MBs with urokinase [243, 244].

There are DDS studies using ultrasound and bubble reagents targeting other CNS diseases. BBBO with MRIgFUS and MBs can deliver and knockdown cholesterol-conjugated siRNA targeting mutant huntingtin in the striatum of Huntington's disease, using HD rat models [43]. Delivery of GDNF plasmids loaded on liposomes into the brain of HD model mice by BBBO using MBs and FUS reportedly prolongs survival and neuroprotection and improves motor symptoms [245].

7.5 Conclusion

In recent years, focused ultrasound irradiation systems have evolved into precisely controllable and safe systems. At the same time, there has been a remarkable evolution of ultrasound-responsive agents as biomaterials. Furthermore, bubble reagents are becoming more sophisticated with the loading of small molecule drugs, proteins, and nucleic acid/gene drugs onto their surface and the addition of targeting by modification with antibodies and peptides.

NBs, which are smaller than MBs, and NDs with liquefied cores have been developed. These trends of progress are expected to be enhanced further. Some of these technologies have advanced to clinical trials. However, most technologies are still in the initial research phase. It is expected that ultrasound devices and their responsive reagents will continue to develop like intermeshing gears. In the future, it is expected that the high safety and spatial controllability of ultrasound will be put to practical use in the curative treatment of malignant tumors or central nervous system diseases (Figure 7.5). Contrary to the vast amount of scientific literature on use of ultrasound for DDS, very few products have actually reached the market using this technology, with the exception of the ultrasound catheter device for sonothrombolysis. This is in part due to the complexity of combining physical acoustic energy with pharmaceutical materials. In addition, compared to conventional DDS products, obtaining FDA approval becomes a more rigorous process for this reason. However, many researchers remain optimistic due to the fact that new discoveries are constantly being made, and if this technology is realized, it would have a tremendous impact on patients in need of a cure.

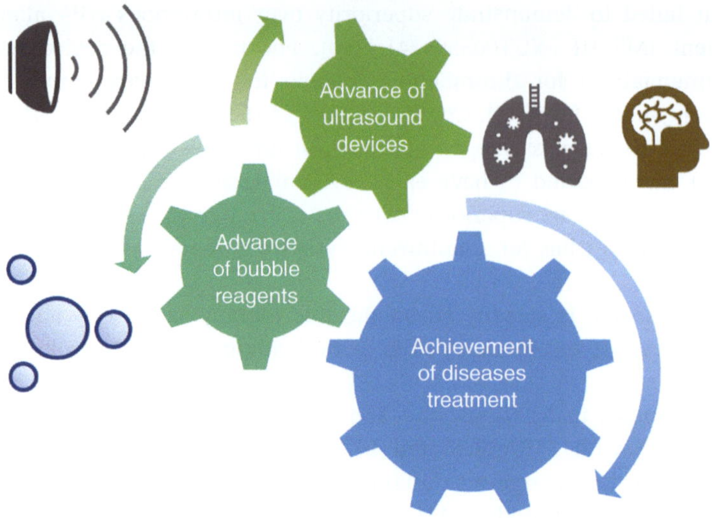

Figure 7.5 Schematic of future developments in ultrasound-responsive materials and ultrasound devices. The advance of "ultrasound devices" and "bubble reagents" meshing together as two cogs in a wheel will finally achieve "treatment of diseases" in the future.

References

1 Calliada, F., Campani, R., Bottinelli, O. et al. (1998). Ultrasound contrast agents: basic principles. *European Journal of Radiology* 27: S157–S160.

2 De Jong, N. (1997). Physics of microbubble scattering. In: *Advances in Echo Imaging Using Contrast Enhancement* (ed. N.C. Nanda, R. Schlief, and B.B. Goldberg), 39–64. Dordrecht, Netherlands: Springer.

3 Gramiak, R. and Shah, P.M. (1968). Echocardiography of the aortic root. *Investigative Radiology* 3 (5): 356–366.

4 Kremkau, F.W., Gramiak, R., Carstensen, E.L. et al. (1970). Ultrasonic detection of cavitation at catheter tips. *The American Journal of Roentgenology, Radium Therapy, and Nuclear Medicine* 110 (1): 177–183.

5 Feigenbaum, H., Stone, J.M., Lee, D.A. et al. (1970). Identification of ultrasound echoes from the left ventricle by use of intracardiac injections of indocyanine green. *Circulation* 41 (4): 615–621.

6 Ziskin, M.C., Bonakdarpour, A., Weinstein, D.P. et al. (1972). Contrast agents for diagnostic ultrasound. *Investigative Radiology* 7 (6): 500–505.

7 Roelandt, J. (1982). Contrast echocardiography. *Ultrasound in Medicine & Biology* 8 (5): 471–492.

8 Meltzer, R.S. and Roelandt, J. (1982). *Contrast Echocardiography*. The Hague/Boston/London: Martinus Nijhoff Publishers.

9 Valdes-cruz, L.M. and Sahn, D.J. (1984). Ultrasonic contrast studies for the detection of cardiac shunts. *Journal of the American College of Cardiology* 3 (4): 978–985.

10 Reid, C.L., Kawanishi, D.T., McKay, C.R. et al. (1983). Accuracy of evaluation of the presence and severity of aortic and mitral regurgitation by contrast 2-dimensional echocardiography. *The American Journal of Cardiology* 52 (5): 519–524.

11 Matsuda, Y. and Yabuuchi, I. (1986). Hepatic tumors: US contrast enhancement with CO_2 microbubbles. *Radiology* 161 (3): 701–705.

12 Feinstein, S.B., Ten Cate, F.J., Zwehl, W. et al. (1984). Two-dimensional contrast echocardiography. I. In vitro development and quantitative analysis of echo contrast agents. *Journal of the American College of Cardiology* 3 (1): 14–20.

13 Frinking, P., Segers, T., Luan, Y. et al. (2020). Three decades of ultrasound contrast agents: a review of the past, present and future improvements. *Ultrasound in Medicine & Biology* 46 (4): 892–908.

14 Unger, E., Shen, D., Fritz, T. et al. (1994). Gas-filled lipid bilayers as ultrasound contrast agents. *Investigative Radiology* 29 (Suppl 2): S134–S136.

15 Schneider, M., Broillet, A., Bussat, P. et al. (1997). Gray-scale liver enhancement in VX2 tumor-bearing rabbits using BR14, a new ultrasonographic contrast agent. *Investigative Radiology* 32 (7): 410–417.

16 Schneider, M., Anantharam, B., Arditi, M. et al. (2011). BR38, a new ultrasound blood pool agent. *Investigative Radiology* 46 (8): 486–494.

17 Schneider, M., Arditi, M., Barrau, M.B. et al. (1995). BR1: a new ultrasonographic contrast agent based on sulfur hexafluoride-filled microbubbles. *Investigative Radiology* 30 (8): 451–457.

18 Hedrick, W.R. and Hykes, D.L. (1991). Biological effects of ultrasound: II. Interactions of ultrasound with matter. *Journal of Diagnostic Medical Sonography* 7 (4): 194–196.

19 Minnaert, M. (1933). XVI. On musical air-bubbles and the sounds of running water. *The London, Edinburgh, and Dublin Philosophical Magazine and Journal of Science* 16 (104): 235–248.

20 Leighton, T.G. (1997). *The Acoustic Bubble*. San Diego, London: Academic Press.

21 Suslick, K.S. (1989). The chemical effects of ultrasound. *Scientific American* 260 (2): 80–87.

22 Wu, J. and Nyborg, W.L. (2008). Ultrasound, cavitation bubbles and their interaction with cells. *Advanced Drug Delivery Reviews* 60 (10): 1103–1116.

23 Paliwal, S. and Mitragotri, S. (2006). Ultrasound-induced cavitation: applications in drug and gene delivery. *Expert Opinion on Drug Delivery* 3 (6): 713–726.

24 Miller, M.W., Miller, D.L., and Brayman, A.A. (1996). A review of in vitro bioeffects of inertial ultrasonic cavitation from a mechanistic perspective. *Ultrasound in Medicine & Biology* 22 (9): 1131–1154.

25 Birkin, P.R., Offin, D.G., Vian, C.J.B. et al. (2011). Investigation of noninertial cavitation produced by an ultrasonic horn. *The Journal of the Acoustical Society of America* 130 (5): 3297–3308.

26 Holland, C.K. and Apfel, R.E. (1990). Thresholds for transient cavitation produced by pulsed ultrasound in a controlled nuclei environment. *The Journal of the Acoustical Society of America* 88 (5): 2059–2069.

27 Collis, J., Manasseh, R., Liovic, P. et al. (2010). Cavitation microstreaming and stress fields created by microbubbles. *Ultrasonics* 50 (2): 273–279.

28 Kodama, T. and Tomita, Y. (2000). Cavitation bubble behavior and bubble–shock wave interaction near a gelatin surface as a study of in vivo bubble dynamics. *Applied Physics B* 70 (1): 139–149.

29 Postema, M., van Wamel, A., ten Cate, F.J. et al. (2005). High-speed photography during ultrasound illustrates potential therapeutic applications of microbubbles. *Medical Physics* 32 (12): 3707–3711.

30 Tachibana, K. and Tachibana, S. (1995). Albumin microbubble echo-contrast material as an enhancer for ultrasound accelerated thrombolysis. *Circulation* 92 (5): 1148–1150.

31 Tachibana, K., Uchida, T., Ogawa, K. et al. (1999). Induction of cell-membrane porosity by ultrasound. *The Lancet* 353 (9162): 1409.

32 Greenleaf, W.J., Bolander, M.E., Sarkar, G. et al. (1998). Artificial cavitation nuclei significantly enhance acoustically induced cell transfection. *Ultrasound in Medicine & Biology* 24 (4): 587–595.

33 Kooiman, K., Roovers, S., Langeveld, S.A.G. et al. (2020). Ultrasound-responsive cavitation nuclei for therapy and drug delivery. *Ultrasound in Medicine & Biology* 46 (6): 1296–1325.

34 Moosavi Nejad, S., Hosseini, S.H., Akiyama, H. et al. (2011). Optical observation of cell sonoporation with low intensity ultrasound. *Biochemical and Biophysical Research Communications* 413 (2): 218–223.

35 Song, J.H., Moldovan, A., and Prentice, P. (2019). Non-linear acoustic emissions from therapeutically driven contrast agent microbubbles. *Ultrasound in Medicine & Biology* 45 (8): 2188–2204.

36 Unger, E.C. (1997). SonoPoration™ and gene delivery with acoustically active carriers. *European Journal of Ultrasound* 6: S2.

37 Yu, H. and Chen, S. (2014). A model to calculate microstreaming-shear stress generated by oscillating microbubbles on the cell membrane in sonoporation. *Bio-medical Materials and Engineering* 24 (1): 861–868.

38 Lentacker, I., De Cock, I., Deckers, R. et al. (2014). Understanding ultrasound induced sonoporation: definitions and underlying mechanisms. *Advanced Drug Delivery Reviews* 72: 49–64.

39 Omata, D., Hagiwara, F., Munakata, L. et al. (2020). Characterization of brain-targeted drug delivery enhanced by a combination of lipid-based microbubbles and non-focused ultrasound. *Journal of Pharmaceutical Sciences* 109 (9): 2827–2835.

40 McConnell, H.L. and Mishra, A. (2022). Cells of the blood-brain barrier: an overview of the neurovascular unit in health and disease. *Methods in Molecular Biology* 2492: 3–24.

41 Wang, F., Wei, X.X., Chang, L.S. et al. (2021). Ultrasound combined with microbubbles loading BDNF retrovirus to open BloodBrain barrier for treatment of Alzheimer's disease. *Frontiers in Pharmacology* 12: 615104.

42 Lin, C.Y., Lin, Y.C., Huang, C.Y. et al. (2020). Ultrasound-responsive neurotrophic factor-loaded microbubble-liposome complex: preclinical investigation for Parkinson's disease treatment. *Journal of Controlled Release* 321: 519–528.

43 Burgess, A., Huang, Y., Querbes, W. et al. (2012). Focused ultrasound for targeted delivery of siRNA and efficient knockdown of Htt expression. *Journal of Controlled Release* 163 (2): 125–129.

44 Fan, C.H., Chang, E.L., Ting, C.Y. et al. (2016). Folate-conjugated gene-carrying microbubbles with focused ultrasound for concurrent blood-brain barrier opening and local gene delivery. *Biomaterials* 106: 46–57.

45 De Cock, I., Zagato, E., Braeckmans, K. et al. (2015). Ultrasound and microbubble mediated drug delivery: acoustic pressure as determinant for uptake via membrane pores or endocytosis. *Journal of Controlled Release* 197: 20–28.

46 Deng, J., Huang, Q., Wang, F. et al. (2012). The role of caveolin-1 in blood-brain barrier disruption induced by focused ultrasound combined with microbubbles. *Journal of Molecular Neuroscience* 46 (3): 677–687.

47 Ye, D., Yuan, J., Yue, Y. et al. (2021). Focused ultrasound-enhanced delivery of intranasally administered anti-programmed cell death-ligand 1 antibody to an intracranial murine glioma model. *Pharmaceutics* 13 (2): e190.

48 Zhigang, W., Zhiyu, L., Haitao, R. et al. (2004). Ultrasound-mediated microbubble destruction enhances VEGF gene delivery to the infarcted myocardium in rats. *Clinical Imaging* 28 (6): 395–398.

49 Taniyama, Y., Tachibana, K., Hiraoka, K. et al. (2002). Local delivery of plasmid DNA into rat carotid artery using ultrasound. *Circulation* 105 (10): 1233–1239.

50 Zhang, L., Sun, Z., Ren, P. et al. (2017). Localized delivery of shRNA against PHD2 protects the heart from acute myocardial infarction through ultrasound-targeted cationic microbubble destruction. *Theranostics* 7 (1): 51–66.

51 Ghanem, A., Steingen, C., Brenig, F. et al. (2009). Focused ultrasound-induced stimulation of microbubbles augments site-targeted engraftment of mesenchymal stem cells after acute myocardial infarction. *Journal of Molecular and Cellular Cardiology* 47 (3): 411–418.

52 Zhong, S., Shu, S., Wang, Z. et al. (2012). Enhanced homing of mesenchymal stem cells to the ischemic myocardium by ultrasound-targeted microbubble destruction. *Ultrasonics* 52 (2): 281–286.

53 Tong, J., Ding, J., Shen, X. et al. (2013). Mesenchymal stem cell transplantation enhancement in myocardial infarction rat model under ultrasound combined with nitric oxide microbubbles. *PLoS One* 8 (11): e80186.

54 Toma, C., Fisher, A., Wang, J. et al. (2011). Vascular endoluminal delivery of mesenchymal stem cells using acoustic radiation force. *Tissue Engineering. Part A* 17 (9–10): 1457–1464.

55 Sun, Z., Xie, Y., Lee, R.J. et al. (2020). Myocardium-targeted transplantation of PHD2 shRNA-modified bone mesenchymal stem cells through ultrasound-targeted microbubble destruction protects the heart from acute myocardial infarction. *Theranostics* 10 (11): 4967–4982.

56 Palumbo, P., Daffinà, J., Bruno, F. et al. (2021). Basics in magnetic resonance guided focused ultrasound: technical basis and clinical application. A brief overview. *Acta Biomed* 92 (S5): e2021403.

57 Weintraub, D. and Elias, W.J. (2017). The emerging role of transcranial magnetic resonance imaging-guided focused ultrasound in functional neurosurgery. *Movement Disorders* 32 (1): 20–27.

58 McDannold, N., King, R.L., Jolesz, F.A. et al. (2002). The use of quantitative temperature images to predict the optimal power for focused ultrasound surgery: in vivo verification in rabbit muscle and brain. *Medical Physics* 29 (3): 356–365.

59 Cosgrove, G.R., Lipsman, N., Lozano, A.M. et al. (2022). Magnetic resonance imaging-guided focused ultrasound thalamotomy for essential tremor: 5-year follow-up results. *Journal of Neurosurgery* 1: 1–6.

60 Lescrauwaet, E., Vonck, K., Sprengers, M. et al. (2022). Recent advances in the use of focused ultrasound as a treatment for epilepsy. *Frontiers in Neuroscience* 16: 886584.

61 Baek, H., Lockwood, D., Mason, E.J. et al. (2022). Clinical intervention using focused ultrasound (FUS) stimulation of the brain in diverse neurological disorders. *Frontiers in Neurology* 13: 880814.

62 Kilroy, J.P., Klibanov, A.L., Wamhoff, B.R. et al. (2012). Intravascular ultrasound catheter to enhance microbubble-based drug delivery via acoustic radiation force. *IEEE Transactions on Ultrasonics, Ferroelectrics, and Frequency Control* 59 (10): 2156–2166.

63 Kaya, M., Toma, C., Wang, J. et al. (2012). Acoustic radiation force for vascular cell therapy: in vitro validation. *Ultrasound in Medicine & Biology* 38 (11): 1989–1997.

64 Straub, J.A., Chickering, D.E., Church, C.C. et al. (2005). Porous PLGA microparticles: AI-700, an intravenously administered ultrasound contrast agent for use in echocardiography. *Journal of Controlled Release* 108 (1): 21–32.

65 Goldberg, B.B., Liu, J.B., Burns, P.N. et al. (1993). Galactose-based intravenous sonographic contrast agent: experimental studies. *Journal of Ultrasound in Medicine* 12 (8): 463–470.

66 Kwan, J.J. and Borden, M.A. (2012). Lipid monolayer collapse and microbubble stability. *Advances in Colloid and Interface Science* 183–184: 82–99.

67 Kurosaki, T., Kawakami, S., Higuchi, Y. et al. (2014). Development of anionic bubble lipopolyplexes for efficient and safe gene transfection with ultrasound exposure in mice. *Journal of Controlled Release: Official Journal of the Controlled Release Society* 176: 24–34.

68 Wang, D.S., Panje, C., Pysz, M.A. et al. (2012). Cationic versus neutral microbubbles for ultrasound-mediated gene delivery in cancer. *Radiology* 264 (3): 721–732.

69 Fisher, N.G., Christiansen, J.P., Klibanov, A. et al. (2002). Influence of microbubble surface charge on capillary transit and myocardial contrast enhancement. *Journal of the American College of Cardiology* 40 (4): 811–819.

70 Nomikou, N., Tiwari, P., Trehan, T. et al. (2012). Studies on neutral, cationic and biotinylated cationic microbubbles in enhancing ultrasound-mediated gene delivery in vitro and in vivo. *Acta Biomaterialia* 8 (3): 1273–1280.

71 Negishi, Y., Endo-Takahashi, Y., and Maruyama, K. (2016). Gene delivery systems by the combination of lipid bubbles and ultrasound. *Drug Discoveries & Therapeutics* 10 (5): 248–255.

72 Omata, D., Negishi, Y., Suzuki, R. et al. (2015). Nonviral gene delivery systems by the combination of bubble liposomes and ultrasound. *Advances in Genetics* 89: 25–48.

73 Wu, S.Y., Chen, C.C., Tung, Y.S. et al. (2015). Effects of the microbubble shell physicochemical properties on ultrasound-mediated drug delivery to the brain. *Journal of Controlled Release* 212: 30–40.

74 Bloch, S.H., Wan, M., Dayton, P.A. et al. (2004). Optical observation of lipid- and polymer-shelled ultrasound microbubble contrast agents. *Applied Physics Letters* 84: 631–633.

75 Bouakaz, A., Versluis, M., and de Jong, N. (2005). High-speed optical observations of contrast agent destruction. *Ultrasound in Medicine & Biology* 31 (3): 391–399.

76 Liu, Y., Yan, J., and Prausnitz, M.R. (2012). Can ultrasound enable efficient intracellular uptake of molecules? A retrospective literature review and analysis. *Ultrasound in Medicine & Biology* 38 (5): 876–888.

77 Hume, D.A., Irvine, K.M., and Pridans, C. (2019). The mononuclear phagocyte system: the relationship between monocytes and macrophages. *Trends in Immunology* 40 (2): 98–112.

78 Khalili, K., Atri, M., Kim, T.K. et al. (2021). Recognizing the role of the reticuloendothelial system in the late phase of US contrast agents. *Radiology* 298 (2): 287–291.

79 Moriyasu, F. and Itoh, K. (2009). Efficacy of perflubutane microbubble-enhanced ultrasound in the characterization and detection of focal liver lesions: phase 3 multicenter clinical trial. *American Journal of Roentgenology* 193 (1): 86–95.

80 Yan, J., Yin, M., Foster, F.S. et al. (2020). Tumor contrast imaging with gas vesicles by circumventing the reticuloendothelial system. *Ultrasound in Medicine & Biology* 46 (2): 359–368.

81 Wrenn, S.P., Dicker, S.M., Small, E.F. et al. (2012). Bursting bubbles and bilayers. *Theranostics* 2 (12): 1140–1159.

82 Wang, S., Hossack, J.A., and Klibanov, A.L. (2018). Targeting of microbubbles: contrast agents for ultrasound molecular imaging. *Journal of Drug Targeting* 26 (5–6): 420–434.

83 Klibanov, A.L. (2009). Preparation of targeted microbubbles: ultrasound contrast agents for molecular imaging. *Medical & Biological Engineering & Computing* 47 (8): 875–882.

84 Wang, Y., Feng, Y., Yang, X. et al. (2022). Diagnosis of atherosclerotic plaques using vascular endothelial growth factor receptor-2 targeting antibody nano-microbubble as ultrasound contrast agent. *Computational and Mathematical Methods in Medicine* 2022: 6524592.

85 Li, J., Zhang, Y., Zou, C. et al. (2022). Binding properties of flowing fibrin-targeted microbubbles evaluated with a thrombus-embedded microchannel. *Lab on a Chip* 22 (12): 2292–2298.

86 Dundas, C.M., Demonte, D., and Park, S. (2013). Streptavidin-biotin technology: improvements and innovations in chemical and biological applications. *Applied Microbiology and Biotechnology* 97 (21): 9343–9353.

87 Hamano, N., Kamoshida, S., Kikkawa, Y. et al. (2019). Development of antibody-modified nanobubbles using Fc-region-binding polypeptides for ultrasound imaging. *Pharmaceutics* 11 (6): 283.

88 Li, Y., Du, M., Fang, J. et al. (2021). UTMD promoted local delivery of miR-34a-mimic for ovarian cancer therapy. *Drug Delivery* 28 (1): 1616–1625.

89 Dong, L., Li, N., Wei, X. et al. (2022). A Gambogic acid-loaded delivery system mediated by ultrasound-targeted microbubble destruction: a promising therapy method for malignant cerebral glioma. *International Journal of Nanomedicine* 17: 2001–2017.

90 Zhang, L., Sun, L., Tang, Q. et al. (2022). Cascade drug delivery through tumor barriers of pancreatic cancer via ultrasound in combination with functional microbubbles. *ACS Biomaterials Science & Engineering* 8 (4): 1583–1595.

91 Lentacker, I., De Geest, B.G., Vandenbroucke, R.E. et al. (2006). Ultrasound-responsive polymer-coated microbubbles that bind and protect DNA. *Langmuir* 22 (17): 7273–7278.

92 Lin, C.W., Fan, C.H., and Yeh, C.K. (2021). The impact of surface drug distribution on the acoustic behavior of DOX-loaded microbubbles. *Pharmaceutics* 13 (12): 2080.

93 Endo-Takahashi, Y., Negishi, Y., Kato, Y. et al. (2012). Efficient siRNA delivery using novel siRNA-loaded bubble liposomes and ultrasound. *International Journal of Pharmaceutics* 422 (1–2): 504–509.

94 Koch, S., Pohl, P., Cobet, U. et al. (2000). Ultrasound enhancement of liposome-mediated cell transfection is caused by cavitation effects. *Ultrasound in Medicine & Biology* 26 (5): 897–903.

95 Sun, R.R., Noble, M.L., Sun, S.S. et al. (2014). Development of therapeutic microbubbles for enhancing ultrasound-mediated gene delivery. *Journal of Controlled Release* 182: 111–120.

96 Panje, C.M., Wang, D.S., Pysz, M.A. et al. (2012). Ultrasound-mediated gene delivery with cationic versus neutral microbubbles: effect of DNA and microbubble dose on in vivo transfection efficiency. *Theranostics* 2 (11): 1078–1091.

97 Tinkov, S., Winter, G., Coester, C. et al. (2010). New doxorubicin-loaded phospholipid microbubbles for targeted tumor therapy: part I–Formulation development and in-vitro characterization. *Journal of Controlled Release* 143 (1): 143–150.

98 Shi, F., Yang, F., He, X. et al. (2016). Inhibitory effect of epirubicin-loaded lipid microbubbles with conjugated anti-ABCG2 antibody combined with therapeutic ultrasound on multiple myeloma cancer stem cells. *Journal of Drug Targeting* 24 (1): 34–46.

99 Shi, F., Li, M., Wang, J. et al. (2018). Induction of multiple myeloma cancer stem cell apoptosis using conjugated anti-ABCG2 antibody with epirubicin-loaded microbubbles. *Stem Cell Research & Therapy* 9 (1): 144.

100 Negishi, Y., Endo-Takahashi, Y., Ishii, K. et al. (2011). Development of novel nucleic acid-loaded bubble liposomes using cholesterol-conjugated siRNA. *Journal of Drug Targeting* 19 (9): 830–836.

101 Unger, E.C., McCreery, T.P., Sweitzer, R.H. et al. (1998). Acoustically active lipospheres containing paclitaxel: a new therapeutic ultrasound contrast agent. *Investigative Radiology* 33 (12): 886–892.

102 Browning, R., Thomas, N., Marsh, L.K. et al. (2021). Ultrasound-triggered delivery of iproplatin from microbubble-conjugated liposomes. *ChemistryOpen* 10 (12): 1170–1176.

103 Kheirolomoom, A., Dayton, P.A., Lum, A.F. et al. (2007). Acoustically-active microbubbles conjugated to liposomes: characterization of a proposed drug delivery vehicle. *Journal of Controlled Release* 118 (3): 275–284.

104 Peyman, S.A., Abou-Saleh, R.H., McLaughlan, J.R. et al. (2012). Expanding 3D geometry for enhanced on-chip microbubble production and single step formation of liposome modified microbubbles. *Lab on a Chip* 12 (21): 4544–4552.

105 De Temmerman, M.L., Dewitte, H., Vandenbroucke, R.E. et al. (2011). mRNA-Lipoplex loaded microbubble contrast agents for ultrasound-assisted transfection of dendritic cells. *Biomaterials* 32 (34): 9128–9135.

106 Zhang, J., Wang, S., Deng, Z. et al. (2018). Ultrasound-triggered drug delivery for breast tumor therapy through iRGD-targeted paclitaxel-loaded liposome-microbubble complexes. *Journal of Biomedical Nanotechnology* 14 (8): 1384–1395.

107 Charalambous, A., Mico, V., McVeigh, L.E. et al. (2021). Targeted microbubbles carrying lipid-oil-nanodroplets for ultrasound-triggered delivery of the hydrophobic drug, combretastatin A4. *Nanomedicine* 36: 102401.

108 Logan, K.A., Nesbitt, H., Callan, B. et al. (2021). Synthesis of a gemcitabine-modified phospholipid and its subsequent incorporation into a single microbubble formulation loaded with paclitaxel for the treatment of pancreatic cancer using ultrasound-targeted microbubble destruction. *European Journal of Pharmaceutics and Biopharmaceutics* 165: 374–382.

109 Wallace, N. and Wrenn, S.P. (2015). Ultrasound triggered drug delivery with liposomal nested microbubbles. *Ultrasonics* 63: 31–38.

110 Bartolomeo, A., Dicker, S., Dierkes, J. et al. (2012). Nesting microbubbles: influence on acoustic activity and image brightness. *Bubble Science, Engineering & Technology* 4 (2): 78–84.

111 Liu, M., Dasgupta, A., Qu, N. et al. (2021). *Strategies to Maximize Anthracycline Drug Loading in Albumin Microbubbles.* ACS Biomaterials Science & Engineering.

112 Chen, H.K., Zhang, S.M., Chang, J.L. et al. (2018). Insonation of systemically delivered cisplatin-loaded microbubbles significantly attenuates nephrotoxicity of chemotherapy in experimental models of head and neck cancer. *Cancers (Basel)* 10 (9): 311.

113 Lee, S., Kim, J.H., Moon, H. et al. (2020). Combined treatment of sorafenib and doxorubicin-loaded microbubble-albumin nanoparticle complex for hepatocellular carcinoma: a feasibility study. *PLoS One* 15 (12): e0243815.

114 Shohet, R.V., Chen, S., Zhou, Y.T. et al. (2000). Echocardiographic destruction of albumin microbubbles directs gene delivery to the myocardium. *Circulation* 101 (22): 2554–2556.

115 Cochran, M.C., Eisenbrey, J., Ouma, R.O. et al. (2011). Doxorubicin and paclitaxel loaded microbubbles for ultrasound triggered drug delivery. *International Journal of Pharmaceutics* 414 (1–2): 161–170.

116 Delaney, L.J., Eisenbrey, J.R., Brown, D. et al. (2021). Gemcitabine-loaded microbubble system for ultrasound imaging and therapy. *Acta Biomaterialia* 130: 385–394.

117 Liu, M., Dasgupta, A., Koczera, P. et al. (2020). Drug loading in poly(butyl cyanoacrylate)-based polymeric microbubbles. *Molecular Pharmaceutics* 17 (8): 2840–2848.

118 Zullino, S., Argenziano, M., Stura, I. et al. (2018). From micro- to nano-multifunctional theranostic platform: effective ultrasound imaging is not just a matter of scale. *Molecular Imaging* 17: 1536012118778216.

119 Tachibana, K. and Kida, H. (2021). *Recent Trends in Application of Encapsulated Ultrafine Bubbles in Medicine*, 1e, 215–236. New York: Jenny Stanford Publishing.

120 ISO (2017). 20480-1:2017 Fine bubble technology — General principles for usage and measurement of fine bubbles — Part 1: Terminology.

121 Stokes, G.G. (1851). On the effect of the internal friction of fluids on the motion of pendulums. *Transactions of the Cambridge Philosophical Society* 9: 8–27.

122 Takahashi, M. (2005). ζ potential of microbubbles in aqueous solutions: electrical properties of the gas–water interface. *The Journal of Physical Chemistry B* 109 (46): 21858–21864.

123 Nirmalkar, N., Pacek, A.W., and Barigou, M. (2018). On the existence and stability of bulk nanobubbles. *Langmuir* 34 (37): 10964–10973.

124 Kida, H., Nishimura, K., Ogawa, K. et al. (2020). Nanobubble mediated gene delivery in conjunction with a hand-held ultrasound scanner. *Frontiers in Pharmacology* 11.

125 Watanabe, A., Sheng, H., Endo, H. et al. (2019). Echographic and physical characterization of albumin-stabilized nanobubbles. *Heliyon* 5 (6): e01907.

126 Lafond, M., Watanabe, A., Yoshizawa, S. et al. (2018). Cavitation-threshold determination and rheological-parameters estimation of albumin-stabilized nanobubbles. *Scientific Reports* 8 (1): 7472.

127 Batchelor, D.V.B., Abou-Saleh, R.H., Coletta, P.L. et al. (2020). Nested nanobubbles for ultrasound-triggered drug release. *ACS Applied Materials & Interfaces* 12 (26): 29085–29093.

128 Hu, C., Jiang, D., Wu, M. et al. (2020). Ultrasound-mediated nanobubble destruction (UMND) facilitates the delivery of VEGFR2-targeted CD-TK-loaded cationic nanobubbles in the treatment of bladder cancer. *Journal of Cancer Research and Clinical Oncology* 146 (6): 1415–1426.

129 Wu, M., Zhao, H., Guo, L. et al. (2018). Ultrasound-mediated nanobubble destruction (UMND) facilitates the delivery of A10-3.2 aptamer targeted and siRNA-loaded cationic nanobubbles for therapy of prostate cancer. *Drug Delivery* 25 (1): 226–240.

130 Akbaba, H., Erel-Akbaba, G., Kotmakçı, M. et al. (2020). Enhanced cellular uptake and gene silencing activity of survivin-siRNA via ultrasound-mediated nanobubbles in lung cancer cells. *Pharmaceutical Research* 37 (8): 165.

131 Yu, Z., Wang, Y., Xu, D. et al. (2020). G250 antigen-targeting drug-loaded nanobubbles combined with ultrasound targeted nanobubble destruction: a potential novel treatment for renal cell carcinoma. *International Journal of Nanomedicine* 15: 81–95.

132 Wang, J.P., Yan, J.P., Xu, J. et al. (2019). Paclitaxel-loaded nanobubble targeted to pro-gastrin-releasing peptide inhibits the growth of small cell lung cancer. *Cancer Management and Research* 11: 6637–6649.

133 Fang, K., Wang, L., Huang, H. et al. (2021). Therapeutic efficacy and cardio-protection of nucleolin-targeted doxorubicin-loaded ultrasound nanobubbles in treating triple-negative breast cancer. *Nanotechnology* 32 (24): 245102.

134 Peng, Y., Zhu, L., Wang, L. et al. (2019). Preparation of nanobubbles modified with a small-molecule CXCR4 antagonist for targeted drug delivery to tumors and enhanced ultrasound molecular imaging. *International Journal of Nanomedicine* 14: 9139–9157.

135 Yao, H., Zhang, L., Yan, S. et al. (2022). Low-intensity pulsed ultrasound/nanomechanical force generators enhance osteogenesis of BMSCs through microfilaments and TRPM7. *Journal of Nanobiotechnology* 20 (1): 378.

136 Endo-Takahashi, Y., Kurokawa, R., Sato, K. et al. (2021). Ternary complexes of pDNA, neuron-binding peptide, and PEGylated polyethyleneimine for brain delivery with nano-bubbles and ultrasound. *Pharmaceutics* 13 (7): 1003.

137 Yang, H., Shen, X., Yan, J. et al. (2018). Charge-reversal-functionalized PLGA nanobubbles as theranostic agents for ultrasonic-imaging-guided combination therapy. *Biomaterials Science* 6 (9): 2426–2439.

138 Li, T., Zhou, J., Zhang, C. et al. (2018). Surface-engineered nanobubbles with pH-/light-responsive drug release and charge-switchable behaviors for active NIR/MR/US imaging-guided tumor therapy. *NPG Asia Materials* 10 (11): 1046–1060.

139 Tayier, B., Deng, Z., Wang, Y. et al. (2019). Biosynthetic nanobubbles for targeted gene delivery by focused ultrasound. *Nanoscale* 11 (31): 14757–14768.

140 Jugniot, N., Massoud, T.F., Dahl, J.J. et al. (2022). Biomimetic nanobubbles for triple-negative breast cancer targeted ultrasound molecular imaging. *Journal of Nanobiotechnology* 20 (1): 267.

141 Li, H., Wang, Z., Zhang, J. et al. (2019). Enhanced shRNA delivery by the combination of polyethylenimine, ultrasound, and nanobubbles in liver cancer. *Technology and Health Care* 27 (S1): 263–272.

142 Prabhakar, A. and Banerjee, R. (2019). Nanobubble liposome complexes for diagnostic imaging and ultrasound-triggered drug delivery in cancers: a theranostic approach. *ACS Omega* 4 (13): 15567–15580.

143 Kida, H., Feril, L.B., Irie, Y. et al. (2022). Influence of nanobubble size distribution on ultrasound-mediated plasmid DNA and messenger RNA gene delivery. *Frontiers in Pharmacology* 13: 855495.

144 Lea-Banks, H., O'Reilly, M.A., and Hynynen, K. (2019). Ultrasound-responsive droplets for therapy: a review. *Journal of Controlled Release* 293: 144–154.

145 Aliabouzar, M., Kumar, K.N., and Sarkar, K. (2018). Acoustic vaporization threshold of lipid-coated perfluoropentane droplets. *The Journal of the Acoustical Society of America* 143 (4): 2001.

146 Gorelikov, I., Martin, A.L., Seo, M. et al. (2011). Silica-coated quantum dots for optical evaluation of perfluorocarbon droplet interactions with cells. *Langmuir* 27 (24): 15024–15033.

147 Giesecke, T. and Hynynen, K. (2003). Ultrasound-mediated cavitation thresholds of liquid perfluorocarbon droplets in vitro. *Ultrasound in Medicine & Biology* 29 (9): 1359–1365.

148 Schad, K.C. and Hynynen, K. (2010). In vitro characterization of perfluorocarbon droplets for focused ultrasound therapy. *Physics in Medicine and Biology* 55 (17): 4933–4947.

149 Maples, D., McLean, K., Sahoo, K. et al. (2015). Synthesis and characterisation of ultrasound imageable heat-sensitive liposomes for HIFU therapy. *International Journal of Hyperthermia* 31 (6): 674–685.

150 Shpak, O., Verweij, M., Vos, H.J. et al. (2014). Acoustic droplet vaporization is initiated by superharmonic focusing. *Proceedings of the National Academy of Sciences of the United States of America* 111 (5): 1697–1702.

151 Cao, Y., Chen, Y., Yu, T. et al. (2018). Drug release from phase-changeable nanodroplets triggered by low-intensity focused ultrasound. *Theranostics* 8 (5): 1327–1339.

152 Rapoport, N., Payne, A., Dillon, C. et al. (2013). Focused ultrasound-mediated drug delivery to pancreatic cancer in a mouse model. *The Journal of Therapeutic Ultrasound* 1: 11.

153 Rapoport, N., Kennedy, A.M., Shea, J.E. et al. (2010). Ultrasonic nanotherapy of pancreatic cancer: lessons from ultrasound imaging. *Molecular Pharmaceutics* 7 (1): 22–31.

154 Rapoport, N.Y., Kennedy, A.M., Shea, J.E. et al. (2009). Controlled and targeted tumor chemotherapy by ultrasound-activated nanoemulsions/microbubbles. *Journal of Controlled Release* 138 (3): 268–276.

155 Dayton, P.A., Zhao, S., Bloch, S.H. et al. (2006). Application of ultrasound to selectively localize nanodroplets for targeted imaging and therapy. *Molecular Imaging* 5 (3): 160–174.

156 Rapoport, N., Gao, Z., and Kennedy, A. (2007). Multifunctional nanoparticles for combining ultrasonic tumor imaging and targeted chemotherapy. *Journal of the National Cancer Institute* 99 (14): 1095–1106.

157 Gao, D., Xu, M., Cao, Z. et al. (2015). Ultrasound-triggered phase-transition cationic nanodroplets for enhanced gene delivery. *ACS Applied Materials & Interfaces* 7 (24): 13524–13537.

158 Gao, D., Gao, J., Xu, M. et al. (2017). Targeted ultrasound-triggered phase transition nanodroplets for Her2-overexpressing breast cancer diagnosis and gene transfection. *Molecular Pharmaceutics* 14 (4): 984–998.

159 Ishijima, A., Minamihata, K., Yamaguchi, S. et al. (2017). Selective intracellular vaporisation of antibody-conjugated phase-change nano-droplets in vitro. *Scientific Reports* 7: 44077.

160 Zhang, X., Hu, J., Zhao, G. et al. (2017). PEGylated PLGA-based phase shift nanodroplets combined with focused ultrasound for blood brain barrier opening in rats. *Oncotarget* 8 (24): 38927–38936.

161 Fix, S.M., Novell, A., Yun, Y. et al. (2017). An evaluation of the sonoporation potential of low-boiling point phase-change ultrasound contrast agents in vitro. *The Journal of Therapeutic Ultrasound* 5: 7.

162 Chen, C.C., Sheeran, P.S., Wu, S.Y. et al. (2013). Targeted drug delivery with focused ultrasound-induced blood-brain barrier opening using acoustically-activated nanodroplets. *Journal of Controlled Release* 172 (3): 795–804.

163 Bérard, C., Desgranges, S., Dumas, N. et al. (2022). Perfluorocarbon nanodroplets as potential nanocarriers for brain delivery assisted by focused ultrasound-mediated blood-brain barrier disruption. *Pharmaceutics* 14 (7): 1498.

164 Zhang, N., Li, J., Hou, R. et al. (2017). Bubble-generating nano-lipid carriers for ultrasound/CT imaging-guided efficient tumor therapy. *International Journal of Pharmaceutics* 534 (1–2): 251–262.

165 Wan, Z., Li, C., Gu, J. et al. (2021). Accurately controlled delivery of temozolomide by biocompatible UiO-66-NH(2) through ultrasound to enhance the antitumor efficacy and attenuate the toxicity for treatment of malignant glioma. *International Journal of Nanomedicine* 16: 6905–6922.

166 Vaupel, P. (2008). Hypoxia and aggressive tumor phenotype: implications for therapy and prognosis. *The Oncologist* 13 (Suppl 3): 21–26.

167 Kizaka-Kondoh, S., Inoue, M., Harada, H. et al. (2003). Tumor hypoxia: a target for selective cancer therapy. *Cancer Science* 94 (12): 1021–1028.

168 Lacerda, Q., Rochani, A., Oeffinger, B. et al. (2022). Tumoral oxygenation and biodistribution of lonidamine oxygen microbubbles following localized ultrasound-triggered delivery. *International Journal of Pharmaceutics* 625: 122072.

169 Xu, Y., Liu, R., Yang, H. et al. (2022). Enhancing photodynamic therapy efficacy against cancer metastasis by ultrasound-mediated oxygen microbubble destruction to boost tumor-targeted delivery of oxygen and renal-clearable photosensitizer micelles. *ACS Applied Materials & Interfaces* 14 (22): 25197–25208.

170 Liu, C., Jin, Y., and Fan, Z. (2021). The mechanism of warburg effect-induced chemoresistance in cancer. *Frontiers in Oncology* 11: 698023.

171 Luo, W., Wen, G., Yang, L. et al. (2017). Dual-targeted and pH-sensitive doxorubicin prodrug-microbubble complex with ultrasound for tumor treatment. *Theranostics* 7 (2): 452–465.

172 Lv, Y., Hao, L., Hu, W. et al. (2016). Novel multifunctional pH-sensitive nanoparticles loaded into microbubbles as drug delivery vehicles for enhanced tumor targeting. *Scientific Reports* 6: 29321.

173 Gao, F., Wu, J., Niu, S. et al. (2019). Biodegradable, pH-sensitive hollow mesoporous organosilica nanoparticle (HMON) with controlled release of pirfenidone and ultrasound-target-microbubble-destruction (UTMD) for pancreatic cancer treatment. *Theranostics* 9 (20): 6002–6018.

174 De Palma, M., Biziato, D., and Petrova, T.V. (2017). Microenvironmental regulation of tumour angiogenesis. *Nature Reviews. Cancer* 17 (8): 457–474.

175 Zhang, C., Huang, P., Zhang, Y. et al. (2014). Anti-tumor efficacy of ultrasonic cavitation is potentiated by concurrent delivery of anti-angiogenic drug in colon cancer. *Cancer Letters* 347 (1): 105–113.

176 Jing, Y., Xiu-Juan, Z., Hong-Jiao, C. et al. (2019). Ultrasound-targeted microbubble destruction improved the antiangiogenic effect of Endostar in triple-negative breast carcinoma xenografts. *Journal of Cancer Research and Clinical Oncology* 145 (5): 1191–1200.

177 Wang, Z. (2017). ErbB receptors and cancer. *Methods in Molecular Biology* 1652: 3–35.

178 Liao, A.H., Chou, H.Y., Hsieh, Y.L. et al. (2015). Enhanced therapeutic epidermal growth factor receptor (EGFR) antibody delivery via pulsed ultrasound with targeting microbubbles for glioma treatment. *Journal of Medical and Biological Engineering* 35 (2): 156–164.

179 Hirabayashi, F., Iwanaga, K., Okinaga, T. et al. (2017). Epidermal growth factor receptor-targeted sonoporation with microbubbles enhances therapeutic efficacy in a squamous cell carcinoma model. *PLoS One* 12 (9): e0185293.

180 Sun, L., Zhang, J., Xu, M. et al. (2022). Ultrasound microbubbles mediated sonosensitizer and antibody co-delivery for highly efficient synergistic therapy on HER2-positive gastric cancer. *ACS Applied Materials & Interfaces* 14 (1): 452–463.

181 Han, Y., Liu, D., and Li, L. (2020). PD-1/PD-L1 pathway: current researches in cancer. *American Journal of Cancer Research* 10 (3): 727–742.

182 Kim, D., Lee, S.S., Moon, H. et al. (2020). PD-L1 targeting immune-microbubble complex enhances therapeutic index in murine colon cancer models. *Pharmaceuticals (Basel)* 14 (1): 6.

183 Fernández, M., Javaid, F., and Chudasama, V. (2018). Advances in targeting the folate receptor in the treatment/imaging of cancers. *Chemical Science* 9 (4): 790–810.

184 Modi, S.J. and Kulkarni, V.M. (2019). Vascular endothelial growth factor receptor (VEGFR-2)/KDR inhibitors: medicinal chemistry perspective. *Medicine in Drug Discovery* 2: 100009.

185 Ingram, N., McVeigh, L.E., Abou-Saleh, R.H. et al. (2020). Ultrasound-triggered therapeutic microbubbles enhance the efficacy of cytotoxic drugs by increasing circulation and tumor drug accumulation and limiting bioavailability and toxicity in normal tissues. *Theranostics* 10 (24): 10973–10992.

186 Liu, Z., Wang, F., and Chen, X. (2008). Integrin α(v)β(3)-targeted cancer therapy. *Drug Development Research* 69 (6): 329–339.

187 Li, J., Zhou, P., Xu, H. et al. (2019). Antitumor activity of integrin α(V)β(3) antibody conjugated-cationic microbubbles in liver cancer. *Translational Cancer Research* 8 (3): 899–908.

188 Dayton, P.A., Pearson, D., Clark, J. et al. (2004). Ultrasonic analysis of peptide- and antibody-targeted microbubble contrast agents for molecular imaging of alphavbeta3-expressing cells. *Molecular Imaging* 3 (2): 125–134.

189 Beekers, I., Langeveld, S.A.G., Meijlink, B. et al. (2022). Internalization of targeted microbubbles by endothelial cells and drug delivery by pores and tunnels. *Journal of Controlled Release* 347: 460–475.

190 Kondo, E., Iioka, H., and Saito, K. (2021). Tumor-homing peptide and its utility for advanced cancer medicine. *Cancer Science* 112 (6): 2118–2125.

191 Ludwig, B.S., Kessler, H., Kossatz, S. et al. (2021). RGD-binding integrins revisited: how recently discovered functions and novel synthetic ligands (re-)shape an ever-evolving field. *Cancers (Basel)* 13 (7): 1711.

192 Willmann, J.K., Kessler, H., Kossatz, S. et al. (2008). Dual-targeted contrast agent for US assessment of tumor angiogenesis in vivo. *Radiology* 248 (3): 936–944.

193 Warram, J.M., Sorace, A.G., Saini, R. et al. (2011). A triple-targeted ultrasound contrast agent provides improved localization to tumor vasculature. *Journal of Ultrasound in Medicine* 30 (7): 921–931.

194 Yuan, H.X., Wang, W.P., Wen, J.X. et al. (2018). Dual-targeted microbubbles specific to integrin αVβ3 and vascular endothelial growth factor receptor 2 for ultrasonography evaluation of tumor angiogenesis. *Ultrasound in Medicine & Biology* 44 (7): 1460–1467.

195 Matsumura, Y. and Maeda, H. (1986). A new concept for macromolecular therapeutics in cancer chemotherapy: mechanism of tumoritropic accumulation of proteins and the antitumor agent smancs. *Cancer Research* 46 (12 Pt 1): 6387–6392.

196 Duan, L., Yang, L., Jin, J. et al. (2020). Micro/nano-bubble-assisted ultrasound to enhance the EPR effect and potential theranostic applications. *Theranostics* 10 (2): 462–483.

197 Rapoport, N. (2012). Phase-shift, stimuli-responsive perfluorocarbon nanodroplets for drug delivery to cancer. *Wiley Interdisciplinary Reviews: Nanomedicine and Nanobiotechnology* 4 (5): 492–510.

198 Mantovani, F., Collavin, L., and Del Sal, G. (2019). Mutant p53 as a guardian of the cancer cell. *Cell Death and Differentiation* 26 (2): 199–212.

199 Chang, S., Guo, J., Sun, J. et al. (2013). Targeted microbubbles for ultrasound mediated gene transfection and apoptosis induction in ovarian cancer cells. *Ultrasonics Sonochemistry* 20 (1): 171–179.

200 Nande, R., Greco, A., Gossman, M.S. et al. (2013). Microbubble-assisted p53, RB, and p130 gene transfer in combination with radiation therapy in prostate cancer. *Current Gene Therapy* 13 (3): 163–174.

201 Zarogoulidis, P., Darwiche, K., Sakkas, A. et al. (2013). Suicide gene therapy for cancer – current strategies. *Journal of Genetic Syndromes and Gene Theraphy* 4: 16849.

202 Zhang, H., Li, Y., Rao, F. et al. (2020). A novel UTMD system facilitating nucleic acid delivery into MDA-MB-231 cells. *Bioscience Reports* 40 (2): BSR20192573.

203 Zhao, S., Xie, J., Zhao, C. et al. (2021). Ultrasound-targeted microbubble destruction enhances the inhibitive efficacy of miR-21 silencing in HeLa cells. *Medical Science Monitor* 27: e923660.

204 Goel, H.L. and Mercurio, A.M. (2013). VEGF targets the tumour cell. *Nature Reviews Cancer* 13 (12): 871–882.

205 Florinas, S., Kim, J., Nam, K. et al. (2014). Ultrasound-assisted siRNA delivery via arginine-grafted bioreducible polymer and microbubbles targeting VEGF for ovarian cancer treatment. *Journal of Controlled Release* 183: 1–8.

206 Xu, Q., Sun, T., Tian, H. et al. (2013). Ultrasound-mediated vascular endothelial growth factor C (VEGF-C) gene microbubble transfection inhibits growth of MCF-7 breast cancer cells. *Oncology Research* 20 (7): 297–301.

207 Chan, D.A. and Giaccia, A.J. (2010). PHD2 in tumour angiogenesis. *British Journal of Cancer* 103 (1): 1–5.

208 Li, A., Zhang, Y., Wang, Z. et al. (2019). The roles and signaling pathways of prolyl-4-hydroxylase 2 in the tumor microenvironment. *Chemico-Biological Interactions* 303: 40–49.

209 Liao, Y., Luo, H., He, Z. et al. (2019). A combination of UTMD-mediated HIF-1α shRNA transfection and TAE in the treatment of hepatic cancer. *BioMed Research International* 2019: 1937460.

210 Sun, S., Xu, Y., Fu, P. et al. (2018). Ultrasound-targeted photodynamic and gene dual therapy for effectively inhibiting triple negative breast cancer by cationic porphyrin lipid microbubbles loaded with HIF1α-siRNA. *Nanoscale* 10 (42): 19945–19956.

211 Zhang, L., Sun, Z., Ren, P. et al. (2015). Ultrasound-targeted microbubble destruction (UTMD) assisted delivery of shRNA against PHD2 into H9C2 cells. *PLoS One* 10 (8): e0134629.

212 Garg, H., Suri, P., Gupta, J.C. et al. (2016). Survivin: a unique target for tumor therapy. *Cancer Cell International* 16: 49.

213 Bae, Y.J., Yoon, Y.I., Yoon, T.J. et al. (2016). Ultrasound-guided delivery of siRNA and a chemotherapeutic drug by using microbubble complexes: in vitro and in vivo evaluations in a prostate cancer model. *Korean Journal of Radiology* 17 (4): 497–508.

214 Abbas, R. and Larisch, S. (2020). Targeting XIAP for promoting cancer cell death-the story of ARTS and SMAC. *Cell* 9 (3): 663.

215 Wang, P., Yin, T., Li, J. et al. (2016). Ultrasound-responsive microbubbles for sonography-guided siRNA delivery. *Nanomedicine* 12 (4): 1139–1149.

216 Sobhani, N. and Samadani, A.A. (2021). Implications of photodynamic cancer therapy: an overview of PDT mechanisms basically and practically. *Journal of the Egyptian National Cancer Institute* 33 (1): 34.

217 Zhao, R., Liang, X., Zhao, B. et al. (2018). Ultrasound assisted gene and photodynamic synergistic therapy with multifunctional FOXA1-siRNA loaded porphyrin microbubbles for enhancing therapeutic efficacy for breast cancer. *Biomaterials* 173: 58–70.

218 Liu, H.L., Hua, M.Y., Chen, P.Y. et al. (2010). Blood-brain barrier disruption with focused ultrasound enhances delivery of chemotherapeutic drugs for glioblastoma treatment. *Radiology* 255 (2): 415–425.

219 McDannold, N., Zhang, Y., Supko, J.G. et al. (2019). Acoustic feedback enables safe and reliable carboplatin delivery across the blood-brain barrier with a clinical focused ultrasound system and improves survival in a rat glioma model. *Theranostics* 9 (21): 6284–6299.

220 Zhang, D.Y., Dmello, C., Chen, L. et al. (2020). Ultrasound-mediated delivery of paclitaxel for glioma: a comparative study of distribution, toxicity, and efficacy of albumin-bound versus cremophor formulations. *Clinical Cancer Research* 26 (2): 477–486.

221 Wei, H.J., Upadhyayula, P.S., Pouliopoulos, A.N. et al. (2021). Focused ultrasound-mediated blood-brain barrier opening increases delivery and efficacy of etoposide for glioblastoma treatment. *International Journal of Radiation Oncology, Biology, Physics* 110 (2): 539–550.

222 Liu, H.L., Hsu, P.H., Lin, C.Y. et al. (2016). Focused ultrasound enhances central nervous system delivery of bevacizumab for malignant glioma treatment. *Radiology* 281 (1): 99–108.

223 Tran, V.L., Novell, A., Tournier, N. et al. (2020). Impact of blood-brain barrier permeabilization induced by ultrasound associated to microbubbles on the brain delivery and kinetics of cetuximab: an immunoPET study using (89)Zr-cetuximab. *Journal of Controlled Release* 328: 304–312.

224 Sheybani, N.D., Breza, V.R., Paul, S. et al. (2021). ImmunoPET-informed sequence for focused ultrasound-targeted mCD47 blockade controls glioma. *Journal of Controlled Release* 331: 19–29.

225 Park, E.-J., Zhang, Y.-Z., Vykhodtseva, N. et al. (2012). Ultrasound-mediated blood-brain/blood-tumor barrier disruption improves outcomes with trastuzumab in a breast cancer brain metastasis model. *Journal of Controlled Release: Official Journal of the Controlled Release Society* 163 (3): 277–284.

226 Chen, P.Y., Hsieh, H.Y., Huang, C.Y. et al. (2015). Focused ultrasound-induced blood-brain barrier opening to enhance interleukin-12 delivery for brain tumor immunotherapy: a preclinical feasibility study. *Journal of Translational Medicine* 13: 93.

227 Alkins, R., Burgess, A., Ganguly, M. et al. (2013). Focused ultrasound delivers targeted immune cells to metastatic brain tumors. *Cancer Research* 73 (6): 1892–1899.

228 Idbaih, A., Canney, M., Belin, L. et al. (2019). Safety and feasibility of repeated and transient blood-brain barrier disruption by pulsed ultrasound in patients with recurrent glioblastoma. *Clinical Cancer Research* 25 (13): 3793–3801.

229 Idbaih, A., Ducray, F., Stupp, R. et al. (2021). A phase I/IIa study to evaluate the safety and efficacy of blood-brain barrier (BBB) opening with the SonoCloud-9 implantable ultrasound device in recurrent glioblastoma patients receiving IV carboplatin. *Journal of Clinical Oncology* 39 (15_suppl): 2049–2049.

230 Mainprize, T., Lipsman, N., Huang, Y. et al. (2019). Blood-brain barrier opening in primary brain tumors with non-invasive MR-guided focused ultrasound: a clinical safety and feasibility study. *Scientific Reports* 9 (1): 321.

231 Dubey, S., Heinen, S., Krantic, S. et al. (2020). Clinically approved IVIg delivered to the hippocampus with focused ultrasound promotes neurogenesis in a model of Alzheimer's disease. *Proceedings of the National Academy of Sciences of the United States of America* 117 (51): 32691–32700.

232 Hsu, P.H., Lin, Y.T., Chung, Y.H. et al. (2018). Focused ultrasound-induced blood-brain barrier opening enhances GSK-3 inhibitor delivery for amyloid-beta plaque reduction. *Scientific Reports* 8 (1): 12882.

233 Leinenga, G., Koh, W.K., and Götz, J. (2021). A comparative study of the effects of Aducanumab and scanning ultrasound on amyloid plaques and behavior in the APP23 mouse model of Alzheimer disease. *Alzheimer's Research & Therapy* 13 (1): 76.

234 Lipsman, N., Meng, Y., Bethune, A.J. et al. (2018). Blood-brain barrier opening in Alzheimer's disease using MR-guided focused ultrasound. *Nature Communications* 9 (1): 2336–2336.

235 Mehta, R.I., Carpenter, J.S., Mehta, R.I. et al. (2021). Blood-brain barrier opening with MRI-guided focused ultrasound elicits meningeal venous permeability in humans with early Alzheimer disease. *Radiology* 298 (3): 654–662.

236 Karakatsani, M.E., Wang, S., Samiotaki, G. et al. (2019). Amelioration of the nigrostriatal pathway facilitated by ultrasound-mediated neurotrophic delivery in early Parkinson's disease. *Journal of Controlled Release* 303: 289–301.

237 Long, L., Cai, X., Guo, R. et al. (2017). Treatment of Parkinson's disease in rats by Nrf2 transfection using MRI-guided focused ultrasound delivery of nanomicrobubbles. *Biochemical and Biophysical Research Communications* 482 (1): 75–80.

238 Zhang, H., Sierra, C., Kwon, N. et al. (2018). Focused-ultrasound mediated anti-alpha-synuclein antibody delivery for the treatment of Parkinson's disease. In: *2018 IEEE International Ultrasonics Symposium (IUS)*. Kobe, Japan, 1–4.

239 Gasca-Salas, C., Fernández-Rodríguez, B., Pineda-Pardo, J.A. et al. (2021). Blood-brain barrier opening with focused ultrasound in Parkinson's disease dementia. *Nature Communications* 12 (1): 779.

240 Palesch, Y.Y., Yeatts, S.D., Tomsick, T.A. et al. (2015). Twelve-month clinical and quality-of-life outcomes in the interventional management of stroke III trial. *Stroke* 46 (5): 1321–1327.

241 Tran, D., Hays, N., Shah, A. et al. (2021). Ultrasound-assisted catheter directed thrombolysis for pulmonary embolus during extracorporeal membrane oxygenation. *Journal of Cardiac Surgery* 36 (8): 2685–2691.

242 Alexandrov, A.V., Tsivgoulis, G., Köhrmann, M. et al. (2019). Endovascular equipoise shift in a phase III randomized clinical trial of sonothrombolysis for acute ischemic stroke. *Therapeutic Advances in Neurological Disorders* 12: 1756286419860652–1756286419860652.

243 Zhu, Y., Guan, L., and Mu, Y. (2016). Combined low-frequency ultrasound and urokinase-containing microbubbles in treatment of femoral artery thrombosis in a rabbit model. *PLoS One* 11 (12): e0168909.

244 Guan, L., Wang, C., Yan, X. et al. (2020). A thrombolytic therapy using diagnostic ultrasound combined with RGDS-targeted microbubbles and urokinase in a rabbit model. *Scientific Reports* 10 (1): 12511.

245 Lin, C.Y., Tsai, C.H., Feng, L.Y. et al. (2019). Focused ultrasound-induced blood brain-barrier opening enhanced vascular permeability for GDNF delivery in Huntington's disease mouse model. *Brain Stimulation* 12 (5): 1143–1150.

8

Acoustic Levitation and Acoustic Holograms

Tatsuki Fushimi[1,2] *and Yoichi Ochiai*[1,2,3]

[1] *University of Tsukuba, Institute of Library, Information and Media Science, Kasuga 1-2, Tsukuba, Ibaraki, Japan*
[2] *University of Tsukuba, R&D Center for Digital Nature, Kasuga 1-2, Tsukuba, Ibaraki, Japan*
[3] *Pixie Dust Technologies, Inc. Misakicho 2-20-5 Chiyoda, Tokyo, Japan*

8.1 Introduction

Levitation is an unnatural and unintuitive technology. Nevertheless, it is gaining increasing attention for its application in the real world and opening new worlds of opportunity. For instance, levitation is used in technologies previously considered science fiction, such as maglev trains, quantum technologies, and volumetric displays. An object can be suspended via various methods, such as those based on electrostatics, magnetism, optics, or acoustics [1]. However, only acoustic forces can manipulate a levitation target, regardless of its material properties.

This chapter focuses on levitation techniques based on acoustophoretic forces (forces generated by acoustics) and discusses the basics of acoustic levitation, types of acoustic levitators, numerical simulations, and ways to optimize the levitators for specific purposes.

8.1.1 History of Acoustic Levitation

The history of the acoustic levitator begins in the twentieth century with Lord Rayleigh, who laid the initial foundation for acoustic radiation pressure [2, 3]. The levitation of objects with acoustic radiation forces was credited to Bücks and Müller in 1933, who demonstrated the levitation of liquid droplets (alcohol) using a standing wave [4]. As shown in Figure 8.1, a quartz crystal and reflector created a standing wave field in the space, and the alcohol droplets were levitated within that volume. The standing wave created stable acoustic traps, and the droplets were levitated at the nodes of the standing wave. The levitated objects were equidistant from each other as multiple nodes of the standing wave existed between the gaps (and each node became a trap).

One notable application of acoustic levitation, which truly took off in the sky – the usage of acoustic levitator in the Space Shuttle mission – was demonstrated by

Acoustic Technologies in Biology and Medicine, First Edition.
Edited by Adem Ozcelik, Ryan Becker, and Tony Jun Huang.

Figure 8.1 Levitation of alcohol droplets as demonstrated by Bücks & Müller. Source: Bücks and Müller [4]/Reproduced with permission from Springer Nature.

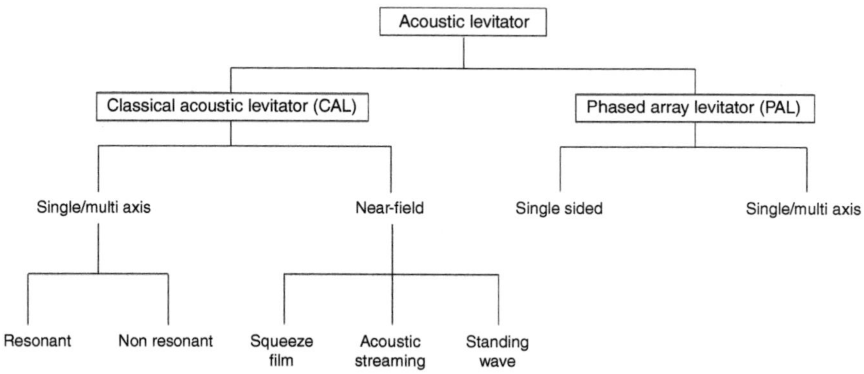

Figure 8.2 Categories and types of acoustic levitators available to date, including classical acoustic levitators (CALs) and phased-array levitators (PALs).

astronauts Dr. Taylor G. Wang and Dr. Eugene H. Trinh in 1985 and 1992 [5–9]. They used a closed resonant chamber with an audible frequency source (100–4000 Hz) [10], and they observed the spin of droplets and their oscillations in microgravity condition.

Since then, a wide range of acoustic levitators has been developed (see Figure 8.2). These levitators can be classified into two classes: classical acoustic levitators (CALs) with audible or Langevin transducers and phased array levitators (PALs) with a group of less-powerful small ultrasonic transducers.

8.1.1.1 Classical Acoustic Levitator

A CAL is characterized by the use of a few powerful transducers (typically Langevin transducers). It can be operated in either a single-/multi-axis or near-field configuration. The single-/multi-axis configuration enables the levitation of the object at least a few wavelengths away from the transducers using standing waves generated by transmitted and reflected waves. By contrast, the near-field configuration enables the levitation of the object at much closer distances using near-field effects.

Single-/Multi-Axis CAL

Single/multi-axis CALs can be further classified into resonant and nonresonant acoustic levitators. A resonant acoustic levitator requires the separation distance between the transducers or reflectors to be an integer multiple of a half-wavelength. This is because the resonant acoustic levitator obtains a high-pressure amplitude field by generating a standing wave based on multiple reflected waves from an opposing transducer/reflector. By contrast, a nonresonant acoustic levitator allows placing the reflector at an arbitrary distance, as it does not rely on volume resonance to obtain a high-pressure amplitude field.

In 1975, R.R. Whymark studied the characteristics and limitations of single-axis acoustic levitator, as shown in Figure 8.3; it comprised a transducer and reflector. The entire system was mounted on a rotatable jig so that it could be rotated, as shown in Figure 8.3. In general, as a resonant CAL can achieve a high-pressure amplitude, it can also obtain a high acoustic radiation force, thus allowing it to levitate dense materials, such as liquid mercury or iridium [12].

A multi-axis acoustic levitator can be constructed using multiple Langevin transducers. In the multi-axis configuration, the transducer axes can be orthogonal to each other [13] or parallel to each other (to increase the working area and transport between the transducers) [14]. The volume of the levitated object is an important factor in the stability of levitated object in a resonant CAL. By increasing the object volume relative to the resonance volume, the presence of the levitated object "detunes" the resonance, thereby causing the acoustic radiation force to act as an effective "negative damping" force in the field [15, 16].

A nonresonant CAL, as shown in Figure 8.4, does not require the separation distance between the reflector and transducer to be an integer multiple of a half-wavelength; additionally, it uses a limited number of reflected waves. Although this system was proposed in 1978 [18], the development of nonresonant CALs has been limited.

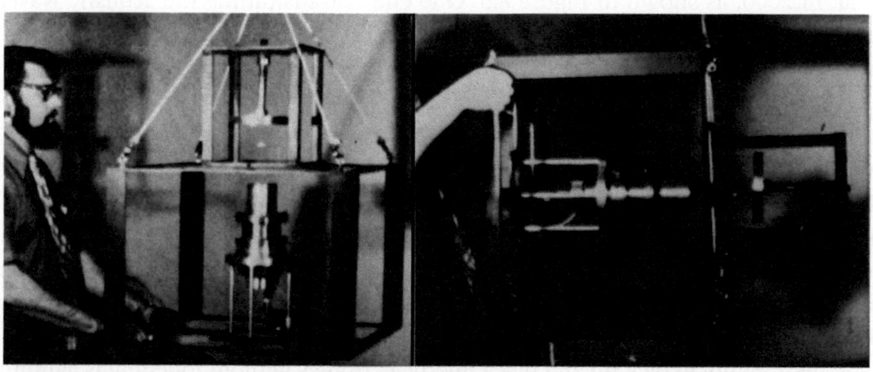

Figure 8.3 Single-axis acoustic levitator demonstrated by Whymark. It is mounted on a rotatable jig. Source: Reproduced from Whymark [11], with permission from Elsevier.

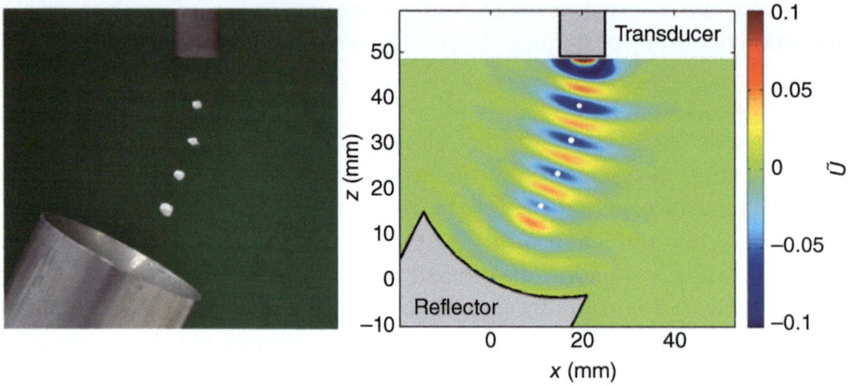

Figure 8.4 Nonresonant acoustic levitator using a single transducer and reflector. Since the transducer is small, the reflection is limited. Source: Reproduced from [17], with permission of AIP Publishing LLC.

Near-Field CAL

The near-field is defined as "region close to a source where the sound pressure and acoustic particle velocity are not in phase" and a region within "a wavelength of the sound or equal to three times the largest dimensions of the source" [19]. Near-field is rich with interesting phenomena, and three types of near-field levitators are possible: squeeze film, acoustic streaming, and standing wave.

Squeeze Film

Hashimoto et al. demonstrated squeeze film levitation in 1995, and as the name suggests, it achieves levitation by squeezing fluids between the vibrating surface and the levitated object [20]. The squeezing effect causes the time-averaged pressure to increase between the clearance and the surroundings [21]. Squeeze film levitation can levitate objects much heavier than single/multi-axis acoustic levitator, and Hashimoto et al. demonstrated levitation of up to 0.1 kg using their setup [20].

Noncontact transportation of levitated objects was also demonstrated by Ueha et al. in 2000, as shown in Figure 8.5. One end is transmitting the acoustic wave, and the other end is setup to receive the energy. This causes the plate to transmit a traveling wave along its axis, and the levitated object can follow the flow of energy.

Acoustic Streaming

Acoustic streaming is the "time-averaged streaming that is induced in any fluid flow... dominated by its fluctuating components" [22]. Although the setup is similar to that of the squeeze film levitator, the acoustic streaming generated by the transducer carries the air out of the clearance between the transducer and levitated object, thus generating a time-averaged pressure between the gaps. The development of near-field acoustic streaming-based levitators has been limited in the past; however, their ability to pick up and place microelectronics without touching is gaining renewed interest in the field [23].

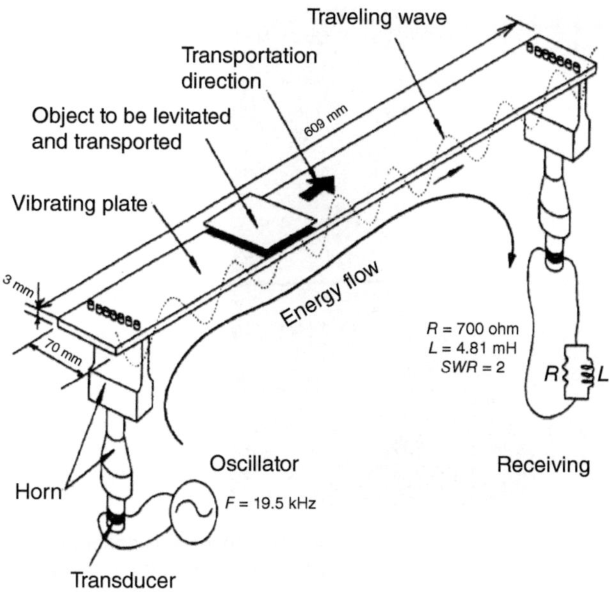

Figure 8.5 Squeeze film CAL designed by Ueha et al. By generating a traveling wave, the squeeze film levitated object can be made to follow its propagation axis. Source: Ueha et al. (2000)/Reproduced with permission of Elsevier.

Standing Wave

Standing wave near-field levitation is different from that of the single-/multi-axis resonant acoustic levitator in that the levitated object itself becomes the reflector. The standing wave generated in the cavity between the levitated object and the transducer generates a sufficient acoustic radiation force to levitate the object in midair. For example, a large expanded polystyrene sphere with a mass of 1.46 g and diameter of 50 mm can be levitated in midair using standing wave near-field levitation, as shown in Figure 8.6 [24]. Levitations of planar objects with lengths exceeding the wavelength were also demonstrated in 2017 [25].

8.1.1.2 Phased Array Levitator (PAL)

PALs represent a more recent innovation in acoustic levitation and are characterized by a group of low-power ultrasonic transducers. These ultrasonic transducers are often used for distance sensors in modern automobiles. Although a wide range of ultrasonic frequencies (ranging from 30 to 300 kHz) is available, the most common frequency is 40 kHz, with a wavelength of 8.6 mm in air.

In a PAL, each transducer is driven by independent drive signals, thus allowing each transducer to have a time offset and, therefore, a phase difference between each transducer. This can be achieved using a microcontroller [26] or a field-programmable gate array (FPGA) board [27, 28]. Further, acoustic holograms can be encoded with the amplitude specifications of phased arrays [29], or the amplitude and phase holograms can be combined [30]. Numerous PALs, ranging

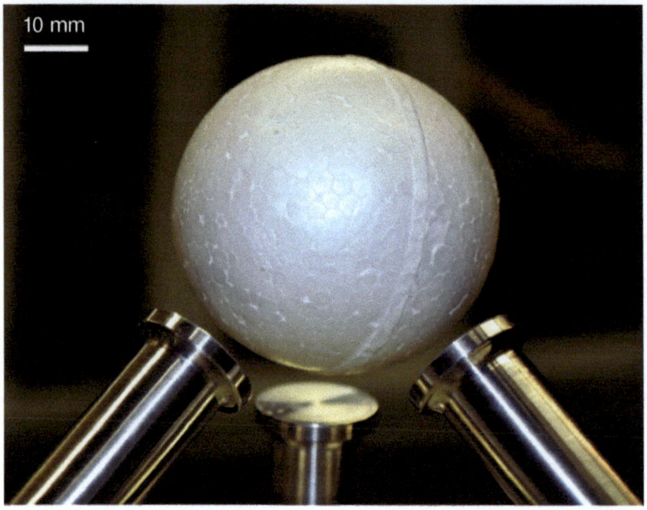

Figure 8.6 Near-field standing wave acoustic levitator. A standing wave is generated between the transducer and sphere and creates sufficient force with stability to levitate object in mid-air. Source: Reproduced from [24], with the permission of AIP Publishing LLC.

from a simple specimen-holding device [31, 32] to dynamically configurable device [26, 27, 33] are available to the public. These devices are easy and relatively cheap (<1500 US$) to manufacture, thus lowering the barrier to entry into the field.

Single/Multi-Axis Phased Array Levitator

In 2014, Ochiai et al. [28] proposed a multi-axis PAL, and Seah et al. [34] proposed a PAL with a Bessel beam. The multi-axis configuration is as shown in Figure 8.7, where four sets of phased arrays are used to focus acoustic energy within a volume. PAL is commonly operated solely in a non-resonance mode, and this enables the acoustic field to rapidly change the positions of levitated objects (up to $8.75\,\mathrm{ms}^{-1}$) [35]; in ideal conditions [36], the speed of the levitated particles can exceed $10\,\mathrm{ms}^{-1}$. However, the upper limit may be imposed not by the aerodynamic drag force or acoustic radiation force but by the bandwidth of the acoustic transducer used in the array [37]. One of the largest single-axis PALs to date was demonstrated by Inoue et al. in 2019 [38]; it comprised 1992 transducers and achieved the levitation of a Mie regime sphere (diameter, 3.5 λ) and octahedron (diagonal length, 5.9 λ).

Single-Sided PAL

Using a single-sided acoustic levitator for the manipulation of objects is beneficial as it enables a wider angle of access to observe the levitated object. The development of the single-sided PAL was driven by the development of an acoustic hologram optimizer for PALs (see Figure 8.8) in 2015 by Marzo et al. [39]. Three acoustic traps

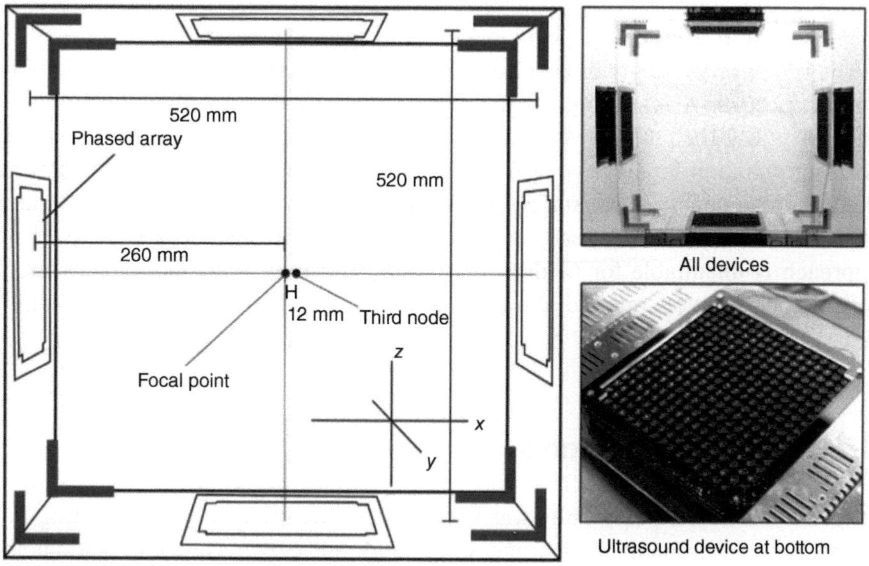

Figure 8.7 Multi-axis PAL by Ochiai et al. Source: Reproduced from Ochiai et al. [28], PLOS ONE/CC BY 4.0.

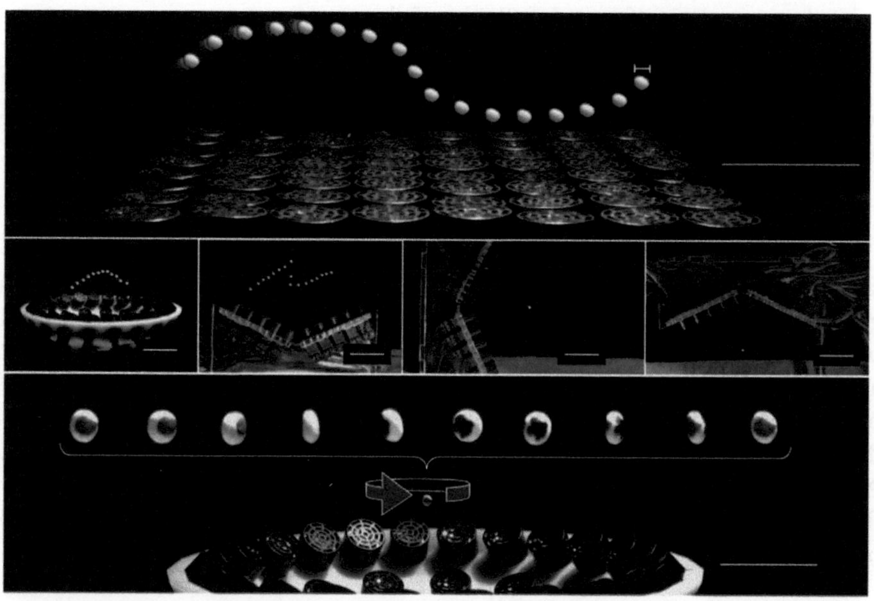

Figure 8.8 Single-sided acoustic levitator demonstrated by Marzo et al. Source: Reproduced from Marzo et al. [39], Springer Nature/CC BY 4.0.

(twins, bottle, and vortex traps) were identified in the process; these traps were used in single-/multi-axis acoustic levitators to stably levitate objects [39].

The levitation of Mie-regime objects using acoustic vortices [41] was first demonstrated in 2018. A scattering calculation via the boundary element method was necessary in 2019 [38]; however, recent advances in the analytical solutions for acoustic radiation forces have opened up more efficient approaches to calculating the forces. In general, acoustic radiation force in a single-sided acoustic levitator should not be calculated based on Gor'kov potential. This is because the Gor'kov approach is unsuitable for fields with stronger traveling wave modes. Instead, a wave decomposition method should be used to calculate the acoustic radiation force [42].

8.2 Acoustic Holograms

Originally, acoustic holography was used to describe methods for estimating the near-field measurements from a measured field; however, recently, acoustic holograms have been used to describe the methods for encoding three-dimensional (3D) acoustic fields into 2D fields. This is an important concept in the operation of PALs; moreover, the identification of a suitable acoustic hologram often requires optimization. However, acoustic hologram for Rayleigh regime spheres ($r \ll \frac{\lambda}{10}$) can be calculated using a straightforward method, as follows:

$$\phi_{final} = \phi_{focus} + \phi_{trap}$$

where ϕ_{focus} and ϕ_{trap} are the acoustic holograms for the focus and trap, respectively; and ϕ_{trap} is an acoustic hologram for encoding different types of traps (Table 8.1), such as twin, bottle, or vortex traps [39].

In the Table 8.1, $\boldsymbol{x_t} = (x_t, y_t, z_t)$ is the transducer position.

One common factor in these acoustic traps (Table 8.1) concerns the phase singularities in the trap. Bottle, vortex, and twin traps have point, line, and plane phase singularities, respectively; additionally, levitation with acoustic beams (such as Bessel beams) also has singularities at their cores [41] A singularity is defined as a point at which "a mathematical quantity is not defined". It represents a point where the pressure amplitude is zero with no phase information. A singularity appears as a sudden jump in phase; essentially, the derivative of phase is infinite at the singularity. As the acoustic radiation force is a function of the acoustic pressure and velocity field, singularities help enhance the acoustic radiation force applied to the levitated particle.

Practice has revealed that these traps can be translated within the work volume by simply adding the trap signatures to the acoustic focus. The singular acoustic focus can be identified as follows:

$$\phi_{focus} = -\frac{2\pi f_0}{c_0}[d(x_f, x_t) - d(0, x_f)]$$

Table 8.1 Types of acoustic traps.

Name	Trap hologram	Trap signature (white = 0, black = π)	Phase singularity
Bottle	$\phi_{trap}(x_t, y_t) = \begin{cases} 0, & r < T \\ \pi, & r \geq T \end{cases}$ where $r = \sqrt{x_t^2 + y_t^2}$, $T = 0.8(k/2)$ k is the width of transducer array	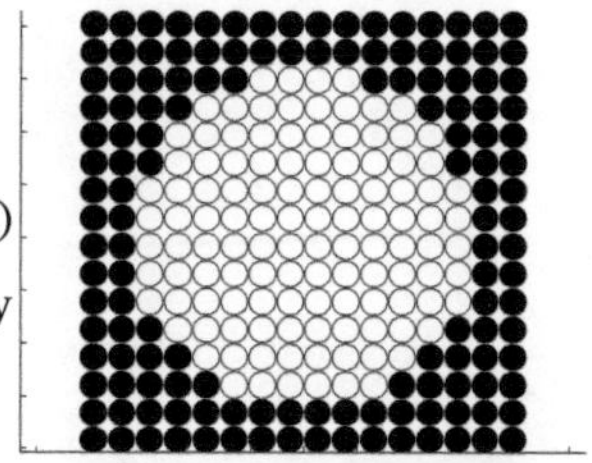	Point
Vortex	$\phi_{trap}(x_t, y_t) = atan2(x_t, y_t)$	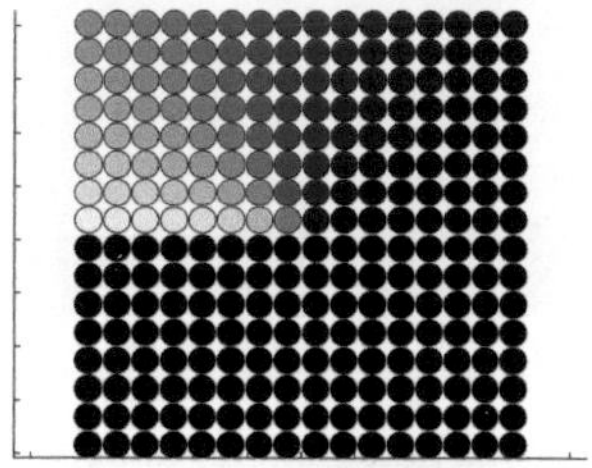	Line
Twin	$\phi_{trap}(x_t, y_t) = \begin{cases} 0, & x_t < 0 \\ \pi, & x_t \geq 0 \end{cases}$	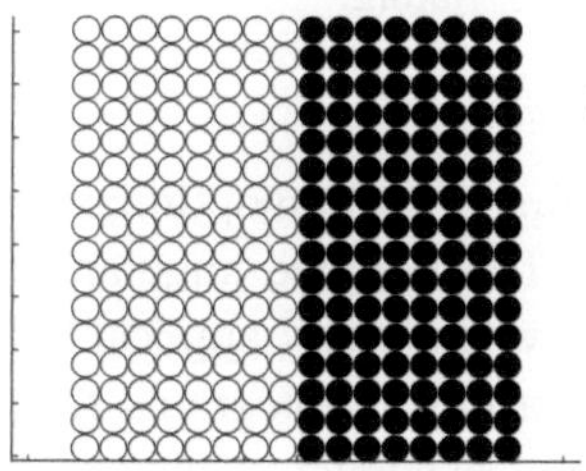	Plane

where $d(x_1, x_2)$ is the Euclidean distance between the points x_1 and x_2, and x_f is the focal point. Note that the practical applications of acoustic holograms often require more complex acoustic fields (for example, multiple acoustic foci or varying acoustic energies between two focal points) compared with simple analytical solutions such as those for focal points or beams (for example, Bessel, and Airy). Hence, acoustic holograms optimization method is important.

However, the optimization of acoustic hologram is a mathematically challenging problem, as the objective functions are often real-valued nonconvex functions of complex variables (i.e. f: $C \mapsto R$). For example, the targets of acoustic holograms may be amplitudes or phases at the target position. To evaluate such objectives, the wave needs to be propagated with an input guess for acoustic holograms; moreover, the arguments or magnitude of the field must be identified at this point. This process converts the complex functions to real values; additionally, as the objective function is nonconstant, "the derivatives of any order are not defined in its domain" (non-holomorphic) [43].

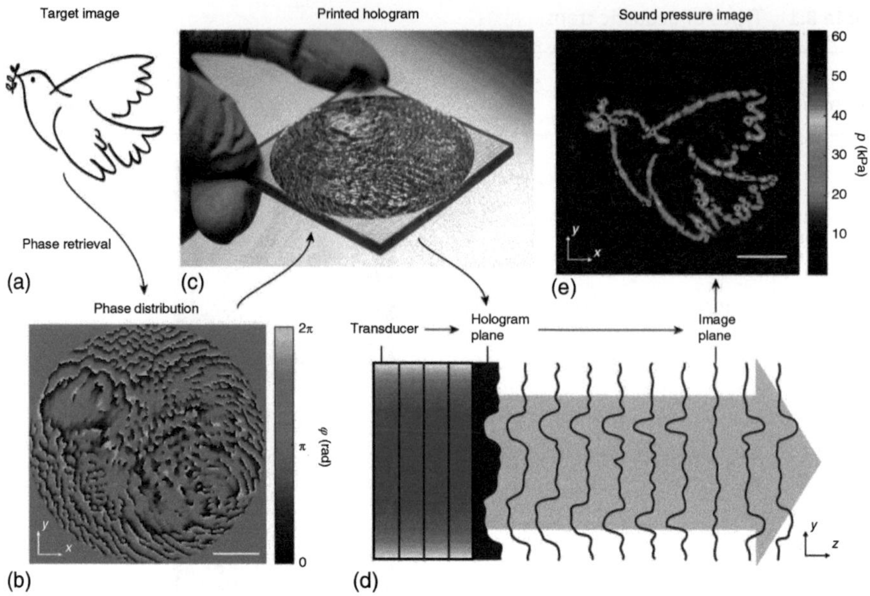

Figure 8.9 Acoustic hologram for underwater developed by Melde et al. Source: Melde et al. [44]/Springer Nature.

One of the first hologram optimization approaches was developed by Melde et al. in 2016 [44]. The iterative angular spectrum approach (IASA) is demonstrated in optics [45] was directly applicable in acoustic hologram optimization. In this approach, the acoustic holograms are physically encoded into an additively manufactured plate, as shown in Figure 8.9. The spatial distribution of the thickness represents the phase at each point as follows:

$$\Delta t(x,y) = \frac{\phi(x,y)}{k_m - k_h}$$

where $\Delta t(x, y)$, $\phi(x, y)$, and k are the thickness distribution of plate, phase, and wavenumber for the medium (m) and hologram (h), respectively.

Phase plates can be used in underwater acoustics as the impedance mismatch between the fluid medium and the phase plate is low. Despite the progress in material sciences, finding suitable materials (with a density and speed of sound as low as those of air) for phase plates in midair acoustics is challenging.

The best alternative to phase plates in airborne ultrasonics is the use of metamaterial bricks, as shown in Figure 8.10. This was proposed by Memoli et al. in 2017 [46]. This method guides the acoustic wave through the bricks to encode physical delays in the acoustic signals. It is designed to transmit a specified frequency (e.g. 40 kHz) without transmission losses; however, it can be expanded to include amplitude control [47].

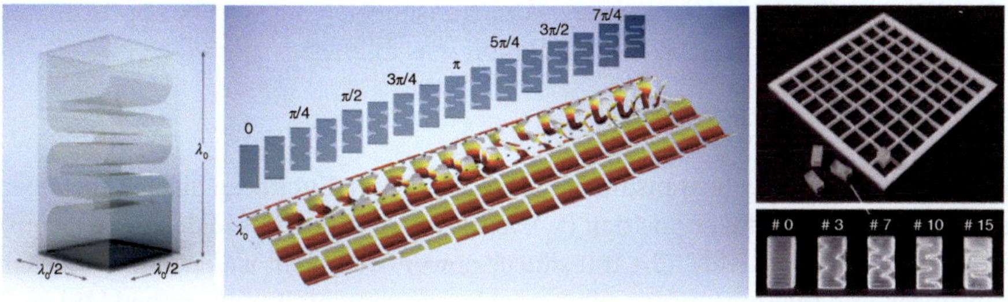

Figure 8.10 Metamaterial bricks developed by Memoli et al. Source: Reproduced from Memoli et al. [46], Springer Nature/CC BY 4.0.

Acoustic holograms are of three types: (A) phase, (B) amplitude, and (C) phase and amplitude acoustic holograms. Further, the aim of the optimizer' can be set to modify the (i) phase, (ii) amplitude, or (iii) phase and amplitude at the target point. Thus, an optimizer that has a phase-only hologram (A) and manipulates both phase and amplitude (iii) at the target point can be designated as "A:iii" optimizer.

8.3 Numerical Simulation of Acoustic Levitator

The efficient design and operation of acoustic levitation require a good numerical model. Acoustic waves are invisible to naked eye; thus, prior knowledge of the expected outcome is critical in debugging the experimental system. Numerical simulations also play a significant role in the optimization of an acoustic hologram. To be useful, the model must be fast, efficient, and accurate.

8.3.1 Pressure Field Calculation

Numerically predicting the pressure field emitted by the developed device helps in developing and optimizing the setups for specific purposes. This section presents an analytical approach to predicting the pressure field within the acoustic levitators.

8.3.1.1 Huygens' Approach

Huygens' approach to calculating the acoustic pressure field is one of the most common approaches for midair acoustic levitators. Simply put, the resulting pressure field is the sum of the pressure contributions from each transducer in the array, as follows:

$$p_{tot} = \sum_{t=0}^{T} p_t$$

where T is the number of transducers. For a transducer, which can be approximated as a piston source transducer, the calculation is as follows:

$$p_t = \frac{A_v}{r} D(\theta_t) e^{-j(kr(x,x_t)-\phi_t)}$$

where A_v, r, and ϕ_t are the transducer power coefficient, Euclidean distance between a point (x) and transducer position (x_t), and user specified phase delay (i.e. acoustic hologram), respectively. The transducer power coefficient is dependent on each transducer, and the transducer power must be experimentally determined [48]. Reference transducer power coefficients for common transducers are available in the supplementary material of TinyLev and coworkers [31]; however, it should be noted that the transducer power also depends on the driving signal (i.e. sine or square wave) and could vary between each device. It is therefore best to measure the average transducer power coefficient with a calibrated microphone in each device to get results with numerical simulation.

$D(\theta_t) = 1$ $(\theta_t = 0)$, $\frac{2J_1(ka\ \sin(\theta_t))}{ka\ \sin(\theta_t)}$ (otherwise) is the far-field directivity function for the piston source [49], where θ is the angle from the normal to the point in space. A useful approach for calculating this angle is to define the orientation of the transducer with a unit vector (n_t) and use the cosine rule, as follows:

$$\theta_t = \cos^{-1}\left(\frac{(x-x_t)\cdot n_t}{|(x-x_t)|\cdot|n_t|}\right)$$

This simplifies calculating the pressure field in cases where the phased-array transducers (PATs) are arranged in a non-flat configuration (i.e. spherical, cylindrical, or face-to-face).

For most of the optimization problem, the transducer position remains invariant and only the specified phase delay changes (i.e. parts within exponential function). Thus, an efficient algorithm can be constructed by generating a lookup table.

8.3.1.2 Spherical Harmonics Expansion

The spherical harmonic decomposition approach to PATs was introduced by Andersson and Ahrens [42] in 2019. Spherical harmonic decomposition is less efficient than Huygens' approach; however, it has the significant advantage of being able to calculate the scattered field from objects present in the field.

This is particularly useful when calculating acoustic radiation force for Mie-regime spheres. In 2021, Zehnter et al. [50] demonstrated optimization with the solution and expanded the solution to include a directivity function. The input pressure field from a PAT can be calculated as follows:

$$p_{in} = \sum_{n=0}^{\infty}\sum_{m=-n}^{n} j_n(kr) S_n^m Y_n^m(\theta, \varphi)$$

where $j_n(kr)$ are the nth-order spherical Bessel functions of the first kind; and (r, θ, φ) are spherical coordinates calculated from the center of the levitated sphere, and where $r = \sqrt{x^2+y^2+z^2}$, $\theta = \arccos\left(\frac{z}{r}\right)$ with $\theta \in [0, \pi]$, and $\varphi = \text{atan2}(y, x)$ with $\varphi \in (-\pi, \pi)$. Notably, there are two types of Bessel functions: spherical Bessel functions and Bessel functions; although they are similar, they have different formulations. $Y_n^m(\theta, \varphi)$ is the spherical harmonic bases. Spherical harmonics are widely implemented in popular packages, such as Scipy. S_n^m denotes the expansion coefficients and can be calculated as follows:

$$S_n^m = \sum_j^J q^j 4\pi ikh_n(kr^j)Y_n^m(\theta^j, \varphi^j)^* D_f(\xi^j)$$

where J is the number of the transducers in the PAT; $h_n(kr^j)$ denotes the nth-order spherical Hankel functions of the first kind; $(r^j, \theta^j, \varphi^j)$ are the spherical coordinates, which are calculated from the center of the levitated sphere to the transducer; $D_f(\xi^j)$ is the piston source directivity function, as mentioned earlier; and ξ^j denotes the angle between the transducer normal and vector $\boldsymbol{r} - \boldsymbol{r^j}$. Further, the scattered field can be calculated as follows:

$$p_s = \sum_{n=0}^{\infty} \sum_{m=-n}^{n} h_n(kr)\hat{S}_n^m Y_n^m(\theta, \varphi)$$

$\hat{S}_n^m$ for a compressible sphere is calculated as follows:

$$\hat{S}_n^m = -\frac{j_n(k_0a)j_n'(k_pa) + Zj_n^{\prime(k_0a)j_n}(k_pa)}{h_n(k_0a)j_n'(k_pa) - Zh_n^{\prime(k_0a)j_n}(k_pa)} S_n^m$$

where the symbol $'$ denotes the derivative of the function; $Z = \frac{\rho_p c_p}{\rho_0 c_0}$ is the relative impedance; the subscripts p and c denote the levitated sphere and surrounding medium respectively; and a denotes the radius of sphere. Thus, the total pressure field is calculated as follows: $p_{tot} = p_{in} + p_s$.

8.3.1.3 Angular Spectrum Method

The angular spectrum approach (ASM) predicts an acoustic pressure field using the principle of acoustic wave diffraction and propagates the spatial frequency components [51] as follows:

$$P(k_x, k_y, z) = P_{in}(k_x, k_y)H_u(k_x, k_y, z)$$

$$H_u(k_x, k_y, z) = \begin{cases} \dfrac{e^{-jz\sqrt{k^2-k_x^2-k_y^2}}}{j\sqrt{k^2-k_x^2-k_y^2}}, & k_x^2 + k_y^2 \le k^2 \\ \dfrac{e^{-z\sqrt{k_x^2+k_y^2-k^2}}}{\sqrt{k_x^2+k_y^2-k^2}}, & k_x^2 + k_y^2 > k^2 \end{cases}$$

where $P_{in}(k_x, k_y)$, $H_u(k_x, k_y, z)$, and $P(k_x, k_y, z)$ are the input pressure field, propagator, and resulting pressure field, respectively. The interchange between the spatial distribution of the pressure field $p(x, y)$ and temporal frequency space $P(k_x, k_y)$ is achieved by a 2D fast Fourier transform (FFT) and inverse FFT.

The ASM is frequently used in underwater acoustics in conjunction with an acoustic lens, but it is less frequently used for midair levitators. This is because the phase modulator in midair (for e.g. metabricks [46, 52]) is at the size of the wavelength and cannot encode much spatial data. However, one successful application of the ASM to acoustic levitation in midair was demonstrated by Melde et al. [44] (optimized acoustic wave reflection from a surface with 300 kHz ultrasound).

In comparison to an integral approach that calculates the acoustic pressure field at each observed point, the 2D FFT significantly increases the calculation speed. In addition, the ASM can easily incorporate attenuation losses in the transmitting media [51]. Although the spectral propagators encounter aliasing errors as $\sqrt{k_x^2 + k_y^2}$ approaches k, this error can be mitigated by restricting the angular frequency a low-pass frequency filter to the propagator. The open-source acoustic toolbox, k-Wave, includes ASM modules (with both angular frequency and attenuation losses), and readers wishing to implement the ASM can easily integrate it into their system [53].

8.3.2 Acoustic Radiation Force

Numerous formulations can be used for the calculation of the acoustic radiation force. Here, we limit ourselves to the formulations actively used for single-/multi-axis PALs (see Table 8.2).

8.3.2.1 Gor'kov

The calculation method by L.P. Gor'kov was published in 1962 in Soviet Physics Doklady [54, 55]. It is the standard method for calculating acoustic radiation and is applicable for a spherical particle in the Rayleigh regime (i.e. $r \ll \lambda$, typically less than 10% of the wavelength). It is formulated as follows.

$$F = -\left(\frac{dU}{dx}, \frac{dU}{dy}, \frac{dU}{dz}\right)$$

$$U = \frac{4\pi}{3}a^3\left[\frac{f_1}{2}\kappa_0\left\langle p_{in}^2\right\rangle - \frac{3f_2}{4}\rho_0\left\langle v_{in}^2\right\rangle\right]$$

Table 8.2 Acoustic radiation force calculation in PALs.

Approach	Particle geometry	Levitator configuration	Advantage	Disadvantage
Gor'kov	Sphere ($r \ll \lambda$)	Standing wave	The simplest, and efficient	Only applicable for spherical and Rayleigh regimes Not suitable for a system with strong traveling waves
Wave expansion	Sphere/ axisymmetric	Traveling/ standing wave	Efficient than finite element. Able to calculate Rayleigh and Mie regime scatterer	Requires calculation until the solution converges. Analytical solutions need to be derived for a specific case

where $f_1 = 1 - \frac{\kappa_p}{\kappa_0}$; $f_2 = \frac{2(\tilde{\rho}-1)}{\tilde{\rho}+1}$; $\tilde{\rho} = \frac{\rho_p}{\rho_0}$. $\kappa = \frac{1}{\rho c^2}$ is the wavenumber; ρ is the density; c is the speed of sound; a is the radius of the sphere; and the subscripts p and c indicate the properties for the particle and surrounding media, respectively. Recently, a group of Rayleigh regime spheres were used for approximating nonspherical objects such as sticks and calculating the acoustic radiation force and torque [56].

The Gor'kov method was shown to be less appropriate for a PAT in a traveling wave configuration (i.e. single sided). Instead, a spherical harmonic approach is more suitable for a transducer array in a single-sided configuration.

8.3.2.2 Spherical Harmonic Approach

Spherical harmonics decomposition, as mentioned earlier, has the added advantage of calculating the acoustic radiation force for Mie-regime targets and arrays in a single-sided configuration. Various studies have developed formulations for various types of acoustic fields and targets; however, here, we limit ourselves to equations directly applicable to PAT.

Using the pressure field calculation using spherical harmonics in Section 8.3.1.2 [42], the approach follows:

$$F_x = \frac{1}{8\rho_0 c_0^2 k^2} \mathrm{Re}\left\{\sum_{n=0}^{\infty}\sum_{m=-n}^{n} \Psi_n A_{nm} \cdot \left(S_{n,m}S^*_{n+1,m+1} - S_{n,-m}S^*_{n+1,-m-1}\right)\right\}$$

$$F_y = \frac{1}{8\rho_0 c_0^2 k^2} \mathrm{Im}\left\{\sum_{n=0}^{\infty}\sum_{m=-n}^{n} \Psi_n A_{nm} \cdot \left(S_{n,m}S^*_{n+1,m+1} + S_{n,-m}S^*_{n+1,-m-1}\right)\right\}$$

$$F_z = \frac{1}{8\rho_0 c_0^2 k^2} \mathrm{Re}\left\{\sum_{n=0}^{\infty}\sum_{m=-n}^{n} \Psi_n B_{nm} \cdot \left(S_{n,m}S^*_{n+1,m}\right)\right\}$$

where $\Psi_n = 2i\left(c_n + c^*_{n+1} + 2c_n c^*_{n+1}\right)$, $A_{nm} = \sqrt{\frac{(n+m+1)(n+m+2)}{(2n+1)(2n+3)}}$, and $B_{nm} = -2\sqrt{\frac{(n+m+1)(n-m+1)}{(2n+1)(2n+3)}}$, with * denoting the complex conjugates.

In 2022, Tang and Huang developed a scheme for calculating acoustic radiation force and torque for axisymmetric irregular objects levitated by PATs [57], and designed an interactive tool called "Soundiation" written in MATLAB [58].

8.4 Acoustic Hologram Optimization

Acoustic hologram optimization is rarely necessary in the context of acoustic radiation force. Acoustic hologram optimization is essential when manipulating more than one scatterer simultaneously (unless they are all on the same axis in a standing wave). Table 8.3 is a list of approaches useful for determining acoustic holograms for multi-focus purposes.

The Eigensolver approach was proposed by Long et al. in 2014 [30]. The problem was framed as a phase eigenproblem and solved using a weighted Tikhonov regularization. An improved version of Eingensolver approach was also published in 2020 [59].

Table 8.3 Acoustic hologram optimization.

Approach	Optimizers	Advantages	Disadvantages
Gerchberg–Saxton	39, 43, 57, 58	• Simple to implement • Great at maximizing amplitudes	• Moderate accuracy
Eigensolver approach	29, 57	• Moderately accurate	• Does not scale well with the number of elements • Inflexible with change in targets
Gradient descent	28, 38, 59	• Flexible objective functions • Scales well with the number of transducers when used with automatic differentiation	• Inefficient when optimized with finite differences
Direct approach	60	• Accurate generation of holograms • Flexible objective functions	• Slow unless tradeoffs are made
Deep learning	61, 62	• Fast execution time • No iteration	• Currently limited to fixed-size elements

The Gerchberg–Saxton (GS) method is an iterative approach that minimizes errors by transmitting and backpropagating the acoustic field between the transducer and target point/plane [60]. In 2018, Marzo and Drinkwater proposed an iterative back propagation (IBP) algorithm incorporating multiple focal point generation with a phase-only acoustic hologram [40]. In 2020, Plasencia et al. proposed the GS-PAT for rapidly calculating acoustic holograms (up to 23 kHz) [59].

Marzo et al. first applied the gradient descent algorithm in 2015 [39] to identify suitable acoustic traps for PALs using the Broyden–Fletcher–Goldfarb–Shanno (BFGS) algorithm. The gradient descent algorithm is a useful approach as it can implement wide range of objective functions. In 2021, Fushimi et al. demonstrated acoustic hologram optimization with stochastic gradient optimization and automatic differentiation (known as Diff-PAT). Automatic differentiation determines the gradient of a function using a computational graph. Unlike the numerical differentiation, it can determine the derivatives with the highest precision and efficiency [61], and hologram optimization with a high accuracy was demonstrated.

The direct approach is a more brute-force method that updates the solution with minor improvements. It has as high accuracy and flexibility as gradient descent [62]; however, its execution time is slower than that of gradient descent unless a suitable tradeoff is specified [63].

Finally, deep learning can be used to generate acoustic holograms, particularly with the recent advancements in machine learning. Lin et al. [64, 65] demonstrated an approach with U-net (a convolutional neural network developed for image segmentation) in 2021. Once trained, the deep learning neural network approach can determine a suitable acoustic hologram without iteration.

8.4.1 Optimization Example with Diff-PAT

In this section, we will describe an example of acoustic hologram optimization with a gradient descent approach, Diff-PAT [66]. Here, we show how every combination of transducer control (phase, amplitude, phase, and amplitude) and target point control (phase, amplitude, phase, and amplitude) can be combined together. To begin the optimization, we define the optimization function as [29]:

$$\underset{\tau}{minimize}\ \mathcal{L}\left(\boldsymbol{\tau}, \boldsymbol{\Omega}, \boldsymbol{x}_t, \boldsymbol{x}_c\right)$$

where $\mathcal{L}, \boldsymbol{\tau}_t, \boldsymbol{\Omega}, \boldsymbol{x}_t,$ *and* $\boldsymbol{x}_c$ are the loss function, acoustic hologram, target specification, transducer, and target point coordinates, respectively. The pressure field can be calculated using Huygens' as defined earlier:

$$p_a(x_c, x_t, A_t, \phi_t) = \sum_{t=0}^{T} \frac{P_{ref}}{d(x_c, x_t)} D(\theta) A_t e^{j(kd(x_c,x_t)+\phi_t)}$$

The definition of $\boldsymbol{\tau}_t$ depends on the control mode of the transducer array. For (A) phase-only control, $\tau_t = \emptyset_t$ with size T and $A_t = 1$ for all transducers. For (B) amplitude-only control, $\tau_t = A_t$ with size T and $\emptyset_t = 0$ for all transducers. For (C) phase and amplitude control $\tau_t = \Psi$ with $\Psi \in \mathbb{C}$ and size T. Since the optimizer cannot optimize complex numbers, the input is split into real and imaginary numbers $\tau_t = [\mathbb{R}(\Psi), \Im(\Psi)]$. In order to force the transducer magnitude A_t to 1, it was normalized as follows: $A_t = \frac{|\tau|}{\max(\tau)}$. The acoustic hologram is updated using a stochastic gradient descent algorithm, Adam which is calculated as follows [67].

$$\tau_t = \tau_{t-1} - \alpha \cdot \frac{\widehat{m_t}}{\sqrt{\widehat{v_t}} + \varepsilon}$$

where $\widehat{m_n} = \frac{m_t}{1-\beta_1^n}$, and $\widehat{v_n} = \frac{v_n}{1-\beta_2^n}$ are bias-corrected first and second raw moment estimates; $m_n = \beta_1 m_{n-1} + (1-\beta_1)\left(\frac{\delta\mathcal{L}}{\delta\tau}\right)$; and $v_n = \beta_2 v_{n-1} + (1-\beta_2)\left(\frac{\delta\mathcal{L}}{\delta\tau}\right)^2$. $\alpha, \beta_1, \beta_2,$ are hyperparameters, and are commonly 0.1, 0.9, 0.999, and 1×10^{-8}, respectively. $\frac{\delta\mathcal{L}}{\delta\tau}$ is the derivative of the loss function with respect to the acoustic hologram, and this can be automatically differentiated on machine learning packages such as TensorFlow, JAX, or PyTorch. The generalized loss function for optimization can be calculated as follows.

$$\mathcal{L} = \sum_{g}^{G} \frac{1}{G}\left(\left(A_a^c \cos\phi_a^c - A_g \cos\phi_g\right)^2 + \left(A_a^c \sin\phi_a^c - A_g \sin\phi_g\right)^2\right)$$

where superscript c denotes the control point, and $A_a^c = |p_a|$, $\phi_g = \text{Arg}(p_a)$. If you wish to (i) only set phase constraints at the target point, $\Omega = \emptyset_g$ with size G, while $A_a = A_g = 1$. For (ii) amplitude only control at the target point, $\Omega = A_g$ with size G, with $\emptyset_g = \emptyset_a = 0$. For (iii) phase and amplitude control at the target point, $\Omega = [A_g, \emptyset_g]$ with size $2G$ is set. An example code to implement the loss function is available in the data availability section of Fushimi et al. [29].

8.5 Applications in Biology and Medicine

Acoustic levitation can levitate and manipulate objects in midair and presents significant potential in the field of biology. Containerless processing minimizes the possibility of contamination. Moreover, the ability to manipulate targets regardless of their material properties is unique to acoustic levitation. Essentially, acoustic levitation can be used to hold specimens in midair for noncontact measurements; further, it can be combined with an experiment automation system to improve the efficiency of lab trials.

8.5.1 Specimen Holding

Noncontact holding of a specimen can minimize damage, contamination, and the effects of holders during measurements. In 2005, Wood et al. demonstrated the remote monitoring of algal cells using Raman spectroscopy [68], Puskar et al. demonstrated the spectroscopy of blood cells in 2007 [69]. Delißen et al. [70] and Seddon et al. [71] observed the agglomeration processes of proteins within a levitated liquid and self-assembly processes of nanomaterials using X-ray facilities in 2008 and 2016, respectively.

The levitation of insects is possible in both single-/multi-axis CALs and PALs. Xie et al. levitated a wide range of small animals, as shown in Figure 8.11. A 16.7 kHz single-axis CAL was used to levitate the animals. The results showed that small animals that live on the ground, in the air, or in water could be acoustically levitated. Zebrafish are a popular choice of fish in medicine, and several studies with single-axis CALs have observed the effects of the growth of zebrafish embryos under states of acoustic levitation [73, 74]. The asymmetry of the levitated object causes an acoustic radiation torque and, thus, instability in the levitated object. Cox et al. demonstrated the levitation of fruit flies and crickets using a single-axis PAL by "locking" the levitated objects using two types of acoustic traps [75].

8.5.2 Experiment Automation

By merging two levitated droplets in midair, experiments can be performed without the use of containers, thereby automating the process and minimizing contamination. Experimental automation with a multi-axis CAL was demonstrated by Foresti et al. [14] in 2013. Two droplets were merged in midair by manipulating the amplitudes of the transducers; in addition, processes such as DNA transfection, were demonstrated.

When an object is levitated within the nodes of a standing wave in a CAL or PAL, the existence of antinodes allows for multiple samples to be levitated simultaneously along its axis, but it renders it difficult to merge two droplets between each node. This challenge can be addressed by oscillating the levitated liquids [76] and forcing them to merge together through the use of a secondary acoustic radiation force.

Moreover, experiments in midair can be performed using the PAL, as shown in Figure 8.12. A standing wave formed between the PAL and reflector can merge

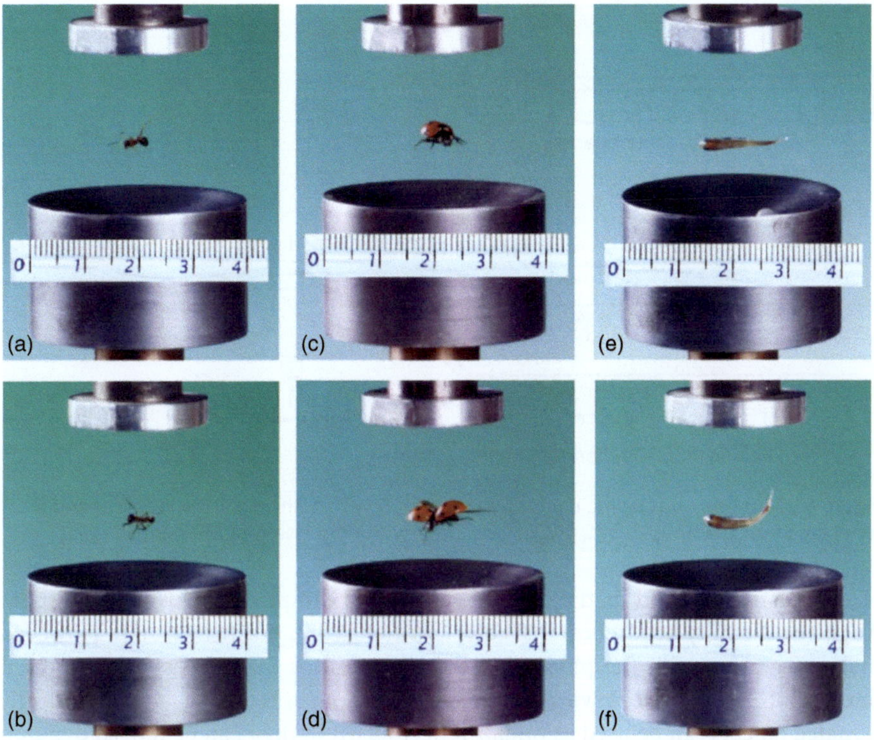

Figure 8.11 Levitation of small animals, as demonstrated by Xie et al. in 2006 [72]. A single-axis CAL with a 16.7 kHz transducer is used. Source: Reproduced from [72], with the permission of AIP Publishing LLC.

the two droplets in midair. Andrade et al. chose to blend two singular-focus holograms in an array rather than determining multi-focus acoustic holograms with optimizers [77].

Currently, both specimen holding and experimental automation approaches are limited by the use of standing waves and their ability to only levitate spheres with a radius smaller than a quarter of the wavelength [11]. This limits the volume of levitation to approximately 40 μl in 40 kHz levitators [31]. However, volumes larger than this (40 μl) are commonly used in biology/medical fields, thus limiting their applications. Additionally, the number of liquid droplets that can be simultaneously levitated is limited [40]. The splitting of acoustically levitated liquids has not yet been demonstrated. Acoustophoretic digital microfluidics with specimens on hydrophobic meshes with traveling waves have been proposed recently to address these issues [78].

8.5.3 3D Display

Another way acoustic levitation can assist in the fields of biology and medicine is through volumetric display. Essentially, 3D displays can display the volumetric data

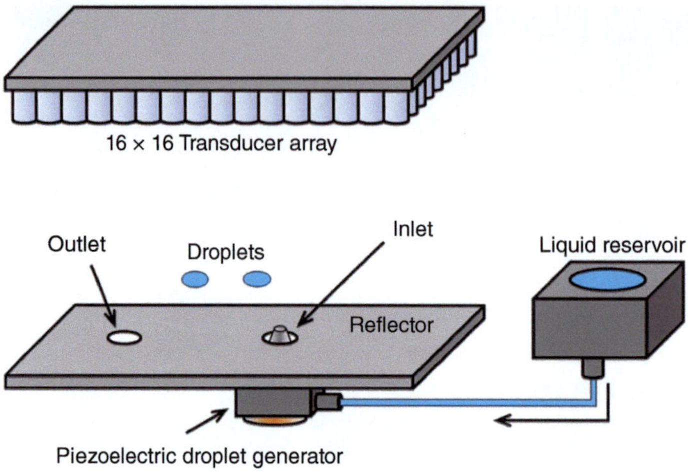

Figure 8.12 Midair ultrasonic microfluidic platform with a phased array levitator and a reflector. Source: Andrade et al. [77]/Reproduced with permission of AIP publishing.

of a given structure and spatial understanding can be improved by using the display. Augmented reality headsets and 3D displays are used in practice [79, 80]. However, their integration with the physical world is limited as the images are often confined within the closure, have a limited field of vision, or require eyewear. Statically levitated objects can be used as midair reflectors [81–84], or can be moved dynamically in midair to create an image, thereby creating a volumetric display in mid-air [35, 37, 81].

8.6 Conclusion and Future Remarks

A wide range of acoustic levitation systems, from CALs to PALs, have been proposed since the inception of the field. Further, field optimization methods for PALs have also been developed, and analytical expressions of the field have been innovated with the aim of relying less on finite element analyses.

However, the field of midair acoustic levitation has several limitations, which must be addressed; these include (i) the reliance on standing wave fields, (ii) transducers, and (iii) limitations in the temporal/spatial resolution of acoustic holograms.

The acoustic levitation of heavy objects still relies on having counter-propagating standing waves, thus limiting the volume of the levitation targets. Therefore, a system that can levitate large volumes of liquids with standing waves or a novel approach for levitating heavy objects using a traveling wave configuration must be developed. Changing acoustic frequency may be another approach to address this; however, a longer wavelength requires a lower acoustic frequency, and the audible frequencies are already close to 40 kHz.

Transducers are another limiting factor in the field. The current transducer specifications are driven by the automobile industry. A wider-bandwidth transducer can allow more rapid control of levitated objects, and wider ranges of transducer frequencies can provide additional tradeoffs between accuracy and power. A new application, which renews the demand for ultrasonic transducers and thus expands the transducer market is essential for continuous innovation in the field.

Finally, both the temporal and spatial resolution of acoustic holograms remain limited. The temporal limitation is partly a limitation of the transducers; however, no other form of active phase modulation has been developed. Active spatial sound modulators (similar to spatial light modulators in optics) can help improve temporal accuracy. Furthermore, the spatial resolution with metamaterial bricks and the achievable width of the tubes before thermoviscous effect takes over are both limited. With a higher temporal and spatial distribution, acoustic fields with more complex topologies can be encoded, and complex phase singularities or acoustic knots could help levitate a broader range of objects in midair.

Acknowledgments

Some parts of this work was supported by JSPS KAKENHI Grant Number JP21K14103. We would like to thank Editage (http://www.editage.com) for editing and reviewing this manuscript for English language. We also acknowledge OpenAI and ChatGPT for its automated generation of abstract and poem.

References

1 Brandt, E.H. (1989). Levitation in Physics. *Science* 243: 349–355.

2 Rayleigh, J.W.S. (1902). XXXIV. On the pressure of vibrations. *The London, Edinburgh, and Dublin Philosophical Magazine and Journal of Science* 3: 338–346.

3 Rayleigh, J.W.S. (1905). XLII. On the momentum and pressure of gaseous vibrations, and on the connexion with the virial theorem. *Philosophical Magazine Series 6* 10: 364–374.

4 Bücks, K. and Müller, H. (1933). Über einige Beobachtungen an schwingenden Piezoquarzen und ihrem Schallfeld. *Zeitschrift für Physik* 84: 75–86.

5 Trinh, E.H., Holt, R.G., and Thiessen, D.B. (1996). The dynamics of ultrasonically levitated drops in an electric field. *Physics of Fluids* 8: 43.

6 Wang, T.G., Trinh, E.H., Croonquist, A.P., and Elleman, D.D. (1986). Shapes of rotating free drops: Spacelab experimental results. *Physical Review Letters* 56: 452–455.

7 Trinh, E., Robey, J., Jacobi, N., and Wang, T. (1986). Dual-temperature acoustic levitation and sample transport apparatus. *Journal of the Acoustical Society of America* 79: 604–612.

8 Trinh, E.H. and Hsu, C.J. (1986). Acoustic levitation methods for density measurements. *The Journal of the Acoustical Society of America* 80: 1757–1761.

9 Trinh, E.H. and Robey, J.L. (1994). Experimental study of streaming flows associated with ultrasonic levitators. *Physics of Fluids* 6: 3567–3579.

10 Wang, T.G., Trinh, E., Rhim, W.-K. et al. (1984). Containerless processing technologies at the jet propulsion laboratory. *Acta Astronautica* 11: 233–237.

11 Whymark, R.R. (1975). Acoustic field positioning for containerless processing. *Ultrasonics* 13: 251–261.

12 Xie, W.J., Cao, C.D., Lü, Y.J., and Wei, B. (2002). Levitation of iridium and liquid mercury by ultrasound. *Physical Review Letters* 89: 1–4.

13 Clough, J., Sracic, M.W., Piombino, D. et al. (2016). Design and prototype of a two-axis acoustic levitator. In: *Volume 13: Acoustics, Vibration, and Wave Propagation*, 1–10. American Society of Mechanical Engineers https://doi.org/10.1115/IMECE2016-66193.

14 Foresti, D., Nabavi, M., Klingauf, M. et al. (2013). Acoustophoretic contactless transport and handling of matter in air. *Proceedings of the National Academy of Sciences* 110: 12549–12554.

15 Rudnick, J. and Barmatz, M. (1990). Oscillational instabilities in single-mode acoustic levitators. *The Journal of the Acoustical Society of America* 87: 81–92.

16 Andrade, M.A.B., Polychronopoulos, S., Memoli, G., and Marzo, A. (2019). Experimental investigation of the particle oscillation instability in a single-axis acoustic levitator. *AIP Advances* 9: 035020.

17 Andrade, M.A.B., Pérez, N., and Adamowski, J.C. (2015). Particle manipulation by a non-resonant acoustic levitator. *Applied Physics Letters* 106: 014101.

18 Rey, C.A. (1982). Acoustic levitation and methods for manipulating levitated objects. *JASA* 72 (1): 300.

19 Hansen, C.H. (2001). Fundamentals of acoustics. *Occupational Exposure to Noise: Evaluation, Prevention and Control* 23–52.

20 Hashimoto, Y., Koike, Y., and Ueha, S. (1995). Acoustic levitation of planar objects using a longitudinal vibration mode. *Journal of the Acoustical Society of Japan (E)* 16: 189–192.

21 Minikes, A., Bucher, I., and Haber, S. (2004). Levitation force induced by pressure radiation in gas squeeze films. *The Journal of the Acoustical Society of America* 116: 217–226.

22 Riley, N. (1998). Acoustic streaming. *Theoretical and Computational Fluid Dynamics* 10: 349–356.

23 Andrade, M.A.B., Ramos, T.S., Adamowski, J.C., and Marzo, A. (2020). Contactless pick-and-place of millimetric objects using inverted near-field acoustic levitation. *Applied Physics Letters* 116: 054104.

24 Andrade, M.A.B., Bernassau, A.L., and Adamowski, J.C. (2016). Acoustic levitation of a large solid sphere. *Applied Physics Letters* 109: 044101.

25 Andrade, M.A.B., Okina, F.T.A., Bernassau, A.L., and Adamowski, J.C. (2017). Acoustic levitation of an object larger than the acoustic wavelength. *The Journal of the Acoustical Society of America* 141: 4148–4154.

26 Marzo, A., Corkett, T., and Drinkwater, B.W. (2018). Ultraino: an open phased-array system for narrowband airborne ultrasound transmission. *IEEE Transactions on Ultrasonics, Ferroelectrics, and Frequency Control* 65: 102–111.

27 Morales, R., Ezcurdia, I., Irisarri, J. et al. (2021). Generating airborne ultrasonic amplitude patterns using an open hardware phased array. *Applied Sciences (Switzerland)* 11: 2981.

28 Ochiai, Y., Hoshi, T., and Rekimoto, J. (2014). Three-dimensional mid-air acoustic manipulation by ultrasonic phased arrays. *PLoS One* 9: e97590.

29 Fushimi, T., Yamamoto, K., and Ochiai, Y. (2021). Target acoustic field and transducer state optimization using Diff-PAT. *AIP Advances* 11: 125007.

30 Long, B., Seah, S.A., Carter, T., and Subramanian, S. (2014). Rendering volumetric haptic shapes in midair using ultrasound. *ACM Transactions on Graphics* 33: 1–10.

31 Marzo, A., Barnes, A., and Drinkwater, B.W. (2017). TinyLev: a multi-emitter single-axis acoustic levitator. *Review of Scientific Instruments* 88: 085105.

32 Marzo, A., Ghobrial, A., Cox, L. et al. (2017). Realization of compact tractor beams using acoustic delay-lines. *Applied Physics Letters* 110: 014102.

33 Suzuki, S., Inoue, S., Fujiwara, M. et al. (2021). AUTD3: scalable airborne ultrasound tactile display. *IEEE Transactions on Haptics* 1–9. https://doi.org/10.1109/TOH.2021.3069976.

34 Seah, S.A., Drinkwater, B.W., Carter, T. et al. (2014). Dexterous ultrasonic levitation of millimeter-sized objects in air. *IEEE Transactions on Ultrasonics, Ferroelectrics, and Frequency Control* 61: 1233–1236.

35 Hirayama, R., Martinez Plasencia, D., Masuda, N., and Subramanian, S. (2019). A volumetric display for visual, tactile and audio presentation using acoustic trapping. *Nature* 575: 320–323.

36 Fushimi, T., Drinkwater, B.W., and Hill, T.L. (2020). What is the ultimate capability of acoustophoretic volumetric displays? *Applied Physics Letters* 116: 244101.

37 Fushimi, T., Marzo, A., Drinkwater, B.W., and Hill, T.L. (2019). Acoustophoretic volumetric displays using a fast-moving levitated particle. *Applied Physics Letters* 115: 064101.

38 Inoue, S., Mogami, S., Ichiyama, T. et al. (2019). Acoustical boundary hologram for macroscopic rigid-body levitation. *The Journal of the Acoustical Society of America* 145: 328–337.

39 Marzo, A., Seah, S.A., Drinkwater, B.W. et al. (2015). Holographic acoustic elements for manipulation of levitated objects. *Nature Communications* 6: 8661.

40 Marzo, A. and Drinkwater, B.W. (2018). Holographic acoustic tweezers. *Proceedings of the National Academy of Sciences* 116: 84–89.

41 Marzo, A., Caleap, M., and Drinkwater, B.W. (2018). Acoustic virtual vortices with tunable orbital angular momentum for trapping of mie particles. *Physical Review Letters* 120: 44301.

42 Andersson, C. and Ahrens, J. (2019). Acoustic levitation from superposition of spherical harmonics expansions of elementary sources: analysis of dependency on wavenumber and order. *IEEE International Ultrasonics Symposium, IUS* 2019 (October): 920–923.

43 Chakravarthula, P., Peng, Y., Kollin, J. et al. (2019). Wirtinger holography for near-eye displays. *ACM Transactions on Graphics* 38: 213.

44 Melde, K., Mark, A.G., Qiu, T., and Fischer, P. (2016). Holograms for acoustics. *Nature* 537: 518–522.

45 Mellin, S.D. and Nordin, G.P. (2001). Limits of scalar diffraction theory and an iterative angular spectrum algorithm for finite aperture diffractive optical element design. http://www.opticsexpress.org/oearchive/.

46 Memoli, G., Caleap, M., Asakawa, M. et al. (2017). Metamaterial bricks and quantization of meta-surfaces. *Nature Communications* 8: 14608.

47 Zhu, Y. and Assouar, B. (2019). Systematic design of multiplexed-acoustic-metasurface hologram with simultaneous amplitude and phase modulations. *Physical Review Mater* 3: 1–5.

48 Fushimi, T. (2020). *Nonlinear Dynamics of Phased Array Levitators*. University of Bristol.

49 Kinsler, L.E., Frey, A.R., Coppens, A.B., and Sanders, J.V. (1982). *Fundamentals of Acoustics*, 3e. Wiley.

50 Zehnter, S., Andrade, M.A.B., and Ament, C. (2021). Acoustic levitation of a Mie sphere using a 2D transducer array. *Journal of Applied Physics* 129 (13): 134901.

51 Zeng, X. and McGough, R.J. (2008). Evaluation of the angular spectrum approach for simulations of near-field pressures. *The Journal of the Acoustical Society of America* 123: 68–76.

52 Zhang, J., Tian, Y., Cheng, Y., and Liu, X. (2020). Acoustic holography using composite metasurfaces. *Applied Physics Letters* 116: 1–6.

53 Treeby, B.E. and Cox, B.T. (2010). k-wave: MATLAB toolbox for the simulation and reconstruction of photoacoustic wave fields. *Journal of Biomedical Optics* 15: 21314.

54 Gor'kov, L.P. (1962). On the forces acting on a small particle in an acoustical field in an ideal fluid. *Soviet Physics Doklady* 6: 773.

55 Bruus, H. (2012). Acoustofluidics 7: the acoustic radiation force on small particles. *Lab on a Chip* 12: 1014–1021.

56 Ezcurdia, I., Morales, R., Andrade, M.A.B., and Marzo, A. (2022). LeviPrint: contactless fabrication using full acoustic trapping of elongated parts. In: *Special Interest Group on Computer Graphics and Interactive Techniques Conference Proceedings*, 1–9. ACM. https://doi.org/10.1145/3528233.3530752.

57 Tang, T. and Huang, L. (2022). Theoretical framework to predict the acoustophoresis of axisymmetric irregular objects above an ultrasound transducer array. *Physical Review E* 105: 055110.

58 Tang, T. and Huang, L. (2022). Soundiation: a software in evaluation of acoustophoresis driven by radiation force and torque on axisymmetric objects. *The Journal of the Acoustical Society of America* 152: 2934–2945.

59 Plasencia, D.M., Hirayama, R., Montano-Murillo, R., and Subramanian, S. (2020). GS-PAT: high-speed multi-point sound-fields for phased arrays of transducers. *ACM Transactions on Graphics* 39: 138:1–138:12.

60 Gerchberg, R.W. (1972). A practical algorithm for the determination of phase from image and diffraction plane pictures. *Optik (Stuttg)* 35: 237–246.

61 Güne, A., Baydin, G., Pearlmutter, B.A., and Siskind, J.M. (2018). Automatic differentiation in machine learning: a survey. *Journal of Machine Learning Research* 18: 1–43.

62 Li, J., Lv, Z., Hou, Z., and Pei, Y. (2020). Comparison of balanced direct search and iterative angular spectrum approaches for designing acoustic holography structure Comparison of balanced direct search and iterative angular spectrum approaches for designing acoustic holography structure. *Applied Acoustics* 175: 107848.

63 Suzuki, S., Fujiwara, M., Makino, Y., and Shinoda, H. (2021). Radiation pressure field reconstruction for ultrasound midair haptics by greedy algorithm with brute-force search. *IEEE Transactions on Haptics* 14: 914–921.

64 Lee, M.H., Lew, H.M., Youn, S. et al. (2022). Deep learning-based framework for fast and accurate acoustic hologram generation. *IEEE Transactions on Ultrasonics, Ferroelectrics, and Frequency Control* 69: 3353–3366.

65 Lin, Q., Wang, J., Cai, F. et al. (2021). A deep learning approach for the fast generation of acoustic holograms. *The Journal of the Acoustical Society of America* 149: 2312–2322.

66 Fushimi, T., Yamamoto, K., and Ochiai, Y. (2021). Acoustic hologram optimisation using automatic differentiation. *Scientific Reports* 11: 12678.

67 Kingma, D.P. and Ba, J.L. (2015). Adam: a method for stochastic optimization. In: *International Conference on Learning Representations (ICLR), ICLR 2015 – Accepted Papers - Main Conference - Poster Presentations*, Ithaca, NY: arXiv.org, vol. 1, 1–15.

68 Wood, B.R., Heraud, P., Stojkovic, S. et al. (2005). A portable Raman acoustic levitation spectroscopic system for the identification and environmental monitoring of algal cells. *Analytical Chemistry* 77: 4955–4961.

69 Puskar, L., Tuckermann, R., Frosch, T. et al. (2007). Raman acoustic levitation spectroscopy of red blood cells and *Plasmodium falciparum* trophozoites. *Lab on a Chip* 7: 1125.

70 Delißen, F., Leiterer, J., Bienert, R. et al. (2008). Agglomeration of proteins in acoustically levitated droplets. *Analytical and Bioanalytical Chemistry* 392: 161–165.

71 Seddon, A.M., Richardson, S.J., Rastogi, K. et al. (2016). Control of nanomaterial self-assembly in ultrasonically levitated droplets. *Journal of Physical Chemistry Letters* 7: 1341–1345. https://doi.org/10.1021/acs.jpclett.6b00449.

72 Xie, W.J., Cao, C.D., Lü, Y.J. et al. (2006). Acoustic method for levitation of small living animals. *Applied Physics Letters* 89: 214102.

73 Li, L., Gu, N., Dong, H. et al. (2020). Analysis of the effects of acoustic levitation to simulate the microgravity environment on the development of early zebrafish embryos. *RSC Advances* 10: 44593–44600.

74 Sundvik, M., Nieminen, H.J., Salmi, A. et al. (2015). Effects of acoustic levitation on the development of zebrafish, Danio rerio, embryos. *Scientific Reports* 5: 13596.

75 Cox, L., Croxford, A., Drinkwater, B.W., and Marzo, A. (2018). Acoustic lock: position and orientation trapping of non-spherical sub-wavelength particles in midair using a single-axis acoustic levitator. *Applied Physics Letters* 113: 054101.

76 Brotton, S.J. and Kaiser, R.I. (2020). Controlled chemistry via contactless manipulation and merging of droplets in an acoustic levitator. *Analytical Chemistry* 92: 8371–8377.

77 Andrade, M.A.B., Camargo, T.S.A., and Marzo, A. (2018). Automatic contactless injection, transportation, merging, and ejection of droplets with a multifocal point acoustic levitator. *Review of Scientific Instruments* 89: 125105.

78 Koroyasu, Y., Nguyen, T., Sasaguri, S. et al. (2023). *Microfluidic platform using focused ultrasound passing through hydrophobic meshes with jump availability.* PNAS Nexus. https://doi.org/10.1093/pnasnexus/pgad207.

79 Kim, Y., Kim, H., and Kim, Y.O. (2017). Virtual reality and augmented reality in plastic surgery: a review. *Archives of Plastic Surgery* 44: 179–187. https://doi.org/10.5999/aps.2017.44.3.179.

80 Klein, M.I., Lio, C.H., Grant, R. et al. (2009). A mental workload study on the 2d and 3d viewing conditions of the da Vinci surgical robot. In: *Proceedings of the Human Factors and Ergonomics Society*, vol. 2, 1186–1190. Human Factors an Ergonomics Society Inc.

81 Ochiai, Y., Hoshi, T., and Rekimoto, J. (2014). Pixie dust: graphics generated by levitated and animated objects in computational acoustic-potential field. *ACM Transactions on Graphics* 33: 85.

82 Sahoo, D.R., Takuto Nakamura, Asier Marzo et al. (2016). JOLED: a midair display based on electrostatic rotation of levitated janus objects. In: *Proceedings of the 29th Annual Symposium on User Interface Software and Technology - UIST '16*, 437–448. ACM Press. https://doi.org/10.1145/2984511.2984549.

83 Uno, Y., Qiu, H., Sai, T. et al. (2018). Luciola: a millimeter-scale light-emitting particle moving in mid-air based on acoustic levitation and wireless powering. *Proceedings of the ACM on Interactive, Mobile, Wearable and Ubiquitous Technologies* 1: 1–17.

84 Morales, R., Marzo, A., Subramanian, S., and Martínez, D. (2019). LeviProps: Animating levitated optimized fabric structures using holographic acoustic tweezers. In: *Proceedings of the 32nd Annual ACM Symposium on User Interface Software and Technology*, 651–661. https://doi.org/10.1145/3332165.3347882.

9

Application of Ultrasonic Waves in Bioparticle Manipulation and Separation

M. Bülent Özer[1] *and Barbaros Çetin*[2]

[1] *Department of Mechanical Engineering, Middle East Technical University, 06800 Ankara, Türkiye*
[2] *Department of Mechanical Engineering, Microfluidics & Lab-on-a-chip Research Group, İ.D. Bilkent University, 06800 Ankara, Türkiye*

9.1 Introduction

In this chapter, we discuss the principles of using acoustic (specifically ultrasonic) waves to manipulate biological particles in microfluidic systems. Microfluidic systems have enabled advances in the manipulation of micro/nano-scale particles and cells. With improved fabrication methods and technologies, the features that can be manufactured on a microfluidic device have become comparable to the size of the micro/nanoparticles and cells. These improvements lead to superior manipulation accuracy and hence enable the detection of minor nuances in particles and cells, which has evolved innovative separation technologies based on these minor differences.

Due to the scope of this book, we focus on the implementation of acoustic waves in bioparticle manipulation and separation. However, to paint a clearer picture of acoustic methods, the resemblance and differences of this method compared to other physical field manipulation methods are discussed in Section 9.2. Section 9.3 focuses on the physical implementation of acoustofluidic devices and describes the components of these devices. Section 9.4 derives and introduces the fundamental theoretical relations, with a particular focus on the underlying assumptions used in analytical formulae. Section 9.5 summarizes the numerical solutions employed in solving the governing partial differential equations. This section is organized such that the chronological evolution of numerical solution approaches for each component of acoustofluidic devices is explained. Section 9.6 contains a detailed summary of the implementation of acoustofluidic devices in bioparticle manipulation and separation. The summary is organized such that the applications targeting different bioparticles (such as cancer cells, bacteria and blood components) are grouped under different subsections. These studies are presented chronologically and in a detailed table that also classifies them according to the originating research group, type of acoustofluidic device, operating frequency, etc. In Section 9.7 of the chapter, design recommendations for a successful acoustofluidic product

Acoustic Technologies in Biology and Medicine, First Edition.
Edited by Adem Ozcelik, Ryan Becker, and Tony Jun Huang.

launch and commercialization in light of medical device regulatory and marketing needs are presented. Several excellent review articles [1–7], are available in the literature regarding the different aspects of acoustofluidic technologies; however, we believe this book chapter is a comprehensive overview of acoustofluidics on bioparticle manipulation, including theory, simulation methods, applications as well as commercialization aspects.

9.2 Bioparticle Manipulation

In microfluidic applications, as particles flow within a microchannel network, they are exposed to certain physical effects that result in an action (force) on them. These forces can originate due to different physical fields in a microfluidic system, such as electrical, acoustic, magnetic, flow fields, or forces between molecules of these bioparticles due to their high affinity for specific biochemical reactions. These aforementioned fields/forces can be utilized to manipulate or separate bioparticles from each other.

9.2.1 Hydrodynamic Bioparticle Manipulation

Hydrodynamic manipulation methods utilize the interaction of bioparticles with the flow field in order to manipulate/separate certain types of bioparticles of interest from a mixture. Bioparticle manipulation is performed using the inertial forces on the particles through specially designed channel geometries and topologies. In hydrodynamic methods, since drag and inertial forces have size and shape dependence, the differentiation of bioparticles can be achieved based on size difference and deformability. Especially with the introduction of specially designed (i.e. slanted and/or anisotropic) obstacle and post structures and/or contraction/expansion within a microchannel network, capturing and isolation of bioparticles through some filter structures and separation, sorting, and focusing through size/shape-based lateral movement of bioparticles can be achieved [8, 9]. Since the manipulation of bioparticles is based only on the flow field, hydrodynamic methods are classified as passive methods. Although utilizing hydrodynamic effects may require three-dimensional (3D) channel structures, which may be challenging for lithography-based fabrication techniques, with the involvement of 3D printing, laser engraving, and mechanical machining-based fabrication techniques, the creation of the desired channel geometry with 3D structures is becoming feasible. These improvements make hydrodynamic techniques quite favorable, especially for high-throughput applications. Sample preparation may be required since the concentration of the particles may affect the flow field and result in undesired flow patterns within the microchannels. As an example, dilution of the whole blood may be needed depending on the application. Since the manipulation of the particles is based on size and/or shape, the sensitivity of the manipulation is low when the bioparticles of interest have a similar size and/or shape.

9.2.2 Immunological (Antigen–Antibody Reaction) Bioparticle Manipulation

Immunological manipulation methods employ the reactions between antibodies–antigens expressed on the surface of bioparticles of interest, which lead to the binding of specific antigens with antibodies. Specific antibodies can be implanted on the surface of the post structures within a microchannel network and/or on the surface of the microchannel walls to trap/capture target bioparticles with the right antigens. Alternatively, antibodies can also be coated on the surfaces of micro/nanoparticles, which attach themselves to the target bioparticles [10]. Since labeling of micro/nanoparticles can be tailored to possess different size, acoustic, or magnetic properties, target bioparticles that are labeled with micro/nanoparticles can further be manipulated under the effect of the corresponding external field of acoustic, magnetic, and/or flow. Therefore, immunological methods can be implemented by themselves or in conjunction with other bioparticle manipulation methods as well.

9.2.3 Electrokinetic Bioparticle Manipulation

Electrokinetic manipulation methods employ electrical forces such as electrophoresis (EP) and dielectrophoresis (DEP) to manipulate bioparticles [8]. EP is the movement of the electrically charged particles in an electrical field due to the Coulombic body force (electrophoretic force) acting on the particles because of their surface charge and is a well-developed manipulation technique used in conventional lab environments (e.g. capillary EP to separate DNA and proteins). The movement of particles in a nonuniform electric field is due to the interaction of the particle's dipole field and the spatial gradient of the electric field [11]. DEP force has a dependence on particle size, dielectric properties of the particle, and the medium. Moreover, in the case of an AC field, DEP force also becomes a function of the frequency of the AC field. More specifically, DEP response of bioparticles is also affected by the morphology and chemical composition of the internal matrix of the bioparticle as well as its size, which may lead to superior selectivity at a certain AC field frequency range for a given buffer solution. Both negative and positive forces can be generated with different configurations of the electrodes and/or the microchannel network and by switching the polarity and/or frequency of the electric field for electrokinetic manipulation. The electrokinetic response of bioparticles strongly depends on dielectric properties of the buffer solution. In the case of biological buffer solutions with relatively large electrical conductivity, severe Joule heating-related issues can be observed (e.g. heating of the medium, which may change the physiological state of bioparticles, electrothermal effects, which may severely affect the flow field, etc.) [12] and/or necessitate the use of electrodes with fractal structure for desired performance [13].

9.2.4 Magnetophoretic Bioparticle Manipulation

Magnetic bioparticle manipulation utilizes forces due to the interaction of bioparticles with magnetic field lines. However, for a successful manipulation, there needs to be a significant difference between the magnetic susceptibility of bioparticles and suspending medium. Usually, such a significant magnetic susceptibility difference does not exist between the bioparticles and the suspension medium since both are predominantly water based. A possible way to increase the magnetic susceptibility of bioparticles is to attach superparamagnetic spheres to them through antigen–antibody reaction [14, 15]. Alternatively, the magnetic susceptibility of suspending medium can also be increased through the inclusion of nanomagnetic materials in the suspension medium [16–18]. The magnetophoretic bioparticle separation with microsphere attachment on the bioparticles can have high selectivity due to the antibody–antigen reactions, and the separation process can be quite fast with a magnetic field by a permanent magnet. On the other hand, microspheres covered with antibodies can be quite expensive. Increasing the magnetic susceptibility of a suspension fluid with nanomagnetic particles may require extensive biocompatibility studies to understand the possible effects of magnetic nanoparticles on the bioparticles.

9.2.5 Acoustophoretic Bioparticle Manipulation

In addition to flow, electric, and magnetic fields, another physical field that can generate forces on a bioparticle in a suspension medium is the acoustic field. An acoustic field generated in the suspension medium causes acoustic radiation forces on bioparticles and drag forces on the suspending fluid due to acoustic streaming. The acoustic field can be generated inside the suspension medium through two different approaches. One approach is to generate the acoustic waves through elastic waves in the chip material where the microchannel is embedded (the chip material surrounds the channel) [19]. The other method is to generate surface acoustic waves (SAWs) on a piezoelectric actuator and use these elastic waves to generate acoustic waves inside a microfluidic channel [20]. The details of these two approaches will be investigated in Section 9.3.

9.2.6 Unification of Field Manipulation Methods

Three microfluidic bioparticle manipulation methods described above rely on an external energy source generating a physical field inside the microchannel. These generated fields are the electrical, magnetic, and acoustic fields. Electrodes, magnets, and piezoelectric actuators are used to generate the dielectric, magnetic and acoustic fields, respectively. As shown in Figure 9.1, particles of two different sizes and/or physical properties can be separated since different amounts of dielectric, magnetic, and acoustic radiation forces affect them.

The expressions of the forces due to different types of generated fields are shown in Figure 9.1a–c. When the force terms are investigated, one can see that all force expressions are composed of three terms. These terms are volume, contrast, and field gradient terms. Simple investigation of the force formulae shows that all three methods generate forces that are linearly proportional to the volume of the particles (shown in the formulae in blue). All the force expressions contain a contrast term

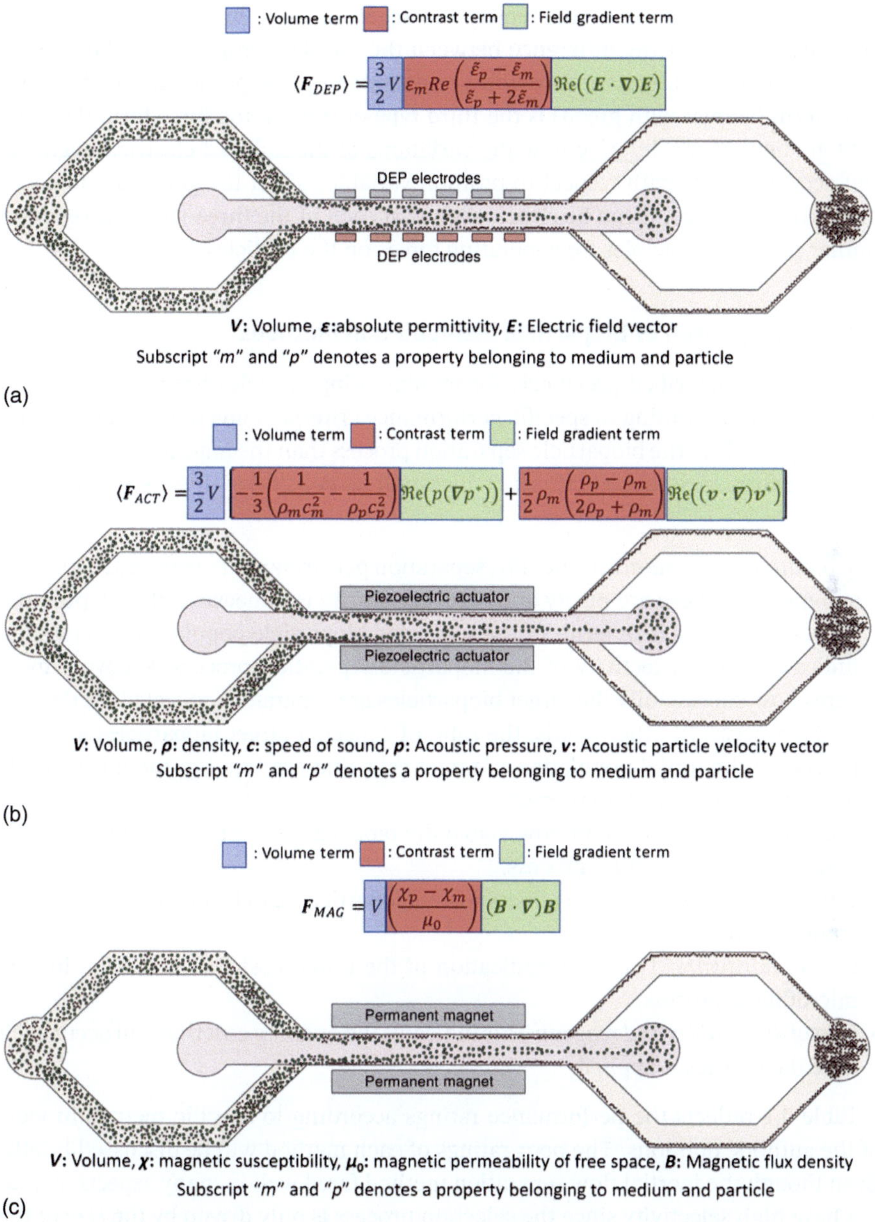

Figure 9.1 Representative architectures and governing equations for particle manipulation with dielectrophoretic, acoustic, and magnetic approaches.

(shown in red) proportional to the difference between the particles' physical properties and the medium properties where the particles are suspended. These related physical properties are electric permittivity for dielectrophoretic, magnetic susceptibility for magnetophoretic, and density and speed of sound for the acoustophoretic methods. The larger the difference between the physical properties of the particle and the medium, the more force will be generated on the particles. The field gradient term (shown with green) is the third type of term in the force formulae. The field gradient terms describe how the variations of the induced electric, magnetic, and acoustic fields with respect to position variables affect the generated force on the particles. Hence, it can be clearly seen that each of the three methods relies on similar physical principles for generating forces on the particles.

9.2.7 Comparison of Bioparticle Manipulation Methods

Each method described previously for manipulating particles has advantages and disadvantages according to specific performance criteria. Some of these criteria are more applicable to the bioparticle separation process than the manipulation process. These performance-related criteria, which will be considered in this study, are as follows:

- *Separation performance*: Generally, separation performance is evaluated using two sub-metrics. These are purity and recovery. Purity is a measure of how pure the processed sample is in terms of the separated bioparticle population. It is a good indicator of the selectivity of the bioparticle separation process. Recovery measures how successfully the target bioparticles are separated (collected) at the end of the process. In other words, the ratio of collected target bioparticles after the process to the initial population of target cells that are input to the microfluidic manipulation/separation process.
- *Viability*: It is a measure of how viable the bioparticles are after the microfluidic manipulation/separation process.
- *Throughput*: It is a measure of how many bioparticles can be processed in a given amount of time.
- *Cost of hardware*: This is an indication of the initial cost that is needed for the microfluidic process.
- *Cost of disposables and reagent*: This indicates the ongoing cost that will occur each time the process is applied.

Table 9.1 reflects the performance ratings according to specific metrics in view of the authors' opinions. The poor ratings of each method will be discussed briefly. Even though the inertial flow separation method fared well in many aspects, it does not have high selectivity since the selection process is only driven by the size of the bioparticle. Electrokinetic separation can have problems with the viability of cells due to the electric field imposed on the cells as well as the Joule heating effects. Also, the cost of hardware can be high, especially due to the cost of electrode fabrication. Also, throughput values remain on the low end, primarily due to the weak dielectrophoretic force generated on the particles. The magnetic manipulation approach

Table 9.1 Performance comparison of different bioparticle manipulation methods.

	Inertial	Electrokinetic	Magnetic	Antigen–Antibody	Acoustophoretic
Separation Perf.	−1	1	0	1	0
Viability	1	−1	−1	1	1
Throughput	1	−1	−1	0	0
Cost of hardware	1	−1	0	1	0
Cost of disposables and reagents	1	1	1	−1	1

Rating: −1: Poor, 0: Mediocre, 1: Good.

needs careful evaluation due to magnetic materials interfacing with live cells. Also, the literature shows that throughput values are at the low end when nano-magnetic materials are used to change the magnetic behavior of the medium. Immunological manipulation has essential separation performance benefits, but the reagents required for the immunological reactions are expensive.

9.3 General Architecture of Acoustofluidic Devices

The general architecture of acoustophoretic devices may differ depending on whether the device is a bulk acoustic wave (BAW) device or a SAW device. In general, two components are common for both types of acoustophoretic devices. One of them is the actuator, and the other is the conduit in which bioparticles in a buffer solution are processed. BAW and SAW device architectures will be discussed in detail in Sections 9.3.1 and 9.3.2.

9.3.1 BAW Device Architecture

BAW devices are generally composed of the following components: a piezoelectric actuator that generates the elastic waves, a chip material where the channel geometry is etched, and a layer of material that completes a closed conduit (acts as a lid) for the fluid flow. These components are demonstrated in Figure 9.2a.

9.3.1.1 Piezoelectric Actuator

A piezoelectric actuator in a BAW device is employed to generate elastic waves in the chip material, which are mainly longitudinal and flexural waves. In BAW devices, mostly lead zirconate titanate (PZT) types of piezoelectric actuators are used due to their capability to generate high-amplitude vibrations. Among the PZT family, PZT-4 is mostly preferred due to its thermal stability and high quality factor at high frequencies. The PZTs in the BAW devices are operated at or around their thickness mode resonances. Hence, the thickness of the PZT constrains the frequency bandwidth at which the BAW device can efficiently operate.

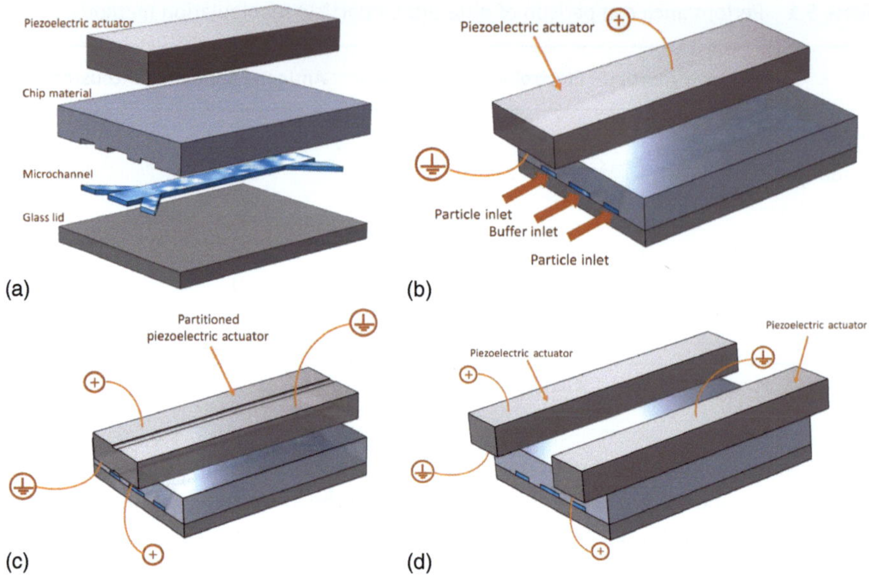

Figure 9.2 (a) Components of a BAW device, (b) shifted PZT configuration, (c) partitioned electrode configuration, and (d) two piezoelectric actuator configuration.

9.3.1.2 Chip Material

The chip substrate contains the microchannel pattern on its surface. Therefore, the chip material in a BAW device is in contact with the buffer solution, where bioparticles are suspended. Furthermore, elastic waves generated by the piezoelectric actuator are transmitted to the chip material, and elastic waves inside the chip material induce acoustic waves in the fluid flowing in the microchannel structure. These two functions of the chip materials set two essential requirements for the selection of the chip materials. The chip material should not be a material, which may harm the bioparticles suspended in the buffer solution (especially in cell applications). As for the second requirement, the chip material should not significantly dampen the elastic waves so that strong acoustic resonances can be effectively formed inside the microchannel. The materials of choice are generally materials with high acoustic impedance. These materials have low damping (high quality factor) and high acoustic impedance (such as silicon or glass) compared to the fluid flowing in the channel. This large acoustic impedance difference causes high reflections at the microchannel-chip interface, resulting in high-amplitude acoustic standing waves. On the other hand, materials with low acoustic impedances (such as polymers) are also implemented as chip materials due to easy fabrication, low cost, and good biocompatibility. A detailed discussion on the performance of different chip materials in BAW devices can be found in [21].

The microchannel's shape and geometry can differ depending on whether the bioparticles need to be manipulated only (such as focusing) or separated from a complex mixture of micro/nanoscale engineered and/or bioparticles. A common shape of choice for microchannels is the trifurcated channel structure, where the particles enter the main channel from side channels, as shown in Figure 9.2a. The flow of buffer fluid from the center channel pushes the particles to the side walls of the channel. In this type of channel and flow structure, acoustic radiation force aims to move the particles to the center of the channel. The particles flowing slowly (because of the flow gradient due to no slip condition) in a narrow region near the channel walls are moved towards the center of the channel under the action of the acoustophoretic force.

9.3.1.3 Lid Material

As shown in the bottom portion of Figure 9.2a, lid material creates a closed conduit for the fluid flow. It also acts as a barrier for acoustic wave reflections, which can contribute to the amplification of the acoustic standing wave inside the microchannel. In practice, the preferred material of choice is glass. Its transparency favors the optical imaging process of the bioparticles, and its high acoustic impedance compared to water results in reflected acoustic waves.

9.3.1.4 Device Assembly and Critical Dimensions

The whole BAW device assembly is designed to generate a high-amplitude resonant wave in the channel. In order to achieve acoustic resonance, the microchannel width needs to be adjusted to either a quarter [22] or the half wavelength [23] of the acoustic wave in the fluid.

In a BAW device, the location of the piezoelectric actuator on the chip geometry and the electrode pattern determine whether the microparticles move toward the center of the channel or toward the side walls of the channel. Most acoustophoretic devices in which commonly used trifurcated geometry is utilized aim to manipulate the bioparticles toward the center of the channel. As shown in Figure 9.2b, the piezoelectric actuator is placed asymmetrically on the chip material in most applications. The top and bottom surfaces are the live and ground electrodes, respectively. In this configuration, at some resonance frequencies, the particles are directed toward the center of the channel, whereas at other resonance frequencies, they are directed toward the channel side walls. The theoretical analysis of this behavior can be found elsewhere [24]. In order to achieve bioparticle motion towards the center of the microchannel at all resonance frequencies, a partitioned electrode pattern is needed with the electrode polarities, as shown in Figure 9.2c. The partitioned electrode pattern is explained and used for both acoustically soft and acoustically hard chip materials in [21]. A similar response that results in bioparticles' motion toward the channel center can also be achieved with two identical piezoelectric actuators driven out-of-phase, as shown in Figure 9.2d.

In the assembly of BAW devices, PZT is attached to the chip material, and the chip is attached to the glass lid through bonding. Generally, cyanoacrylate is preferred as a bonding agent due to its fast curing and acceptable thermal stability. Alternatively, some studies also implemented ultrasonic coupling gel between the chip and piezoelectric actuator rather than using a permanent bonding agent [25]. In some studies where polymer chips are used, thermofusion is used to permanently connect the chip and the lid components [26].

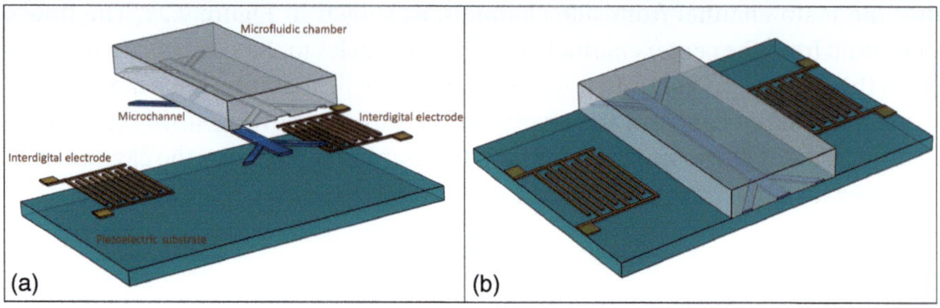

Figure 9.3 (a) Components of a SAW acoustofluidic device with two IDTs, and (b) assembled SAW device.

The configurations shown in Figure 9.2 represent only some of the designs in the literature. Especially in the earlier years of acoustofluidic bioparticle manipulation, different layered BAW device configurations were also used where some of which are explained in [27].

9.3.2 SAW Device Architecture

At the highest level, the working principle of a SAW device is similar to that of a BAW device. The elastic waves generated by a piezoelectric actuator induce acoustic waves inside a microchannel, causing acoustic radiation forces and acoustic streaming drag forces on the bioparticles used for manipulation and/or separation. The components of a SAW device are shown in Figure 9.3a. These components consist of a piezoelectric actuator, an interdigital electrode and a microfluidic chamber. The assembled SAW device is presented in Figure 9.3b.

9.3.2.1 Piezoelectric Actuator

Similar to BAW devices, SAW devices also employ piezoelectric actuators for the generation of elastic waves. The types of waves in SAW devices are generally Rayleigh waves, which travel predominantly on the surface of the piezoelectric substrate.

The type of piezoelectric actuator is also different from BAW systems. BAW devices generally utilize piezoelectric ceramics, whereas SAW systems utilize single-crystal piezoelectric actuators such as $LiNbO_3$ (Lithium Niobate), $LiTaO_3$ (Lithium Tantalate), or quartz. However, the 128° YX-cut $LiNbO_3$ crystal is the most commonly preferred piezoelectric actuator type. In SAW systems, the excitation frequency does not have to correspond to the resonance frequency of the piezoelectric actuator. Therefore, unlike BAW devices, in SAW devices the selected geometry of the piezoelectric actuator does not highly affect the wave generation performance as long as the thickness of the piezoelectric actuator is significantly larger than the wavelength generated by the IDT. This flexibility in the geometry selection of the piezoelectric actuator is an important feature and benefit of SAW devices.

Another significant difference between SAW and BAW devices is that the piezoelectric actuator is in contact with the buffer solution in most SAW applications. This contact may lead to additional sterilization and disposability requirements on the piezoelectric actuators, especially for applications where the processed bioparticles need to be further used, such as separated cells being delivered to patients in therapeutic applications.

9.3.2.2 Interdigital Electrodes (IDT)

In SAW devices, the actuation of piezoelectric actuators is accomplished by the IDTs. The SAWs are generated due to the interdigitated nature of the electrodes. A general form of the interdigital electrode is shown in Figure 9.3. Even though the shape of the IDT depicted in Figure 9.3 is the most common one, there can be other shapes for the IDT geometry, including spiral-shaped (for focused SAW) [28], slanted-shaped (for increased bandwidth) [29], and chirped IDT [30]. The dimensions of the fingers of one branch of the electrode determine the surface acoustic wavelength and the excitation frequency of the electrical signal that should be provided to the electrodes. Unlike the BAW devices, the required excitation frequency is not determined by the piezoelectric actuator dimensions but by the dimensions of the IDT. The IDT fingers' pitch determines the wavelength; the width of an IDT finger is generally set to a quarter of a wavelength and which is also the gap between two successive IDT fingers.

The electrode materials are generally gold or silver. Chromium is also used to assist in the deposition of the IDT onto the piezoelectric actuator surface. Generally, IDT fabrication includes photolithography and e-beam evaporation processes. The number of IDT electrodes in a SAW device can vary significantly. A device with a single IDT generates a traveling SAW. A second identical IDT can also be used on the same actuator, where two SAWs with the same wavelength interfere with each other resulting in standing surface acoustic waves (SSAWs). The devices which utilize SSAW can have more stable wave patterns with higher wave amplitudes. Also, more than two IDT configurations exist, especially in cases where continuous flow manipulation is not used. In recent studies, four similar IDTs have been implemented to create 2D and 3D patterns of particles and cells [31, 32].

The utilization of microfabrication methods allows the possibility of using high frequencies as the excitation frequency. This ability to achieve and expose bioparticles to high frequencies has benefits and disadvantages. The benefits include the possibility of increasing selectivity during the separation of bioparticles. The ability of acoustic waves to differentiate bioparticles from each other is related to the bioparticle's dimensions and its mechanical features compared to the wavelength of the incident acoustic wave. Therefore, bioparticles with slight differences may generate different amplitude reflections when exposed to small-wavelength acoustic waves in comparison to their sizes. Hence, exposing bioparticles to high frequencies improves the wave's ability to differentiate bioparticles with slight differences in mechanical and geometric properties.

The disadvantage of high operation frequencies is the inherent difficulty of generating and sustaining large-amplitude waves at high frequencies. Due to this restriction, the acoustic pressure amplitude of SAW systems is generally lower than that of their BAW counterparts, resulting in lower flow rates and throughputs, which overshadow the improved selectivity of the SAW process.

9.3.2.3 Microfluidic Chamber

The microfluidic chamber is the fluid path where the microchannel and the fluid with the bioparticles flow through, similar to the chip in the BAW device. In non-continuous flow applications, it is the chamber where the suspension fluid and the bioparticles are contained. Unlike a BAW device, this component does not have to transmit the elastic waves generated by the piezoelectric actuator to the fluid since the piezoelectric actuator is in direct contact with the fluid in SAW devices. Therefore, the microfluidic chamber does not have to resonate. Due to this reason, materials with poor acoustic transmission properties, such as poly-dimethylsiloxane (PDMS), are widely used as microfluidic chamber material. The width of the channel can be half wavelength or longer, depending on the targeted locations of the manipulated bioparticles.

In several studies, the preferred choice of microfluidic chamber material was PDMS due to its low cost and easier manufacturing (molding or microfabrication). However, the literature also includes some studies where the chamber material is glass [33, 34]. The use of acoustically hard materials results in higher acoustic wave amplitudes inside the fluid due to reflected waves at the microfluidic chamber walls.

9.3.2.4 Device Assembly and Critical Dimensions

In a SAW device, the IDT is already deposited on the piezoelectric actuator at the microfabrication step. Generally, the main step is to bond the microfluidic chamber to the piezoelectric actuator. Since the microfluidic chamber material is commonly PDMS, plasma treatment of the PDMS and the piezoelectric surface simplifies the bonding process. When acoustically hard materials such as pyrex, glass, or silicon are used, then ultraviolet (UV)- cured bonding agents are widely used.

The dimensional sensitivity of SAW devices is less compared to BAW counterparts since there is no requirement to form resonant waves in electromechanical, structural, and acoustic domains. The most critical dimensions of a SAW acoustophoretic device are the dimensions of the IDT, since they determine the wavelength of the SAW generated at the surface of the piezoelectric material. If there are multiple IDTs used for the generation of SSAWs, the symmetric deposition of the IDT on the piezoelectric actuator can become important for standing wave generation.

Also, the microfluidic chamber width is affected by the acoustic wavelength that is induced by the IDT(s). The dimension of the microfluidic chamber is determined by the number of nodal lines (collection planes for the bioparticles) targeted in the device.

9.3.3 Comparison of BAW and SAW Devices

In Sections 9.3.1 and 9.3.2, general device architecture and components of BAW and SAW devices were explained, and also some requirements and common practices for the design, assembly, and fabrication of these devices are detailed. It is almost impossible to favor BAW or SAW acoustophoretic devices over each other, but depending on the application, one device architecture can be more advantageous than the other.

Table 9.2 A comparison table for the advantages and disadvantages of BAW and SAW devices.

	Advantages	Disadvantages
BAW devices	• When polymers are used as chip material, conventional manufacturing methods can be used • Suitable for high throughput applications • Expensive components do not contact the fluid and the bio-particles	• High sensitivity to chip and actuator geometry and material properties • Difficulties in high-frequency applications
SAW devices	• Ability to operate at a wide radio frequency range • Increased selectivity • Ability to perform single-cell analysis conveniently	• Higher piezoelectric actuator cost • High electrode cost • Requires clean room fabrication • Lower throughput

Table 9.2 summarizes some of the advantages and disadvantages discussed throughout this section.

9.4 Governing Equations in Acoustic Bioparticle Manipulation

Similar to other microfluidic applications, Navier–Stokes equations are the starting point for deriving the acoustophoretic bioparticle manipulation equations. Two main forces are responsible for the motion of the bioparticles in an acoustophoretic device. These forces are the acoustic radiation force and the drag forces acting on the particle due to acoustic streaming flow inside the microfluidic channel. It is a general assumption that bioparticles are generally not significantly affected by the gravitational forces (because of their density), and the bioparticles considered in this study are large enough not to be affected by Brownian forces. There are two different approaches to the calculation of the acoustophoretic variables in a fluid domain. These approaches include calculation of first-order acoustic variables and the calculation of second-order acoustic variables with perturbation methods in Sections 9.4.1 and 9.4.2, respectively.

9.4.1 First-Order Acoustic Field Variables

First-order acoustic field variables are calculated from the Navier–Stokes equations and the thermodynamic equation of state. Equations (9.1) and (9.2) show the

Navier–Stokes equations, which are the continuity equation (mass conservation) and the force balance equation (conservation of momentum) for a fluid medium.

$$\frac{\partial \rho_m}{\partial t} = -\boldsymbol{\nabla} \cdot (\rho_m \boldsymbol{v}) \tag{9.1}$$

$$\rho_m \frac{\partial \boldsymbol{v}}{\partial t} = -\nabla p - \rho_m (\boldsymbol{v} \cdot \boldsymbol{\nabla})\boldsymbol{v} + \eta \nabla^2 \boldsymbol{v} + \beta \eta \boldsymbol{\nabla}(\boldsymbol{\nabla} \cdot \boldsymbol{v}) \tag{9.2}$$

In the above equations, "ρ_m" is the density of the fluid (medium), "p" is the pressure in the fluid, "η" and "β" are the dynamic shear and bulk viscosities of the fluid, respectively.

The acoustic variables in the above two equations are "p," "v," and "ρ." Their first-order variations are considered, which assumes orders of magnitude smaller variations added to the mean values of these variables, which are their values in the static case.

$$\rho_m = \rho_{m0} + \rho_{m1} \tag{9.3a}$$

$$p = p_0 + p_1 \tag{9.3b}$$

$$\boldsymbol{v} = \boldsymbol{v}_1 \tag{9.3c}$$

For the third perturbation equation, the fluid is assumed to be quiescent initially. Therefore, the perturbation includes only the first-order perturbation. After substituting Eq. (9.3a)–(9.3c) in the first two equations and disregarding the terms that are higher than the first order, the following two equations are obtained.

$$\frac{\partial \rho_{m_1}}{\partial t} = -\rho_{m_0} (\boldsymbol{\nabla} \cdot \boldsymbol{v}_1) \tag{9.4}$$

$$\rho_{m0} \frac{\partial \boldsymbol{v_1}}{\partial t} = -\nabla p_1 + \eta \nabla^2 \boldsymbol{v_1} + \beta \eta \boldsymbol{\nabla}(\boldsymbol{\nabla} \cdot \boldsymbol{v_1}) \tag{9.5}$$

The thermodynamic equation of state is an equation that relates the pressure in a compressible fluid to the density of the fluid. There are different approaches to defining this relationship. The most practical approach assumes that acoustic wave propagation has faster dynamics than thermal energy flow. So, by the time an acoustic wave generates compression and rarefaction in a control volume, the energy does not have time to flow in and out of the control volume; hence, the process can assumed to be adiabatic. If the losses during the passage of the acoustic wave are ignored, the acoustic process can assumed to be isentropic, which results in the following thermodynamic state equation given below:

$$p_1 = c_m^2 \rho_{m_1} \tag{9.6}$$

The term "c_m" is the isentropic speed of sound in the fluid medium. After several steps of algebraic manipulations, which are shown in Figure 9.4, one can obtain the following equation, which relates the first-order variations of the density, pressure, and acoustic particle velocity.

$$\frac{\partial^2 \rho_{m1}}{\partial t^2} = \nabla^2 p_1 - (1 + \beta)\eta \nabla^2 (\boldsymbol{\nabla} \cdot \boldsymbol{v}_1) \tag{9.7}$$

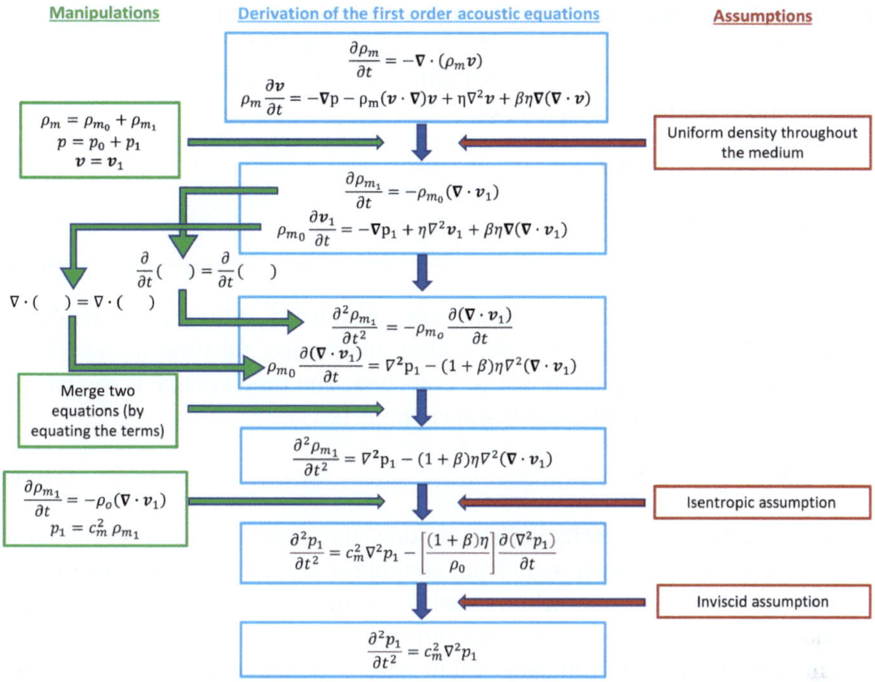

Figure 9.4 The derivation schematic of the first-order acoustic field with the assumptions involved.

To solve for these first-order acoustic variables, it is desirable to have the equations in terms of a single variable only, which is traditionally chosen to be the acoustic pressure "p_1." By writing the acoustic particle velocity in Eq. (9.7) in terms of "ρ_{m_1}" using Eq. (9.4) and replacing density terms with the acoustic pressure terms using the thermodynamic equation of state Eq. (9.6), one can obtain the following equation:

$$\frac{\partial^2 p_1}{\partial t^2} = c_m^2 \nabla^2 p_1 + \left[\frac{(1+\beta)\eta}{\rho_0}\right] \frac{\partial(\nabla^2 p_1)}{\partial t} \tag{9.8}$$

When the harmonic excitation of the fluid medium is considered, the second term on the right-hand side of Eq. (9.8) is orders of magnitude smaller than the first term; hence, it can be ignored. The first-order acoustic pressure for an inviscid fluid is given below.

$$\frac{\partial^2 p_1}{\partial t^2} = c_m^2 \nabla^2 p_1 \tag{9.9}$$

The above equation is also called the acoustic wave equation, which, when solved, can be employed in calculating the second-order acoustic variables.

9.4.2 Second-Order Acoustic Field Variables

The first-order acoustic variables have zero mean values; therefore, the net force acting on the particles due to the first-order acoustic field is zero. But it is known that bioparticles experience force and motion when exposed to an acoustic field. A second-order perturbation is needed to calculate a steady acoustic radiation force acting on a particle due to a first-order acoustic field. The perturbation expressions for a second-order field are as follows:

$$\rho_m = \rho_{m0} + \rho_{m1} + \rho_{m2} \tag{9.10a}$$

$$p = p_0 + p_1 + p_2 \tag{9.10b}$$

$$v = v_1 + v_2 \tag{9.10c}$$

The above perturbation equations are substituted into the Navier–Stokes equations given in Eqs. (9.1) and (9.2). After substitution, the time average of each term in the resulting equation will be evaluated. The time average is shown with the "$< >$" sign and its expression is shown in Eq. (9.11).

$$\langle X \rangle = \frac{1}{\tau} \int_0^{\tau} x(t)dt \tag{9.11}$$

Under the harmonic excitation assumption, the time average of all first-order terms will be zero, bringing significant simplification to the resulting expression. Further simplifications are achieved with the time derivative of time-averaged second-order acoustic variables being zero since the time-harmonic components become zero due to time averaging and the steady components become zero when time derivatives are applied. Finally, the following expression can be obtained with the assumption of inviscid flow.

$$\langle \boldsymbol{\nabla} p_2 \rangle = -\left\langle \rho_1 \frac{\partial \boldsymbol{v_1}}{\partial t} \right\rangle - \rho_{m0} \langle (\boldsymbol{v_1} \cdot \boldsymbol{\nabla}) \boldsymbol{v_1} \rangle \tag{9.12}$$

Using the fact that first-order acoustic velocity is a gradient field and utilizing some calculus identities shown in Figure 9.5, one can obtain Eq. (9.13) for the second-order acoustic pressure in terms of first-order acoustic variables.

$$\langle p_2 \rangle = \frac{1}{2\rho_{m_0} c_m^2} \langle p_1^2 \rangle - \frac{1}{2} \rho_{m_0} \left\langle |\boldsymbol{v_1}|^2 \right\rangle \tag{9.13}$$

9.4.3 Acoustic Radiation Force on a Particle

Once the expression for the second-order pressure is obtained, one can calculate the acoustic radiation force on a particle by integrating the acoustic pressure and the contribution of the momentum flux term over the particle geometry, which leads to the Eq. (9.14). Intermediate derivation steps are shown in Figure 9.5.

$$\boldsymbol{F}_{rad} = -\int_{\Omega} \left\{ \left(\frac{1}{2\rho_{mo} c_m^2} \langle p_1^2 \rangle - \frac{1}{2} \rho_{mo} \left\langle |v_1|^2 \right\rangle \right) \boldsymbol{n} - \rho_{mo} (\boldsymbol{n} \cdot \boldsymbol{v_1}) \boldsymbol{v_1} \right\} da \tag{9.14}$$

Equation (9.14) is a general representation and applies to any geometry without size and geometric shape restrictions. On the other hand, the calculation of the acoustic field variables "p_1" and "v_1" requires the calculation of the scattered first-order acoustic field due to the existence of the particle. In order to calculate

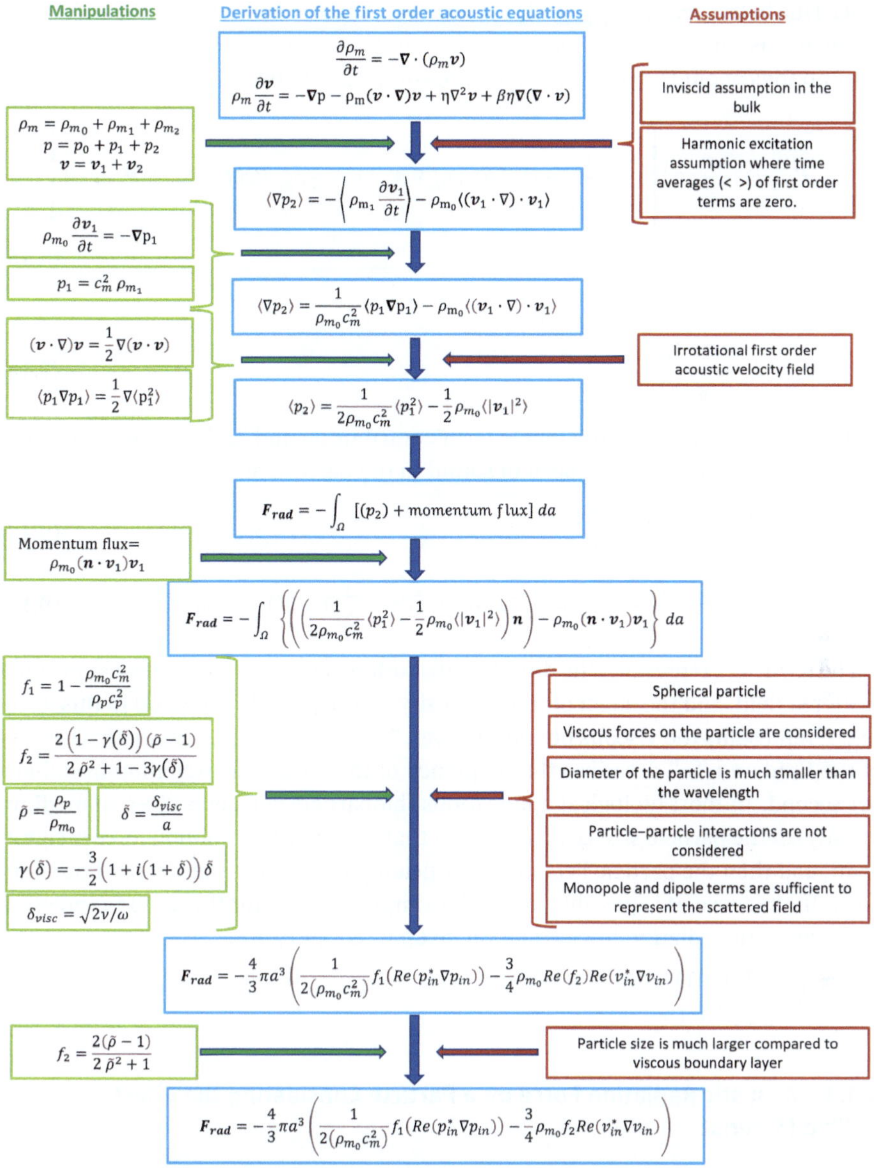

Figure 9.5 The derivation schematic of the second-order acoustic field with the assumptions involved.

the scattered field from a single spherical particle, a multi-pole expansion using monopole and dipole terms can be utilized. Further details of the expansion terms can be found in Bruus [35]. The analytical prediction of the scattered force due to a point-like (compared to acoustic wavelength) particle enables one to represent the acoustic radiation force in terms of the incident field without the need to consider the scattering of the incident field, which generally may require numerical analysis tools. Due to this reason, "p_1" and "v_1" terms are replaced with the incident acoustic variables, as shown in Eq. (9.15). This simplification is possible under certain assumptions about the particle, which are shown in Figure 9.5.

$$\boldsymbol{F}_{rad} = -\frac{4}{3}\pi a^3 \left(\frac{1}{2\left(\rho_{m_0} c_m^2\right)} f_1 \left(Re\left(p_{in}^* \boldsymbol{\nabla} p_{in} \right) \right) - \frac{3}{4}\rho_{m_0} Re(f_2) Re\left(\boldsymbol{v}_{in}^* \boldsymbol{\nabla} \cdot \boldsymbol{v}_{in} \right) \right) \tag{9.15}$$

where

$$f_1 = 1 - \frac{\rho_{m_0} c_m^2}{\rho_p c_p^2} \tag{9.16}$$

In Eq. (9.16) "f_1" is the monopole term contribution and "ρ_p" and "c_p" represent the particle density and the speed of sound in the particle material.

$$f_2 = \frac{2(1-\gamma(\tilde{\delta}))(\tilde{\rho}-1)}{2\,\tilde{\rho}^2 + 1 - 3\gamma(\tilde{\delta})} \tag{9.17}$$

$$\tilde{\rho} = \frac{\rho_p}{\rho_{m_0}}, \tilde{\delta} = \frac{\delta_{visc}}{a}, \delta_{visc} = \sqrt{2\nu/\omega}, \gamma(\tilde{\delta}) = -\frac{3}{2}(1 + i(1+\tilde{\delta}))\tilde{\delta} \tag{9.18}$$

The term "f_2" represents the dipole contribution of the multipole expansion of the scattered field, and the effect of fluid viscosity on the particle shows up in this term. "δ_{visc}" represents the viscous boundary layer, "ν" represents the kinematic viscosity of the fluid and "ω" describes the frequency of the harmonic excitation in radians per second. In most biological applications (bioparticle density is close to medium density), if the particle size is large enough that the acoustic radiation force is significant, and then the particle becomes significantly larger than the viscous boundary layer. In that case, it is possible to use the simplified form for the dipole expression in the acoustic radiation force equation given in Eq. (9.19).

$$f_2 = \frac{2(\tilde{\rho}-1)}{2\,\tilde{\rho}^2 + 1} \tag{9.19}$$

9.4.4 Acoustic Radiation Force on a Particle Considering the Effect of Chip Material

The calculation of the acoustic radiation force on a particle is explained in detail in the Section 9.4.3. One of the fundamental assumptions in these calculations is that the chip material does not affect the resonance conditions inside the channel. The requirement for resonance is generally assumed to be the frequency that results in fitting the half-acoustic wave along the width of the microchannel. In literature, some studies show that this assumption for resonance conditions inside a microchannel provides a good approximation when acoustically hard materials are used as chip materials, such as silicon and glass. On the other hand, in studies that use polymer materials, the resonance frequency estimates do not reflect the experimentally observed resonances. When the channel is fabricated from acoustically soft

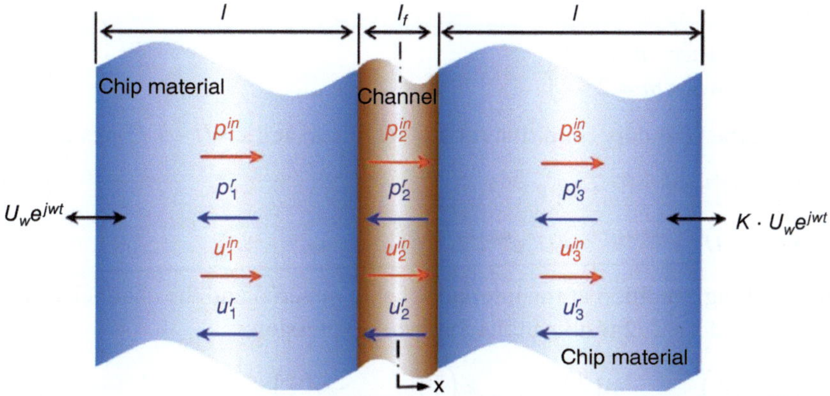

Figure 9.6 1D representation of the acoustic chip–channel interaction considered in the study of Özer and Çetin. Source: Özer and Çetin [24]/Acoustical Society of America.

chip materials, the resonance of the chip may affect the resonance conditions inside the microchannel. Recently, Özer and Çetin [24] attempted to formulate the effect of the chip material on the acoustic resonance inside the channel. The following assumptions were used in the formulations:

- Acoustic attenuation is not accounted for.
- The system is assumed to be infinitely long along the length direction, and no acoustic reflections occur along that direction, as shown in Figure 9.6.
- The acoustofluidic chip is excited from outer boundaries only.
- If the wall on the left side of the figure moves with harmonic amplitude of U_w, the wall on the other side (right side) moves with an amplitude of KU_w where K is a real number ranging between $-1 \leq K \leq 1$ (no phase difference between channel walls).
- It is assumed that longitudinal waves dominate the response.

As shown in Figure 9.6, the chip has a width "l" and the channel width is "l_f." The chip walls are assumed to have a harmonic oscillation at a frequency "ω." The outer chip wall vibration amplitude is "U_w" for one of the walls and "KU_w" for the other wall. The "$p_1{}^{in}$", "$p_3{}^{in}$" are the incident acoustic wave field inside the chip material on both sides of the channel. "$p_1{}^{r}$" "$p_3{}^{r}$" are the reflected acoustic wave fields in the chip material at both sides of the channel. The same naming convention is true for the acoustic wave velocities in the chip material. "$p_2{}^{in}$", "$p_2{}^{r}$" the incident and reflected acoustic pressure wave fields inside the channel, and a similar naming convention is used for the acoustic particle velocity "u." The acoustic wave variables are shown below with the complex notation:

$$u_m^{in}(x,t) = U_m^{in}(x)\, e^{j(\omega t - k_m x)} \tag{9.20a}$$

$$u_m^{r}(x,t) = U_m^{r}(x)\, e^{j(\omega t + k_m x)} \tag{9.20b}$$

$$p_m^{in}(x,t) = P_m^{in}(x)\, e^{j(\omega t - k_m x)} \tag{9.20c}$$

$$p_m^{r}(x,t) = P_m^{r}(x)\, e^{j(\omega t + k_m x)} \tag{9.20d}$$

where,

$$k_m \begin{Bmatrix} k \text{ when } m = 1,3 \\ k_f \text{ when } m = 2 \end{Bmatrix} \tag{9.21}$$

Matching the boundary conditions at $x = -(l + l_f/2)$ and $x = (l + l_f/2)$ results in:

$$U_w = U_1^{in}\, e^{jk(l+l_f/2)} + U_1^r\, e^{-jk(l+l_f/2)} \tag{9.22}$$

$$KU_w = U_3^{in}\, e^{-jk(l+l_f/2)} + U_3^r\, e^{jk(l+l_f/2)} \tag{9.23}$$

The matching condition at the interface of chip material and the channel located at $x = -l_f/2$ and $x = l_f/2$ results in the following expression:

$$U_1^{in} e^{jk\left(\frac{l_f}{2}\right)} + U_1^r e^{-jk\left(\frac{l_f}{2}\right)} = U_2^{in} e^{jk_f\left(\frac{l_f}{2}\right)} + U_2^r e^{-jk_f\left(\frac{l_f}{2}\right)} \tag{9.24}$$

$$U_2^{in} e^{-jk_f\left(\frac{l_f}{2}\right)} + U_2^r e^{jk_f\left(\frac{l_f}{2}\right)} = U_3^{in} e^{-jk\left(\frac{l_f}{2}\right)} + U_3^r e^{jk\left(\frac{l_f}{2}\right)} \tag{9.25}$$

Equating the acoustic pressure on each side of the boundaries $x = -l_f/2$ and $x = l_f/2$ results in the following equations:

$$\frac{Z_f}{Z_c} = \hat{Z} = \frac{U_1^{in} e^{jk\left(\frac{l_f}{2}\right)} - U_1^r e^{-jk\left(\frac{l_f}{2}\right)}}{U_2^{in} e^{jk_f\left(\frac{l_f}{2}\right)} - U_2^r e^{-jk_f\left(\frac{l_f}{2}\right)}} \tag{9.26}$$

$$\frac{Z_c}{Z_f} = \frac{1}{\hat{Z}} = \frac{U_2^{in} e^{-jk_f\left(\frac{l_f}{2}\right)} - U_2^r e^{jk_f\left(\frac{l_f}{2}\right)}}{U_3^{in} e^{-jk\left(\frac{l_f}{2}\right)} - U_3^r e^{jk\left(\frac{l_f}{2}\right)}} \tag{9.27}$$

where Z_f and Z_c denote infinite domain acoustic impedances of the fluid and the chip materials, respectively.

Solving Eqs. (9.22)–(9.27) for six unknowns, which are incident and reflected velocity fields of the three media, namely the channel region and chip regions at two sides of the channel ($U_1{}^{in}$, $U_1{}^r$, $U_2{}^{in}$, $U_2{}^r$, $U_3{}^{in}$, $U_3{}^r$), in terms of channel wall velocities "U_w" and "KU_w." Utilizing channel acoustic velocity variables $U_2{}^{in}$ and $U_2{}^r$, after several steps of algebraic manipulations, one can write the acoustic velocity field inside the channel as:

$$u(x) = U_2^{in} e^{-jk_f x} + U_2^r e^{jk_f x} \tag{9.28}$$

$$\frac{u(x)}{U_w} = \frac{(1+K)}{2\Gamma_{in}} \cos\cos(k_f x) - \frac{(1-K)}{2\Gamma_{out}} \sin\sin(k_f x) \tag{9.29}$$

Similarly, using the acoustic impedance of the fluid medium, one can obtain the acoustic pressure inside the channel as:

$$p(x) = U_2^{in} Z_f e^{-jk_f x} - U_2^r Z_f e^{jk_f x} \tag{9.30}$$

$$\frac{p(x)}{-jZ_f U_w} = \frac{(1+K)}{2\Gamma_{in}} \sin\sin(k_f x) + \frac{(1-K)}{2\Gamma_{out}} \cos\cos(k_f x) \tag{9.31}$$

"Γ_{in}" and "Γ_{out}" in the above equations is derived as:

$$\Gamma_{in} = \cos\cos(kl)\cos\cos\left(\frac{k_f l_f}{2}\right) - \hat{Z}\sin(kl)\sin\left(\frac{k_f l_f}{2}\right) \tag{9.32}$$

$$\Gamma_{out} = \cos\cos(kl)\sin\sin\left(\frac{k_f l_f}{2}\right) + \hat{Z}\sin(kl)\cos\left(\frac{k_f l_f}{2}\right) \tag{9.33}$$

As it can be seen from Eqs. (9.29) and (9.31), the acoustic pressure and acoustic particle velocity expressions inside the microchannel are represented in terms of the chip width "l" and the channel wave number "k" which is a function of the acoustic wave propagation inside the chip material. The acoustic radiation force on a particle inside the channel is shown in Eqs. (9.34) and Eqs. (9.35a)–(9.35d) as explained in [24].

$$F_{rad} = -\pi a^3 \rho_f k_f U_w^2 \Phi \Gamma_{chip}^{(1)} \sin(2k_f x + \phi) \tag{9.34}$$

$$\Gamma_{chip}^{(1)} = \left(\frac{1+K}{\Gamma_{in}}\right)^2 - \left(\frac{1-K}{\Gamma_{out}}\right)^2 \tag{9.35a}$$

$$\Gamma_{chip}^{(2)} = \frac{2(1-K^2)}{\Gamma_{in}\Gamma_{out}} \tag{9.35b}$$

$$\phi = \arctan\left(\Gamma_{chip}^{(2)}/\Gamma_{chip}^{(1)}\right) \tag{9.35c}$$

$$\Phi = \frac{1}{12}f_1 + \frac{1}{8}f_2 \tag{9.35d}$$

As seen from Eqs. (9.34) and (9.35), one can obtain the acoustic radiation force accounting for the acoustic properties of the chip material. Careful examination of the Eqs. (9.34) and (9.35a) shows that acoustic radiation can reach very high values as "Γ_{in}" or "Γ_{out}" tends to zero. Equation (9.35a) shows that the $\Gamma_{chip}^{(1)}$ expression, which shows up in the acoustic radiation force Eq. (9.34) is also dependent on the value of "K." If the "K" value is "1," which means both outer channel walls move in the same direction then the second term of the Eq. (9.35a) becomes zero and the "Γ_{in}" becomes the only contributing term to acoustic radiation force. Also, "Γ_{in}" shows that the particles will move toward the center of the channel due to its positive sign. In the other situation, where the outer channel walls move in the opposite direction ($K = -1$) then, the negative sign coming from the "Γ_{out}" causes particles to move towards the channel walls. Therefore the destination of the particles is determined by the movement of the outer channel walls with respect to each other.

Also, examining Eq. (9.32) shows that when "$\hat{Z}$" is very small (i.e. chip acoustic impedance is much larger compared to fluid acoustic impedance) and the channel width ("l_f") is half the wavelength of the acoustic wavelength in the fluid, then "Γ_{in}" term in Eq. (9.32) approaches zero regardless of the chip width "l." Hence for acoustically hard chip materials ($\hat{Z} \approx 0$) the chip width does not have a strong effect on the channel acoustic resonance. However, the same is not true for acoustically soft materials ($\hat{Z} \approx 1$) therefore for polymer chip materials, the chip width affects acoustic resonance. The aforementioned conclusions for a polymer chip are experimentally observed results and are discussed in Silva et al. [36]. Their experimental observations of channel resonance being affected by chip geometry in a polystyrene chip can be explained with the equations derived in this section.

9.5 Simulation of Acoustophoretic Bio-Particle Manipulation

Numerical simulation of acoustophoresis is a rather complex and challenging task. The difficulty mainly arises from various coupled physical domains and orders of magnitude differences in time scales at which the physical variables need to be calculated. The general approach that needs to be applied for a numerical solution is similar for both the BAW and SAW numerical simulations. On the other hand, the frequencies used in SAW applications are generally higher; therefore, the computational cost of simulations is usually higher. The components that need to be modeled and the required physical analyses are shown in Figure 9.7. Even though this numerical simulation flow chart contains the components of a BAW device, this general approach is mostly valid for SAW devices as well.

9.5.1 Simulation of Piezoelectric Actuators

For the motion of the bioparticles, power input to the acoustofluidic system is required. This power is provided to the system by a power amplifier that excites the piezoelectric actuator. The electrical field, as well as the elastic wave motion, needs to be accurately simulated. To simulate the resulting motion, piezoelectric constitutive relationships need to be coupled with the elastic wave equations of the solid elastic domain. The equations for piezoelectric constitutive relations as well as the stress and strain relations in the solid domain, are provided in several different references [37–40].

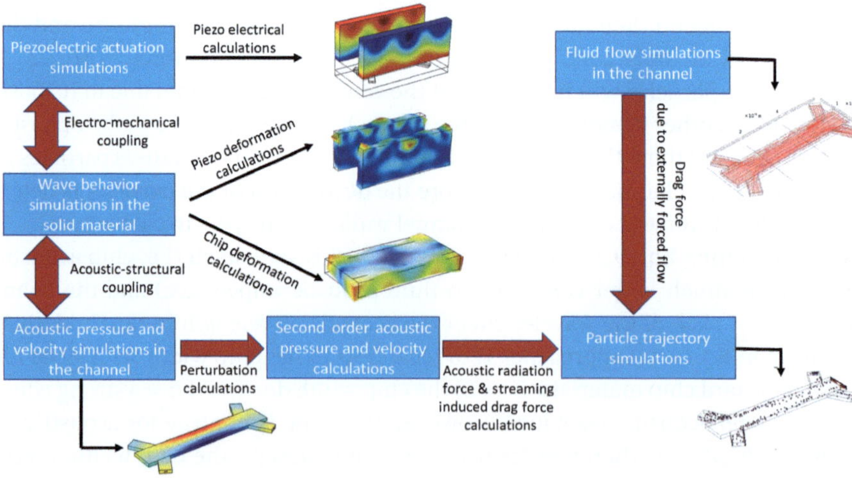

Figure 9.7 Numerical simulation scheme of a BAW acoustofluidic device.

Due to the high complexity of the acoustophoretic simulation process, in the initial years of the numerical simulation efforts, several studies did not account for the piezoelectric material dynamics [41, 42]. Some of these studies assumed a uniform wall motion of the chip material in BAW studies. This approach results in a more uniform acoustic pressure distribution in the microchannel due to complicated flexural wave patterns that do not affect the boundary motion of the acoustic domains. In SAW studies, the modeling of the piezoelectric material is computationally more expensive due to significantly higher excitation frequencies, a much more complicated electrode structure that excites the piezoelectric material as well as the larger geometric dimensions of the piezoelectric actuator compared to the channel structure. This difficulty is overcome by analytically defined displacement or velocity fields, which provide motion to the fluid field where acoustic waves propagate [42, 43]. As the computational power of the computers and the capabilities of the commercial finite element software packages increased, numerical simulations of the piezoelectric materials became possible and more practical. The simulations initially started with the 2D domain simulations [37, 44], which extended into 3D simulations [38–40] over time.

9.5.2 Numerical Simulations of the Elastic Material Surrounding the Channel

The piezoelectric actuators transmit the elastic wave energy into the chip material through the physical interface between the actuator and the chip material in devices with BAW architecture. In SAW devices, the power from the actuator is directly transmitted to the fluid and the channel material that surrounds the fluid. The elastic material that covers the fluid in either configuration must be coupled with two different physics fields. These two physical domains are the electromechanical domain due to the elastic material's physical interface with the piezoelectric actuator and the coupling with the acoustic domain due to its large interface area with the fluid.

The chip material over the channel structure covers a much larger area than the channel structure; therefore, it may require significant processing power to fully simulate this domain, and difficulties may arise due to coupling of solid mechanics domain with the electromechanical and acoustic fields. Hence, similar to the piezoelectric actuator modeling, some studies chose not to include the chip material in the numerical simulations in 2D [43] and 3D [45]. Overtime, as chip material is included in the numerical models, due to the large area it covers, 2D models were preferred. The mesh size that needs to be used is an important parameter in finite element simulations. In finite element simulations, 5–10 finite elements per wavelength are a generally accepted norm for acoustic numerical simulations. However, for domains of solid elastic materials, the mesh density needs to be at least three times higher than the mesh density in the acoustic medium to resolve the shear and flexural waves that have lower wavelengths compared to the longitudinal waves. The choice of chip material is rather crucial regarding the impact on the computation power required for the numerical simulation. Since acoustically hard materials such as silicon and glass have high acoustic wave propagation speeds and hence longer wavelengths, they require lower mesh densities for accurate simulation. In that sense, the most demanding material is PDMS, commonly used in SAW acoustophoretic devices. It has low acoustic wave propagation speed close to that of water, and if the shear waves are to be resolved, its mesh requirement is more demanding than the mesh density requirements for water. In situations where the mesh density requirements start to challenge the computational capabilities, an obvious solution is to analyze the problem in 2D. However, one can extend the numerical simulation to 3D without the full burden of the computational cost. The perfectly matched layers (PMLs) are used to extend the problem to 3D without the need for meshing the entire geometry of the solid elastic material. PML regions allow one to just model a portion of the geometry when reflections from boundaries are believed to be weak. This weak reflection assumption is particularly valid for highly lossy materials such as PDMS [40], but is also used for other chip materials such as glass as well as piezoelectric actuators [40, 46].

9.5.3 Simulation of Fluid Flow

The fluid domain inside the microchannel interfaces with the chip and the lid materials in the BAW acoustophoretic devices. In SAW devices, the fluid interfaces with a piezoelectric actuator and solid elastic material around the fluid. Generally, the motion induced by the chip material or the piezoelectric actuator (in SAW devices) at the channel walls is used for the calculation of the first-order acoustic pressures and velocities inside the microchannel. As in the previous components, the models for the microchannels evolved from simple 2D numerical models toward 3D full models for both SAW and BAW device numerical simulations. Since the microchannel can extend to long distances along certain spatial directions, several studies used PML to decrease the size of the meshed geometry, hence the computational burden [45, 47, 48].

The first-order field is calculated by using the numerical solution of the acoustic wave equation shown in Eq. (9.9). This equation is numerically solved mainly through the application of finite element analysis. Sometimes the boundary element method (BEM) is employed in acoustophoresis problems, but these were limited to the calculation of the scattered field from a spherical/non spherical particles or cells but not for the solution of the first-order acoustic field [49–51].

After the calculation of the first-order fields, one can calculate the second-order acoustic variable through the second-order perturbation of the Navier–Stokes equations. In order to calculate the second-order acoustic pressure, which is necessary to calculate the acoustic radiation force on a microparticle, the calculation of the first-order field variables is enough, as shown in Eq. (9.13). The force generated by the acoustic radiation force may not be the only dominant force on a microparticle, especially if the particle size is small. In that case, the acoustic flow (second-order acoustic velocity field with a steady component), generally called the acoustic streaming generated by the first-order acoustic fields, may become significant. In order to calculate the acoustic streaming field, one needs to account for the viscous effects on the fluid since the streaming in the bulk of the fluid is driven by the vibrating boundaries of the microchannel, where the viscous effects are significant. Therefore, applying the second-order perturbation to Navier–Stokes equations is required without neglecting the bulk and shear viscosity terms. Also, at the boundaries of the fluid domain, there is also a thermal boundary layer in addition to the viscous boundary layer. This thermal boundary layer and the first-order temperature variations of the acoustic field may also contribute to acoustic streaming. In that case, one needs to solve the thermo-viscous first-order acoustic equations, and the thermo-viscous terms of the first-order variables affect the second-order acoustic solutions as derived in Muller et al. [52] and Muller and Bruus [53].

Generally, for the numerical solution of the second-order acoustic fields, which results in the acoustic streaming velocity, the finite element method has almost always been used in the literature. The second-order perturbation of the Navier–Stokes equation results in partial differential equations where additional terms appear in the standard Navier–Stokes equations. Therefore, in order to implement these additional terms to finite element solutions, they need to be added to the Navier–Stokes equations either as a source term or weak contributions, as explained in [39, 54, 55]. Another difficulty in obtaining the acoustic streaming field is the very small size of the viscous and thermal boundary layers and the requirement that the solution mesh resolve these layers with several elements. Due to this requirement, the number of finite elements needs to be drastically increased, increasing the computational burden. This problem is even more pronounced in SAW systems where the frequencies are high; hence, the boundary layer thicknesses are very small (in the order of nanometers). A remedy to this problem is offered in some studies [45, 47, 56] where an approximation to the time-averaged acoustic streaming velocities is calculated for the boundaries, and they act as a boundary condition for the bulk of the domain, resulting in the acoustic streaming in the bulk of the fluid.

9.5.4 Simulation of Particle Motion

The microparticles in the microchannel are subjected to two kinds of forces in acoustofluidic devices. These are the forces due to acoustic pressure and acoustic velocity fields resulting in an acoustic radiation force on the particle and/or the drag forces due to flow. The drag force acting on the particle may be due to acoustic flow resulting from the second-order acoustic velocity field inside the microchannel or an external flow due to a powered device such as a pump. The trajectories of the particles can also be affected by gravity or other mechanical body forces, but for small particles, it is common practice to neglect them. Also, in practice, the manipulated bioparticles are large enough to ignore effects such as van der Wall's forces and random walk.

Particle trajectory calculations are one reason why orders of magnitude time scale differences exist in the acoustophoretic numerical simulation process. Due to high excitation frequency, the elastic and acoustic wave motion occurs in the order of microseconds. Because of the weak, steady second-order acoustic field forces, particle motion occurs in the order of seconds. This vast time scale difference results in difficulties in time simulations. This difficulty is overcome by performing simulations in the frequency domain for wave motion and in the time domain for particle motion simulations.

The size of the particles is essential for choosing the correct approach in particle trajectory simulations. When small-diameter particles ($<1\,\mu m$) are considered, the acoustophoretic radiation force becomes very small, and the particle's trajectory is determined by acoustic streaming. On the contrary, acoustic streaming has a weak effect on the trajectory of large particles. Also, trajectory simulations of large particles may conflict with the Rayleigh limit, which requires particle size to be significantly smaller than the acoustic wavelength ($ka \ll 1$). For instance, a HeLa cell with a $40\,\mu m$ diameter manipulated in a SAW device operating at 25 MHz may violate the Rayleigh limit; hence, Eq. (9.15) may not be used to calculate the acoustic radiation force on the particle. Also, Eq. (9.15) cannot be used to calculate the acoustic radiation force for nonspherical particles. In the case of an arbitrarily shaped particle, Eq. (9.14) needs to be utilized with the scattered first-order acoustic field calculated by the numerical simulation [47, 57, 58].

It should also be remembered that the acoustic radiation force expression in Eq. (9.15) is derived for a single spherical scatterer. If the bioparticle concentration is high, then the scattering from other particles may disturb the incident acoustic field. This disturbed acoustic field may result in inaccurate estimations of acoustic radiation force since the assumed incident field in the equation will not be correct. This situation can certainly be the case for whole blood numerical simulation, where cell concentrations are pretty high. Some studies accounted for the particle–particle acoustic interactions through analytical methods [59, 60] or numerical methods [58]. Also, a similar argument can be extended to the effect of particle–particle interactions on fluid flow. Two close-moving bioparticles may affect each other's motions, and in high-concentration cases, one may need to account for the particle–particle interactions for fluid flow. Some studies have investigated the effect of particle–particle interactions with analytical methods and with finite element method [61, 62].

The choice of the numerical simulation method can also be important when particle–particle interactions are modeled. When the microparticles and the fluid domain are meshed in the finite element method to account for particle–particle interactions, mesh quality issues may arise in the finite element approach. The problem may occur when the particles get very close to each other during the numerical simulation. Since the minute distance between the particles needs to be meshed due to the requirement of meshing the fluid domain in finite element methods, poor mesh quality and numerical instabilities may occur. The BEM can be immune to such problems since when the fluid domain is simulated with BEM, the domain does not have to be meshed; only the boundaries of the domain need to be meshed. BEM has been successfully implemented for the simulation of particle motion in microchannels [63, 64] even for electrokinetic applications [65, 66]. Even though the lack of necessity of domain mesh is a significant benefit, the computation is still challenging since the system matrices for a boundary element formulation are fully populated, unlike the sparse matrix structure of FEM.

Another issue that can be important in the numerical simulation of trajectories of the particles is assigning single geometric and material constants to populations of microparticles. This is especially true when acoustophoretic separation of bioparticles is simulated since sizes and material properties vary widely for bioparticles. When single-value properties of geometric dimensions and material properties are used for bioparticles, perfect separations of bioparticle populations can be observed, which rarely happens in real-life biological applications. Hence in some studies [41], variations in sizes and material/acoustic properties are considered and statistical distributions are applied in particle property assignments.

9.6 Acoustofluidic Devices in Biological and Medical Applications

There has been vast literature on the acoustofluidic manipulation of micro/nanoscale biological particles for biomedical applications. Owing to the relatively insensitive nature of acoustofluidic techniques to pH, salt concentration, and other media properties, both BAW and SAW platforms have been implemented for the manipulation and separation of biological particles. Although BAW devices offer simple device architecture and high throughput (which is essential for clinical applications [8]), they often require thermal management due to excessive heating during the operation and may not be viable for the manipulation of submicron particles. On the other hand, SAW devices offer high precision at the price of relatively complex device architecture and the capability of nanoscale particle manipulation with relatively low throughput. The references given in this section of the chapter are organized and listed in Table 9.3, along with fundamental information on each study.

9.6.1 Applications Regarding Lipid Particles

Following the fundamental studies with microscale engineered particles, one of the pioneering applications of acoustofluidic biological particle manipulation is the separation of lipid particles from RBCs to avoid any possible embolization problem after cardiac surgery. Laurell's group developed a BAW device [67–69], with a silicon chip for the separation of lipid particles from RBCs at different levels of RBC concentration (i.e. hematocrit) with a sample flow rate of 300 μl/min [68, 69] and 500 μl/min [67]. In these studies, RBCs are pushed toward the center stream, and on the other hand, lipid particles are pushed toward the channel walls due to the difference in acoustic contrast factors. In an initial study, to mimic lipid particles, a mixture of blood and milk was processed, and the separation of blood and milk was demonstrated at a sample rate of 300 μl/min with 80% recovery rate for lipid particles and 70% RBC removal [68]. In the following studies, lipid particles were mimicked by phospholipid-stabilized emulsion of triolein, and they achieved 82% recovery rate

Table 9.3 Summary of the literature on acoustofluidic particle/cell manipulation.[a)]

	Author (Year)	Application	Method	Frequency (MHz)			Sample flowrate (ul/min)	Performance	Miscellaneous details
Lipid particles	Jonsson et al. [67]	Separate lipid particles and RBCs	BAW	1.97–2.03	Silicon	Silicon	500	Recovery rate of 82% for lipid particles and 80% for RBC (hematocrit ranging between 5% and 30%)	A chip with eight parallel channels was used
	Petersson et al. [68]	Separation of lipid particles from blood and milk	BAW	~2	Silicon	Glass	300	Recovery rate of 70% for RBCs Removal rate of 80% for lipid particle	Single inlet/multiple outlet chip
	Petersson et al. [69]	Separation of lipid particles from RBC	BAW	Not specified	Silicon	Silicon	300	Recovery rate of 70% for RBCs Removal rate of 80% for lipid particle	Single-inlet/multiple outlet chip Hematocrit level of 2.5%
	Grenvall et al. [70]	Separation of lipid particles in raw milk.	BAW	~2	Silicon	Glass	10	Lipid content depletion above 90%	Two-inlet/two-outlet chip

Cell wash	Petersson et al. [71]	Carrier medium exchange	BAW	~2	Silicon	Glass	170	Recovery rate of at least 95% of for RBCs Removal rate of up to 98% for the contaminants	Two-inlet/ two-outlet chip
	Augustsson et al. [72]	Buffer exchange RBC	BAW	~2	Silicon	Glass	50	Recovery rate of 58% for RBCs Removal rate of 98.3% for the original buffer	Single-inlet/ single outlet with eight cross-flow junctions
	Li et al. [73]	Cell washing	Tilted SSAW	19.4–19.6	$LiNbO_3$	PDMS	1	Purity of 97% for WBCs washing	Triple-inlet/ two-outlet chip
	Li et al. [74]	Transfer of inflammatory cells from liquefied sputum	Tilted SSAW	19.54	$LiNbO_3$	PDMS	1	Recovery rate of 83.9% for cells	Triple-inlet/ two-outlet chip
Plasma	Lenshof et al. [75]	Whole blood plasmapheresis	BAW	~2	Silicon	Glass	80	Cell count of 3.65×10^9 cells/l for whole blood (approved plasma quality by Council of Europe for plasma transfusion)	Plasmapheresis was followed by PSA detection
	Adams et al. [76]	Separation blood cells from whole blood for high-quality plasma	BAW	0.93–0.95	Glass	Glass	16.7	Transfer ratio of 95% for blood cells in whole blood	Two inlet/outlet chip with PDMS gaskets and temperature control unit
	Petersson et al. [77]	Hematocrit determination	BAW	1.8–2.1	Glass	Glass	2000	Hematocrit measurement with error less than 1% point for whole blood in 20 s	Single inlet single exit device with image processing unit

(*continued overleaf*)

Table 9.3 (Continued)

	Author (Year)	Application	Method	Frequency (MHz)			Sample flowrate (ul/min)	Performance	Miscellaneous details
Platelets	Petersson et al. [19]	Separation of platelets from RBC	BAW	~2	Silicon	Glass	20	Recovery rate of 99% for platelets Removal rate of 80% for RBCs	Two-inlet/ multiple-outlet chip
	Nam et al. [78]	Separation of platelets from whole blood	SSAW	7.54	$LiNbO_3$	PDMS	0.25	Recovery rate of 98% for platelets Removal rate of 99% for RBCs in whole blood	Two-inlet/ two-outlet chip
	Chen et al. [79]	Separation of platelets from whole blood	BAW	0.23–0.24	Stainless steel	Stainless steel	10 000	Recovery rate of 86% for platelets Removal rate of 88% for WBCs/RBCs in whole blood	Two-inlet/ two-outlet chip
	Zhang et al. [80]	RBC and platelet separation	spiral SAW	40	$LiNbO_3$	N/A	N/A	Purity of 93% for RBCs Purity of 84% for platelets	Droplet size: 1.5 ul 152° Y-rotated $LiNbO_3$
WBCs	Dykes et al. [81]	Removal of platelets from PBPC products for retransfusion	BAW	~2	Silicon	Glass	20	Removal rate of 98% for WBCs Removal rate of 89% for platelets	Two-inlet/ two-outlet chip
	Urbansky et al. [82]	Separation of mononuclear cells from blood	BAW	~5/~2[b)]	Silicon	Silicon	5	Separation eff. of 43% (87%) with purity of 54% (53%) for WBCs (for MNCs) Removal rate of >99.95% for RBCs	Two-stage, two-inlet/ two-outlet chip
	Lissandrello et al. [83]	Purification of lymphocytes	BAW		plastic	plastic	30–100	Purity of 59% and recovery rate of 37% in whole blood for lymphocytes	Single inlet/ two-outlet chip with temperature control unit

Cancer cells	Augustsson et al. [84]	Separation of DU145 and PC3 and LNCaP from WBCs	BAW	~5/~2[b)]	Silicon	Glass	70	Removal rate of >94% (79%) for DU145 and PC3 (LNCaP) with a depletion of 96%for WBCs	Two-stage, two-inlet/ two-outlet chip with temp. control unit
	Magnusson et al. [85]	Separation of MC7 from WBCs	BAW	~5/~2[b)]	Silicon	Glass	75	Recovery rate of 86% (62%) for CTCs at high recovery mode (high purity mode) with a depletion of >99.5% for WBCs	Two-stage, fully automated device with temp. control unit
	Cushing et al. [86]	Reducing WBC background in cancer cell separation products	BAW	1.89–2.09	Silicon	Glass	100	Separation efficiencies of 98.6% (MCF-7) and 99.7% (DU145) with cancer cell recoveries of 89.8% (MCF-7) and 85.0% (DU145)	Two-stage, two-inlet/ two-outlet chip with temp. control unit
	Antfolk et al. [87]	Separation of CTCs from WBCs	BAW	~4.5/~2[b)]	Silicon	Glass	100	Recovery rate of 95% for CTCs Removal rate of 97.8% for WBCs	Two stage, single-inlet device with temperature control unit
	Ding et al. [88]	Sorting MCF-7 human breast cancer cells from non- malignant leukocytes	Tilted SSAW	~19.4	$LiNbO_3$	PDMS	2	Recovery rate of %71 with purity of 84% for MC7 cells	Triple-inlet/ two-outlet chip
	Li et al. [89]	Separation of CTCs from WBCs	Tilted SSAW	~19.57	$LiNbO_3$	PDMS	20	Recovery rate of >83% for CTCs Removal rate of about 90% for WBCs	Triple-inlet/ two-outlet chip
	Wu et al. [90]	Separation of CTC from whole blood	Tilted SSAW	19.9	$LiNbO_3$	PDMS and glass	125	Recovery rate of >86% for CTCs	Triple-inlet/ two-outlet chip

(*continued overleaf*)

Table 9.3 (Continued)

	Author (Year)	Application	Method	Frequency (MHz)			Sample flowrate (ul/min)	Performance	Miscellaneous details
Miscellaneous cells	Yang et al. [91]	Separation of viable and nonviable mammalian cells (MCF-7)	BAW	2.04	Silicon	Fused silica	6.7–200	Recovery rate of 91% for viable cells Removal of 78% for non-viable cells	Two-inlet/two-outlet chip
	Zalis et al. [92]	Separation of viable MCF-7 Separation of viable neuronal cells	BAW	4.93/2[b)]	Silicon	Glass	100	Recovery rate of 85% (75%) for viable MCF7 (neuronal) cells with removal rate of 82% (96%) for dead cells	Two-stage, two-inlet/two-outlet chip with temp. control unit
	Thévoz et al. [93]	Cell cycle phase synchronization in mammalian cells	BAW	~2	Silicon	Glass	50	84% G1 phase synchronization	Two-inlet/two-outlet chip
	Nam et al. [94]	Density-based separation of monodisperse encapsulated cells	SSAW	3.94	$LiNbO_3$	PDMS	8	Recovery rate of 97% for large-cell-quantity beads Removal rate of 99% for empty beads	Triple-inlet/multiple outlet chip
	Evander et al. [95]	Trapping of neural stem cells	BAW	12.4	Glass	Epoxy	1	No performance quantification	Two-inlet/single-outlet device with 3 orthogonal junctions
	Norris et al. [96]	Trapping of sperm cells	BAW	11.6–12.1	Glass	PZT	1	No performance quantification	Multiple-inlet/two-outlet device
	Hammarström et al. [97]	RBC trapping (×10 diluted whole blood)	BAW	10.8	Glass capillary		40	Only visual assessment	Single-inlet/single-outlet chip operated at aspirate/dispense mode

Bacteria	Ai et al. [98]	Separation of bacteria from peripheral blood mononuclear cells	SSAW	13.02	$LiNbO_3$	PDMS	0.5	Purity of 96% for bacteria and 91% for peripheral blood mononuclear cells	Two-inlet/two-outlet chip
	Li et al. [99]	Separation of bacteria from blood	tilted SSAW	19.54	$LiNbO_3$	PDMS	1	Purity of 96% for bacteria and 96% for RBCs	Triple-inlet/ two-outlet chip
	Silva et al. [36]	Separation of bacteria from whole blood	BAW	0.632	Plastic	Plastic	110	Yield of 34% for bacteria Removal rate of 95% for RBCs	Single inlet/ two-outlet chip with optimized performance
	Dow et al. [100]	Separation of bacteria from whole blood	BAW	0.94	Plastic	Plastic	50	Yield of 45–60% for bacteria Removal rate of >85% for RBCs	Two-inlet/two-outlet chip with temperature control unit
	Ohlsson et al. [101]	Bacteria separation blood	BAW	1.99	Silicon	Glass	400	Recovery rate of 90% for bacteria Removal rate of 99.9% for RBCs	Two-inlet/ two-outlet chip with temperature control unit
	Hammarström et al. [102]	Trapping of nanoparticles and bacteria	BAW	4	Glass capillary		10	Capture efficiency of 95% for bacteria at low concentration	10 μm PS seed particles were used
	Hammarström et al. [118]	Purification of bacteria	BAW	4	Glass capillary		50	Capture efficiency of 95% for bacteria	Automated device operated at aspirate/ dispense-mode
	Ohlsson et al. [103]	Separation of bacteria from RBCs	BAW	1.96	Glass capillary		80	Recovery rate of 91% for bacteria Removal rate of 99% for RBCs	Single-inlet/ two-outlet device with temperature control unit
	Ngamsom et al. [104]	Isolation of microflora	BAW	1.30	Glass	Glass	10	Recovery rate of 60–90% for pathogens Removal rate of 99.8% for RBCs	Raw chicken, beef and blood samples

(*continued overleaf*)

Table 9.3 (Continued)

	Author (Year)	Application	Method	Frequency (MHz)			Sample flowrate (ul/min)	Performance	Miscellaneous details
Nanoscale (bio)particles	Evander et al. [105]	Trapping of platelet-derived microparticles (nano vesicles)	BAW	4	Glass capillary		10	Recovery rate of about 30% in undiluted and 60% in diluted plasma for microparticles	12 μm PS seed particles were used
	Ku et al. [106]	Enrichment of EVs from biological fluids using acoustic trapping	BAW	4	Glass capillary		15	Trapping efficiency of 1–100% for nanoparticles depending on size	12 μm PS seed particles were used/automated platform
	Collins et al. [107]	Separation of 300 and 500 nm particles	TSAW	49	$LiNbO_3$	PDMS	0.45/1.8	Separation eff. of 79%/87% for 500 nm particles	Tunable separation by size
	Seghal and Kirby [108]	Separation of 100 and 300 nm particles	SSAW	48.8	$LiNbO_3$	PDMS	0.2	Separation efficiency of 86% for 300 nm particles and 98% 100 nm particles	Triple-inlet/two-outlet chip with Bragg reflectors
	Collins et al. [28]	Manipulation of nanoparticles	Focused TSAW	633	$LiNbO_3$	PDMS	0.45	84% of 300 nm, 62% of 500 nm, and 5% of 100 nm particles were manipulated	—
	Lee et al. [109]	Purification of extracellular MVs	SSAW	38.5	$LiNbO_3$	PDMS	0.43	Recovery rate of >80% for exosomes and >90% for other MVs	Two-inlet/triple-outlet chip
	Wu et al. [110]	separation of nanoparticles	tilted SSAW	33.1	$LiNbO_3$	PDMS	5	Recovery rate of about 90% for nanoparticles	Two-inlet/two-outlet chip
	Wu et al. [111]	Cell removal + isolation of exosomes from other MVs	Tilted SSAW	19.6/39.4[b)]	$LiNbO_3$	PDMS	4	Recovery rate of 82% with purity of 82% for exosome and removal rate of 99.999% for RBCs	Two-stage, multiple-inlet/multiple outlet device

Shi et al. [112]	1D and 2D patterning of bacteria and bRBC (acoustic tweezer)	SSAW		$LiNbO_3$	PDMS	N/A		
Ding et al. [113]	2D manipulation of cells and organisms	SSAW		$LiNbO_3$	PDMS	N/A		
Collins et al. [31]	2D single-cell patterning	SSAW				N/A		
Guo et al. [114]	2D cell patterning	SSAW				N/A		
Guo et al. [32]	3D manipulation of cells	SSAW		$LiNbO_3$	PDMS	N/A		
Ahmed et al. [115]	Rotational manipulation of single cells and organisms	BAW		Glass	PDMS	N/A		
Ozcelik et al. [116]	Rotational manipulation of single cells and organisms	BAW	0.005/0.07–0.1	Glass	PDMS	N/A		Single-inlet/ single-outlet chip with sharp structures
Zhang et al. [117]	Rotational manipulation of organisms	SSAW		$LiNbO_3$	PDMS	5		

a) Color coding is used to demonstrate the work of Laurell's group (blue) and Huang's group (gray).
b) For two-stage devices, the frequency of each stage is specified in order.

for lipid particles with an 80% RBC removal for the range of hematocrit levels of 5–30% [67], and 80% recovery of triglyceride emulsions with more than 60% RBC removal [69]. A similar idea was also employed for the separation of lipid particles in raw milk. Unlike the previous studies, the concentration of lipid particles is much higher than that of the blood samples, and the low velocity of the lipid particles close to channel wall results in clogging problems after certain duration of operation. To circumvent this issue, two-inlet and two-exit acoustofluidic chip was proposed to be operated in two different operation modes for enrichment and depletion of lipid particles, keeping the protein and lactose content the same as the original [70]. The mixture was loaded from a center inlet together with sheath flows from two side inlets. In lipid enrichment mode, lipid particles are focused at the center antinode, whereas RBCs are focused at pressure nodes on both sides of the centerline with a clear separation from channel walls. The lipid-rich outlet was analyzed by FTIR, and a lipid concentration of 3.5 was reported. In lipid depletion mode, the pressure node was placed at the centerline, and antinodes were placed on both sides of the centerline. As the sample mixture flowed at the centerline, the lipid particles (with negative contrast factor) moved towards the antinodes. The efficient removal of lipid emulsion from milk without any effect on the lactose or protein composition was demonstrated through multiangle light scattering measurements.

9.6.2 Applications Regarding Cell Wash

In some biomedical applications, it is desirable to transfer cell groups of interest from one buffer medium to another (i.e. buffer medium exchange). One good example is the complete removal of blood plasma from RBCs when the plasma contains high levels of inflammatory or coagulation factors. The same BAW principle introduced in the aforementioned studies (i.e. focusing of RBCs in the center stream) was also implemented to transfer RBCs from their original medium to a clean plasma with a sample flow rate of 170 μl/min. At least 95% of the RBCs were collected in clean blood plasma, with 98% removal of the contaminants [71]. Huang's group introduced a SAW device with tilted SAW architecture to deflect the particle trajectory in the lateral direction [88]. They implemented tilted SAW architecture for the purification of WBCs from lysed blood samples (i.e. transferring WBCs into a clean buffer solution), which is quite essential in many immunological and clinical diagnostic applications [73]. A washing efficiency of above 97% was demonstrated with microbeads, and WBCs were isolated from lysed blood samples with a purity of >97% at a sample flow rate of 1 μl/min. The same SAW architecture was also utilized to transfer inflammatory cells in liquefied human sputum to avoid the negative effects of dithiothreitol present in the samples as a result of the liquefaction process [74]. 84% cell recovery was reported at a sample flow rate of 1.0 μl/min.

A silicon BAW device with a prefocusing section followed by a section with multiple lateral flow streams was also proposed to accomplish acoustofluidic cell wash (i.e. buffer medium exchange) [72]. At a sample flow rate of 50 μl/min, 58% of RBCs (at hematocrit level of about 2%) were successfully removed with a washing efficiency of more than 98%.

9.6.3 Applications Regarding Separation of Blood Components

Blood, which is composed of plasma (about 55% of blood fluid) with various proteins, antibodies, hormones, molecules etc., red blood cells (RBCs, 6–8 μm in diameter, about 4–6 × 10^{12} cells/ml), white blood cells (WBCs, 12–15 μm in diameter, about 4–11 × 10^9 cell/ml), platelets (1–5 μm in diameter, 100–400 × 10^9 cells/ml), is an essential biological fluid which encompasses rich information about the physiological and chemical state of the body. Each constitute has a specific role for the body and possesses vital information for many diagnostic and therapeutic applications. The size, density, and compressibility dependence of acoustophoretic force enable the manipulation and separation of blood components from a background medium.

9.6.3.1 Plasma Separation

Alternative to conventional centrifuge-based (which may rupture and/or activate blood cells due to high shear forces induced by centrifugal forces and lead to the contamination of plasma with intracellular content) and microfluidic filtration-based (which are prone to clogging) techniques, plasmapheresis was achieved in a single-inlet, multiple-exit BAW device with a whole blood flow rate of 80 μl/min [75]. The produced plasma was further processed with a subsequent immunoassay for prostate-specific antigen (PSA) detection. Almost 100% removal of RBCs was reported over a range of hematocrit with an output rate of 10 μl/min for the cell-free plasma containing an RBC level of $<6.0 \times 10^9$ cells/ml (which requires the removal of more than 99.9% of the total RBC count), which complies with the benchmark regulations of the Council of Europe. Later, Adams et al. [76] introduced a temperature-controlled BAW device for the removal of RBCs from whole blood. The temperature- controlled nature of the device enabled the use of elevated power intensity, boosting the processing rate up to 33.3 μl/min. They stated 95% and 82% RBC removal at a processing rate of 16.7 and 33.3 μl/min, respectively. The separation of plasma and blood cells is also quite important to determine the hematocrit level of blood. An image acquisition and processing unit were also integrated into a single inlet, single exit BAW device to determine the hematocrit measurements of whole blood [77]. This time, the chip was fabricated out of borosilicate glass rather than silicon to be able to have visual access to the main channel. The device was capable of determining the hematocrit level within 20 seconds with 1% point accuracy on a 200 ml whole blood sample.

9.6.3.2 Platelet Separation

Being one of the blood components, platelets are involved in many essential physiological processes such as wound healing, liver regeneration, and tumor metastasis. In many therapeutic applications, platelets are the main ingredient for chemotherapy, thrombocytopenia secondary to bone marrow neoplasms, and platelet function defects [79]. The conventional technique for the separation of platelets from whole blood is centrifugation. However, centrifugation has some drawbacks since it may activate platelets due to high shear stress induced on them and affect their integrity and functionality. To overcome the adverse effects of centrifugation, several researchers have proposed acoustofluidic techniques for the separation of platelets from the blood. A two-inlet/two-outlet silicon BAW device was employed for the separation of platelets from RBCs, utilizing both density and size differences [19]. 99% recovery of platelets was achieved with 92% removal of RBCs at a sample flow rate of 20 μl/min. However, cesium chloride solution, which may have some adverse effects, was added to the sample to tune the medium density. Nam et al. [78] demonstrated a SAW device, which can separate platelets from whole blood. At a low sample rate of 0.25 μl/min, 98% platelet recovery with 99% RBC removal was accomplished. Chen et al. [79] proposed a stainless steel BAW device for the separation of platelets from both RBCs/WBCs with high purity and throughput. Whole blood was processed at a rate of 10^3 μl/min with a more than 80% recovery of platelets and more than 80% RBC/WBC removal. To achieve successful operation, they used impedance-matched buffers. They also assessed the integrity of platelets following the process, and it was shown that proposed BAW device damaged the platelets much less than the conventional centrifuge technique. Very recently, Zhang et al. [80] proposed a novel SAW device with an omnidirectional spiral IDT to induce planar recirculating acoustic streaming in a droplet, which creates a centrifuge within the droplet, and demonstrated the separation of platelets and RBCs in 1.5 μl sessile droplet. Unlike many SAW devices, which 127.86° Y-rotated LN, they used a 152° Y-rotated cut LN that is superior to supporting SAW in an arbitrary direction. They demonstrated that omnidirectional IDT designs induce higher rotational speeds compared to those of standard planar IDTs. They accomplished the separation of RBCs and platelets with a purity of 83% and 97%, respectively.

9.6.3.3 Separation of WBCs

WBCs composed of monoctytes, lymphocytes, and granulates play a critical role in the human immune system and are important in many diagnostic and therapeutic applications; therefore, purification of WBCs and even subpopulations is crucial in a variety of clinical and research applications. Laurell's group applied the two-inlet/two-outlet silicon BAW architecture to separate WBCs and deplete platelets from peripheral blood progenitor cells (PBPCs) products, which would be transfused to the PBPC donor without any tuning of the density of the medium [81]. WBCs were separated with a recovery rate of 98% and 89% removal rate of platelets at a sample rate of 20 μl/min. In a later study [82], mononuclear cells (MNCs, i.e. monocytes and lymphocytes) were separated from diluted whole blood in a BAW device, which had a similar architecture to previous studies [84, 85]. The performance of the acoustofluidic separation was also compared against that of two standard methods, namely Ficoll separation and RBC lysis. To improve the acoustic differentiation of acoustic contrast factor among WBCs and RBCs, blood samples are diluted with 80% stock isotonic percoll. Over 99.97% RBC depletion was observed for all three methods, and 68% recovery of MNCs was achieved with acoustofluidic, which was higher than that of Ficoll separation (51%), but lower than RBC lysis (79%). However, acoustofluidic technique yielded higher purity (53%) than that of RBC lysis (27%), with significant granulocyte depletion at a sample flow rate of 100 μl/min with 5% blood concentration (which is equivalent to 5 μl/min of whole blood). Lissandrello et al. [83] introduced a single-inlet, plastic BAW device for the purification of lymphocytes. Three different samples, whole blood, buffy coat, and leukapheresis products were tested at high purity and high recovery modes. The sample flow rate ranged between 30 and 100 μl/min, and lymphocyte purity values ranging between 59% and 97% were reported at a recovery rate ranging between 25% and 43% for different samples, leukapheresis product having the superior performance and whole blood having the worst performance. They also suggested two-stage separation procedure to enhance lymphocyte purity (88%) and RBC depletion (increased from 24% to 84%).

9.6.4 Applications Regarding Cancer Cells

Circulating tumor cells (CTCs), which are the tumor cells circulating in the peripheral blood either shed from a primary tumor or distant metastases, are essential ingredients for cancer biology. Isolating CTCs from peripheral blood is a critical step for further genetic and physiological profiling to understand the role of CTCs in cancer metastasis and prognosis. However, the main challenge in the isolation process is the scarcity of CTC in peripheral blood. Typically, 1–100 CTCs/ml are present in peripheral blood, surrounded by billions of RBCs and millions of WBCs. Immunolabeling is the common technique for enriching of CTCs from peripheral blood, which relies on the presence of specific surface markers. Label-free isolation has been an active research area for the last couple of decades. Acoustofluidics techniques have also been utilized to overcome the challenges of effective isolation of CTCs. Laurell's group utilized the two-inlet/two-outlet silicon BAW architecture by incorporating a pre-alignment unit (prior to the second inlet) followed by a separation unit to separate CTCs from WBCs [84–86]. In a pioneering study [84], the samples were prepared by spiking tumor cells from three prostate cancer cell-lines (DU145, PC3, LNCaP) in RBC-lysed blood for both fixed and non-fixed cells. They reported CTC recovery of 93.6–97.9% with purity of 97.4–98.4% for fixed cells and tumor cell recovery of 72.5–93.9% with purity of 76.9–99.7% for non-fixed cells at a sample flow rate of 70 μl/min. Cell viability and proliferation were also assessed following the acoustophoretic process for viable tumor cells. Ten times diluted blood spiked with about 10^5 tumor cells/ml (which is well above the actual concentration of CTCs in peripheral blood) was used in the experiments. Following the proof-of-concept experiments, Magnusson et al. [85] employed an automated version of the same device for clinical-scale samples. Fifty breast cancer (MCF7) cell lines were spiked into 10 ml (×2 dilution) RBC-lysed blood. In the high-recovery mode, tumor cells were isolated with 87% recovery and 99.4% WBC removal rate. On the other hand, tumor cells were isolated with 62% recovery and 99.9% WBC removal rate in the high-purity mode. Even though 99.9% WBC removal is achievable with acoustophoresis in CTC isolation, it still leads to about 1 CTC/1000 WBCs. Such high background may interfere with molecular analysis that would take place following the isolation. To further reduce the WBC background in CTC isolation process as a second step for their tumor cell separation devices [84, 85], the same group proposed immunolabeling of WBCs for negative selection [86]. Negative acoustic contrast elastomeric particles activated with CD45 antibodies specifically bind to WBCs to amplify the acoustic contrast difference between CTCs and WBCs. The efficacy of the technique was demonstrated through prostate cancer (DU145) and breast cancer (MCF7) cell lines in RBC-lysed blood. The samples were prepared as a 10^3 tumor cells/10^5 WBCs. DU145 cells were isolated with an 84% recovery and 85% WBC removal, which leads to a final ratio of 1 CTC/2 WBCs. Similarly, MCF7 cells were isolated with a 86% recovery, 89% WBC removal and a final ratio of 1 CTC : 1.3 WBCs. For both cell lines, the sample flow rate was 100 μl/min. A single-inlet version of the BAW structure used in the aforementioned studies was also implemented for the isolation of DU145 cells from WBCs [87]. This single-inlet version removes the necessity of matching the acoustic properties of the multiple buffers loaded into the chip, simplifies the fluidic components, and enhances the stability of flow conditions. At a sample rate of 100 μl/min, 95% tumor cell recovery and 98% WBC removal were reported. As an alternative to BAW devices, Huang's group [88–90] applied tilted SAW architecture for the isolation of CTCs in a series of papers. In an early study [88], they demonstrated the separation of MCF-7 cells from WBCs at a sample flow rate of 2 μl/min. They run their experiments with 1 ml RBC-lysed blood ($\sim 3 \times 10^6$ WBCs/ml) spiked with $\sim 3 \times 10^5$ MCF-7 cells, and 71% MCF-7 recovery with 84% purity was reported. By tuning the flow conditions, they also achieved 98% purity at a recovery rate of 20%. Following this study, through rigorous simulations and optimization, the SAW device was operated at a higher sample flow rate of 20 μl/min (which is relatively high for typical SAW devices and comparable with BAW devices). The performance of the device was assessed through the isolation of multiple cancer cell lines: breast cancer cells (MCF-7), cervical cancer cells (HeLa), and prostate cancer cells (UACC903M-GFP and LNCaP). To mimic the scarcity of CTCs 50–1000 cancer cells were spiked into 1 ml of WBC solution, which was obtained by directly lysing 1 ml of human whole blood ($3\text{–}6 \times 10^6$ WBCs/ml). Recovery rate of more than 83% was accomplished for all cell lines with about 90% WBC removal. In a later study [90], same group further improved the sample flow rate to 125 μl/min by introducing two design improvements: (i) placement of a glass layer at the top wall of the microchannel to amplify the acoustic energy density within the channel via reflected acoustic waves from the glass wall, and (ii) inclusion of divider wall to locally decrease the velocity of the cells within the channel. They tested the device for the isolation of prostate cancer lines (PC-3, LNCaP), HeLa and MCF-7 cells. An average recovery rate of above 86% was observed for all samples. Isolation of CTCs was also demonstrated for the clinical samples (whose cell counts ranged between 0.93 and 400 CTCs/ml) collected from patients with metastatic prostate cancer, followed by some detailed characterization of the phenotypic heterogeneity of CTCs.

9.6.5 Applications Regarding Miscellaneous Cells

In addition to manipulation of blood components, acoustofluidics has been utilized for the separation of different cell populations. Yang et al. [91] present a silicon

BAW device with two-inlet architecture for the sorting of viable and nonviable mammalian cells based on the size difference. MCF-7 breast tumor cells at a cell concentration of 10^6 cells/ml were used to assess the performance of the device. Sample flow rates ranging between 6.7 and 200 μl/min were tested. At a flow rate of 6.7 μl/min, 91% recovery of viable cells with a 60% purity and 78% removal of nonviable cells were reported. At the highest flow rate of 200 μl/min, purity of 80% was achieved with only 30% recovery and 97% removal of nonviable cells. The BAW device with pre-alignment unit previously developed by Laurell's group [84] was also employed for the separation of viable and nonviable MCF-7, mouse neuroblastoma (N2a) and human embryonic stem cells (hESCs) [92]. Although they reported lack of measurable size differences among viable and nonviable MCF-7 cells, as reported in Yang et al. [91], they were able to separate the cells. A successful separation of viable MCF-7 cells with 98% purity and 19% recovery at high purity mode and with 73% purity and 85% recovery at high recovery mode was demonstrated at a sample flow rate of 100 μl/min. For N2a cells, the effects of different death-inducing stimuli, namely induced osmotic shock, treatment with staurosporine (STS, a well-known apoptosis inducer isolated from *Streptomyces staurosporeus*) and treatment with dimethyl sulfoxide (DMSO, a chemical highly toxic to cells), and slow-progressing apoptotic cell death by starvation, were investigated. In all four treatments, similar size distributions among viable and nonviable cells were observed, yet successful separations were achieved due to the change in cell stiffness and compressibility, which eventually affects the acoustic contrast factor. At an optimum voltage, 75% and 56% recovery of viable cells were obtained with a purity of 95% and 90%, and 97% and 93% of removal of nonviable cells for osmotic shock induced stimuli and STS treatment, respectively. In the case of DMSO treatment, the separation efficiency obtained with STS-treated cells was not achieved, and highly variable results with low purity of viable cells were observed. A similar performance was reported for starving cells with osmotic shock and STS cases. In the case of hESCs, 42% recovery was stated with a 88% purity and 93% removal of nonviable cells at an optimum voltage. Thévoz et al. [93] introduced a two-inlet, silicon BAW device to purify cells in a particular phase from an asynchronous mixture based on size difference of cells at different phases. MDA-MB-231 human breast ductal carcinoma cell line was used, 84% synchronization of specific phase of interest (i.e. 84% purity), which was only 59% for the initial sample, was accomplished at a sample flow rate of 50 μl/min. Nam et al. [94] presented a SAW device to separate uniformly sized cell-encapsulating beads containing different quantities of cells according to the bead density (i.e. according to the number of encapsulated cells in a bead). Alginate beads with 150.7 μm average diameter encapsulating P19 mouse embryonic carcinoma stem cells were separated based on their cell content by controlling the lateral displacement of the beads. Large-cell-quantity alginate beads were successfully separated from small-cell-quantity alginate and empty beads with a recovery rate of 97% and a removal rate of 99% for the small-cell-quantity alginate and empty beads at a sample flow rate of 8 μl/min.

Sometimes controlled trapping or holding of cells of interest in a flow is desired for some set of chemical and/or microfluidic unit operations. In a series of papers,

Laurell's group introduced several acoustofluidic devices for cell trapping [95–97]. First, Evander et al. [95] presented a glass BAW device with miniature ultrasonic transducers to trap single cells as well as larger cell clusters against a flow of 1 μl/min. The microfluidic channels were fabricated by direct etching in a glass reflector cover; thus, polymer gaskets, which may require photolithography based fabrication and undergo swelling during the operations of acoustofluidic devices, were avoided. Stable operations over six hours were succeeded. Viability tests and temperature measurements were also performed to assess the possible adverse effects of acoustic trapping. Similar device architecture was also applied to trap sperm cells at a sample flow rate of 1 μl/min in the presence of epithelial cell lysate to direct male and female fractions to separate outlets for forensic DNA analysis of samples obtained from sexual assault evidence al. [96]. In a following study, Hammarstrom et al. [97] presented an acoustic trapping platform using commercially available borasilicate glass capillaries with a rectangular cross-section, which totally avoids fabrication techniques that rely on relatively high costs either for precision machining or microfabrication facilities. They demonstrated successful trapping of RBCs at a flow rate of 40 μl/min for ×10 diluted whole blood.

9.6.6 Application Regarding Bacteria

Pathogenic bacteria present in bloodstream are an important public health concern and may cause sepsis in some severe cases. Rapid and reliable identification of pathogens is an essential step for efficient treatment. The common practice for detecting bacteria in blood samples is blood culture, which is labor-intensive, time-consuming, and may have limited sensitivity in some instances. Although different culture-independent, relatively faster methods, such as direct polymerase chain reaction (PCR) detection, have been proposed, the presence of blood components may potentially increase the background noise and affect the sensitivity and specificity of the detection. In pathogen identification, the reliability of detection process can be significantly improved by an isolation/purification step for bacteria to remove the remaining blood components. Similar to the previous applications, acoustofluidic techniques also offer several advantages; therefore, several studies have focused on the acoustofluidic separation of bacteria from blood samples.

Since acoustic force is proportional to size, the major challenge for the manipulation of bacteria is their size, which is around a micron or sub-micron. In one of the pioneering studies, Ai et al. [98] introduced a two-inlet/two-outlet SAW device with two IDTs to form standing SAW whose pressure nodes were located near the two sidewalls of the microchannel. The sample is fed from a center inlet and acoustic force was utilized to migrate the larger PBMCs toward pressure nodes, smaller *Escherichia coli* (*E. coli*) bacteria remained undisturbed and flowed out with the center stream. At a sample flow rate of 0.5 μl/min, the separation of *E. coli* was accomplished with 96% purity. Huang's group utilized the previously developed tilted SAW architecture [73, 74, 88, 89] for the separation of *E. coli* from RBCs [99]. For a sample with *E. coli* and RBC concentration of 4.4×10^7 per ml and 1.32×10^8

per ml, *E. coli* was separated from the sample with a purity of 96%, and RBCs were collected with a purity of 96%. They also demonstrated the separation and detection of *E. coli* from whole blood, and concluded that downstream detection of bacteria significantly benefits from the separation unit prior to the detection. Silva et al. [36] proposed a single-inlet, plastic BAW device for the separation of bacteria from whole blood. Whole blood diluted in PBS to 5% hematocrit spiked with 10^5 *Pseudomonas aerouginosa* cells/ml were tested. A thorough optimization was performed for the geometrical dimensions (recall channel width is correlated with the frequency of the acoustic signal for BAW devices), flow rate and actuation voltage, and bacteria were separated with 34% recovery and 98% RBC removal rate at a sample flow rate of 110 μl/min. In a following study, Dow et al. [100] further revised the design of the plastic device with two-inlet configuration and assessed BAW device using three clinically relevant species of bacteria: *Pseudomonas aeruginosa*, *Staphylococcus aureus*, and *E. coli*. At a concentration level of 10^6 CFU/ml, all three species were reported to be separated at a recovery rate ranging between 45% and 60%, with more than 85% RBC removal at a sample flow rate of 50 μl/min. By changing the blood anticoagulant and operating the chip in a single-inlet mode (i.e. no sheath flow), they were also successful in isolating *P. aeruginosa* at clinically relevant concentration (120 CFU/ml) with 68% recovery and 92% RBC removal. They also showed that the limit of detection of the luminescent bacteriophage assay used for detection could be significantly improved by employing acoustic purification. Laurell's group applied the two-inlet/two-outlet silicon BAW architecture used in their previous studies [84–86, 92] for the separation of bacteria from whole blood, but this time the pre-alignment unit was utilized only hydrodynamically rather than acoustically [101]. They specifically addressed the fact that the presence of blood cells and plasma proteins in whole blood increases the acoustic impedance of the blood sample stream relative to a regular buffer that was used for hydrodynamic alignment. Therefore, the acoustic impedance of the buffer solution used for hydrodynamic alignment needed to be matched with that of whole blood for a successful separation. They optimally operated the device for bacteria recovery, blood cell removal, and throughput. They achieved 99.7% bacteria recovery with >99.9% blood cell removal at a sample flow rate of 400 μl/min (with ×100 dilution) for high bacterial recovery mode, 75% bacteria recovery with >99.99% blood cell removal at a sample flow rate of 100 μl/min (with ×5 dilution) for high blood removal mode, and 90% bacteria recovery with >99% blood cell removal at a sample flow rate of 400 μl/min (with ×5 dilution) for high throughput mode.

Acoustic trapping of bacteria was also studied in the literature. Actually, fluid drag both from acoustic streaming and channel flow needs to be dominated by the acoustic radiation forces, which scale with size, for a successful trapping. However, considering the size of bacteria (single micron or submicron), trapping of bacteria using acoustic forces is challenging if not impossible. Laurell's group implemented their previously proposed acoustic trapping platform [97] for the acoustic trapping of bacteria [102]. They showed that at certain concentration ($>5 \times 10^6$ bacteria/ml), trapping of *E. coli* is possible utilizing particle–particle acoustic interactions (i.e.

secondary acoustic forces) rather than utilizing primary acoustic forces. In addition, they also accomplished trapping of *E. coli* at lower concentrations by preloading larger polystyrene seed particles to enhance the secondary acoustic forces. At a flow rate of 10 μl/min and bacteria concentration of 1.1×10^5 bacteria/ml, trapping was reported with a capture efficiency of 95%. In a following study, Hammarstrom et al. [118] employed the seed particle-assisted acoustic trapping platform for the extraction of purified bacteria proteins obtained by bacteria lysis followed by a reverse-phase solid-phase extraction. The entire process was used for a sample preparation step for matrix assisted laser desorption/ionization time-of-flight mass spectrometry for the identification of *E. coli*. In this platform, trapping was demonstrated against a flow rate of 50 μl/min with silica microbeads rather than polymeric ones. Moreover, they implemented kerfed transducers for continuous optimization of the driving frequency in the acoustic trap, which also led to monitoring of the trapping as it takes place, thereby facilitating quality assurance. The entire sample handling was integrated into standard laboratory robotics. The platform provided accurate identification in 12 out of 12 cases of *E. coli* positive blood cultures with a score better than a standard method. However, the sensitivity of matrix-assisted laser desorption ionization time-of-flight mass spectrometry is affected by the sensitivity of blood culture. Laurell's group integrated their BAW architecture with the trapping platform to separate and enrich bacteria to be detected in a PCR chip [103]. First, the separation unit was tested with a sample of washed blood cells and spiked *E. coli* at a final concentration of 3×10^8 bacteria/ml, and recovery rate of 91% with a blood cell removal of 99.1% was reported at a sample flow rate of 80 μl/min. Then, bacteria enrichment was tested for *E. coli* suspension of 22.2×10^6 bacteria/ml at a sample flow rate of 10 μl/min using 12 μm polystyrene seed particles. At a latter step, the platform was also tested for citrated blood from healthy volunteers spiked with known amounts of *P. putida* bacteria and 57 clinical samples for the detection of *E. coli*, *S. aureus*, and *St. pneumoniae* results of which were compared against standard blood culture. For the citrated blood samples, the detection was successful for concentrations as low as (i.e. limit of detection) 1000 *P. putida*/ml citrated blood. In the case of clinical samples, 7 out of 57 samples were positive, and only two of the samples were detected by the proposed system.

Pathogenic bacteria present in food are as important as the ones in the blood stream for public health and safety. Ngamsom et al. [104] introduced a glass BAW device as a sample preparation step to enhance the performance of subsequent detection systems such as flow cytometry and q-PCR. The undesired large particles were removed from the intrinsic microflora of food and blood samples. The system was successfully tested with samples of chicken and minced beef spiked with *Salmonella typhimurium* at a concentration of 10^3 CFU/ml and a sample flow rate of 10 μl/min, and pathogen recovery of 60–90% was obtained. *S. typhimurium*-contaminated blood samples were also tested at a concentration of 10^7 CFU/ml, 99.8% removal rate of RBCs, but a recovery rate of 10% (which was attributed to the adsorption/specific binding of pathogen to the RBCs) was reported. Despite low recovery rate of the pathogen, the number of pathogens was found to be sufficient for further downstream analyses.

9.6.7 Applications Regarding Nanoscale (Bio)Particles

Manipulation of nanoscale bioparticles whose diameters are typically less than 1 μm has gained significant attention from researchers. Some of the target bioparticles are viruses and extracellular vesicles (EVs) such as exosomes and microvesicles. Since EVs potentially carry variety of molecular markers, which belong to their originating cells, such as DNA, RNA, and proteins, they can serve as target biomarkers in the diagnosis and prognosis of various diseases. Although vast amounts of EVs are present in circulation (>10^{12} vesicles/ml), isolation of EVs is a challenging process due to their nanoscale size and presence in complex media. Acoustofluidics can also be utilized for the manipulation of nanoscale particles; however, the major challenge associated with the acoustic force is the scaling of the primary acoustic particle forces and the hydrodynamic drag force resulting from the induced acoustic streaming with size. The typical frequencies used for BAW devices limit their application to ~10 μm order particles. Alternatively, SAW devices, which typically operate 10–1000 MHz, may be a viable option for the manipulation of nanoscale particles. However, regardless of the technique, the manipulation of nanoscale particles may become even more challenging as the particle size reduces.

As successful trapping of bacteria was demonstrated in a BAW device through secondary acoustic forces (i.e. particle–particle acoustic interactions), Laurell's group implemented the same platform for the capture of 110 nm polystyrene particles at a concentration of 0.1%wt using 10 μm seed particles at a sample flow rate of 10 μl/min [102]. In the following study, Evander et al. [105] demonstrated successful capture of platelet-derived microparticles in minute sample volumes using 12 μm seed particles at a flow rate of 10 μl/min. The effect of key parameters (sample volume and flow rate, and dilution of plasma, which affect the viscosity and density of the sample) was investigated to characterize and optimize trapping performance. The performance of the platform was also compared to a more standardized differential centrifugation protocol to demonstrate the clinical relevance of the technique. At a sample volume of 10 μl, microparticle recovery as high as 60% was accomplished for the samples with ×5 dilution. Moreover, acoustic trapping exhibited a better recovery compared to differential centrifugal protocols in the case of clinical samples. The overall recovery rate was lower for the clinical samples than that of the cell-free plasma samples, which was attributed to the long storage time of the clinical samples. Later, the acoustic trapping principle was employed for the enrichment of EVs on an automated trapping platform again using 12 μm seed particles for sensitive downstream assays such as qRT-PCR or next-generation sequencing [106]. The device's performance was characterized by processing beads whose sizes ranged from 0.1 to 1 μm at biologically relevant concentrations of 2×10^6 to 4×10^9 particles/ml. Trapping efficiencies as high as 82% and 44% were reported for 1 and 0.3 μm particles, whereas 5% and 1% were reported for 0.2 and 0.1 μm particles, respectively. Considering the abundant presence of EVs in biological fluids (at least 10^9 particles/ml), it was concluded that even trapping efficiencies as low as 1% were sufficient for most downstream analyses. Following the nanoscale particle tests, EVs were successfully enriched from cell culture-conditioned media as well as urine and blood plasma from healthy volunteers. Collins et al. [107] introduced a virtual deterministic lateral displacement (vDLD) device, which uses high-frequency traveling SAW for the sorting of microscale as well as nanoscale particles. They placed IDTs directly at the bottom of the microchannel to avoid any attenuation losses. The IDTs replace the DLD pillars used in hydrodynamic manipulation of particles [8] and generate virtual obstacles, which apply tunable acoustic and DEP forces on the particles. They demonstrated the separation of 300 and 500 nm particles at a flow rate of 0.45 and 1.8 μl/min, and reported separation efficiency of 87% and 79% for 500 nm particles, respectively. Seghal and Kirby [108] proposed a SAW device, which implements Fabry–Perot acoustic resonators in a microfluidic geometry to separate nanoscale particles (more specifically, particles with size less than 0.5 μm). The overall ARF on the particles was enhanced by the use of Fabry–Perot-type acoustic resonators, in which traveling SAWs emanating from a single IDT were reflected from Bragg reflectors, leading to an SSAW field. Similar to the device proposed by Collins et al. [107], IDTs are also placed at the bottom surface of the microchannel. At a sample flow rate of 0.2 μl/min, 300 nm particles were separated from 100 nm particles with a recovery of 86% and 98% of removal rate of 100 nm particles, respectively. Collins et al. [28] utilized the focused traveling SAW approach to robustly manipulate nanoscale particles with strong local acoustic streaming in the vicinity of the focused beam, which led to amplified ARF by recirculating particles through this region. They demonstrated successful manipulation of 0.3 and 0.5 μm particles with this approach at a flow rate of 0.45 μl/min. However, the deterioration of the manipulation performance for 0.5 μm particles due to the trapping in the acoustic streaming vortices was also observed. Lee et al. [109] presented a SAW device using standing waves to separate exosomes and microvesicles in a cell culture medium and erythrocyte-derived vesicles from stored blood units. The performance of the device was first validated using polystyrene beads, and the separation of particles with diameters of 0.19 and 1.0 μm was demonstrated. Smaller particles were isolated with a recovery rate of above 90% and with >90% removal rate of larger particles at a total flow rate of 0.8 μl/min (including sheath flows). Following the particle tests, the separation of exomes and microvesicles was reported with a >80% recovery rate for exomes and >90% recovery rate for microvesicles at a total flow rate of 0.43 μl/min. The acoustic enrichment of microvesicles in packed red blood cell units was also illustrated with similar performance to that of a standard method.

The tilted SAW architecture developed by Huang's group was also utilized for the manipulation of nanoscale particles [110]. Separation of mixtures of 110 and 900 nm particles, 240 and 900 nm, 240 and 700 nm, and 240 and 500 nm was successfully demonstrated with about 90% recovery rate at a sample flow rate of 5 μl/min. In a follow-up study [111], two sequential tilted SAW modules were integrated to isolate exosomes from undilute blood. The first module removed the microscale blood components to enrich EVs, and the second module isolated exomes by removing other EV subgroups. They achieved the isolation of exosomes with a recovery rate of 86% and a purity of 98.4% in the isolation unit at a sample flow rate of 4 μl/min. A blood cell removal rate of over 99.999% was reported for the integrated device with a 4 μl/min whole blood flow rate.

9.6.8 Miscellaneous Applications

The manipulation of large amounts of microscale bioparticles in a continuous flow is a vital task for many biomedical applications, and acoustofluidic manipulation has been implemented successfully for many of these applications, as discussed earlier. Apart from these applications, the manipulation of cells into desired locations to form specific patterns is also crucial in tissue engineering and regenerative medicine studies. SAW-based acoustofluidic techniques, so-called "acoustic tweezers," were developed for bioparticle patterning. Shi et al. [112] presented a SAW device with a pair of parallel and orthogonal IDTs to generate SSAWs in the fluidic chamber. In the parallel configuration, 1D pattern of particles along multiple lines where the pressure nodes (or antinodes) were aligned was demonstrated. In the orthogonal configuration, on the other hand, 2D-patterned particle clusters were formed, and 2D-patterning of bovine RBCs and *E. coli* was successfully performed.

Precise manipulation of individual bioparticles on the cellular scale is also essential for many fundamental biomedical studies. Capability of trapping and patterning individual cells at some desired locations may facilitate studies regarding cell–cell and cell–environment interactions as well as single-cell analysis. Through these studies, the mechanism and process of cell–cell interactions and heterogeneous cellular parameters can be investigated on the basis of individual cells, and the response of the cell(s) to their local environment as well as to any external stimuli can be fully explored. Huang's group [113] introduced a SAW device to demonstrate single cell manipulation in 2D by employing two pairs of orthogonally arranged chirped IDTs. By tuning the frequency of each pair of IDTs, the motion of the cells was controlled in two orthogonal directions independently. They successfully manipulated bovine red blood cells and *Caenorhabditis elegans (C. elegans)* a multicellular organism. *C. elegans* was immobilized along its entire length and even stretched in a SSAW field. Collins et al. [31] introduced a SAW device to create 2D acoustic force field with an inter-nodal spacing, which has the same order of magnitude as the size of cell of interest using two pairs of orthogonal IDTs with high frequency (>100 MHz). They demonstrated active patterning of spatially isolated RBCs and lymphocytes. Individual RBCs were successfully positioned at positions 7.5 and 12.5 μm far away from each other. Around same time, Huang's group [114] developed a similar SAW architecture, which can form thousands of cell assemblies with tunable geometric configurations and control intercellular distance with a resolution of micrometer scale to assess cell–cell interactions. Gap junctional intercellular communication among HEK 293T cells was explored. Later, 2D patterning capability over a planar surface was extended to 3D by utilizing acoustic streaming to control the motion in the height direction by Huang's group [32]. With the help of this 3D manipulation capability, they demonstrated acoustic cell printing of living 3T3 mouse fibroblasts and HeLa cells as a potential application.

Precise rotational manipulation of bioparticles is also desired in variety of applications, such as 3D optical imaging. Ahmed et al. [115] presented a BAW-based acoustofluidic device for the rotational manipulation of cells utilizing microstreaming vortices created by the oscillatory motion (induced by low-cost piezo transducer) of air microbubbles trapped in a microfluidic channel. As a model organism, *C. elegans* was rotated to analyze the morphological features of its nervous and reproductive systems. Although the proposed manipulation with the oscillatory motion of air microbubbles was successful, they observed some variations in acoustic parameters since the microbubble size and geometry might change over a period of operation. Instead, they utilized bare microfluidic channel walls and sharp edge structures for more robust and precise in-plane and out-of-plane rotational manipulation of *C. elegans and* HeLa cells [116]. Rotational speeds as high as 14 000 rpm was achieved (which was about 3000 rpm in their previous study [115]), and they successfully simultaneously image left/right pairs of neurons in the *C. elegans* head, which was quite challenging by standard microscopy techniques. As an alternative to BAW devices, which enable rotation at some specific locations in a static flow and/in one direction, Zhang et al. [117] recently proposed a SAW device that has rotational capability in both directions in a static as well as continuous flow. The device consisted of a pair of IDTs, which are capable of focusing the microorganism upon the application of SSAW or rotating either in clockwise or counter-clockwise direction upon the application of tSAW emitting from either of the IDT. They also demonstrated the rotational manipulation of *C. elegans* in a continuous flow of 5 μl/min, and simultaneous manipulation of multiple was also illustrated for possible higher-throughput 3D imaging applications.

9.7 Commercial and Regulatory Considerations for Acoustofluidic Devices

So far we have discussed the attempts, we have discussed the attempts to understand the physics behind acoustofluidic devices and investigated the proposed methods to improve their performances in bioparticle manipulation. These intended improvements have the general purpose of leading to a commercial medical device. Even though "Performance" is an important part of the commercial success of a medical device, it is not the only ingredient. In order to grant regulatory permissions, for instance, Food and Drug Administration (FDA) of the U.S. requires demonstration of "safe and effectiveness" in a medical device. Even though effectiveness is the main focus of academic studies in medical device research, FDA lists "safety" as the first criterion in the regulatory process. The features that the medical device should satisfy for a safe and effective device may change depending on the intended use of the medical device. Figure 9.8 gives a general list of criteria that can affect a medical device's commercial success and regulatory admission process.

It can be assumed that a commercialized acoustofluidic device will have the "intended use" of the manipulation , separation, and/or enumeration of certain bioparticles. For commercialization in the USA, according to FDA regulations, such a device is most likely to be treated as a "Class II medical device" which is subjected to a "510(k) Premarket Notification" process. This process may require a significant amount of documentation to demonstrate regulatory compliance with some of the criteria listed in Figure 9.8. The criteria listed in Figure 9.8, which are critical

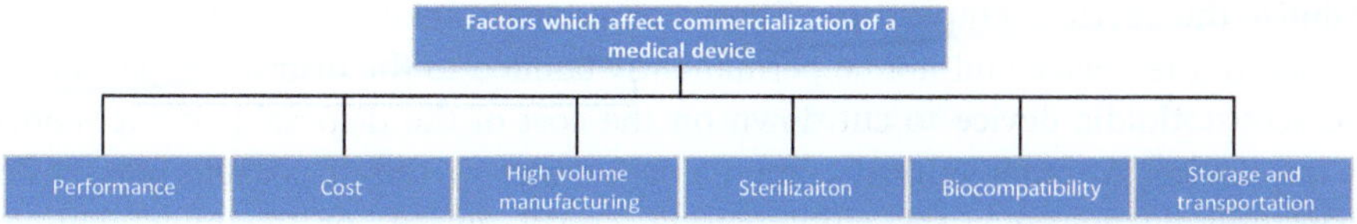

Figure 9.8 The list of factors that may affect the commercialization of an acoustofluidic device.

during the commercialization process, will be discussed in the scope of BAW and SAW devices in this section. The first component, "Performance," is not included since several academic studies focusing on improving the performance of such acoustofluidic devices for many different applications have already been discussed so far.

9.7.1 Cost

It is beneficial and essential to keep in mind that an electro-mechanical biomedical device generally comprises three main components; hardware, software, and disposables and reagents. The most important ingredient in the determination of cost is the cost of disposable sets and reagents. The reasoning behind this is that the hardware is marketed once, but disposable components and reagents need to be manufactured in high volumes since a new set is required for each use. Hence, disposables and reagents are the primary cost and profit sources for medical device manufacturers. Therefore for BAW and SAW devices, the implementation of high-cost components of device architecture in hardware and low-cost components in disposable parts is an essential step.

Keeping in mind the "label-free" nature of acoustofluidics, there will be no reagent component in most of the acoustofluidic applications. It is clear that among the architectural components of BAW devices (shown in Figure 9.2), the most critical component in terms of cost is PZT (unless an expensive chip material is considered). For SAW devices, the most critical components in terms of cost are the single-crystal piezoelectric actuator and the IDT. Therefore, a piezoelectric actuator and the required electrode structure permanently bonded to the disposable part of the device may not be preferable considering the commercialization step of the technology. Unlike most of the configurations in the literature for SAW devices, a beneficial configuration in terms of cost is the engagement of the hardware (including the piezoelectric material and the IDT) on the chip during the process and disengagement once the process is completed to be used with another set of disposables in the next cycle. Such a configuration would prevent the actuator and IDT from being thrown away with the disposable components. Of course, realizing such a configuration requires researchers to focus on the transmission of longitudinal and transverse waves generated in the actuator to the disposable parts of the device (chip material or the microchannel structure) without using bonding agents but, possibly, with the use of possible acoustic coupling gels and preload. A similar argument is also valid for any thermoelectric cooling device to stabilize the device's temperature. These components need to be on the hardware portion of the device and not be permanently bonded to the disposable portion of the acoustofluidic device to cut down on the cost of the disposable components of the device. Another critical component for cost in the disposable part of an acoustofluidic device is the chip material (BAW device) or the microchannel structure (SAW device). A clear beneficial choice in terms of cost would be thermoplastic materials, which unfortunately have generally poor acoustic performance. Silicon, a widely employed material for the chip component of BAW devices, has relatively high material and manufacturing process costs with superior acoustic performance.

9.7.2 High Volume Manufacturing

High-volume manufacturing is crucial and even necessary for manufacturing disposable components. High-volume manufacturing may decrease the disposable cost of medical devices as well as their ability to meet high demand. The most compatible manufacturing methodology for high-volume manufacturing is mold-based manufacturing. Since thermoplastic parts can be manufactured using injection molding methods, low manufacturing costs and high manufacturing volumes are achievable for disposable components made of thermoplastics. An opposite argument can be stated for disposable component materials that require clean-room facilities for production. Reaching high-volume manufacturing for disposable parts that require clean-room facilities would be rather challenging and relatively costly. Therefore, it is highly likely that disposable chips fabricated from silicon materials for BAW devices may have issues once high-volume manufacturing is needed.

9.7.3 Sterilization

Sterilization may be an important requirement for medical devices, depending on their intended use. If no bodily fluids/tissues are contacted or there is no infusion of substance into the body in the intended use of the device, then sterilization may not be a requirement. In addition, if samples of bodily fluids/tissues are to be tested for diagnosis or some infusion to the body is performed with the instrument, such as separated blood components, then sterilization would be an important safety as well as a regulatory issue. A commercial acoustofluidic device would most likely require sterilization, considering the application area of these devices. If a component of a medical device needs to be sterilized, this generally means that after its use (contact with fluids/tissue), it needs to be disposed of. Therefore, it is rather important that the piezoelectric actuators and IDTs are not placed in the sub-components that are sterilized (disposable parts) to decrease the cost of the process. There are several methods for sterilizing of biomedical devices. These methods include high-temperature dry or moist heating, the use of chemicals or gases to kill pathogens, such as ethylene oxide (EtO) gas and vaporized hydrogen peroxide, which are the most widely used ones, and the use of radiation to kill microorganisms, such as gamma and X-rays as well as E-beam methods. Each of these methodologies has its advantages and drawbacks, which are discussed widely in the literature [119–121].

The type of material selected for the disposable set of acoustofluidic devices may have an effect on the preferred sterilization methods. For instance, thermoplastic materials can limit the use of high-temperature sterilization methods since most of these materials cannot handle high temperatures during the sterilization process. On the other hand, high-temperature sterilization methods, which are cost-effective, can be implemented when metals, silicon, or glass are used as the disposable components of the acoustofluidic devices. Moreover, if bonding agents are implemented, it should be ensured that the bonding quality is not affected by high temperatures and moisture during sterilization.

The sterilization methods of EtO and other gases are effective and can be applied to a wide range of materials, but they require stringent controls to ensure that the harmful gases do not affect the environment or people. The radiation methods may be effective but require high capital investment and correct dose determination, which may be quite challenging. More than necessary doses may result in change in color of the thermoplastic materials [120] which may be even more important if an optical detection is intended on-chip. Design solutions for medical devices may also have to consider the requirements of sterilization and staying sterile during the operation of the devices. For instance, the requirements for sterilization of disposable components may affect the way the pumping of the fluid occurs in an acoustofluidic device. If the pump is part of the hardware, then the pump should not be in contact with the fluids in the acoustofluidic device. Therefore, if syringe pumps are employed, any syringes used during pumping need to be part of the disposable set and be sterile. Furthermore, any connections and the filling of the syringes need to be performed using the aseptic technique, which would significantly increase the number of steps taken and the disposable and hygienic components used in order to start the acoustofluidic manipulation/separation process. To circumvent these issues, a disposable sample loading interface [122, 123], which can be integrated to a syringe pump may be a viable option. As an alternative to syringe pumps, peristaltic pumps ensure contact-free pumping on the hardware side since the tubing used in the disposable set is squeezed to pump the fluid. The use of peristaltic pumps results in a convenient way of pumping since the interface between the hardware and the disposable is simple. As a disadvantage, pumping accuracy and consistency are worse than syringe pumps. In the case of peristaltic pumps, it needs to be ensured that acoustofluidic manipulation/separation process is not negatively affected by the pulsatile flow of the peristaltic pump.

Moreover, air management is an important issue, especially if on-line bioparticle separation is employed. On-line bioparticle separation means that bioparticles such as cells are collected from one end of the device while the acoustofluidic device manipulates/separates certain cells and infuses the remaining cells back into the patient from the output port of the device in real-time. Infusion of air into the patient can be lethal; hence, air detectors are needed to prevent the infusion of air. It should be noted that air in microfluidic devices affects the performance of the device severely in a negative way. Also, all the bonded components should be checked for integrity with a pressure test so that air is not pulled into the acoustofluidic device from a region with failed bonds. A bonding failure will also result in a sterilization breach. In that sense, using chemical bonding and welding methods (such as ultrasonic welding) for permanently binding the two components has advantages over a bonding procedure with the use of glues. Like many medical devices, priming an acoustofluidic device is critical since air must be driven entirely out before its use. On the other hand, if there is a concern that air is being pulled in from other ports during priming, then sterilizing filters can be placed in these ports to prevent potential pathogens from entering the device.

9.7.4 Biocompatibility

Biocompatibility requires the materials in the disposable set not to cause any adverse effects on the bioparticles processed in the device. For instance, there can be a biocompatibility issue if "lead" material in PZT comes into contact with the bioparticles. To that effect, the contact between lithium niobate and the bioparticles in SAW devices may need to be investigated, and documentation stating that lithium niobate is not causing any harm to bioparticles may be needed. There are already some studies on the biocompatibility of lithium niobate [124]. A biocompatible material can be applied between the lithium niobate and the fluid in SAW devices, which may also act as an impedance matching layer. The choice of thermoplastics as chip materials in BAW devices and microfluidic channel enclosures in SAW devices can be advantageous since many predicate devices prove the biocompatibility of several polymer materials.

9.7.5 Storage and Transportation Requirements

A commercial acoustofluidic device not only has to function after its fabrication, but it should also endure transportation and storage conditions over its lifetime and be able to function afterwards. Since shelf-life periods are generally long, accelerated life tests are applied to simulate sterilization over shelf-life [ASTM F1980, ISO 11607-1] and transportation conditions over short verification/validation periods [ASTM D47289, ISO 13355:2016]. In order to perform these tests in short amount of time, elevated temperatures can be employed. Special attention should be given to thermoplastic materials and bonding agents since it is known that their performance is affected by temperature during these tests. As connections, luer connections can be preferred to endure high- amplitude vibration transportation-related tests. In addition, a widely used material, PDMS, especially in SAW devices, should not be in a commercial device due to concerns of decay over the shelf-life duration.

9.8 Summary and Outlook

In this chapter of the book, we provide a general overview of the ultrasonic bioparticle manipulation process. A brief description of methods for microfluidic bioparticle manipulation/separation is presented, along with similarities among the forces generated by dielectric, magnetic, and acoustic field forces on bioparticles. We also provide a comparison of different microfluidic bioparticle separation approaches and state the advantages and disadvantages of acoustic methods in comparison to other commonly employed methods. In addition, the functional components of BAW and SAW devices for ultrasonic bioparticle manipulation and the theoretical background required in the analysis of such systems, as well as the numerical simulation approaches utilized in the literature, are discussed in detail. The applications of ultrasonic manipulation methods in micro and nanoscale bioparticles (including different cell types, bacteria, lipids and exosomes) available in the literature are critically discussed. Moreover, the evaluation of the commercialization prospects of ultrasonic bioparticle manipulation from different perspectives such as cost, high-volume manufacturability, biocompatibility, and sterilization is also presented.

The developments in acoustophoresis field in the last decade have led to many successful applications regarding the manipulation of microparticles and cells. We believe the field has matured enough to open a pathway to the commercialization of acoustophoretic devices. Following the discussions we presented in this book chapter, we would like to list some recommendations to improve the possible successful commercialization of acoustofluidic devices as follows:

- Since acoustofluidic technologies are mainly utilized in biomedical and biological fields, it would be required to have disposable sets for medical devices. Considering low manufacturing costs with high manufacturing volume, the use of thermoplastics [26, 36, 125], which enable injection molding, should be implemented as much as possible for both BAW and SAW devices.
- The actuators are mostly not cost-efficient; therefore, it may be important not to include them in the disposable set but rather to have them in the hardware, which will enable their reuse. In this regard, for BAW applications, the actuator may not be bonded to the chip material, but coupling and wave transmission may be ensured with preload on the PZT and the chip material, along with ultrasonic coupling gel layer in between [126]. For SAW applications, a permanent bond between the piezoelectric actuator and the microfluidic chamber needs to be avoided. The bottom of the microfluidic chamber must be covered with a polymeric material such that fluid in the channel does not contact the SAW piezoelectric actuator. The transmission of surface waves may be achieved with the applied preload on the piezoelectric actuator and polymer layer, along with coupling gel in between. Even though wave transmission is possible hinder, this kind of architecture can allow the piezoelectric actuator and the IDT to be on the hardware portion of the medical device rather than the disposable portion of the medical device [30, 127].
- In the case where thermoplastic materials are used, the temperature control of the acoustofluidic platforms may be quite essential. For BAW applications, temperature control may be maintained through contact with PZT [128] such that the temperature control unit stays on the hardware side.

- For BAW devices, the use of bonding agents may be minimized by the implementation of thermofusion [26] or ultrasonic welding in the case where the lid and chip materials are manufactured out of thermoplastic materials.
- Standardization of the fluidic components may be quite important. The implementation of standard luer connections and the use of peristaltic pumping or a disposable sample loading interface [122, 123] may be beneficial for a user-friendly fluidic interface.

References

1 Ding, X., Li, P., Lin, S.-C.S. et al. (2013). Surface acoustic wave microfluidics. *Lab Chip* 13: 3626–3649.

2 Ozcelik, A., Rufo, J., Guo, F. et al. (2018). Acoustic tweezers for the life sciences. *Nature Methods* 15 (12): 1021–1028.

3 Ozcelik, A., Rich, J., and Huang, T.J. (2022). Fundamentals and applications of acoustics in microfluidics. In XiuJun (James) Li. In: *Multidisciplinary Microfluidic and Nanofluidic Lab-on-a-Chip* (ed. P.C.H. Li and C. Yang), 297–321. Elsevier.

4 Wu, M., Ozcelik, A., Rufo, J. et al. (2019). Acoustofluidic separation of cells and particles. *Microsystems and Nanoengineering* 5: 32.

5 Yeo, L.Y. and Friend, J.R. (2014). Surface acoustic wave microfluidics. *Annual Review of Fluid Mechanics* 46 (1): 379–406.

6 Zhang, P., Bachman, H., Ozcelik, A., and Huang, T.J. (2020). Acoustic microfluidics. *Annual Review of Analytical Chemistry (Palo Alto, Calif.)* 13 (1): 17.

7 Huang, Y., Das, P.K., and Bhethanabotla, V.R. (2021). Surface acoustic waves in biosensing applications. *Sensors and Actuators Reports* 3: 100041.

8 Cetin, B., Özer, M.B., and Solmaz, M.E. (2014). Microfluidic bio-particle manipulation for biotechnology. *Biochemical Engineering Journal* 92: 63–82.

9 Warkiani et al. (2014). Slanted spiral microfluidics for the ultra-fast, label-free isolation of circulating tumor cells'. *Lab on a Chip* 14 (1): 128.

10 Miltenyi, S., Müller, W., Weichel, W., and Radbruch, A. (1990). High gradient magnetic cell separation with MACS. *Cytometry: The Journal of the International Society for Analytical Cytology* 11 (2): 231–238.

11 Çetin, B. and Li, D. (2011). Dielectrophoresis in microfluidics technology. *Electrophoresis* 32 (18): 2410–2427.

12 Cetin, B. and Li, D. (2008). Effect of Joule heating on electrokinetic transport. *Electrophoresis* 29 (5): 994–1005.

13 Koklu, A., Sabuncu, A.C., and Beskok, A. (2016). Rough gold electrodes for decreasing impedance at the electrolyte/electrode interface. *Electrochimica Acta* 205: 215–225.

14 Hardwick, R.A., Kulcinski, D., Mansour, V. et al. (1992). Design of large-scale separation systems for positive and negative immunomagnetic selection of

cells using superparamagnetic microspheres. *Journal of Hematotherapy* 1 (4): 379–386.

15 Hornig-Do, H.T., Günther, G., Bust, M. et al. (2009). Isolation of functional pure mitochondria by superparamagnetic microbeads. *Analytical Biochemistry* 389 (1): 1–5.

16 Munaz, A., Shiddiky, M.J., and Nguyen, N.T. (2018). Magnetophoretic separation of diamagnetic particles through parallel ferrofluid streams. *Sensors and Actuators B: Chemical* 275: 459–469.

17 Zhao, W., Zhu, T., Cheng, R. et al. (2016). Label-free and continuous-flow ferrohydrodynamic separation of HeLa cells and blood cells in biocompatible ferrofluids. *Advanced Functional Materials* 26 (22): 3990–3998.

18 Zhao, W., Cheng, R., Lim, S.H. et al. (2017). Biocompatible and label-free separation of cancer cells from cell culture lines from white blood cells in ferrofluids. *Lab on a Chip* 17 (13): 2243–2255.

19 Petersson, F., Åberg, L., Swärd-Nilsson, A.M., and Laurell, T. (2007). Free flow acoustophoresis: microfluidic-based mode of particle and cell separation. *Analytical Chemistry* 79 (14): 5117–5123.

20 Lin, S.C.S., Mao, X., and Huang, T.J. (2012). Surface acoustic wave (SAW) acoustophoresis: now and beyond. *Lab on a Chip* 12 (16): 2766–2770.

21 Açıkgöz, H.N., Karaman, A., Şahin, M.A. et al. (2023). Assessment of silicon, glass, FR4, PDMS and PMMA as a chip material for acoustic particle and cell manipulation in microfluidics. *Ultrasonics* 129: 106911.

22 Martin, S.P., Townsend, R.J., Kuznetsova, L.A. et al. (2005). Spore and micro-particle capture on an immunosensor surface in an ultrasound standing wave system. *Biosensors and Bioelectronics* 21 (5): 758–767.

23 Hawkes, J.J. and Coakley, W.T. (2001). Force field particle filter, combining ultrasound standing waves and laminar flow. *Sensors and Actuators B: Chemical* 75 (3): 213–222.

24 Özer, M.B. and Çetin, B. (2021). An extended view for acoustofluidic particle manipulation: scenarios for actuation modes and device resonance phenomenon for bulk-acoustic-wave devices. *The Journal of the Acoustical Society of America* 149 (4): 2802–2812.

25 Binkley, M.M., Cui, M., Li, W. et al. (2019). Design, modeling, and experimental validation of an acoustofluidic platform for nanoscale molecular synthesis and detection. *Physics of Fluids* 31 (8): 082007.

26 Mueller, A., Lever, A., Nguyen, T.V. et al. (2013). Continuous acoustic separation in a thermoplastic microchannel. *Journal of Micromechanics and Microengineering* 23 (12): 125006.

27 Lenshof, A., Evander, M., Laurell, T., and Nilsson, J. (2012). Acoustofluidics 5: building microfluidic acoustic resonators. *Lab on a Chip* 12 (4): 684–695.

28 Collins, D.J., Ma, Z., Han, J., and Ai, Y. (2017). Continuous micro-vortex-based nanoparticle manipulation via focused surface acoustic waves. *Lab Chip* 17 (1): 91–103.

29 Sinha, S. and Kumar, A. (2013). Slanted finger inter-digital transducers for the design of improved performance small shape factor mid-bandwidth SAW filters. In: *IEEE MTT-S International Microwave and RF Conference*, 1–4. IEEE.

30 Qian, J., Ren, J., Liu, Y. et al. (2020). A two-chip acoustofluidic particle manipulation platform with a detachable and reusable surface acoustic wave device. *Analyst* 145 (23): 7752–7758.

31 Collins, D.J., Morahan, B., Garcia-Bustos, J. et al. (2015). Two-dimensional single-cell patterning with one cell per well driven by surface acoustic waves. *Nature Communications* 6 (1): 1–11.

32 Guo, F., Mao, Z., Chen, Y. et al. (2016). Three-dimensional manipulation of single cells using surface acoustic waves. *Proceedings of the National Academy of Sciences of the United States of America* 113 (6): 1522–1527.

33 Langelier, S.M., Yeo, L.Y., and Friend, J. (2012). UV epoxy bonding for enhanced SAW transmission and microscale acoustofluidic integration. *Lab on a Chip* 12 (16): 2970–2976.

34 Skov, N.R. and Bruus, H. (2016). Modeling of microdevices for SAW-based acoustophoresis—a study of boundary conditions. *Micromachines* 7 (10): 182.

35 Bruus, H. (2012). Acoustofluidics 7: the acoustic radiation force on small particles. *Lab on a Chip* 12 (6): 1014–1021.

36 Silva, R., Dow, P., Dubay, R. et al. (2017). Rapid prototyping and parametric optimization of plastic acoustofluidic devices for blood–bacteria separation. *Biomedical Microdevices* 19: 70.

37 Hsu, J.C. and Chao, C.L. (2020). Full-wave modeling of micro-acoustofluidic devices driven by standing surface acoustic waves for microparticle acoustophoresis. *Journal of Applied Physics* 128 (12): 124502.

38 Şahin, M.A., Çetin, B., and Özer, M.B. (2020). Investigation of effect of design and operating parameters on acoustophoretic particle separation via 3D device-level simulations. *Microfluidics and Nanofluidics* 24 (1): 1–18.

39 Skov, N.R., Bach, J.S., Winckelmann, B.G., and Bruus, H. (2019). 3D modeling of acoustofluidics in a liquid-filled cavity including streaming, viscous boundary layers, surrounding solids, and a piezoelectric transducer. *AIMS Mathematics* 4 (1): 99–111.

40 Skov, N.R., Sehgal, P., Kirby, B.J., and Bruus, H. (2019). Three-dimensional numerical modeling of surface-acoustic-wave devices: acoustophoresis of micro-and nanoparticles including streaming. *Physical Review Applied* 12 (4): 044028.

41 Büyükkoçak, S., Özer, M.B., and Cetin, B. (2014). Numerical modeling of ultrasonic particle manipulation for microfluidic applications. *Microfluidics and Nanofluidics* 17 (6): 1025–1037.

42 Nama, N., Barnkob, R., Mao, Z. et al. (2015). Numerical study of acoustophoretic motion of particles in a PDMS microchannel driven by surface acoustic waves. *Lab on a Chip* 15 (12): 2700–2709.

43 Devendran, C., Albrecht, T., Brenker, J. et al. (2016). The importance of travelling wave components in standing surface acoustic wave (SSAW) systems. *Lab on a Chip* 16 (19): 3756–3766.

44 Lickert, F., Ohlin, M., Bruus, H., and Ohlsson, P. (2021). Acoustophoresis in polymer-based microfluidic devices: modeling and experimental validation. *The Journal of the Acoustical Society of America* 149 (6): 4281–4291.
45 Marefati, S., Ghassemi, M., and Ghazizadeh, V. (2022). Investigation of effective parameters on streaming-induced acoustophoretic particle manipulation in a microchannel via three-dimensional numerical simulation. *Physics of Fluids* 34 (1): 012008.
46 Bodé, W.N. and Bruus, H. (2021). Numerical study of the coupling layer between transducer and chip in acoustofluidic devices. *The Journal of the Acoustical Society of America* 149 (5): 3096–3105.
47 Hahn, P., Leibacher, I., Baasch, T., and Dual, J. (2015). Numerical simulation of acoustofluidic manipulation by radiation forces and acoustic streaming for complex particles. *Lab on a Chip* 15 (22): 4302–4313.
48 Leibacher, I., Dietze, W., Hahn, P. et al. (2014). Acoustophoresis of hollow and core-shell particles in two-dimensional resonance modes. *Microfluidics and Nanofluidics* 16 (3): 513–524.
49 Sepehrirahnama, S., Oberst, S., Chiang, Y.K., and Powell, D. (2021). Acoustic radiation force and radiation torque beyond particles: effects of nonspherical shape and Willis coupling. *Physical Review E* 104 (6): 065003.
50 Wijaya, F.B., Mohapatra, A.R., Sepehrirahnama, S., and Lim, K.M. (2016). Coupled acoustic-shell model for experimental study of cell stiffness under acoustophoresis. *Microfluidics and Nanofluidics* 20 (5): 1–15.
51 Wijaya, F.B., Sepehrirahnama, S., and Lim, K.M. (2018). Interparticle force and torque on rigid spheroidal particles in acoustophoresis. *Wave Motion* 81: 28–45.
52 Muller, P.B., Barnkob, R., Jensen, M.J.H., and Bruus, H. (2012). A numerical study of microparticle acoustophoresis driven by acoustic radiation forces and streaming-induced drag forces. *Lab on a Chip* 12 (22): 4617–4627.
53 Muller, P.B. and Bruus, H. (2014). Numerical study of thermoviscous effects in ultrasound-induced acoustic streaming in microchannels. *Physical Review E* 90 (4): 043016.
54 Ley, M.W. and Bruus, H. (2017). Three-dimensional numerical modeling of acoustic trapping in glass capillaries. *Physical Review Applied* 8 (2): 024020.
55 Tribler, P.M. (2015). Acoustic streaming in microchannels: the trinity of analytics, numerics and experiments. Ph.D. Thesis, Technical University of Denmark.
56 Lee, C.P. and Wang, T.G. (1989). Near-boundary streaming around a small sphere due to two orthogonal standing waves. *The Journal of the Acoustical Society of America* 85 (3): 1081–1088.
57 Glynne-Jones, P., Mishra, P.P., Boltryk, R.J., and Hill, M. (2013). Efficient finite element modeling of radiation forces on elastic particles of arbitrary size and geometry. *The Journal of the Acoustical Society of America* 133 (4): 1885–1893.
58 Zareei, S.M., Sepehrirahnama, S., Jamshidian, M., and Ziaei-Rad, S. (2022). Three-dimensional numerical simulation of particle acoustophoresis: COMSOL implementation and case studies. *Engineering with Computers* 39 (1): 1–16.
59 Doinikov, A.A. (2001). Acoustic radiation interparticle forces in a compressible fluid. *Journal of Fluid Mechanics* 444: 1–21.

60 Silva, G.T. and Bruus, H. (2014). Acoustic interaction forces between small particles in an ideal fluid. *Physical Review E* 90 (6): 063007.

61 Baasch, T., Leibacher, I., and Dual, J. (2017). Multibody dynamics in acoustophoresis. *The Journal of the Acoustical Society of America* 141 (3): 1664–1674.

62 Baasch, T. and Dual, J. (2018). Acoustofluidic particle dynamics: beyond the Rayleigh limit. *J Acoust Soc Am* 143 (1): 509–519.

63 Karakaya, Z., Baranoğlu, B., Çetin, B., and Yazıcı, A. (2015). A parallel boundary element formulation for tracking multiple particle trajectories in Stoke's flow for microfluidic applications. *CMES Computer Modeling in Engineering and Sciences* 104 (3): 227–249.

64 Topuz, A., Baranoglu, B., and Çetin, B. (2021). A multi-domain direct boundary element formulation for particulate flow in microchannels. *Engineering Analysis with Boundary Elements* 132 (1): 221–230.

65 Atay, A., Beskok, A., and Cetin, B. (2022). DC-electrokinetic behavior of colloidal cylinder(s) in the vicinity of a conducting wall. *Electrophoresis* 43 (12): 1263–1274.

66 Cetin, B., Oner, S.D., and Baranoglu, B. (2017). Modeling of dielectrophoretic particle motion: point particle vs finite-sized particle. *Electrophoresis* 38: 1407–1418.

67 Jönsson, H., Holm, C., Nilsson, A. et al. (2004). Particle separation using ultrasound can radically reduce embolic load to brain after cardiac surgery. *The Annals of Thoracic Surgery* 78 (5): 1572–1577.

68 Petersson, F., Nilsson, A., Holm, C. et al. (2004). Separation of lipids from blood utilizing ultrasonic standing waves in microfluidic channels. *Analyst* 129 (10): 938–943.

69 Petersson, F., Nilsson, A., Holm, C. et al. (2005a). Continuous separation of lipid particles from erythrocytes by means of laminar flow and acoustic standing wave forces. *Lab Chip* 5 (1): 20–22.

70 Grenvall, C., Augustsson, P., Folkenberg, J.R., and Laurell, T. (2009). Harmonic microchip acoustophoresis: a route to online raw milk sample precondition in protein and lipid content quality control. *Analytical Chemistry* 81 (15): 6195–6200.

71 Petersson, F., Nilsson, A., Jönsson, H., and Laurell, T. (2005b). Carrier medium exchange through ultrasonic particle switching in microfluidic channels. *Analytical Chemistry* 77 (5): 1216–1221.

72 Augustsson, P., Åberg, L.B., Swärd-Nilsson, A.M.K., and Laurell, T. (2009). Buffer medium exchange in continuous cell and particle streams using ultrasonic standing wave focusing. *Microchim Acta* 164: 269–277.

73 Li, S., Ding, X., Mao, Z. et al. (2015a). Standing surface acoustic wave (SSAW)-based cell washing. *Lab Chip* 15 (1): 331–338.

74 Li, R., Ren, L., Huang, P.H. et al. (2016). Acoustofluidic transfer of inflammatory cells from human sputum samples. *Analytical Chemistry* 88 (11): 5655–5661.

75 Lenshof, A., Ahmad-Taudin, A., Järås, K. et al. (2009). Acoustic whole blood plasmapheresis chip for prostate specific antigen microarray diagnostics. *Analytical Chemistry* 81 (15): 6030–6037.

76 Adams, J.D., Ebbesen, C.L., Barnkob, R. et al. (2012). High-throughput, temperature-controlled microchannel acoustophoresis device made with rapid prototyping. *Journal of Micromechanics and Microengineering* 22 (7): 1–8.

77 Petersson, K., Jakobsson, O., Ohlsson, P. et al. (2018). Acoustofluidic hematocrit determination. *Analytica Chimica Acta* 1000: 199–204.

78 Nam, J., Lim, H., Kim, D., and Shin, S. (2011). Separation of platelets from whole blood using standing surface acoustic waves in a microchannel. *Lab Chip* 11 (19): 3361–3364.

79 Chen, Y., Wu, M., Ren, L. et al. (2016). High-throughput acoustic separation of platelets from whole blood. *Lab Chip* 16 (18): 3466–3472.

80 Zhang, N., Zuniga-Hertz, J.P., Zhang, E.Y. et al. (2021). Microliter ultrafast centrifuge platform for size- based particle and cell separation and extraction using novel omnidirectional spiral surface acoustic waves. *Lab Chip* 21 (5): 904–915.

81 Dykes, J., Lenshof, A., Åstrand-Grundström, I.B. et al. (2011). Efficient removal of platelets from peripheral blood progenitor cell products using a novel micro-chip based acoustophoretic platform. *PLoS One* 6 (8): 1–10.

82 Urbansky, A., Ohlsson, P., Lenshof, A. et al. (2017). Rapid and effective enrichment of mononuclear cells from blood using acoustophoresis. *Scientific Reports* 7 (1): 1–9.

83 Lissandrello, C., Dubay, R., Kotz, K.T., and Fiering, J. (2018). Purification of lymphocytes by acoustic separation in plastic microchannels. *SLAS Technology* 23 (4): 352–363.

84 Augustsson, P., Magnusson, C., Nordin, M. et al. (2012). Microfluidic, label-free enrichment of prostate cancer cells in blood based on acoustophoresis. *Analytical Chemistry* 84 (18): 7954–7962.

85 Magnusson, C., Augustsson, P., Lenshof, A. et al. (2017). Clinical-scale cell-surface-marker independent acoustic microfluidic enrichment of tumor cells from blood. *Analytical Chemistry* 89 (22): 11954–11961.

86 Cushing, K., Undvall, E., Ceder, Y. et al. (2018). Reducing WBC background in cancer cell separation products by negative acoustic contrast particle immuno-acoustophoresis. *Analytica Chimica Acta* 1000: 256–264.

87 Antfolk, M., Antfolk, C., Lilja, H. et al. (2015). A single inlet two-stage acoustophoresis chip enabling tumor cell enrichment from white blood cells. *Lab Chip* 15 (9): 2102–2109.

88 Ding, X., Peng, Z., Lin, S.C.S. et al. (2014). Cell separation using tilted-angle standing surface acoustic waves. *Proceedings of the National Academy of Sciences of the United States of America* 111 (36): 12992–12997.

89 Li, P., Mao, Z., Peng, Z. et al. (2015b). Acoustic separation of circulating tumor cells. *Proceedings of the National Academy of Sciences of the United States of America* 112 (16): 4970–4975.

90 Wu, M., Huang, P.H., Zhang, R. et al. (2018). Circulating tumor cell phenotyping via high-throughput acoustic separation. *Small* 14 (32): 1–10.

91 Yang, A.H.A. and Soh, H.T. (2012). Acoustophoretic sorting of viable mammalian cells in a microfluidic device. *Analytical Chemistry* 84 (24): 10756–10762.

92 Zalis, M.C., Reyes, J.F., Augustsson, P. et al. (2016). Label-free concentration of viable neurons, hESCs and cancer cells by means of acoustophoresis. *Integrative Biology* 8 (3): 332–340.

93 Thévoz, P., Adams, J.D., Shea, H. et al. (2010). Acoustophoretic synchronization of mammalian cells in microchannels. *Analytical Chemistry* 82 (7): 3094–3098.

94 Nam, J., Lim, H., Kim, C. et al. (2012). Density-dependent separation of encapsulated cells in a microfluidic channel by using a standing surface acoustic wave. *Biomicrofluidics* 6 (2): 1–10.

95 Evander, M., Johansson, L., Lilliehorn, T. et al. (2007). Noninvasive acoustic cell trapping in a microfluidic perfusion system for online bioassays. *Analytical Chemistry* 79 (7): 2984–2991.

96 Norris, J.V., Evander, M., Horsman-Hall, K.M. et al. (2009). Acoustic differential extraction for forensic analysis of sexual assault evidence. *Analytical Chemistry* 81 (15): 6089–6095.

97 Hammarström, B., Evander, M., Barbeau, H. et al. (2010). Non-contact acoustic cell trapping in disposable glass capillaries. *Lab Chip* 10 (17): 2251–2257.

98 Ai, Y., Sanders, C.K., and Marrone, B.L. (2013). Separation of *Escherichia coli* bacteria from peripheral blood mononuclear cells using standing surface acoustic waves. *Analytical Chemistry* 85 (19): 9126–9134.

99 Li, S., Ma, F., Bachman, H. et al. (2017). Acoustofluidic bacteria separation. *Journal of Micromechanics and Microengineering* 27 (1): 1–8.

100 Dow, P., Kotz, K., Gruszka, S. et al. (2018). Acoustic separation in plastic microfluidics for rapid detection of bacteria in blood using engineered bacteriophage. *Lab Chip* 18 (6): 923–932.

101 Ohlsson, P., Petersson, K., Augustsson, P., and Laruell, T. (2018). Acoustic impedance matched buffers enable separation of bacteria from blood cells at high cell concentrations. *Scientific Reports* 8 (1): 1–11.

102 Hammarström, B., Laurell, T., and Nilsson, J. (2012). Seed particle-enabled acoustic trapping of bacteria and nanoparticles in continuous flow systems. *Lab Chip* 12 (21): 4296–4304.

103 Ohlsson, P., Evander, M., Petersson, K. et al. (2016). Integrated acoustic separation, enrichment, and microchip polymerase chain reaction detection of bacteria from blood for rapid sepsis diagnostics. *Analytical Chemistry* 88 (19): 9403–9411.

104 Ngamsom, B., Martinez, M.J.L., Raymond, J.C. et al. (2016). On-chip acoustophoretic isolation of microflora including *S. typhimurium* from raw chicken, beef and blood samples. *Journal of Microbiological Methods* 123: 79–86.

105 Evander, M., Gidlöf, O., Olde, B. et al. (2015). Non-contact acoustic capture of microparticles from small plasma volumes. *Lab Chip* 15 (12): 2588–2595.

106 Ku, A., Lim, H.C., Evender, M. et al. (2018). Acoustic enrichment of extracellular vesicles from biological fluids. *Analytical Chemistry* 90 (13): 8011–8019.

107 Collins, D.J., Alan, T., and Neild, A. (2014). Particle separation using virtual deterministic lateral displacement (vDLD). *Lab Chip* 14 (9): 1595–1603.

108 Sehgal, P. and Kirby, B.J. (2017). Separation of 300 and 100 nm particles in Fabry–Perot acoustofluidic resonators. *Analytical Chemistry* 89 (22): 12192–12200.

109 Lee, K., Shao, H., Weissleder, R., and Lee, H. (2015). Acoustic purification of extracellular microvesicles. *ACS Nano* 9 (3): 2321–2327.

110 Wu, M., Mao, Z., Chen, K. et al. (2017a). Acoustic separation of nanoparticles in continuous flow. *Advanced Funtional Materials* 27 (14): 1–8.

111 Wu, M., Ouyang, Y., Wang, Z. et al. (2017b). Isolation of exosomes from whole blood by integrating acoustics and microfluidics. *Proceedings of the National Academy of Sciences of the United States of America* 114 (40): 10584–10589.

112 Shi, J., Ahmed, D., Mao, X. et al. (2009). Acoustic tweezers: patterning cells and microparticles using standing surface acoustic waves (SSAW). *Lab Chip* 9 (20): 2890–2895.

113 Ding, X., Lin, S.C.S., Kiraly, B. et al. (2012). On-chip manipulation of single microparticles, cells, and organisms using surface acoustic waves. *Proceedings of the National Academy of Sciences of the United States of America* 109 (28): 11105–11109.

114 Guo, F., Li, P., French, J.B. et al. (2015). Controlling cell–cell interactions using surface acoustic waves. *Proceedings of the National Academy of Sciences of the United States of America* 112 (1): 43–48.

115 Ahmed, D., Ozcelik, A., Bojanala, N. et al. (2016). Rotational manipulation of single cells and organisms using acoustic waves. *Nature Communications* 7: 1–11.

116 Ozcelik, A., Nama, N., Huang, P.H. et al. (2016). Acoustofluidic rotational manipulation of cells and organisms using oscillating solid structures. *Small* 12 (37): 5120–5125.

117 Zhang, J., Yang, S., and Chen, C. (2019). Surface acoustic waves enable rotational manipulation of *Caenorhabditis elegans*. *Lab Chip* 19: 984–992.

118 Hammerström, B., Nilson, B., Laruell, T. et al. (2014). Acoustic trapping for bacteria identification in positive blood cultures with MALDI-TOF MS. *Analytical Chemistry* 86 (21): 10560–10567.

119 Mubarak, M.T., Ozsahin, I., and Ozsahin, D.U. (2019). Evaluation of sterilization methods for medical devices. In: *2019 Advances in Science and Engineering Technology International Conferences (ASET)*, 1–4. IEEE.

120 Silindir, M. and Özer, A.Y. (2009). Sterilization methods and the comparison of E-beam sterilization with gamma radiation sterilization. *FABAD Journal of Pharmaceutical Sciences* 34 (1): 43.

121 Yavuz, C., Oliaei, S.N.B., Cetin, B., and Yesil-Celiktas, O. (2016). Sterilization of PMMA microfluidic chips by various techniques and investigation of material characteristics. *The Journal of Supercritical Fluids* 107: 114–121.

122 Atay, A., Topuz, A., Sarıarslan, B. et al. (2021). Flow rate-controlled pipetting for microfluidics: second-generation flexible hydraulic reservoir (FHRv2). *Microfluidics and Nanofluidics* 25 (1): 1–10.

123 Hatipoğlu, U., Çetin, B., and Yıldırım, E. (2018). A novel zero-dead-volume sample loading interface for microfluidic devices: flexible hydraulic reservoir (FHR). *Journal of Micromechanics and Microengineering* 28 (9): 097001.

124 Carville, N.C., Collins, L., Manzo, M. et al. (2015). B Biocompatibility of ferroelectric lithium niobate and the influence of polarization charge on osteoblast proliferation and function. *Journal of Biomedical Materials Research Part A* 103 (8): 2540–2548.

125 Dubay, R., Lissandrello, C., Swierk, P. et al. (2019). Scalable high-throughput acoustophoresis in arrayed plastic microchannels. *Biomicrofluidics* 13 (3): 034105.

126 Evander, M., Lenshof, A., Laurell, T., and Nilsson, J. (2008). Acoustophoresis in wet-etched glass chips. *Analytical Chemistry* 80 (13): 5178–5185.

127 Zhao, S., Wu, M., Yang, S. et al. (2020). A disposable acoustofluidic chip for nano/microparticle separation using unidirectional acoustic transducers. *Lab on a Chip* 20 (7): 1298–1308.

128 Augustsson, P., Barnkob, R., Wereley, S.T. et al. (2011). Automated and temperature-controlled micro-PIV measurements enabling long-term-stable microchannel acoustophoresis characterization. *Lab on a Chip* 11 (24): 4152–4164.

10

Acoustic Biosensors

Alper Şişman[1,2,3], Paddy French[1], Ayşe Ogan[4], Erdal Korkmaz[5], Abbas A. Husseini[6], Ali M. Yazdani[2,7], and Johan Meyer[1]

[1] Delft University of Technology, Faculty of Electrical Engineering, Mathematics and Computer Science, 2628CD, Delft, the Netherlands
[2] Marmara University, Faculty of Engineering, 34854, Istanbul, Turkey
[3] Sabanci University Nanotechnology and Application Center (SUNUM), Sabanci University, 34956, Istanbul, Turkey
[4] Marmara University, Faculty of Science, 34722, Istanbul, Turkey
[5] The Hague University of Applied Sciences, Technology-Innovation & Society Faculty, Electrical and Electronic Engineering Department, Smart Sensor Systems Research Group, 2521EN, Delft, the Netherlands
[6] Istanbul Gelisim University Vocational School, Biomedical Department, Life Science and Biomedical Engineering Application and Research Center, 34310, Istanbul, Turkey
[7] Nisantasi University Vocational School, Mechanics Department, 34398, Istanbul, Turkey

10.1 Introduction

An acoustic biosensor has two main components: (i) A physical sensing mechanism that converts the interaction between the biomolecules into a measurable signal, and (ii) A sensitive surface to recognize a biologic material [1]. In this chapter, these two components will be discussed in detail and the acoustic biosensor applications will be given along with future projections.

Many physical sensing mechanisms are in application like optical based [2], electrochemical (e.g. ohmic) based [3], and mechanical waves based [4, 5] (Figure 10.1). All these physical domains have their own advantages for various applications in terms of sensitivity, cost, ease of use, accessibility, etc. Also, several domains can be used together to enable more optimized solutions [6]. Acoustic biosensors use mechanical waves that vibrate a medium to affect the biomaterial attached. These waves may create vibrations only on a surface or the vibrations may affect the whole material (body). The wave type defines the vibration propagating through: (i) Bulk acoustic waves (BAWs) propagate through the whole structure, (ii) In surface acoustic waves, the vibrations only propagate at the surface of the material.

10.1.1 Bulk Acoustic Wave (BAW) Mode

BAWs propagate in any media unless they are interrupted by boundaries or by interfaces. These waves can be decomposed into infinite plane waves propagating along

Acoustic Technologies in Biology and Medicine, First Edition.
Edited by Adem Ozcelik, Ryan Becker, and Tony Jun Huang.

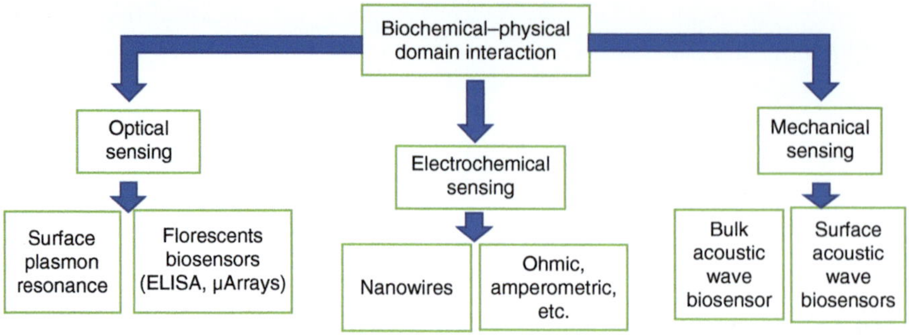

Figure 10.1 The biosensor classification according to physical–biochemical operation principle.

an arbitrary direction g within the solid. The properties of these waves are determined by the relationships between the propagation direction and the constitutive properties of the medium. For an excited pure-bulk wave, the propagation direction is normal to or along the polarization vector. The polarization and propagation directions are the same for longitudinal bulk waves. The other wave type is known as a shear bulk wave, which has a polarization vector normal to the propagation direction (Figure 10.2). For an isotropic medium, the bulk wave speed for longitudinal and shear waves can be seen in (10.1) as c_L and c_S respectively. Here, λ and μ represent Lamé's first and second parameters for the medium and ρ shows the material density.

$$c_L = \sqrt{\frac{2\mu + \lambda}{\rho}}, \quad c_S = \sqrt{\frac{\mu}{\rho}} \tag{10.1}$$

Bulk waves propagating in three-dimensional (3D) media satisfy the general wave equation:

$$\frac{d^2u}{dt^2} = c^2\left(\frac{d^2u}{dx^2} + \frac{d^2u}{dy^2} + \frac{d^2u}{dz^2}\right) \tag{10.2}$$

where "$u(x,y,z,t)$" represents the wave intensity in the space and "c" shows the wave speed [7].

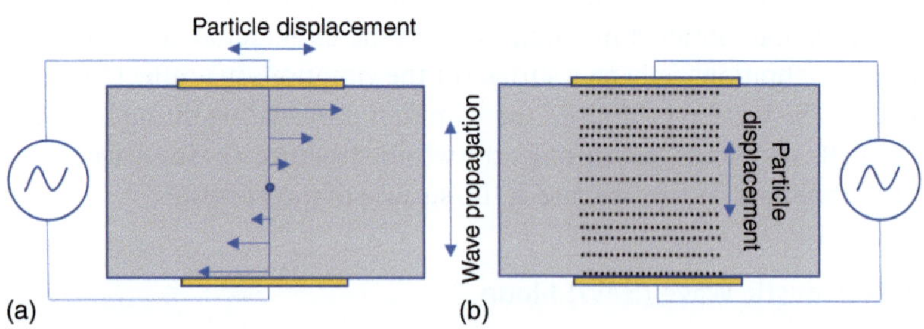

Figure 10.2 Bulk acoustic wave modes. (a) Shear bulk wave mode and (b) longitudinal bulk wave mode.

The operating frequency of BAW devices is directly dependent on the thickness of the material the BAW propagates through. Higher frequency requires a thinner design, and since decreased thickness decreases the mechanical stability of the device, the maximum operation frequency is limited to around 50 MHz [8]. Higher operation frequencies enable enhanced sensitivity, which leads to considerable attention to another wave mode: surface-guided waves.

10.1.2 Surface Guided Acoustic Wave (SGAW) Modes

Surface acoustic waves are acoustic waves (plastic deformations) traveling along the surface of a medium. Acoustic waves can be excited by applying a localized mechanical impact to a solid medium, and they generally travel in different modes in all directions into the medium. However, in some modes, the wave energy is confined to the surface, penetrating the material only in the order of magnitude of several wavelengths. Of these surface-bound waves, the most commonly known types are Rayleigh, shear horizontal (SH), and plate waves. A line source on a plane surface of an elastic solid medium can excite acoustic waves into the bulk and along the surface. The created BAWs have a longitudinal (L) component, shear vertical (SV) component, and SH component. As the L and SV components cannot satisfy the boundary condition, their radiation pattern is not uniform in all directions but directional as illustrated in Figure 10.3. The L and SV components couple with each other through the boundary condition giving rise to an eigenmode that confines most wave energy to within one wavelength of the surface and is evanescent into the bulk. For the Rayleigh mode, the surface particles move in an elliptic orbit normal to the surface normal to the surface (Figure 10.4) and the created wave attenuates rapidly as it moves away from the surface (1–2λ). This wave mode is excitable on all solids, including piezoelectric materials. The Rayleigh-mode surface acoustic wave (SAW) has a first overtone that is sometimes referred to as the Sezawa-mode. This type of SAW has very high phase velocity and can therefore be used to obtain a high operating frequency device. As the Sewaza-mode is an overtone of the Rayleigh-mode, it also has a L and a SV component causing both Rayleigh and Sewaza modes to have large coupling to any liquid loading [12].

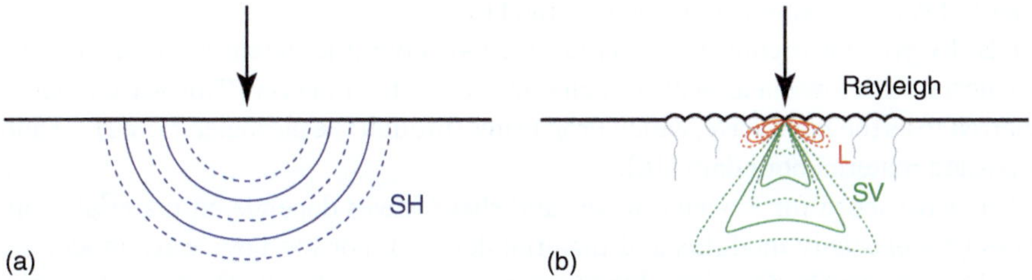

Figure 10.3 Linear source plane excitation for surface and bulk wave propagation modeling in an isotropic solid. (a) Propagation of SH waves along the surface and into the bulk. (b) Coupling of SV and L modes to a Rayleigh mode SAW. Both (a) and (b) are simultaneously at play but drawn separately for clarity. Source: Adapted from Hashimoto and Hashimoto [9]; Krylov [10]; Andreykiv et al. [11].

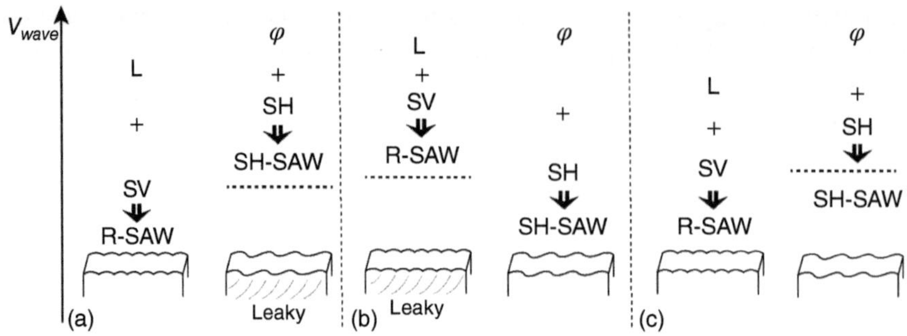

Figure 10.4 (a) and (b) shows a leaky SH-SAW and a leaky Rayleigh-SAW, respectively. (c) The coupling of the modes is not arising a leaky wave. Relation of SH, SV L, and φ wave velocity and the effect when they couple. When the coupled wave causing a SAW is faster than any present shear BAW component, a leaky SAW is created. Source: Adapted from [9].

The SH-type BAW does comply with the boundary condition and can therefore spread evenly in all directions in the bulk for isotropic solids as shown in Figure 10.3a. So, no SH-SAW-mode exists on these isotropic materials. For some anisotropic solids with piezoelectric properties, the SH-type wave can couple to the electric potential φ induced due to deformation. This can confine the wave energy to the surface giving rise to some shear modes around the surface (e.g. SH-type SAW) [13]. The SH-type BAW has a higher propagation velocity than the SH-type SAW. A higher electromechanical coupling (strength of piezoelectric effect) causes a larger difference in velocity. Also, the higher the difference in velocity, the more the wave energy is confined to the surface [9]. When the resulting SAW has a higher velocity than the remaining shear BAW component (SH or SV), energy from the SAW is allowed to leak into the bulk creating a leaky SAW. This is illustrated in Figure 10.3 cases (a) and (b) where a leaky SH-SAW and a leaky Rayleigh-SAW are shown respectively. An interdigital transducer (IDT) can excite a BAW that does not enter the material uniformly. Then the wave is skimming the surface giving rise to a surface skimming bulk wave (SSBW) or shallow bulk acoustic wave (SBAW). This wave type does not have its energy concentrated at the surface and is consequently a BAW [14]. When using proper mass grating structures, spaced half a wavelength, the wave can be captured to the surface. This is known as the surface transverse wave (STW) [5] as was discovered by Auld [15].

It is also possible to confine the SH surface wave energy to the surface by depositing a material with low shear wave velocity on the excited material. This wave mode is referred to as the Love wave, which propagates through the piezoelectric surface and deposited material boundary [16].

The wave speed for Rayleigh waves and shear waves depends on material properties (the elasticity modulus and material density). For the Love wave mode, the speed is not only the function of the material properties, but it also depends on the thickness of the deposited material on the substrate.

Another SAW mode occurs when the material thickness is thinner than the wavelength of the excited surface wave. These types of mediums can be regarded as two-dimensional (2D) mediums since their thickness is smaller than a wavelength

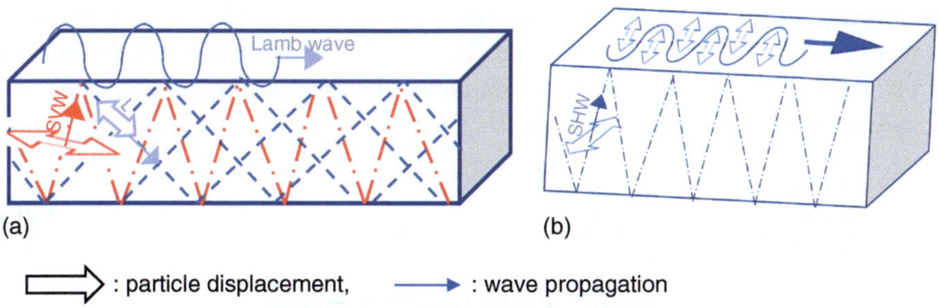

Figure 10.5 The plate wave modes: (a) Lamb wave (b) SH wave.

(e.g. a membrane). A group of surface wave modes can be excited as plate waves since they can propagate among a 2D medium. These modes are Lamb and SH waves as they are illustrated in Figure 10.5 [17]. The Lamb wave mode has two fundamental modes, the symmetric zero order mode S0 and the antisymmetric zero order mode A0. Here the symmetric or extensional S0-mode extends the membrane in the longitudinal direction parallel to wave propagation, and the membrane contracts perpendicular to the direction of wave propagation. This symmetric mode converges to a longitudinal mode when the guide material thickness goes to zero. If the membrane is loaded with a liquid, the S0-mode has low power radiation into the liquid making the lamb wave device suitable for liquid sensing applications [18].

The wave propagation for symmetric and antisymmetric modes is illustrated in Figure 10.6. The wave speed is not only a function of medium parameters, but it depends on the thickness of the medium. Each symmetric and antisymmetric mode has its own propagation speed, resulting in different frequencies for each mode, if their wavelengths are equal. The wave speed for each mode can be plotted as a family of curves according to normalized medium thickness (d/λ), the shown family

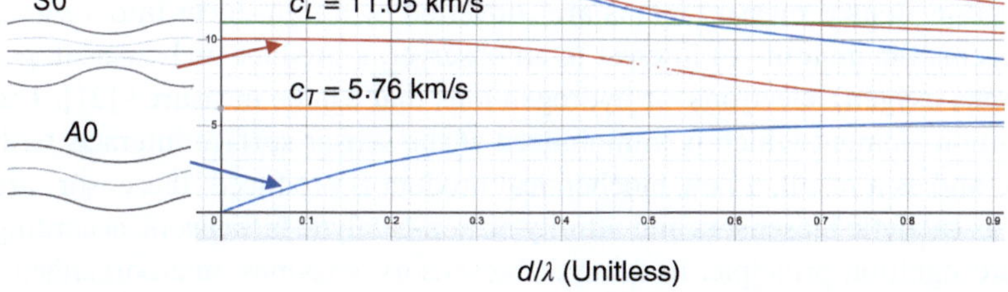

Figure 10.6 Dispersion curves for Lamb waves traveling along an aluminum nitride film, (A0) zeroth order antisymmetric (extensional) mode, and (S0) zeroth order symmetric (flexural) mode.

of curves are known as dispersion curves. These curves can be determined using Rayleigh–Lamb frequency equation set (10.3)–(10.5). The equation set explains the propagation of Lamb waves along a thin plate. The relationship between the phase speed and the normalized thickness (d/λ) for antisymmetric modes is given by,

$$\frac{\tanh\left(\pi\frac{d}{\lambda}\beta\right)}{\tanh\left(\pi\frac{d}{\lambda}\alpha\right)} = \frac{4\alpha\beta}{(1-\beta^2)^2} \tag{10.3}$$

The relationship for symmetric modes is given as follows:

$$\frac{\tanh\left(\pi\frac{d}{\lambda}\beta\right)}{\tanh\left(\pi\frac{d}{\lambda}\alpha\right)} = \frac{(1-\beta^2)^2}{4\alpha\beta} \tag{10.4}$$

$$\alpha^2 = \left(\frac{c_P^2}{c_L^2} - 1\right), \quad \beta^2 = \left(\frac{c_P^2}{c_T^2} - 1\right), \quad \xi = \frac{2\pi}{\lambda}, \quad c_p = \frac{w}{\xi} \tag{10.5}$$

In these equations, "*d*" shows the medium thickness and ξ, λ, w, C_L, C_T, and C_P represent wave number, wavelength, angular frequency, longitudinal wave speed, transversal wave speed, and phase speed. The solution of these equations results in a family of curves, which represents the relationship (c_p vs d/λ) for zeroth and higher-order symmetric and antisymmetric waves (Figure 10.6).

10.2 Biochemical Fundamentals of Sensing

Sensors can be used to measure a great variety of physical, chemical, and biological materials such as proteins, peptides, enzymes, antibodies, nucleic acids, or hormones, chemicals, bacteria or detect gases, light intensity, motion, position, sound, and many others [19]. Biosensors use biochemical mechanisms to detect a bioanalyte in biological samples (blood, saliva, and urine) and in the air, soil, and water [20]. Generally, biosensors are designed to detect very low concentrations of bioanalytes in biological fluids with high sensitivity and specificity. They preferentially detect target biomolecules (proteins, peptides, DNA, RNA, etc.) in small sample volumes and a rapid response time [21]. A biosensor measures biological or chemical reactions by generating signals proportional to the concentration of an analyte in the reaction. Sensor measurements are converted by a transducer into a signal that represents the quantity of interest to an observer; a product such as heat, gas (for example, oxygen), electrons, or hydrogen ions that can be measured [21]. The biorecognition element which is immobilized at the sensor surface interacts with the target, and as a result, a new biochemical product is produced. Biosensors are classified as catalytic biosensors and affinity or non-catalytic biosensors, according to the biorecognition principle. In catalytic biosensors; enzymes, microorganisms, tissues, and whole cells and in case of affinity (non-catalytic) biosensors, antigen/antibody, cell receptors, and nucleic acids, are the key bioreceptors [22]. Immunoglobulins (IgGs) are widely used as detection elements. The diversity in complementarity-determining regions (CDRs) area, which are responsible for the specific antibody–antigen (ab–antigen) interaction, allows the endless supply of Abs with different specificity and binding strength (affinity). IgG derivatives such as antigen-binding fragments (Fabs) and single-chain variable fragments (scFv), which still retain antigen-binding specificity, minimize the Abs and more stable fragments. Camelid Abs and novel antigen receptors are two other Ab classes that are becoming important in biosensor applications because of their superior stabilities and small sizes [23].

10.2.1 Immobilization Strategies of Detection Element

Analyte detection with biosensors is strongly influenced by the preparation of the biosensor surface including choice of sensing layers, coupling methods particularly coupling chemistry, and incubation times for corresponding capture molecules [24]. Detection elements with high analyte specificity and binding strength are required. In the field of biosensors, the most important crucial step is to immobilize bioreceptors onto the transducer. Biosensor stability and sensitivity may vary depending on the immobilization method used [20]. The surface and the recognition sensing component influence the method of immobilization. Ideally, the sensing surface should be stable, robust, and possess a certain lifetime. Surface binding sites must be accessible and able to bind the analyte to be detected and nonspecific binding on the transducer surface has to be inhibited to prevent false-positive signals. In case of Ab immobilization, it is essential to immobilize antibodies on surfaces without changing their binding activities and specificities. Therefore immobilization strategies should be mild. The most commonly used methods for immobilization of analytes are based on noncovalent, covalent immobilization, and affinity bindings [24–26].

10.2.1.1 Noncovalent Immobilization

Sensor surface modification may include immobilization of detection elements directly over a surface by adsorption or entrapment in polymer layers either as 3D matrices or self-assembled monolayers (SAMs) [27–29].

Adsorption

Adsorption is typically regarded as the simplest and most straightforward immobilization technique. It depends on weak electrostatic and/or hydrophobic interactions and Van der Waal forces. Adsorption-based immobilization is a dynamic process that involves ongoing adsorption and desorption from the surface. It is reversible and has no adverse effects on the recognition element such as antigen/antibody, enzyme structure, or functionality [30, 31]. The procedure involves dissolving the recognition element in a solution and exposing the solid support to the solution for a period of time. Following this, the excess recognition element is removed by washing buffers. Polymeric substrates used in traditional adsorption platforms are cheap and stable and easy to manufacture [32]. Surface modification methods are employed to increase the hydrophilicity of microtiter plates which are made of polystyrene, polypropylene, polyethylene, etc. Hydrophilicity can increase antigen/antibody binding and decrease poor binding [33]. However, because of the poor binding, this immobilization is susceptible to biomolecule desorption. Desorption may also be caused by external conditions such as temperature, ionic strength, and pH [25]. To overcome this problem, plasma surface modification is a trusted and inexpensive way to alter physicochemical and mechanical properties of the surface. Microtiter plates are treated with oxygen, nitrogen, and other gas plasmas to create chemical functionalities at the surface, thereby reducing the hydrophobicity of the surface for further chemical treatments and covalent grafting of proteins [32]. *N*-ethyl-*N*-(3-dimethylaminopropyl) carbodiimide (EDC)/*N*-hydroxysuccinimide (NHS) chemistry is generally used to attach IgG or Protein A (with subsequent IgG binding) to the treated surfaces [34, 35].

Three-Dimensional Substrates

3D substrates are attractive as they increase surface area for antibody binding and can minimize steric hindrance that may prevent antigen capture. Porous 3D substrates have been prepared from a variety of materials, including various polymers, silicon, glass slides, metals, and gels [36]. Similarly, agarose or dextran or sol–gels have found considerable applications for improved antibody binding due to their high surface area [37].

Self-Assembled Monolayers

The formation of a SAM provides another means of modifying surface chemistry to promote antibody adsorption or to produce functional groups for subsequent covalent attachments [38].

The majority of molecules which form SAMs consist of a chemical "head" group, which binds to the substrate of interest, as well as an attached tail group, which often possesses a chemically functionalized terminal group. The anchoring head group has an affinity for the surface, while the other tail group enhances protein adsorption or attachment.

10.2.1.2 Covalent Immobilization

Covalent immobilization, on the other hand, is rather powerful since biomolecules like proteins–antigens and antibodies and nucleic acids include functional groups that can be easily modified to bind to polymeric matrices. During immobilization process, stable complexes are formed between the functional groups of the biomolecules and the matrix. The functional groups that are involved in the reaction are the amine, carboxyl, thiol, and carbohydrate moieties [39]. Covalent immobilization of other biomolecules, including DNA and aptamers, is commonly utilized in immunosensors for specific antigen capture [40]. However, a major drawback of covalent bonding is the significant loss of protein activity that can occur after the immobilization process. Amino acids with reactive primary amine and carboxyl side chains can be targeted for covalent attachment. Amine and carboxyl coupling is commonly attained with carbodiimide chemistry that utilizes EDC combined with succinimidyl esters (such as NHS) [41]. This method is known as

EDC/NHS coupling and results in robust amide bond formation. Cross-linking with glutaraldehyde or other bifunctional agents (e.g. glyoxal, hexamethylenediamine) is another well-known approach used in biosensors in which biomolecules are immobilized [42].

10.2.1.3 Bioaffinity Bindings

Biochemical affinity reactions offer a gentle oriented immobilization of analytes, providing an important advantage over other immobilization techniques. Proteins may be gently immobilized using biochemical affinity processes, which have significant benefits over conventional immobilization methods. Additionally, not only is directed and homogenous attachment achieved, but it is also possible to separate proteins and use the same surface repeatedly. Biotin–streptavidin interaction, DNA-directed immobilization, Protein A and Protein G binding, Fc-binding peptides and aptamers, nucleotide binding site, and metal affinity are the tools of bioaffinity immobilization.

Avidin–Biotin Interaction

In the specificity of this, affinity interaction permits uniformly oriented protein immobilization. Avidin–biotin bond is one of the strongest noncovalent bonds ($K_d = 101^5 \times M^{-1}$) ever known enabling the use of extreme conditions during biological experiments. Avidin is a tetrameric glycoprotein that is stable over a wide pH and temperature range and soluble in aqueous solutions [43]. It can bind up to four biotin molecules. The most popular biotinylation agent for amine groups is the biotin NHS ester. The avidin/biotin system offers numerous advantages; widespread availability of biotinylated proteins, separate irradiation and immobilization steps, avoiding direct exposure of proteins to UV light that might damage them, and the possibility of controlling surface density of binding sites by carrying out a partial photolysis step. Aside from being unaffected by pH, temperature, organic solvents, enzymatic proteolysis, and other denaturing agents, bond formation happens relatively quickly. Streptavidin is a closely related tetrameric protein, with similar affinity to biotin, but differing in other aspects, such as molecular weight, amino acid composition, and pI providing enhanced stability and controlled biotin binding.

DNA-Directed Immobilization (DDI)

DDI of proteins is a chemically mild and highly efficient method for generating (micro)structured patterns of proteins on surfaces and highly efficient method for generating (micro) structured patterns of proteins on surfaces. This method requires the binding of short nucleotide sequences to the substrate and to the protein of interest, allowing direct binding between the two and enabling the in situ production of protein arrays, which can be used for multiplexed immunoassays [44]. Additional advantages of DDI include high surface-coating density and possibilities for regeneration of the DNA surface by simple denaturation protocols. However, care must be taken to ensure that the covalent attachment of the DNA to the antibody occurs in a site-direct manner [45–47].

Protein A and Protein G Binding

Protein A and Protein G are small proteins, derived from bacteria which are able to bind with high affinity to the Fc region of IgG antibodies. Protein A, or its derivatives, has been combined with many surface immobilization strategies such as biotin–streptavidin, SAMs, EDC/NHS chemistry, and glutaraldehyde cross-linking [48, 49].

Fc-binding Peptides and Aptamers

Short peptides with specificity to the Fc domain of antibodies have been used to promote oriented immobilization of human, rabbit, goat, and mouse antibodies, with strong selectivity for human IgG1 and IgG2 [50]. The peptide (KHRFNKD) and Fc binding peptide HWRGWVC showed improved IgG binding relative to physical absorption [51]. Aptamers are single-chain DNA, or RNA, oligonucleotides that form complex 3D structures and can specifically immobilize proteins of interest. An RNA aptamer has been shown selectively to bind the Fc portion of human IgG1–IgG4 [52].

Nucleotide-Binding Site

This method utilizes the nucleotide-binding site (NBS) that is located on the Fab variable domain of IgGs to enable capturing of antibody molecules [51].

Metal Affinity

The interaction of histidine and cysteine with metals, particularly copper and zinc, has been utilized for protein fractionation and IgG purification using immobilized metal affinity [53].

In conclusion, biosensors can display superior sensitivity and specificity but they are prone to operating conditions. Because biosensors rely on biological components, their stability may change with the degradation of immobilized biological materials; enzymes or antibodies can lose activity over time [19]. Storage conditions and the immobilization methods can considerably influence the lifespan of a sensor. Biosensors are generally for single use, often have a limited operational range because of temperature, pH, or humidity, and are suitable for point-of-care (POC) applications but they are not suitable for long-term monitoring at present. Preparation of biological samples before application to the sensor, such as micro-dialysis is often necessary and this can increase the complexity of the sensor system. Some compounds can interfere with the sensor readings, particularly biochemical transducers, interfere with readouts, which can lead to improper results [54, 55].

10.3 Bulk Acoustic Wave Biosensors

BAW modality have a broad application spectrum including many sensing applications (e.g. biosensing, gas sensing, etc). The majority of current acoustic biosensor applications are based on BAW biosensors due to the ease of fabrication and low-cost design [56]. They have two main types of biosensing applications: (i) Quartz crystal microbalance (QCM) biosensors, and (ii) Film bulk acoustic resonators (FBARs), in this section, the sensing principles and fabrication details will be discussed.

10.3.1 Quartz Microbalance (QMB) Crystal Biosensors

QCM devices have been widely employed for physical, chemical and biological sensing applications. QCM devices employ a thickness-shear mode (TSM) BAW [56, 57]. They are fabricated using quartz crystal resonators that resonate electromechanically in TSM. The resonant properties of crystal can be influenced by external loading (gravimetric or viscoelastic loading). Shear gradient inside the crystal creates a force that balances the gravimetric force due to the deposited mass on the sensor. The dynamic behavior of the system changes due to the changing mass, so that resonance frequency is shifted. The amount of shift (f_g), can be given by Sauerbrey's equation (10.6) and (10.7) [58, 59].

$$\Delta f_g = -\frac{2f_0^2}{\sqrt{\rho\mu}}\frac{\Delta m}{A} \tag{10.6}$$

$$f_0 = \frac{c}{2t} = \frac{\sqrt{\frac{\mu}{\rho}}}{2t} \tag{10.7}$$

Here, f_0 is the fundamental resonance frequency, Δm is the mass deposited on sensor surface, A is the active electrode area, c is the acoustic wave velocity, ρ is the density of crystal, μ is the shear modulus of the piezoelectric material, and t is the crystal thickness. Crystal thickness defines the wavelength of the generated acoustic wave that has a velocity depending on the medium parameters (density and shear modulus). Since the wave speed is constant, and the wavelength is $\lambda = 2t$, the center frequency depends only on the thickness of the vibrating material. Higher frequencies require smaller material thickness, which makes the fabrication harder (~100–150 μm) [59, 60]. The major drawback of QCM sensors is the limit of operation frequency, that limits the sensitivity of the biosensor. This drawback leads to the development of FBAR biosensors.

QCM biosensor structure has two main blocks, the sensitive area and the resonating structure (Figure 10.7a). The metal electrodes are grown on both sides of a thin piezoelectric material, which can generate shear displacement when the electrical signal with design frequency is applied to the electrodes (Figure 10.7b). The single crystal piezoelectric material should be chosen to produce a TSM BAW. Usually, AT-Cut Quartz is preferred since its crystal orientation can excite TSM mode and it provides good temperature stability, which is another important parameter to consider.

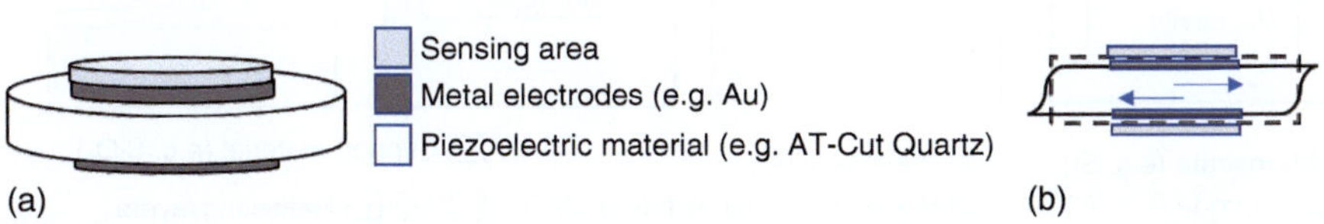

Figure 10.7 (a) QCM biosensor structure, (b) QCM biosensor operation: TSM bulk acoustic wave illustration.

10.3.2 Film Bulk Acoustic Wave (FBAR) Biosensors

The lack of increased sensitivity due to limited crystal thickness is the main problem in the QCM biosensors. The thickness of vibrating material can be decreased dramatically if this layer is formed by depositing the piezoelectric material (instead of using single crystal material). The dramatically decreased piezoelectric layer thickness increases the operation frequency from the order of MHz to sub-GHz frequencies. Increased sensitivity due to high operation frequency makes the FBAR an emerging technology for BAW-based biosensors. FBAR biosensor has various applications that require high mass change sensitivity (e.g. protein level quantification, DNA detection) [61].

The resonator structure has two metal electrodes and a piezoelectric film layer between them. The structure can either be suspended on an air gap (air cavity resonator) or it may be grown on a silicon surface(solidly mounted resonator, SMR). In a SMR structure, a multilayer acoustic reflector structure is used to reflect the wave propagating toward to substrate (Figure 10.8). The reflector layers are quarter wavelength in thickness and the successive layers should have high and low acoustic velocity [62].

Another classification for FBAR devices is according to wave type: (i) Longitudinal mode (L-standing wave) and (ii) Thickness shear wave mode (TSM-standing wave). The propagating wave through the piezoelectric medium depends on the *c*-axis angle of crystal columns deposited on the surface. Longitudinal waves are excited if the columns are deposited normal to the plane [62, 63], and 34.5^{o} tilted *c*-axis results in shear standing waves [64, 65].

Usually, the film is deposited using chemical vapor deposition (e.g. $LiNbO_3$ grown) or sputter (e.g. AlNi ZnO grown) techniques [66, 67], and the film thickness varies from submicrons to a couple of microns [65–67]. The c-axis orientation and the film stress can be adjusted by changing the deposition parameters. After deposition, the orientation should be verified using X-ray diffraction (XRD).

The operation sensitivity and operation frequency calculations are similar to the QCM structure, frequency depends on acoustic the wave speed and the thickness of the piezoelectric layer-metal electrodes structure and given by, [56]. The sensitivity is given by Sauerbrey's formula [57].

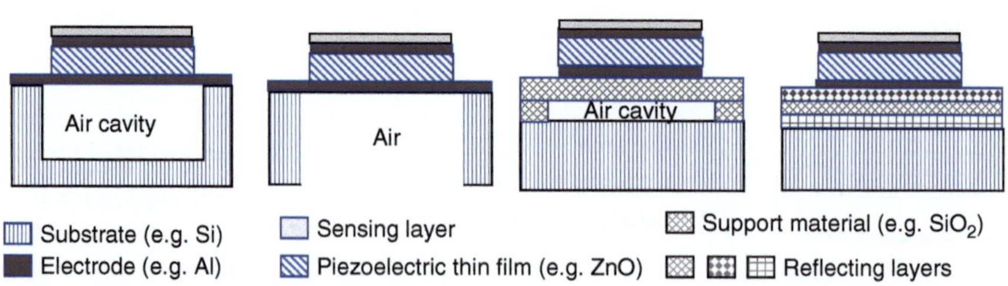

Figure 10.8 The different FBAR device structures: cavity type, hole type, bridge type, and SMR type, respectively.

10.4 Surface Transverse Wave Biosensors

The FBAR devices improve the sensitivity of the QCM by employing a thin piezoelectric film. However, as the frequency increases, their performance is affected by excited lamb modes, which results in spurious resonances and decreased Q-factor [68]. The main problem of BAW resonators is the dependency between operation frequency and piezoelectric film thickness. Surface waves are excited by the electrodes grown on the plane the wave propagates, so the wavelength of the generated acoustic wave is a function of electrode geometry. The fabrication for the SAW devices needs less time and is low cost. The sample is put to the sensitive surface of the SAW device easily, it is not fragile. The surface waves have different modes according to their particle displacement and propagation direction vectors. These modes can be generated on special materials and they have their particular biosensing application (Table 10.1) [69].

In this section, all the biosensor types that employ different surface acoustic wave modes are discussed: (i) The modes that have shear particle displacement (SH-Wave and Lamb Wave) (ii) The modes with out-of-plane particle displacement (Rayleigh mode) and (iii) The modes that propagate on plates (Lamb modes).

10.4.1 SH-Wave and Love Wave Biosensors

A SH mode SAW is a surface wave with only displacement along the plane parallel to the surface of the solid supporting the wave. The displacement is perpendicular to

Table 10.1 Applications, materials, and operation frequencies for different SAW modes.

SAW type	Material	Frequency range	Comments
SH-wave/ love wave	STQuartz (90° Rotated) AT Quartz 36° YX $LiTaO_3$ 64° YX $LiNbO_3$ Love wave must have guiding layer grown on the material: SiO_2, ZnO, PR, PMMA, etc.	100–500 MHz	Low attenuation, proper to detect bioanalytes in liquids or in fluids. Love Wave Mode: Highly Sensitive
Lamb wave	Thin films: ZnO, AlN, PVF membranes, thin-plate structures	Natural materials: <2 MHz, Biosensing: Thin Films: 200 MHz to 2GHz	Low attenuation, Propagates long distances, Liquid sensing applications
Rayleigh wave	ST Quartz 128° YX $LiNbO_3$ 41° YX $LiNbO_3$	2 MHz to 2 GHz, Biosensing: ~100–400 MHz	Highly attenuative, proper to detect biological materials in gas phase, gas sensing

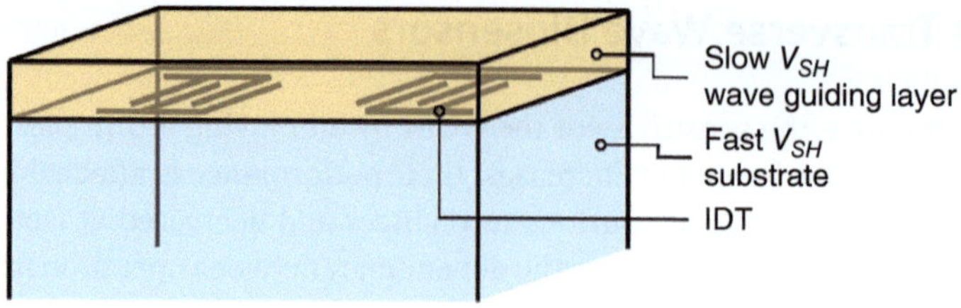

Figure 10.9 Stylistic representation of a basic Love mode device.

the direction of propagation of the wave. It arises from the coupling of the SH-bulk wave and the electric potential in a piezoelectric material [9]. A Love wave is excited when a SH (bulk) wave is excited on the interface with two media of which one medium is pseudo infinitely deep and the second material has finite thickness and has a lower SH wave propagation velocity (Figure 10.9) as was described by Augustus E.H. Love [16]. The wave energy is then majorly trapped in the slow wave velocity material (or waveguide, WG) and only penetrates into the high wave velocity material (or substrate) in the order of magnitude of the wavelength, making it a surface wave mode.

The in-plane displacement of the SH- and Love-mode waves is most commonly applied to sensing applications like viscous loading and chiefly, mass loading. Using a highly selective bioactive layer, the sensor can be made sensitive to anything the bioactive layer is designed to capture due to a change of surface mass upon capture of the target particle. But more generally, the SH- and Love-mode sensors are sensitive to any surface mass density change, making these modes applicable to a wide variety of biosensing applications.

A big advantage of the SH- and Love-mode over other SAW mode sensors, is the suitability of these wave modes for liquid samples. This follows from the limited acoustic coupling to the liquid as a consequence of the lack of any displacement perpendicular to the liquid–solid interface when the SH- or Love-mode wave originates from a pure SH-mode wave. This avoids the wave energy to be radiated into the liquid by mode conversion, limiting the loss mechanism to viscous coupling. High-sensitivity measurements using liquid samples become possible [70].

A further advantage of the Love-mode device is the natural protection of the transducers from the often aggressive sample liquid. The WG layer can be chosen with only a few constraints, so a chemically stable material can be chosen.

In order to excite a SH- or Love-mode wave, any transducer exciting the surface could be applied. A very convenient way, however, is the use of an IDT on a piezoelectric material as originally described by White and Voltmer [71]. The same mechanism can also be used to detect the wave as it has traveled. Using an IDT requires a piezoelectric material to transfer the electrical potential to a mechanical deformation and vice versa.

For Love-mode waves, the material of the WG layer, in combination with its thickness, largely determines the achievable sensitivity and insertion loss. The operating frequency, temperature coefficient of frequency (TCF), and damping are all variable by changing the WG layer thickness.

The operating frequency is between the frequencies following from $f_r = c_s/\lambda$ of the substrate and WG where a thicker WG lowers the frequency. Here, c_S is the shear wave velocity of the medium. The shear wave velocity can be found from the material stiffness matrix and is given by $c_S = \omega/k = (G/\rho)^{1/2}$, with k the wave number, ω the angular frequency, $G = \text{c44}$ the shear wave modulus, equal to the c44 element from the stiffness matrix, and ρ the material density [72]. The bigger the difference between SH-wave velocities of the substrate and WG, the better the wave energy is concentrated at the interface, leading to a better sensitivity for surface changes [73].

The TCF, relating the influence of the ambient temperature to the resonance frequency of the material [18], of the total device, can be tuned as the influence of the WG layer can again be tuned by changing the thickness. By combining a top and bottom layer with opposite TCF, in some cases, it becomes possible to tune to get a nearly zero TCF device [74].

The damping per unit length is largely determined by the WG as the required lower SH-mode wave velocity often comes at the cost of higher tan δ, which is a measure for the conversion of vibrational energy to heat by a material.

Some state of art examples of SH- and Love-mode devices are listed in Table 10.2 to give an indication of the currently often applied materials, applications, and operating frequency. Some state of art examples of SH- and Love-mode devices are listed in Table 10.2 to give an indication of the currently often applied materials, applications, and operating frequency.

Most importantly, there is an optimum top layer thickness where the wave energy is most efficiently captured. At this optimum, the top layer is as thin as possible, so as little as possible wave energy is lost to heat in the top layer, while it is thick enough to still concentrate the wave energy at the interface. When the top layer becomes too thin, the wave energy is not captured at the interface and is allowed to leak into the bulk of the thick material.

This optimum is heavily material and designed operating frequency dependent. It has been determined using many methods among which are coupling of modes theory [80], electric equivalent circuit approaches, finite element models as shown in Figure 10.11, and physical measurements as by Gizeli [74] as shown in Figure 10.10.

Table 10.2 State-of-the-art comparison design parameter comparison.

		State of the Art				
References	**Mode**	**Substrate**	**Waveguide (WG)**	**Biomarker**	***f*0 (MHz)**	**λ (s)**
Alper 2017 [75]	SH	ST-cut Qz	no WG	PSA	16.9	x
Alper 2012 [76]	SH	ST-cut Qz	no WG	Bcl-2	16.8	300
Raimbault 2007 [77]	Love	AT-cut Qz	SiO_2	Viscosity	117.2	40
Johansson 2012 [78]	Love	$LiNbO_3$	PDMS	Flow	40	100
Gizeli 1992 [70]	Love	Y-cut Qz	PMMA	No biomarker	110	45
Tran 2014 [79]	SH	AlN	SiN	No biomarker	211.25	24

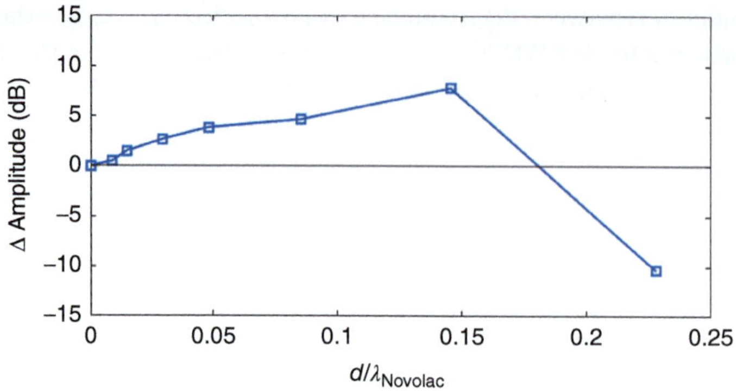

Figure 10.10 Measured change of signal transfer amplitude on a quartz device with a Novolac overlayer as performed by Gizeli et. al. [80]. The Novolac thickness (d) is normalized to the acoustic wavelength in Novolac ($\lambda_{Novolac}$). A clear optimum thickness is apparent where the signal transfer is largest.

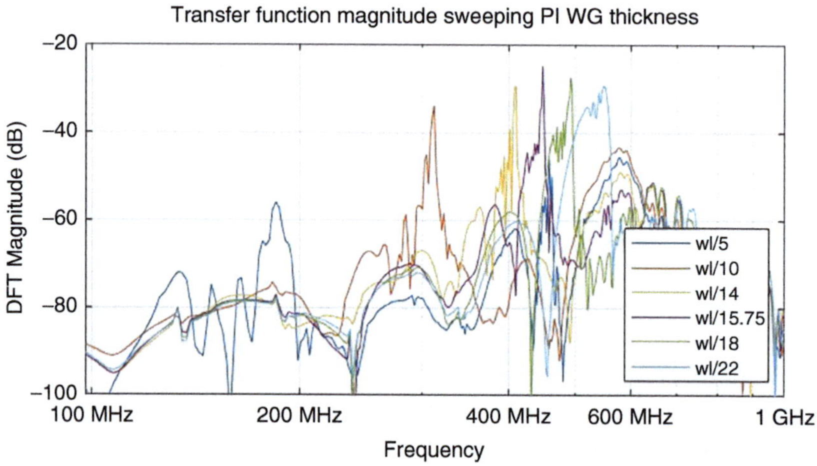

Figure 10.11 Simulated effect of the top layer thickness on the insertion loss and resonant frequency revealing an optimum thickness where the insertion loss is lowest indicating the lowest power loss. This is a COMSOL simulation using quartz and polyimide.

Materials suited and often reported to be used for SH-mode and Love-mode devices are quartz, lithium tantalate ($LiTaO_3$), and lithium niobate ($LiNbO_3$) as these are all capable of supporting SH-mode waves in certain crystal cuts. For example for quartz, several crystal orientations are possible, referred to as crystal cuts. Examples are stable temperature cut quartz (ST-cut Qz), AT-cut, and *Y*-cut quartz.

For the top layer, a lower SH-mode wave velocity than the bottom material is required. It is possible to achieve this with materials such as silicon dioxide (SiO_2) and silicon nitrite (SiN) [21, 23]. Certain metals are suited. But often polymers are applied, such as, but not limited to, polymethyl metacrylate (PMMA), polyimide (PI), and polydimethylsyloxane (PDMS). Polymers generally support much lower SH-wave velocities [16, 22].

10.4.2 Lamb Waves Biosensors

The basic principle of Lamb wave sensors is loading the surface to modulate the speed of the traveling wave by the mass change. In order to make the sensor, we need to be able to generate and detect the wave. When the plate/membrane is made of a piezoelectric material the wave can be generated by interdigitated electrode, with the same structure being used to detect the wave. A range of piezoelectric materials are suitable for this application. Materials include PVF2, PVDF, ZnO, AlN, PZT, GaN, and GaAs.

In 1980, Lamb waves were shown in stretched PVF2 films. Here, the interdigitated transducers were used to generate and detect the Lamb waves. This basic structure is shown in Figure 10.12. Another commonly used structure is a membrane on silicon. This structure is shown in Figure 10.13a. The IDTs and the piezoelectric layer are positioned on a silicon nitride membrane. A thin layer of silicon for the membrane is also often used. The advantage of this configuration is that the biological/chemical environment can be kept on the underside of the membrane, meaning that the electrodes are completely protected. A similar structure, using SOI wafers and AlN piezoelectric layer can be found in [84].

An example of an integrated device can be found in [85, 86]. This device was integrated with a bipolar process where the epi-layer formed the membrane. The piezoelectric layer was ZnO which was sputtered onto the wafer as a post-processing step. The backside of the membrane was exposed to the environment. This was designed to measure liquid properties and did not require any additional coating. The denser the fluid in contact with the membrane, the slower the acoustic wave. The same structure can, with a functionalized surface, be used for a range of biosensors. Such a structure is shown in Figure 10.13b. In this case, an Au/Cr layer is deposited on the underside of the membrane and this is then covered with the biosensitive layer.

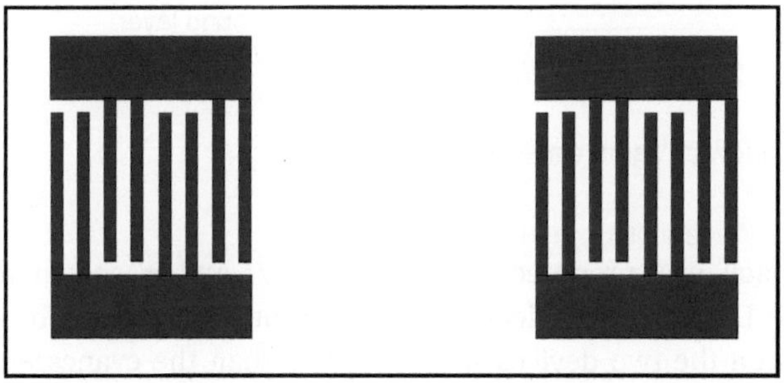

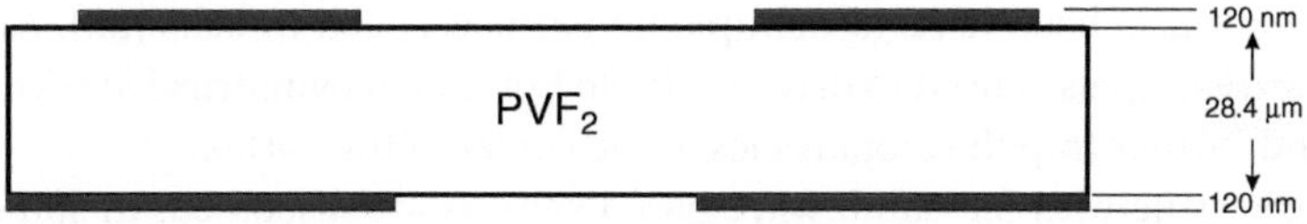

Figure 10.12 The structure where the whole thickness is the piezoelectric layer is illustrated Lamb device is using PVF_2. Source: Adapted from Wagers [81].

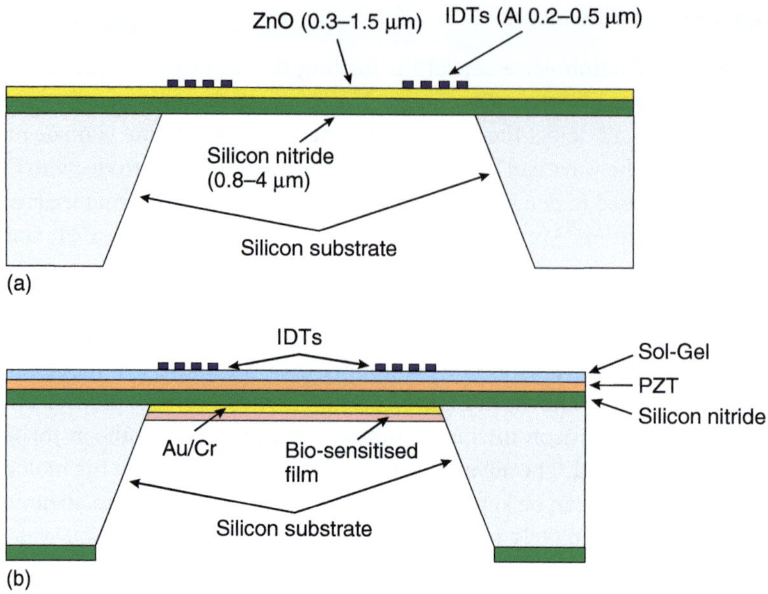

Figure 10.13 (a) Cross section of a Lamb wave device on a silicon nitride membrane. Source: Adapted from Rana et al. [82]. (b) Cross section of Lamb wave device with biosensitive coating on the membrane. Source: Adapted from Yu and Tsai [83].

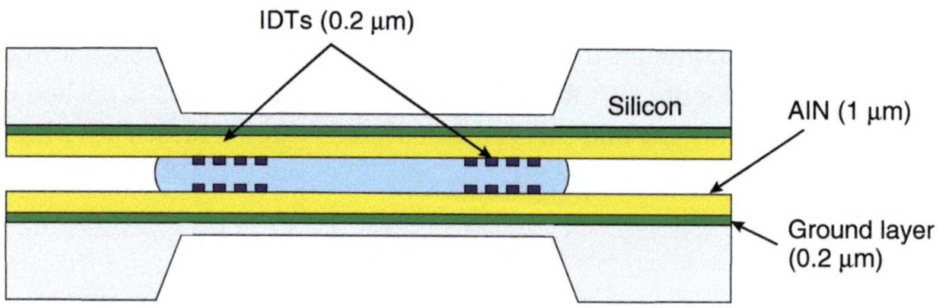

Figure 10.14 Setup of the sandwich Lamb wave device.

A quite different approach was proposed by Li et. al. [87] and is shown in Figure 10.14. This has two Lamb wave devices close to each other, separated by a liquid. The distance between the two devices should be less than the evanescent field penetration depth in the liquid. The two devices interact with each other through the evanescent wave.

Lamb waves have also been used to manipulate particles, such as cells [88]. An array of Lamb wave devices is used to drive multiple horizontal cylindrical vortices in a 1 μl drop and further trap the bioparticles at the center of the vortex.

So far, the basic structures for Lamb wave-based sensors are discussed. In some cases, sufficient information can be extracted without additional layers to the basic structure. Additional layers can be used to trap the cells/bacteria etc. and in all cases,

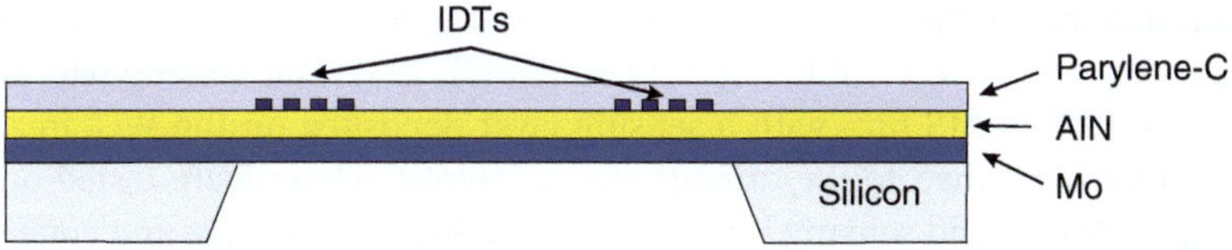

Figure 10.15 Whole blood measurement device using Parylene-C coating layer. Source: Adapted from Kong et al. [89].

the trapped objects cause a slowing of the wave, which can be converted into a frequency shift to quantify. Kong et. al. used Parylene-C as a WG layer on a AlN piezoelectric layer for whole blood measurements. The structure is shown in Figure 10.15. The device was designed to measure the active partial thromboplastic time (aPTT) with enhanced sensitivity. The phase velocity of the Lamb waves within the evanescent wave field is the same as the phase velocity inside the membrane. Blood coagulation is detected by the evanescent wave penetration depth, which finally leads to frequency shifts. This approach was shown to be stable and reliable.

Gao et. al. used Lamb wave devices to detect biomolecules in water [90]. This work used high aspect ratio electrodes to slow down the Lamb waves to lower than the sound of speed in water. This enables high-quality-factor resonance in water and reduces the sensing frequency noise. In this structure, the bioreceptors are placed on the top of the high aspect ratio electrodes. This, along with the confined acoustic energy, is shown in Figure 10.16.

Wei et. al. used a structure similar to Figure 10.15, without the Parylene-C [91]. In this case, the underside of the membrane is functionalized to detect pepsinogen 1 and 2. The SAW device was coated with a polyethylene layer. This PE-Lamb device had a limit of detection (LOD) were 60 pg/ml of PG1 and 30 pg/ml of PG2.

Lemanna et. al. developed an aluminum nitride-based Lamb wave device fabricated on recyclable polyethylene naphthalate [92]. This was applied to the detection of *Escherichia coli* in the food industry. The LOD was 6.54×10^5 CFU/ml, with an estimated mass of a single *E. coli* cell of 6.16×10^{-13} g, which was better than comparable devices in silicon.

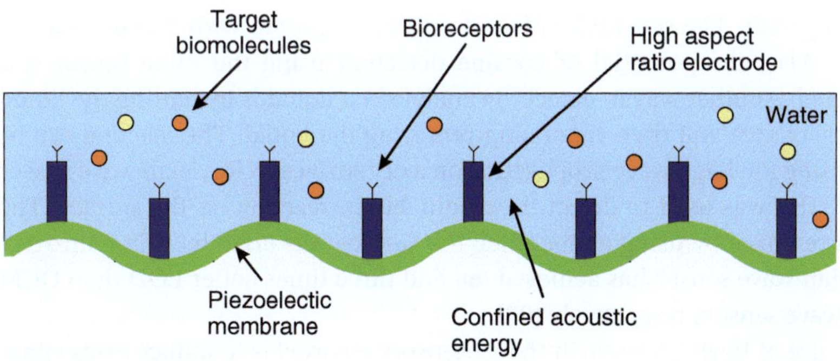

Figure 10.16 High aspect ratio electrodes for Lamb wave devices. Source: Adapted from Gao et al. [90].

Sappati and Bhadra used printed electronics technology to develop a Lamb wave device for low-concentration volatile organic compounds [93]. The sensing part of the device was a PZT-PDMS film with a sensing layer deposited on the WG area. This sensing layer was tuned to the organic gas of interest. These were tested on acetic acid vapor. Acetic acid sensing is vital in wine industry, chocolate and coffee processing, defining the final quality. For concentrations above 85 ppb it can be used for breath analysis. They achieved a detection limit of 10.0 ppm for acetic acid and 0.03 for toluene. The sensitivities were 0.65 kHz/ppm and 150 kHz/ppm for acetic acid and toluene, respectively.

Lamb wave devices represent an effective technique for biosensing. In some cases, this is simply detecting the density or viscosity of a liquid. In some bio-fluids, this is sufficient to determine the concentrations. In other cases, bioreceptors or surface functionalization are used to trap the cells and then load the membrane/WG thus slowing the traveling Lamb wave for quantification of molecule concentration.

10.4.3 Rayleigh Wave Biosensors

The particles move in longitudinal and transverse directions in Rayleigh wave and this displacement results in particle oscillation in an elliptic orbit. The out-of-plane particle displacement reduces the sensing performance for liquid samples [94]. Therefore, the application area for the Rayleigh mode biosensors narrows to samples in gas/vaporized form. Despite this drawback, in some studies, the detection capability in liquid samples was demonstrated. A third application area of Rayleigh wave mode is the enhancement of biosensor performance by inducing extra vibrations.

Gas sensing by using Rayleigh wave biosensors provides high sensitivity compared to BAW-based devices. Since the energy is confined close to surface, in a thin layer for the Rayleigh waves, it provides high sensitivity to surface perturbations [95]. The delay-line area is coated by a sensitive layer (e.g. polymer), which creates perturbations on the surface and it affects the velocity of the propagating Rayleigh wave (Figure 10.17). A vapor-phase detector is reported by [96], which demonstrates a 250 MHz-Rayleigh wave sensor. In this study, brief pulses of analyte vapor were applied to sensitive regions and the analyte concentration was quantified by frequency shift. The sensitivity of 20 Hz/pg was reported with a LOD of a few picograms. Also, the potential of cocaine detection using the same sensor was demonstrated. Another way to detect the analyte in a liquid is by putting the liquid on the sensing area and then vaporizing/removing the liquid. The reaction can be observed using Rayleigh wave propagating on a dry surface. A Rayleigh wave sensor with 1.285 GHz was used to detect the avidin–biotin reaction on the surface. The liquid was removed by using a pipet then the surface was dried by using nitrogen. The Rayleigh wave sensor has achieved ten and three times better LOD than QCM and Love wave sensors respectively [97].

Another use of Rayleigh wave in the biosensors reported is to induce extra vibrations in order to provide better detection [98, 99]. The nonspecific protein binding reduces the measurement performance in biosensors. The actuation is an important

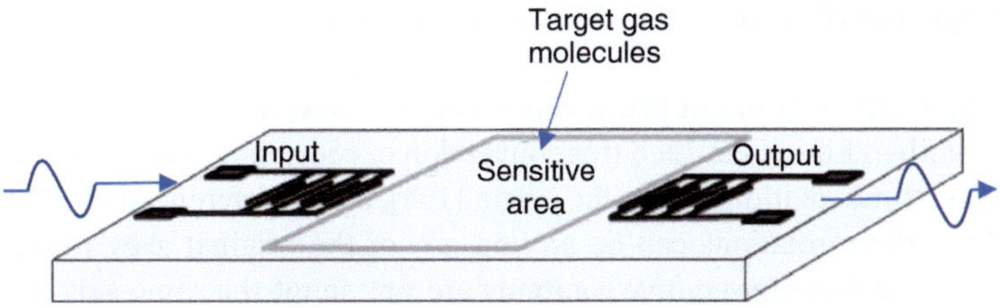

Figure 10.17 A Rayleigh Wave biosensor structure is illustrated.

application area of Rayleigh wave devices and the biosensitive film can be vibrated by using Rayleigh waves to efficiently remove the nonspecifically bounded proteins from the film [98]. The Rayleigh wave was not sensing in this application but it was enhancing the sensor performance. Another study employs two vertically located (to each other) IDT electrode pairs that are generating SH-Wave and Rayleigh-wave modes [99]. The improvement in measurement performance was demonstrated by using the extra vibrations induced by Rayleigh waves.

The sensitivity of Rayleigh wave sensor is directly proportional to operation frequency and inversely proportional to sensing area (10.8) and (10.9).

$$\frac{\Delta c}{c} = -c_m f_0^2 \frac{\Delta m}{A} \tag{10.8}$$

$$f_0 = \frac{c}{\lambda} \tag{10.9}$$

Here, c_m is the mass sensitivity coefficient, f_0 is the center frequency that depends on the wavelength (λ), c is the Rayleigh wave speed, Δm is the change in the mass on the sensitive region, which has the area A and Δc is the change in wave speed during the sensor operation [100].

10.4.4 Crystal Cuts and Axis Orientation

Crystalline materials with piezoelectric properties are anisotropic in nature. Asymmetry in the crystal structure causes polarization when the material is deformed under stress and reciprocally the same is true, a strain is generated when the material experiences an electric field [101]. This piezoelectric effect is dependent on the orientation of the crystalline structure of the material. When a crystalline material is cut into a workable size (for example, when making a wafer), the orientation of the cut can be chosen. As the material is anisotropic, a different cut can have different material properties. When considering piezoelectric materials, the magnitude and direction of the piezoelectric effect varies with cut orientation.

Specifically relevant for SAWs, the TCF is also dependent on cut orientation [102]. These effects cause that for a specific application, a specific cut is beneficial. For example, for quartz, many cut orientations have been investigated and some cuts have proved popular for SAW applications.

The definition of the crystal cut is historically determined by the crystallographic structure of the material and what reference points can be obtained from the crystal.

Even for a single material, multiple definitions can exist as is the case with quartz [103].

For several applications, it might prove convenient to convert one linear coordinate system to another. Euler identified this conversion of coordinate systems can be achieved with just three rotations about the origin [104], as was referenced by Whittaker [105]. These three rotations can be around any of the original axes and the only constraint is that two consecutive rotations are not about the same axis (two possibilities are a X–Y–Z or a Z–X–Z rotation). This leads to 12 ($3 \times 2 \times 2$) possible conversion options that are equivalent but not equal [106]. A commonly used Euler rotation is Z–X–Z, which happens to also be applied in the rotated system function of the finite element method (FEM) computational software product COMSOL.

In this Z–X–Z system, first a rotation about the Z-axis in the XY plane is done. Secondly, a rotation about the new X-axis is performed. Finally, a rotation about the new Z-axis is done. These steps are illustrated in Figure 10.17. To make clear the rotations are about the new axis every time, the Z–X–Z transformation can also be represented by z–x'–z''.

Common symbols for the Euler angles are φ, θ, and ψ. Realize that these symbols do not reveal what Euler rotation set is used. Another set of symbols sometimes used is α, β, γ as used in COMSOL and Figure 10.18. Specifically for SAW applications, the base orientation of the material of interest has to be rotated to its final orientation in the application.

The cut angle reasoned from the point of view of the crystal cutter is generally not coinciding with the Euler angle system convenient for the engineer [103, 107]. This implies caution is required when doing calculations or simulations with a piezoelectric material and the conversion from the material-specific cut angles to Euler angles must be executed with care.

A practical example is the often-applied piezoelectric material quartz. It has a crystalline structure that is built up from unit cells consisting of three SiO_2 molecules that are forming a helix [108]. This helix can be right-hand or left-hand rotating [109, 110]. These, both naturally occurring, variants are referred to as RH-quartz

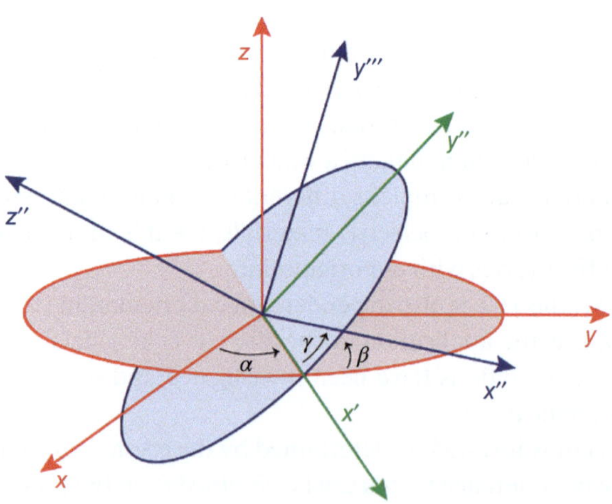

Figure 10.18 Example of the application of Euler angles Z-X-Z sequence by rotating over α (rotate over the xy-plane), β (tilt the xy-plane), γ (rotate over the new xy-plane) respectively.

and LH-quartz, respectively. Then the cuts made in this crystal follow two conventions, the Institute of Radio Engineers 1949 (IRE 1949) [111] and the newer Institute of Electrical and Electronics Engineers 1978 (IEEE 1978) [112]. An example is the ST-cut quartz. This is, for RH-IEEE 1978 quartz, a Y-cut (XZ-cut plane) where the cut plane is 42.75° rotated over the X-axis. To convert this to Z–X–Z Euler angles, no initial rotation over Z is required, then 90° rotation over the Z-axis is required to get a XZ-cut plane and an additional 42.75°Z rotation is required to get to the ST-cut. And no final rotation over Z is done. This leads to Euler angles [0;90+42.75;0] [107].

While for another often-applied material, lithium niobate ($LiNbO_3$), the cut angles are again defined differently. The crystal is oriented along the Cartesian axes according to the IRE 1949 convention [111]. An example of a cut is then the X-cut $LiNbO_3$, which implies the crystal is cut perpendicular to the X-axis (the same can be applied to the Y- and Z-axis). A second naming option for more freedom in cut definition is, for example, XZ 30°-$LiNbO_3$. Here, the cut plane, perpendicular to the first axis (X implies the YZ-plane) is rotated counter clockwise around the second axis (in this case Z). In this notation also any other combination of axes is possible like the XY-, YX-, YZ-, ZX-, and ZY-cut [113]. With this knowledge, the Euler angles can be derived.

10.5 Commercial Biosensors and Trends

Acoustic wave is used for many years in different sectors such as sensors, automotive, aerospace, defense, consumer electronics, healthcare, industrial, and other end-user industries because of its small size, low production cost, rugged nature, high stability, and applications across various end-user industries. The improvements in this field make the market to be expected to reach as high as 5.62 billion by 2026 [114]. In particular, acoustic wave-based biosensors show a great interest in the market promising due to their advantages compared to traditional laboratory techniques. Acoustic wave biosensors are small in size which brings the ability for portable applications, using small sample volumes, and low power consumption. These chrematistics makes acoustic biosensors to be capable of being integrated into microfluidic platforms, and lab-on-a-chip systems. These sensors can also be integrated to work with mobile and smart devices which brings the opportunity to transfer to wearable or in situ device applications [115]. In this chapter, we discuss the potential commercial applications of acoustic wave-based biosensors, their trends, and challenges.

The acoustic wave biosensors mostly have two main parts as the control unit and sensor chips. The control unit is responsible for generating the RF signal and reading the feedback response from the sensor and comparing these two signals.

The Qorvo® Biotechnologies company (Minnesota, US, www.qorvo.com) has developed a POC device with the name of TRUFORMA. This device has a control unit that is as small as a printer (Dimensions 8″X 9″X 4.5″) and uses a disposable test chip (Figure 10.19).

The device shows the result on its touchscreen interface and can also transfer data via wire, wireless, and cellular data connectivity. The sensor chip part is designed so

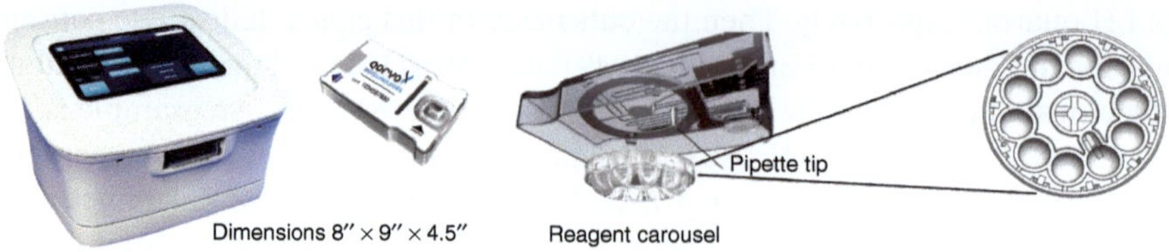

Figure 10.19 The Qorvo® Biotechnologies TRUFORMA product. Source: Qorvo Biotechnologies [116]/Future Science Group.

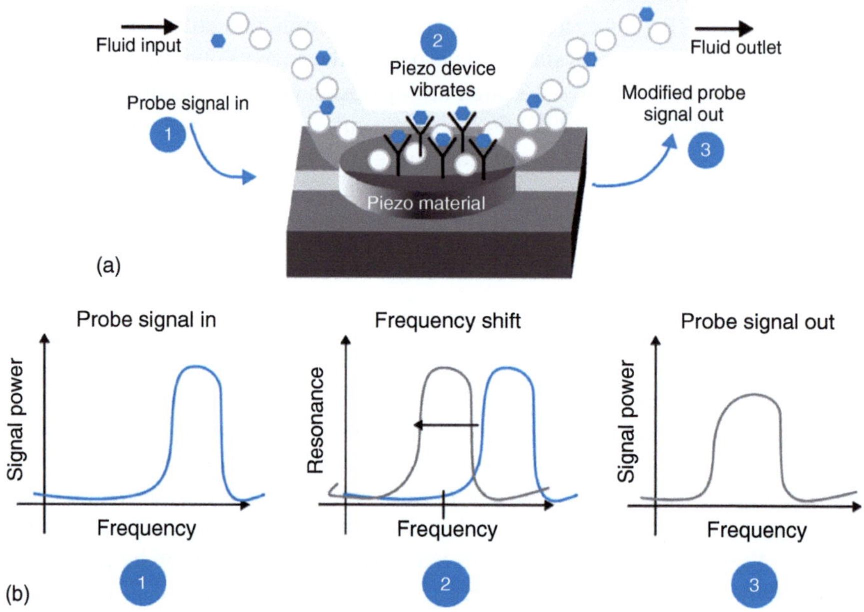

Figure 10.20 Principle of working of the device: (a) the schematic of fluid flow and binding process, and signalization. (b) Frequency shift due to binding process. Source: Fischer et al. [117]/With permission of Future Science Ltd.

as to have great flexibility in assay design as dilution steps, mixing, preincubation and multistep reactions, etc. This device uses the BAW to measure the frequency shift due to the binding process (Figure 10.20) [117]. The fluid with the sample test inside is flowing over a piezoelectric material which is actuated by a radio frequency (RF) signal at the designed resonance frequency. The readout probe reads the output signal from piezoelectric substrate. As the target molecules or biological components bind to the surface of the piezoelectric substrate, the mass of the fluid in study area of the sensor changes. This change in mass would result in a shifting of the provided RF signal. By measuring this signal, the device would report the presence of the target molecules or target biological components (Figure 10.20).

The device is designed as to be able to do molecular and immunoassay tests at the same time on one testing chip. The acoustic wave technology brought the possibility of working with multiple matrix types including whole blood, serum, plasma, saliva, and more. This device proved to be a good alternative to the lab-based test technology for patient care [117].

The AWSensors (Valencia, Spain, www.awsensors.com) has developed the X4 Advanced Multichannel QCMD system (Figure 10.21) which improved the possibility of simultaneous measurement of sensors up to four chips. The device has a main control unit which is responsible for actuating the sensor chip and reading the feedback signal and controlling the temperature of the sensors. The flow over the sensor is controlled by an additional flow control unit, however, the number of flow control can be increased if the flow over a sensor wants to be different than others.

AWSensors Inc. provides three types of sensors to widen the required range of sensor working frequencies. The classical quartz crystal microbalance (QCM) sensors provide working frequency range of 5–10 MHz in two sizes of 1″ and 14 mm diameters. These sensors are coated by Ti/Au, Cr/Au, and SiO_2 as a standard coating to provide higher sensing sensitivity (Figure 10.22b). The high fundamental frequency quartz crystal microbalance HFF QCM sensors are based on ATcut quartz crystal resonators manufactured through inverted mesa technology providing higher frequency ranges as high as 50–150 MHz (Figure 10.22a). The company provides also a Love-mode SAW sensor at 120 MHz to facilitate the easy connection and removing of the sensor (Figure 10.22c).

The working frequency and quality factor of these sensors are very sensitive to the sensor environment. The environment change of the sensor is the result of the mass change of a thin layer or shape of particles adsorbed at the surface of the resonator, or viscosity of the surrounding liquid. These specifications offer a promising technology platform for the development of sensitive, portable, and real-time biosensors in a wide range of applications. These acoustic wave biosensors can be used in nucleic acid systems, lipid systems, protein and immunological systems, polymer systems, and living cell systems.

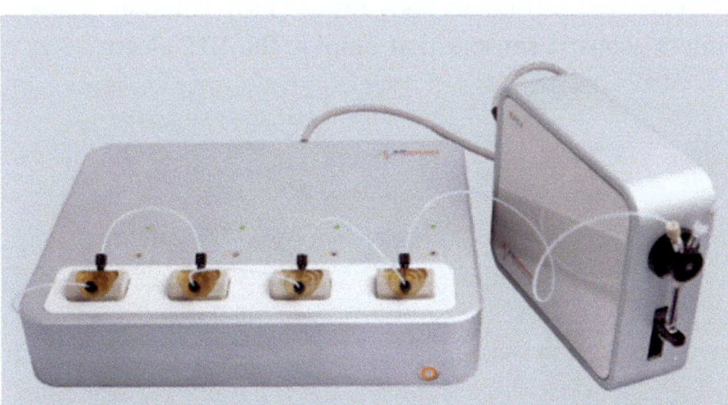

Figure 10.21 The AWSensors X4 Advanced Multichannel QCMD system. Source: AWSensors.

(a) (b) (c)

Figure 10.22 Sensor types provided by AWsensors Inc. (a) HFF QCM, (b) QCM, (c) LOVE-SAW. Source: AWSensors.

Although the experimental results of these devices are not available for all potential applications of acoustic wave biosensors, however, we can study these potential applications through the experimental studies which were done about using acoustic wave biosensors for different applications in the literature. Besides using these sensors as a specific cell isolator, the main difference between these biosensors is the biomarkers used on the surface of these biosensors. The biomarkers used in biosensors to detect the biological components or molecules can be categorized as; antibody–antigen biomarkers, protein biomarkers, genetic material biomarkers, and small molecule biomarkers. The leading motivations to use the biomarkers are the detection of a specific molecule in a small amount of body fluid with high sensitivity. Also, these sensors have the potential of being used in POC applications, which makes them attractive [118]. The application area is so wide including the detection of heart disease, many types of cancer, and many viral infections[118–144]. The detection of biomarkers/biomolecules by using acoustic wave biosensors have remarkable potential to be used in many diagnosis application in near future. The fast response time, high sensitivity, low cost, and ease of use are the key properties making them an important alternative and to put new application areas as well.

New fabrication techniques enable flexible biosensors fabricated on new substrates (Figure 10.23) [18, 145]. These, in turn, could enable lab-on-chips and wearable sensors, which can process human body fluids [146]. Another emerging application field is wireless acoustic sensors that employ the IDT electrode as an antenna [147].

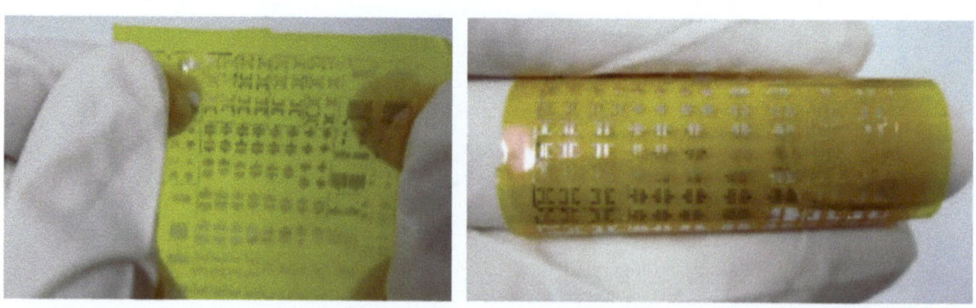

Figure 10.23 Wearable SAW sensor fabricated on a flexible substrate. Source: Jin et al. [145], Springer Nature.

10.6 Conclusion

In this chapter, we provided a detailed view of the application of acoustic waves in the field of biosensing. The acoustic wave modes were described using the biosensing point of view. Two major categories of acoustic wave biosensors, namely BAW and SAW, have been discussed in terms of their structure, material, working principle, and applications. Also, a comprehensive background about biochemistry principles was given to enable better understanding of the operation. The acoustic biosensor types are compared by using their figure of merits and applications. A comparison of the discussed acoustic biosensor branches can be seen in Table 10.3.

Acoustic wave biosensors have extensive application areas in multiple fields (in application perspective), such as clinical and medical research, the early diagnosis of deadly and infectious diseases, and environmental and food safety-related applications. A brief view of the commercially available biosensors was given in the last section. Moreover, the label-free detection, noninvasive application, and biocompatibility properties of these devices enable many challenging applications. The future perspective for the new and emerging application areas has been discussed like monitoring of cell adhesion and spreading, the detection and quantification of specific proteins and other biomolecules.

The advances in microfabrication technology and its reflections on acoustics enable new applications in healthcare and bioengineering. Developments in

Table 10.3 Advantage/disadvantage table for QCM, FBAR and SAW devices.

QCM	FBAR	SAW
Easy Fabrication	Relatively difficult	Easy Fabrication
LOD (Limit of Detection): μg/ml	LOD: femtograms to picograms	LOD: ng/ml to μg/ml
Operation freq. 5 to 50 MHz	Operation freq. sub GHZ to tens of GHz	Operation freq. a few MHz to a few GHz
Susceptibility to noises: good performance	Susceptibility to noises: poor performance	Susceptibility to noises: good performance
Poor sensitivity	Superior sensitivity	Good sensitivity
Commercialized (more than a decade)	—	Commercial products exist
—	CMOS process compatible	Partially CMOS compatibility
Not fragile	Fragile (only for suspended designs)	Not fragile (except lamb wave devices)
Temperature frequency Dependency (TCF)	Temperature frequency Dependency (TCF)	Temperature frequency Dependency (TCF)
Superior (AT-Cut Quartz)	Compensation required	Good (Depends on substrate used)

acoustic biosensors result in considerable advantages over traditional bioassays. The benefits like fast responsiveness, small size, and high sensitivity made BAW/SAW devices promising to show great in developing portable, low-cost biosensors.

References

1 Wijesuriya, D.C. and Rechnitz, G.A. (1993). Biosensors based on plant and animal tissues. *Biosensors and Bioelectronics* 8 (3–4): 155–160.

2 Damborský, P., Švitel, J., and Katrlík, J. (2016). Optical biosensors. *Essays in Biochemistry* 60 (1): 91–100. https://doi.org/10.1042/EBC20150010.

3 Ding, J. and Qin, W. (2020). Recent advances in potentiometric biosensors. *TrAC Trends in Analytical Chemistry* 12 (1): https://doi.org/10.1016/j.trac.2019.115803.

4 Tamayo, J., Kosaka, P.M., Ruz, J.J. et al. (2013). Biosensors based on nanomechanical systems. *Chemical Society Reviews* 42 (3): 1287–1311. https://doi.org/10.1039/C2CS35293A.

5 Länge, K., Rapp, B., and Rapp, M. (2008). Surface acoustic wave biosensors: a review. *Analytical and Bioanalytical Chemistry* 391 (1): 1509–1519. https://doi.org/10.1007/s00216-008-1911-5.

6 Carrascosa, L.G., Moreno, M., Álvarez, M., and Lechuga, L.M. (2006). Nanomechanical biosensors: a new sensing tool. *TrAC Trends in Analytical Chemistry* 25 (3): 196–206. https://doi.org/10.1016/j.trac.2005.09.006.

7 Nayfeh, A.H. (1995). Bulk waves, Chapter 3. In: *Wave Propagation In Layered Anisotropic Media*, North-Holland Series in Applied Mathematics and Mechanics, vol. 39, 31–59. Elsevier https://doi.org/10.1016/S0167-5931(06)80044-6.

8 Länge, K. (2022). Bulk and surface acoustic wave biosensors for milk analysis. *Biosensors* 12 (8): https://doi.org/10.3390/bios12080602.

9 Hashimoto, K.-Y. and Hashimoto, K.-Y. (2000). Surface acoustic wave devices in telecommunications, Chapter 1. In: *Bulk Acoustic and Surface Acoustic Waves*, vol. 116, 1–23. Springer.

10 Krylov, V. (2018). Resonant properties of finite cracks and their acoustic emission spectra, arXiv preprint.

11 Andreykiv, O.Y., Lysak, M.V., Serhiyenko, M., and Skalsky, V.R. (2001). Analysis of acoustic emission caused by internal cracks. *Engineering Fracture Mechanics* 68 (11): 1317–1333.

12 Hadj-Larbi, F. and Serhane, R. (2019). Sezawa saw devices: review of numerical-experimental studies and recent applications. *Sensors and Actuators A: Physical* 292: 169–197.

13 Bleustein, J.L. (1968). A new surface wave in piezoelectric materials. *Applied Physics Letters* 13 (12): 412–413.

14 Yen, K., Lau, K., and Kagiwada, R. (1979). Recent advances in shallow bulk acoustic wave devices. In: *1979 Ultrasonics Symposium*, 776–785. IEEE.

15 Auld, B.A., Gagnepain, J.J., and Tan, M. (1976). Horizontal shear surface waves on corrugated surfaces. *Electronics Letters* 12: 650–651.

16 Love, A. (1911). *Some Problems of Geodynamics*. Cambridge: Cambridge University Press.
17 Kadota, M., Ogami, T., Yamamoto, K. et al. (2010). High-frequency lamb wave device composed of MEMS structure using $LiNbO_3$ thin film and air gap. *IEEE Transactions on Ultrasonics, Ferroelectrics, and Frequency Control* 57 (11): 2564–2571. https://doi.org/10.1109/TUFFc.2010.1722.
18 Fu, Y.Q., Luo, J., Nguyen, N.-T. et al. (2017). Advances in piezoelectric thin films for acoustic biosensors, acoustofluidics and lab-on-chip applications. *Progress in Materials Science* 89: 31–91.
19 D'souza, S.F. (2001a). Immobilization and stabilization of biomaterials for biosensor applications. *Applied Biochemistry and Biotechnology* 96 (1): 225–238.
20 Naresh, V. and Lee, N. (2021). A review on biosensors and recent development of nanostructured materials-enabled biosensors. *Sensors (Basel)* 21 (4): 1109. https://doi.org/10.3390/s21041109. PMID: 33562639; PMCID: PMC7915135.
21 Nguyen, H.H., Lee, S.H., Lee, U.J. et al. (2019). Immobilized enzymes in biosensor applications. *Materials* 12 (1): 121.
22 Rodriguez-Mozaz, S., Marco, M.-P., Lopez de Alda, M.J., and Barceló, D. (2004). Biosensors for environmental monitoring of endocrine disruptors: a review article. *Analytical and Bioanalytical Chemistry* 378 (3): 588–598.
23 Trilling, A.K., Beekwilder, J., and Zuilhof, H. (2013). Antibody orientation on biosensor surfaces: a minireview. *Analyst* 138 (6): 1619–1627.
24 Gruhl, F.J. and Länge, K. (2012). Surface modification of an acoustic biosensor allowing the detection of low concentrations of cancer markers. *Analytical Biochemistry* 420 (2): 188–190.
25 Rusmini, F., Zhong, Z., and Feijen, J. (2007). Protein immobilization strategies for protein biochips. *Biomacromolecules* 8 (6): 1775–1789.
26 Bhardwaj, D., Ansari, M.W., Sahoo, R.K., and Tuteja, N. (2014). Biofertilizers function as key player in sustainable agriculture by improving soil fertility, plant tolerance and crop productivity. *Microbial Cell Factories* 3: 66.
27 Länge, K., Gruhl, F.J., and Rapp, M. (2013). Surface acoustic wave (SAW) biosensors: coupling of sensing layers and measurement. In: *Microfluidic Diagnostics* (ed. G. Jenkins and C.D. Mansfield), 491–505. Springer.
28 Länge, K. and Rapp, M. (2008). Influence of intermediate aminodextran layers on the signal response of surface acoustic wave biosensors. *Analytical Biochemistry* 377 (2): 170–175.
29 Bhalla, N., Chiang, H.-J., and Shen, A.Q. (2018). Cell biology at the interface of nanobiosensors and microfluidics. In: *Methods in Cell Biology*, vol. 148 (ed. D.A. Fletcher, J. Doh, and M. Piel), 203–227. Elsevier.
30 Dugas, V., Elaissari, A., and Chevalier, Y. (2010). Surface sensitization techniques and recognition receptors immobilization on biosensors and microarrays. In: *Recognition Receptors in Biosensors* (ed. M. Zourob), 47–134. New York: Springer.
31 Gao, S., Guisán, J.M., and Rocha-Martin, J. (2022). Oriented immobilization of antibodies onto sensing platforms - a critical review. In: *Analytica Chimica*

Acta, vol. 1189 (ed. N.W. Barnett, W. Buchberger, L. Buydens, et al.), 1–24. Elsevier.

32 North, S.H., Lock, E.H., Cooper, C.J. et al. (2010). Plasma-based surface modification of polystyrene microtiter plates for covalent immobilization of biomolecules. *ACS Applied Materials & Interfaces* 2 (10): 2884–2891.

33 Welch, N.G., Scoble, J.A., Muir, B.W., and Pigram, P.J. (2017). Orientation and characterization of immobilized antibodies for improved immunoassays (Review). *Biointerphases* 12 (2): 02D301.

34 Pâslaru, E., Baican, M.C., Hitruc, E.G. et al. (2014). Immunoglobulin G immobilization on PVDF surface. *Colloids and Surfaces B: Biointerfaces* 115: 139–149.

35 Chan, C.-M., Ko, T.-M., and Hiraoka, H. (1996). Polymer surface modification by plasmas and photons. In: *Surface Science Reports*, vol. 24, 1–2, 1–54.

36 Wang, C. and Feng, B. (2015). Research progress on site-oriented and three-dimensional immobilization of protein. *Molecular Biology* 49 (1): 1–20.

37 Hasan, A. and Pandey, L.M. (2015). Polymers, surface-modified polymers, and self assembled monolayers as surface-modifying agents for biomaterials. *Polymer-Plastics Technology and Engineering* 54 (13): 1358–1378.

38 Acero Sánchez, J.L., Fragoso, A., Joda, H. et al. (2016). Site-directed introduction of disulfide groups on antibodies for highly sensitive immunosensors. *Analytical and Bioanalytical Chemistry* 408 (19): 5337–5346.

39 Smith, C., Haute, M., and Rose, D. (2020). Processing has differential effects on microbiota-accessible carbohydrates in whole grains during ın vitro fermentation. *Applied and Environmental Microbiology* 86: https://doi.org/10.1128/AEM.01705-20.

40 Serafim, B.M., Leitolis, A., Crestani, S. et al. (2016). Electrospinning induced surface activation (EISA) of highly porous PMMA microfiber mats for HIV diagnosis. *Journal of Materials Chemistry B* 4: 6004–6011.

41 Goddard, J.M. and Hotchkiss, J.H. (2007). Polymer surface modification for the attachment of bioactive compounds. *Progress in Polymer Science* 32 (7): 698–725.

42 Ionescu, R.E. (2022). Use of cysteamine and glutaraldehyde chemicals for robust functionalization of substrates with protein biomarkers-an overview on the construction of biosensors with different transductions. *Biosensors (Basel)* 12 (8): 581. https://doi.org/10.3390/bios12080581.

43 Fogel, R., Limson, J., and Seshia, A.A. (2016). Acoustic biosensors. *Essays in Biochemistry* 60 (1): 101–110.

44 Cho, I.-H., Paek, E.-H., Lee, H. et al. (2007). Site-directed biotinylation of antibodies for controlled immobilization on solid surfaces. *Analytical Biochemistry* 365 (1): 14–23.

45 Wacker, R., Schröder, H., and Niemeyer, C.M. (2004). Performance of antibody microarrays fabricated by either DNA-directed immobilization, direct spotting, or streptavidin–biotin attachment: a comparative study. *Analytical Biochemistry* 330 (2): 281–287.

46 Bano, F., Fruk, L., Sanavio, B. et al. (2009). Toward multiprotein nanoarrays using nanografting and DNA directed immobilization of proteins. *Nano Letters* 9 (7): 2614–2618.

47 Yang, Z., Kasprzyk-Hordern, B., Goggins, S. et al. (2015). A novel immobilization strategy for electrochemical detection of cancer biomarkers: DNA-directed immobilization of aptamer sensors for sensitive detection of prostate specific antigens. *Analyst* 140 (8): 2628–2633.

48 Crivianu-Gaita, V., Aamer, M., Posaratnanathan, R.T. et al. (2016). Acoustic wave biosensor for the detection of the breast and prostate cancer metastasis biomarker protein PTHrP. *Biosensors and Bioelectronics* 78: 92–99.

49 Verma, S.K., Amoah, A., Schellhaas, U. et al. (2016). "To catch or not to catch": microcapsule-based sandwich assay for detection of proteins and nucleic acids. *Advanced Functional Materials* 26 (33): 6015–6024.

50 Tsai, C.-W., Jheng, S.-L., Chen, W.-Y., and Ruaan, R.-C. (2014). Strategy of Fc-recognizable peptide ligand design for oriented immobilization of antibody. *Analytical Chemistry* 86 (6): 2931–2938.

51 Yoo, R.-J. and Choi, S.-J. (2016a). Identification of a peptide ligand for antibody immobilization on biosensor surfaces. *BioChip Journal* 10 (2): 88–94.

52 Miyakawa, S., Oguro, A., Ohtsu, T. et al. (2006). RNA aptamers to mammalian initiation factor 4G inhibit cap-dependent translation by blocking the formation of initiation factor complexes. *RNA* 12 (10): 1825–1834.

53 Porath, J. and Olin, B. (1983). Immobilized metal affinity adsorption and immobilized metal affinity chromatography of biomaterials. Serum protein affinities for gel-immobilized iron and nickel ions. *Biochemistry* 22 (7): 1621–1630.

54 McGrath, M.J. and Scanaill, C.N. (2013). *Sensor Technologies: Healthcare, Wellness, and Environmental Applications*. Springer Nature.

55 Zhang, J., Zhang, X., Wei, X. et al. (2021). Recent advances in acoustic wave biosensors for the detection of disease-related biomarkers: a review. *Analytica Chimica Acta* 1164: 338321.

56 Länge, K. (2019). Bulk and surface acoustic wave sensor arrays for multi-analyte detection: a review. *MDPI, Sensors* 19: 5382.

57 Sherrit, S., Bao, X.Q., Bar-Cohen, Y., and Chang, Z. (2003). BAW and SAW sensors for in-situ analysis. In: *Smart Structures and Materials: Smart Sensor Technology and Measurement Systems*, 81–91. SPIE.

58 Rabe, J., Büttgenbach, S., Schröder, J., and Hauptmann, P. (2003). Monolithic miniaturized quartz microbalance array and its application to chemical sensor systems for liquids. *IEEE Sensors Journal* 3: 361–368.

59 Sauerbrey, G. (1959). The use of quartz oscillators for weighing thin layers and for microweighing. *Zeitschrift für Physik* 155 (2): 206–222.

60 Lin, L., Ng, S., Li, F. et al. (2020). MEMS biosensor for monitoring water toxicity based on quartz crystal microbalance. *Kun- Biointerphases* 15: 021006.

61 Vikholm-Lundin, I., Piskonen, R., and Albers, W.M. (2007). Hybridisation of surface immobilised single stranded oligonucleotides and polymer monitored by surface plasmon resonance. *Biosensors and Bioelectronics* 22: 1323–1329.

62 Terzieva, M.D. (2016). Overview of bulk acoustic wave technology and its applications. *Electronıca* 7-8: 15–23.

63 Zhang, H., Pang, W., Kim, E.S., and Yu, H. (2010). Micromachined silicon and polymer probes integrated with film-bulk-acoustic-resonator mass sensors. *Journal of Micromechanics and Microengineering* 20 (12): 125008.

64 Lin, A., Li, Y.J., Wang, L. et al. (2011). Label-free detection of prostate-specific antigen with FBAR-based sensor with oriented antibody immobilization. In: *Ultrasonics Symposium (IUS), 2011, IEEE International*, 797–800. IEEE.

65 Qin, L., Chen, Q., Cheng, H., and Wang, Q.-M. (2010). Analytical study of dual-mode thin film bulk acoustic resonators (FBARs) based on ZnO and AlN films with tilted c-axis orientation. *IEEE Transactions on Ultrasonics, Ferroelectrics, and Frequency Control* 57 (8): 1840–1853. https://doi.org/10.1109/TUFFC.2010.1623.

66 Kadota, M., Suzuki, Y., and Ito, Y. (2011). Label-free detection of prostate-specific antigen with FBAR-based sensor with oriented antibody immobilization. In: *Ultrasonics Symposium (IUS), 2010, IEEE International*, 91–94. IEEE.

67 Ndong, R.O., Sorli, B., Giani, A., and Foucaran, A. (2013). Film bulk acoustic resonator based on zinc oxide thin film. *Advances in Physics Theories and Applications* 23: 31–36.

68 Ruby, R.C., Larson, J.D., Fazzio, R.S., and Feng, C. Performance degradation effects in FBAR filters and resonators due to lamb wave modes. In: *Ultrasonics. Symposium (IUS), 2005, IEEE International*, vol. 2011, 1832–1835. IEEE.

69 Mandal, D. and Banerjee, S. (2022). Surface acoustic wave (SAW) sensors: physics, materials, and applications. *Sensors* 22: 820.

70 Gizeli, E., Stevenson, A., Goddard, N., and Lowe, C. (1992). A novel love-plate acoustic sensor utilizing polymer overlayers. *IEEE Transactions on Ultrasonics, Ferroelectrics, and Frequency Control* 39 (5): 657–659.

71 White, R.M. and Voltmer, F.W. (1965). Direct piezoelectric coupling to surface elastic waves. *Applied Physics Letters* 7 (12): 314–316.

72 Auld, B. (1973). *Acoustic Fields and Waves in Solids, ser. Acoustic Fields and Waves in Solids*, vol. 1. Wiley.

73 Wang, Z., Cheeke, J., and Jen, C. (1994). Sensitivity analysis for love mode acoustic gravimetric sensors. *Applied Physics Letters* 64 (22): 2940–2942.

74 Jakoby, B., Bastemeijer, J., and Vellekoop, M.J. (2000). Temperature-compensated love-wave sensors on quartz substrates. *Sensors and Actuators A: Physical* 82 (1-3): 83–88.

75 Sisman, A., Gur, E., Ozturk, S. et al. (2017). A low-cost biomarker-based saw-biosensor design for early detection of prostate cancer. *Procedia Technology* 27: 248–249, biosensors 2016.

76 Onen, O., Sisman, A., Gallant, N.D. et al. (2012). A urinary bcl-2 surface acoustic wave biosensor for early ovarian cancer detection. *Sensors* 12 (6): 7423–7437.

77 Raimbault, V., Rebière, D., Dejous, C. et al. (2007). High viscosity sensing using a love wave acoustic platform combined with a pdms microfludic chip. *ECS Transactions* 4 (1): 73.

78 Johansson, L., Enlund, J., Johansson, S. et al. (2012). Surface acoustic wave induced particle manipulation in a pdms channel—principle concepts for continuous flow applications. *Biomedical Microdevices* 14 (2): 279–289.

79 Tran, A. (2014). Aln piezoelectric films for sensing and actuation, Master's thesis, TU Delft, the Netherlands.

80 Gizeli, E., Bender, F., Rasmusson, A. et al. (2003). Sensitivity of the acoustic waveguide biosensor to protein binding as a function of the waveguide properties. *Biosensors and Bioelectronics* 18 (11): 1399–1406.

81 Wagers, R.S. (1980). Low-temperature Lamb wave propagation in PVF2. *Journal of Applied Physics* 51 (5797).

82 Lowe, M.J.S. (2002). Wave propagation guided waves in structures. In: *Encyclopedia of Vibration* (ed. S. Braun), 1551–1559.

83 Yu, J.-C. and Tsai, J.-X. (2006). Design and Manufacturing of Lamb-Wave Biosensors Using PZT on SiN Thin Film.

84 Li, C., Kong, H., Tang, Y. et al. (2018). Liquid sound velocity and density decoupling on a compact lamb wave sensor by a two-port local resonating method. *IEEE Sensors Journal* 18: 1384–1389. https://doi.org/10.1109/JSEN.2017.2781691.

85 Vellekoop, M.J., Libking, G.W., Sarro, P.M., and Venema, A. (1994). Evaluation of liquid properties using a silicon Lamb wave sensor. *Sensors and Actuators* A43: 175–180.

86 Vellekoop, M.J., Lubking, G.W., and Venema, A. (1994). Acoustic-wave based monolithic microsensors. In: *Proceedings Ultrasonics Symposium*, 565–574.

87 Li, F., Manceau, J.-F., Wu, Y., and Bastien, F. (2008). Measurements of evanescent wave in a sandwich Lamb wave sensor. *Applied Physics Letters* 93: 174101.

88 Zhang, H., Tang, Z., Wang, Z. et al. (2017). Bioparticle manipulations using lamb wave resonator array. In: *2017 IEEE International Ultrasonics Symposium (IUS)*, 1–4. IEEE.

89 Kong, H., Li, C., Guo, Z. et al. (2019). Sensitivity improved with Parylene-C passivized on Lamb wave sensor for aPTT measurement through monitoring whole blood reaction. *Sensors and Actuators* B285: 479–486.

90 Gao, F., Al-Qahtani, A.M., Khelif, A. et al. (2021). Towards acoustic radiation free lamb wave resonators for high-resolution gravimetric biosensing. *IEEE Sensors Journal* 21 (3): 2725–2733.

91 Zhang, W., Li, C., Kong, H. et al. (2019). Label-free detection of pepsinogen 1 and 2 by polyethylene coating Lamb microfluidic device. *Biosensors and Bioelectronics* 129: 231–237.

92 Lamanna, L., Rizzi, F., Bhethanabotla, V.R., and De Vittorio, M. (2020). Conformable surface acoustic wave biosensor for E-coli fabricated on PEN plastic film. *Biosensors and Bioelectronics* 163: 112164.

93 Sappati, K.K. and Bhadra, S. (2021). Printed acoustic sensor for low concentration volatile organic compound monitoring. *IEEE Sensors Journal* 21 (8): 9808–9818.

94 Calabrese, G.S., Wohltjen, H., and Roy, M.K. (1986). A study of SAW delay line behavior in liquids. In: *IEEE 1986 Ultrasonics Symposium*, Williamsburg, VA, USA, 607–610. https://doi.org/10.1109/ULTSYM.1986.198809.

95 Caliendo, C., Verona, E., and D'Amico, A. (1992). Surface acoustic wave (SAW) gas sensors. In: *Gas Sensors* (ed. G. Sberveglieri), 281–306. Dordrecht: Springer.

96 Hunt, W.D., Stubbs, D.D., and Lee, S.-H. (2003). Time-dependent signatures of acoustic wave biosensors. *Proceedings of the IEEE* 91 (6): 890–901. https://doi.org/10.1109/JPROC.2003.813566.

97 Agostini, M., Greco, G., and Cecchini, M. (2018). A Rayleigh surface acoustic wave (R-SAW) resonator biosensor based on positive and negative reflectors with sub-nanomolar limit of detection. *Sensors and Actuators B* 4005 (17): 31221–31222.

98 Richardson, M., Das, P.K., Morrill, S. et al. (2022). Removal of non-specifically bound proteins using Rayleigh waves generated on ST-quartz substrates. *Sensors* 22: 4096.

99 Yatsuda, H. and Kogai, T. (2006). 3F-3 liquid sensor using SAW and SH-SAW on quartz. In: *2006 IEEE Ultrasonics Symposium*, Vancouver, BC, Canada, 552–555. https://doi.org/10.1109/ULTSYM.2006.143.

100 Wang, S.-H., Shen, C.-Y., Huang, H.-M., and Shih, Y.-C. (2014). Rayleigh surface acoustic wave sensor for ppb-level nitric oxide gas sensing. *Sensors and Actuators A: Physical* 216: 237–242.

101 Middelhoek, S., Audet, S., and French, P. (2020). *Silicon Sensors and Actuators*. Delft Academic Press.

102 Jakoby, B. and Vellekoop, M. (1998). Reducing the temperature sensitivity of love-wave sensors. In: *1998 IEEE Ultrasonics Symposium. Proceedings* (Cat. No. 98CH36102), vol. 1, 447–450. IEEE.

103 Bond, W. (1943). Chapter II: Methods for specifying quartz crystal orientation and their determination by optical means. *The Bell System Technical Journal* 22 (2): 224–262.

104 Euler, L. (1776). Formvlae generales pro translatione qvacvnqve corporvm rigidorvm. *Novi commentarii acedemiae scientairum imperialis petropolitane* XX: 189.

105 Whittaker, E.T. (1964). *A Treatise on the Analytical Dynamics of Particles and Rigid Bodies*. CUP Archive.

106 Pio, R. (1966). Euler angle transformations. *IEEE Transactions on Automatic Control* 11 (4): 707–715.

107 Lam, C., Wang, C.Y., and Wang, S. (2010). A review of the recent development of temperature stable cuts of quartz for saw applications. In: *Proceedings of the Fourth International Symposium on Acoustic Wave Devices for Future Mobile Communication Systems*, Chiba, Japan, 3–5.

108 Bragg, W.L. and Gibbs, R.E. (1925). The structure of α and β quartz. *Proceedings of the Royal Society of London. Series A, Containing Papers of a Mathematical and Physical Character* 109 (751): 405–427.

109 Huggins, M.L. (1922). The crystal structure of quartz. *Physical Review* 19 (4): 363.

110 Bragg, W.H. and Bragg, W.L. (1915). *X Rays and Crystal Structure*. G. Bell and Sons Ltd.

111 IRE-Standards-Committee (1949). Standards on piezoelectric crystals. In: *Proceedings of the I.R.E.*, 1378–1395.

112 ANSI (1978). IEEE standard on piezoelectricity (ANSİ/İEEE Std 176-1978).

113 Weisand, R. and Gaylord, T. (1985). Lithium niobate: Summary of physical properties and crystal structure. *Applied Physics A* 37 (4): 191–203.

114 Mordor Intelligence LLP (2022). Surface Acoustic Wave Sensors Market – Growth, Trends, COVID-19 Impact, and Forecasts (2021–2026). https://www.reportlinker.com/p06192761/Surface-Acoustic-Wave-Sensors-Market-Growth-Trends-COVID-19-Impact-and-Forecasts.html?utm_source=GNW (accessed 29 December 2022).

115 Durmuş, N.G., Lin, R.L., Kozberg, M. et al. (2014). Acoustic-based biosensors. In: *Encyclopedia of Microfluidics and Nanofluidics* (ed. D. Li), 1–15. Boston, MA: Springer US https://doi.org/10.1007/978-3-642-27758-0_10-2.

116 Qorvo Biotechnologies (2023). Qorvo Biotechnologies Receives FDA Emergency Use Authorization (EUA) for Rapid COVID-19 Antigen Testing at the Point of Care. https://www.qorvo.com/newsroom/news/2022/https//www.qorvo.com/newsroom/news/2022/qorvo-biotechnologies-receives-fda-eua-for-rapid-covid-19-antigen-testing-at-point-of-care (accessed 11 January 2023).

117 Fischer, S.K. et al. (2020). Evaluation of two point of care technologies for measuring monoclonal antibody therapeutic - a concentrations in blood. *Bioanalysis* 12 (20): 1449–1458. https://doi.org/10.4155/bio-2020-0193.

118 Gray, E.R. et al. (2018). Ultra-rapid, sensitive and specific digital diagnosis of HIV with a dual-channel SAW biosensor in a pilot clinical study. *npj Digital Medicine* 1 (1): 1. https://doi.org/10.1038/s41746-018-0041-5.

119 Bakhshpour, M., Piskin, A.K., Yavuz, H., and Denizli, A. (2019). Quartz crystal microbalance biosensor for label-free MDA MB 231 cancer cell detection via notch-4 receptor. *Talanta* 204: 840–845. https://doi.org/10.1016/j.talanta.2019.06.060.

120 Zhou, B., Hao, Y., Chen, S., and Yang, P. (2019). A quartz crystal microbalance modified with antibody-coated silver nanoparticles acting as mass signal amplifiers for real-time monitoring of three latent tuberculosis infection biomarkers. *Microchimica Acta* 186 (4): 212. https://doi.org/10.1007/s00604-019-3319-7.

121 Chunta, S., Suedee, R., Boonsriwong, W., and Lieberzeit, P.A. (2020). Biomimetic sensors targeting oxidized-low-density lipoprotein with molecularly imprinted polymers. *Analytica Chimica Acta* 1116: 27–35. https://doi.org/10.1016/j.aca.2020.04.017.

122 Guha, A. et al. (2020). Direct detection of small molecules using a nanomolecular imprinted polymer receptor and a quartz crystal resonator driven at a fixed frequency and amplitude. *Biosensors & Bioelectronics* 158: 112176. https://doi.org/10.1016/j.bios.2020.112176.

123 Pohanka, M. (2019). Piezoelectric immunosensor for the determination of C-reactive protein. *International Journal of Electrochemical Science* 8470–8478. https://doi.org/10.20964/2019.09.02.

124 Yang, H. et al. (2019). Quartz crystal microbalance detection of poly(ADP-ribose) polymerase-1 based on gold nanorods signal amplification. *Analytical Chemistry* 91 (17): 11038–11044. https://doi.org/10.1021/acs.analchem.9b01366.

125 Park, H.J. and Lee, S.S. (2019). QCM sensing of miR-21 by formation of microRNA–DNA hybrid duplexes and intercalation on surface-functionalized pyrene. *Analyst* 144 (23): 6936–6943. https://doi.org/10.1039/C9AN01645G.

126 Rahman, N.A.A.A., Ma'Radzi, A.H., and Zakaria, A. (2019). Fabrication of quartz crystal microbalance with pegylated lipopolymer for detection of non-invasive lung cancer biomarker. *Materials Today: Proceedings* 7: 632–637. https://doi.org/10.1016/j.matpr.2018.12.054.

127 Chi, L., Xu, C., Li, S. et al. (2020). In situ amplified QCM immunoassay for carcinoembryonic antigen with colorectal cancer using horseradish peroxidase nanospheres and enzymatic biocatalytic precipitation. *Analyst* 145 (18): 6111–6118. https://doi.org/10.1039/D0AN01399D.

128 Hwang, S.S. et al. (2019). Detection of amyloid β oligomers toward early diagnosis of Alzheimer's disease. *Analytical Biochemistry* 566: 40–45. https://doi.org/10.1016/j.ab.2018.09.011.

129 Wu, J.-G., Wei, S.-C., Chen, Y. et al. (2018). Critical study of the recognition between C-reactive protein and surface-immobilized phosphorylcholine by quartz crystal microbalance with dissipation. *Langmuir* 34 (3): 943–951. https://doi.org/10.1021/acs.langmuir.7b02724.

130 Suthar, J., Parsons, E.S., Hoogenboom, B.W. et al. (2020). Acoustic immunosensing of exosomes using a quartz crystal microbalance with dissipation monitoring. *Analytical Chemistry* 92 (5): 4082–4093. https://doi.org/10.1021/acs.analchem.9b05736.

131 Liu, J. et al. (2020). A microfabricated thickness shear mode electroacoustic resonator for the label-free detection of cardiac troponin in serum. *Talanta* 215: 120890. https://doi.org/10.1016/j.talanta.2020.120890.

132 Peng, J. et al. (2020). Detection of cardiac biomarkers in serum using a micro-electromechanical film electroacoustic resonator. *Journal of Micromechanics and Microengineering* 30 (7): 075011. https://doi.org/10.1088/1361-6439/ab8dd3.

133 Wajs, E., Rughoobur, G., Burling, K. et al. (2020). A novel split mode TFBAR device for quantitative measurements of prostate specific antigen in a small sample of whole blood. *Nanoscale* 12 (17): 9647–9652. https://doi.org/10.1039/D0NR00416B.

134 Tang, P., Wang, Y., Huo, J., and Lin, X. (2018). Love wave sensor for prostate-specific membrane antigen detection based on hydrophilic molecularly-imprinted polymer. *Polymers* 10 (5): 5. https://doi.org/10.3390/polym10050563.

135 Ji, J., Yang, C., Zhang, F. et al. (2019). A high sensitive SH-SAW biosensor based 36° Y-X black $LiTaO_3$ for label-free detection of *Pseudomonas Aeruginosa*. *Sensors and Actuators B: Chemical* 281: 757–764. https://doi.org/10.1016/j.snb.2018.10.128.

136 Ji, J. et al. (2020). An aptamer-based shear horizontal surface acoustic wave biosensor with a CVD-grown single-layered graphene film for high-sensitivity detection of a label-free endotoxin. *Microsystems & Nanoengineering* 6 (1): 1. https://doi.org/10.1038/s41378-019-0118-6.

137 Kano, K. et al. (2018). Evaluation of SH-SAW biosensor in whole blood. In: *2018 IEEE International Ultrasonics Symposium (IUS)*, 1–4. https://doi.org/10.1109/ULTSYM.2018.8580094.

138 Jandas, P.J., Luo, J., Prabakaran, K. et al. (2020). Highly stable, love-mode surface acoustic wave biosensor using Au nanoparticle-MoS_2-rGO nano-cluster doped polyimide nanocomposite for the selective detection of carcinoembryonic antigen. *Materials Chemistry and Physics* 246: 122800. https://doi.org/10.1016/j.matchemphys.2020.122800.

139 Rana, L., Gupta, R., Tomar, M., and Gupta, V. (2018). Highly sensitive Love wave acoustic biosensor for uric acid. *Sensors and Actuators B: Chemical* 261: 169–177. https://doi.org/10.1016/j.snb.2018.01.122.

140 Šetka, M. et al. (2020). Love wave sensors based on gold nanoparticle-modified polypyrrole and their properties to ammonia and ethylene. *Sensors and Actuators B: Chemical* 304: 127337. https://doi.org/10.1016/j.snb.2019.127337.

141 Kidakova, A., Boroznjak, R., Reut, J. et al. (2020). Molecularly imprinted polymer-based SAW sensor for label-free detection of cerebral dopamine neurotrophic factor protein. *Sensors and Actuators B: Chemical* 308: 127708. https://doi.org/10.1016/j.snb.2020.127708.

142 Greco, G., Agostini, M., and Cecchini, M. (2020). Ultra-high-frequency love surface acoustic wave device for real-time sensing applications. *IEEE Access* 8: 112507–112514. https://doi.org/10.1109/ACCESS.2020.3003076.

143 Jupe, A. et al. (2018). Development of a piezoelectric flexural plate-wave (FPW) biomems-sensor for rapid point-of-care diagnostics. In: *Nanostructured Materials for the Detection of CBRN*, Dordrecht, 199–212. https://doi.org/10.1007/978-94-024-1304-5_15.

144 Kong, H. et al. (2019). Sensitivity improved with Parylene-C passivized on Lamb wave sensor for aPTT measurement through monitoring whole blood reaction. *Sensors and Actuators B: Chemical* 285: 479–486. https://doi.org/10.1016/j.snb.2019.01.053.

145 Jin, H., Zhou, J., He, X. et al. (2013). Flexible surface acoustic wave resonators built on disposable plastic film for electronics and lab-on-a-chip applications. *Scientific Reports* 3: 2140.

146 Sharma, P., Imtiaz, S.A., and Rodriguez-Villegas, E. (2019). Acoustic sensing as a novel wearable approach for cardiac monitoring at the wrist. *Scientific Reports* 9: 20079. https://doi.org/10.1038/s41598-019-55599-5.

147 Hallil, H., Dejous, C., Hage-Ali, S. et al. (2021). Passive resonant sensors: trends and future prospects. *IEEE Sensors Journal*, 15p.

11

Acoustic Micro/Nanorobots in Medicine

Murat Kaynak[1,2], *Amit Dolev*[1], *and Mahmut S. Sakar*[1]

[1] *Institute of Mechanical Engineering, Ecole Polytechnique Fédérale de Lausanne, Lausanne 1015, Switzerland*
[2] *Center for Engineering in Medicine and Surgery, Massachusetts General Hospital, Harvard Medical School, 51 Blossom Street, Boston, MA 02114, USA*

11.1 Introduction

Sound is a vibration that propagates as a mechanical pressure wave through biological tissues. The use of sound in medicine started a long time ago. Auscultation is the act of listening to sounds within the body for diagnostic purposes. Since the early nineteenth century, doctors have used stethoscopes to listen to internal sounds generated through the presence of disturbed flow, such as eddies and turbulence, by frictional rubs or other sources with high accelerations (e.g. cardiac valve opening). Sonography, the use of ultrasound (>20 kHz) to produce images, is now regularly used to monitor the developing baby's health, guide interventional procedures such as biopsies, visualize blood flow through a vessel, and diagnose diseases associated with almost all organs, including the heart (i.e. echocardiography), urinary tract, and kidneys. The speed of sound wave propagation – approximately 1.5 km/s in soft tissue – allows ultrasound to operate with millisecond temporal precision. Sonography generates images several centimeters into tissue with spatial precision on the order of a few millimeters. These favorable properties, along with its relatively low cost, high portability, and safety, have made sonography one of the most widely used technologies in clinical medicine.

Applications of acoustics, the interdisciplinary science of sound, not only empower the clinician with the capability to record diagnostic signals and see visually hidden pathologies, but also provide the opportunity to "operate" inside living systems. For medical imaging, ultrasound has a frequency range of about 1–15 MHz, whereas therapeutic applications typically use a frequency of ~1 MHz or less. Sound waves can be focused to concentrate energy in a small volume, analogous to focusing light with a magnifying glass. Focused ultrasound can penetrate deep into the body and deposit energy in various shapes and sizes, generating a wide range of mechanical and thermal effects that can be used for treatment. Some effects, like tissue destruction, are permanent, while others, such

Acoustic Technologies in Biology and Medicine, First Edition.
Edited by Adem Ozcelik, Ryan Becker, and Tony Jun Huang.

as neuromodulation, are transitory and reversible. Thermal ablation has been the most widely explored clinical application of focused ultrasound. There are a variety of conditions that can be treated using this technology, including tumors in the prostate, breast, and liver; brain disorders such as essential tremor and Parkinson's disease; and neuropathic pain associated with bone metastasis. The thermal ablation of tissue is conducted with high-intensity ultrasound at power levels usually exceeding 600 W/cm^2, whereas ultrasound imaging is performed at significantly lower levels ($<0.2\,W/cm^2$). The mechanical effects of focused ultrasound can also be used for the precise destruction of tissue. Histotripsy is based on the use of inertial cavitation, the nucleation, growth, and destruction of tiny air bubbles. At high enough acoustic intensities with short pulse duration, inertial cavitation releases a shockwave, with amplitudes exceeding 10 000 atmospheres depending on the bubble size, capable of destroying cell membranes and even liquefying or annihilating cells. The same properties that make ultrasound attractive for imaging make it appealing for neuromodulation, including the ability of sound waves to penetrate deep into tissues and achieve high spatiotemporal precision. Transcranial-focused ultrasound has become a new non-invasive mode of regional brain stimulation that can be used to stimulate or suppress neural activity.

This chapter describes the acoustic forces acting on microscale structures submerged in physiological media. Microrobotics research aims to harness forces generated on acoustically excited structures to create injectable or ingestible medical devices that can be transported and operated remotely inside the body. Such mechanical devices are envisioned to be used for minimally invasive surgery, targeted drug delivery, and the collection of liquid or solid biopsy samples. This approach is fundamentally different from another application area of sound waves – ultrasound-powered implantable miniaturized electronic devices. In such devices, a piezoelectric crystal sits on the chip and converts ultrasound waves into electrical energy. Bypassing the electrical circuitry and directly using sound waves to do mechanical work leads to the creation of significantly smaller devices. Notably, without piezo crystals and accompanying metal traces, the whole device can be manufactured from biocompatible and soft materials. The devices can be made mobile and actuated to deform in three-dimensional (3D). These novel features open new avenues for the use of acoustic micro/nanorobots in biology and medicine.

The rest of the chapter is organized as follows. A brief theoretical background on acoustical forces is provided in Section 11.2. It focuses on primary and secondary acoustic radiation forces along with acoustic streaming-induced drag and thrust forces. In Section 11.3, we discuss acoustical manipulation methods, namely acoustic tweezers that can be utilized in medicine. In Section 11.4, we describe state-of-the-art micro-nanorobotic devices. We critically discuss the design, fabrication, actuation, and applications of these robots. Furthermore, we review the current in vivo applications of acoustic microrobotic systems in Section 11.5. In Section 11.6, we discuss acoustic robots in medicine and share our perspective.

11.2 Theoretical Background

11.2.1 Introduction to Acoustics

In this section, we introduce a non-exhaustive derivation of linear acoustic equations. In Section 11.2.2, we discuss nonlinear acoustic phenomena, which require considering additional terms in the governing equations. The objective is to familiarize the reader with the physics behind acoustic waves' propagation and wave–matter interactions. For a detailed derivation, the interested reader is referred to the relevant literature [1–4].

To describe lossless sound propagation in a homogenous fluid, we need three equations: (i) mass conservation, (ii) momentum conservation, and (iii) thermodynamic state. The conservation of mass requires that for a fixed volume, the mass flux through the surfaces equals the time rate of change of the volume's mass. The conservation law can be written using volume and surface integrals. Using Gauss's theorem, and since the obtained result is valid for arbitrary volumes, the mass conservation (or continuity) equation is

$$\frac{\partial \rho}{\partial t} + \nabla \cdot (\rho \mathbf{v}) = 0 \tag{11.1}$$

where t is the time, ρ is the mass density, $\mathbf{v}$ is the fluid velocity vector, and $\nabla\bullet$ is the divergence operator.

The conservation of momentum for a fluid is derived from the conservation law of classical continuum mechanics, which states that the total force exerted on a fluid particle equals its mass times the acceleration of its center of mass. In general, gravity should be considered, but it can be neglected [1] when considering acoustic disturbances at frequencies higher than g/c_0, where g is the gravitational acceleration and c_0 is the speed of sound. For water, it equates to frequencies lower than 0.01 Hz. In the following derivation, the surface forces are assumed to be normal, implying negligible viscosity, and the conservation of momentum equation is

$$\rho \frac{D\mathbf{v}}{Dt} = -\nabla p, \quad \frac{D\mathbf{v}}{Dt} = \frac{\partial \mathbf{v}}{\partial t} + (\mathbf{v} \cdot \nabla \mathbf{v}) \tag{11.2}$$

where p is the pressure and the operator D/Dt is the material derivative.

The thermodynamic state of the fluid can be described by some definite relation

$$p = p(\rho, s) \tag{11.3}$$

where s is the specific entropy. Sound is neither an adiabatic nor an isothermal process, but for most cases of interest, it can be considered adiabatic, which is the same as $Ds/Dt = 0$. Therefore, Eq. (11.3) can be simplified to

$$p = p(\rho, s_0) \tag{11.4}$$

where s_0 is a constant, representing the specific entropy throughout the medium.

Linear acoustics considers the disturbances as a small perturbation to the ambient state of the fluid, which can be described by its ambient values, p_0, ρ_0, and $\mathbf{v}_0$. Hence, the fluid state can be described as:

$$p = p_0 + p_1, \rho = \rho_0 + \rho_1, \mathbf{v} = \mathbf{v}_0 + \mathbf{v}_1 \tag{11.5}$$

Figure 11.1 Illustration of a spherical wave that resembles a planar wave far from the source.

Here, subscript 1 indicates a small linear perturbation to the state. If the medium through which the sound propagates is homogenous, it means that the ambient quantities are position- independent, if in addition, they are time-independent and $\mathbf{v}_0 = 0$, the medium is quiescent. This idealization is a sufficient approximation for the description of most acoustic phenomena. Substituting Eq. (11.5) into the previously derived equations where the medium is assumed to be homogenous and quiescent and considering only first-order terms (zero-order terms equate to zero), the linear acoustic equations are

$$\frac{\partial \rho_1}{\partial t} + \rho_0 \nabla \cdot \mathbf{v}_1 = 0, \rho_0 \frac{\partial \mathbf{v}_1}{\partial t} = -\nabla p_1, p_1 = c_0^2 \rho_1, c_0^2 = \left(\frac{\partial p}{\partial \rho}\right)_0 \tag{11.6}$$

Equation (11.6) can be combined to derive the wave equation, which takes the form

$$\nabla^2 p_1 - \frac{1}{c_0^2} \frac{\partial^2 p_1}{\partial t} = 0 \tag{11.7}$$

where, operator ∇ [2] is the Laplacian, which is the divergence of the gradient.

In the idealized case described thus far, the particles are displaced in parallel to the direction of propagation of the acoustic wave; therefore, it is called a longitudinal wave. This contrasts with transverse waves, such as the wave propagating on a string, where the particles are displaced perpendicularly to the direction of propagation. In both cases, it is important to note that the perturbation traveling from the source is transferred by the wave and not by the particles, which are barely displaced locally [1, 2, 4]. Most acoustic waves spread out spherically from a source [1, 4], but for simplicity, we begin with waves propagating in a single Cartesian direction – plane waves. These waves are useful in understanding many acoustic concepts and can be used for approximating acoustic phenomena far from the source (see Figure 11.1). Plane waves propagating in a single Cartesian direction s (i.e. $s = \mathbf{n} \bullet \mathbf{x}$, where $\mathbf{n}$ is a unit vector) do not disturb the medium in the perpendicular plane. Therefore, all the field variables change only along s. For example, $p_1 = p_1(s,t)$, so its gradient, ∇p_1, has only one component. As a result, Eq. (11.7) can be simplified to the following 1D wave equation:

$$\frac{\partial^2 p_1}{\partial s^2} - \frac{1}{c^2} \frac{\partial^2 p_1}{\partial t^2} = 0 \tag{11.8}$$

The general solution of Eq. (11.8) is a superposition of two arbitrary functions $f(t - s/c_0)$ and $b(t + s/c_0)$, as follows

$$p_1 = f(t - s/c_0) + b(t + s/c_0) \tag{11.9}$$

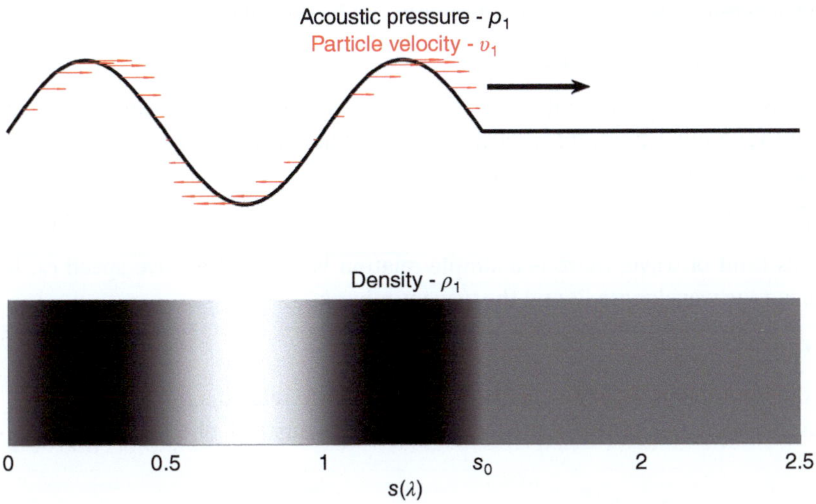

Figure 11.2 Sound wave propagation in a homogenous, quiescent medium. Local and instantaneous pressure, particle velocity (top), and density (bottom) in the medium as the wave propagates.

In addition, for the velocity field component in the s direction, one obtains:

$$v_1 = \frac{1}{\rho_0 c_0}[f(t - s/c_0) - b(t + s/c_0)] \tag{11.10}$$

It can be understood that f describes a forward-traveling wave in the $+s$ direction, while b describes a backward traveling wave in the $-s$ direction, both with speed c_0. The factor, $\rho_0 c_0$, is called the characteristic impedance of the medium. For simplicity, one can assume a forward- traveling wave (i.e. $b = 0$) in a medium that was at rest before $t = 0$ (i.e. $f = 0$ for $t < 0$), and is $p_A \sin(\omega t)$ afterward, as shown in Figure 11.2. The sinusoidal waveform travels from the source and reaches location s_0 after $t_0 = s_0/c_0$. Right before t_0, the particles have positive velocity in the direction of propagation, as indicated by the red arrows. The wave compresses the fluid, and, as a result, ρ_1 increases. Since the pressure is temporarily larger to the left of s_0, its gradient points in the negative direction, and the particles are being accelerated in the opposite direction. These two processes continue until a maximum value is obtained for p_a, at $t = \pi/2\omega$. Later, the pressure and compression begin to decrease and reach the ambient values at the pressure node at $t = \pi/\omega$. The process is then reversed, and minimum values are obtained at $t = 3\pi/2\omega$ when the pressure is minimal. During the second half of the cycle, the pressure and density decrease below the ambient values, leading to fluid rarefaction, where the particle velocity is in the opposite direction to that of the wave propagation. It is important to note that if the time integral of f is zero, the fluid particles remain in the same place after the transmission of the acoustic wave.

Usually, when discussing acoustics, waves of constant frequencies are considered. At constant frequencies, the variables oscillate sinusoidally with time throughout the field. Therefore, at any point, the pressure can be described as follows:

$$p_1 = p_A \cos(\omega t - \varphi) = \mathrm{Re}\{\hat{p}e^{-i\omega t}\} \tag{11.11}$$

where p_A is the pressure amplitude, ω is the angular frequency, φ is the phase constant, $\hat{p}$ is the complex pressure amplitude, and Re denotes the real part. The complex pressure in Eq. (11.11) can be expressed as

$$\hat{p} = p_A e^{i\varphi}, e^{i\varphi} = \cos(\varphi) + i\sin(\varphi) \tag{11.12}$$

For this kind of wave, there is a simple relation between the wave speed c_0, its frequency f, its wavelength λ, and the wavenumber k.

$$c_0 = \lambda f = \frac{\omega}{k} \tag{11.13}$$

where the frequency f and the angular frequency ω are related as $\omega = 2\pi f$.

For a constant frequency, the time average of acoustical phenomena is computed as:

$$X_{av} = \langle X \rangle = \frac{1}{T}\int_{t_0}^{t_0+T} X dt \tag{11.14}$$

Here, X can be any quantity, the angled brackets represent the time-averaging operator, and T is either an integral number of wave periods or an infinite time integral. To compute the time average of a product of quantities oscillating at the same frequency, the complex notation leads to

$$(XY)_{av} = \langle XY \rangle = \frac{1}{2}\mathrm{Re}\left\{\hat{X}\hat{Y}^*\right\} \tag{11.15}$$

where the asterisk denotes the complex conjugate.

11.2.2 Time-Averaged Acoustically Induced Forces

Acoustically induced forces are associated with nonlinear phenomena occurring as the sound propagates through the medium and interacts with the boundaries [1, 3]. In Section 11.2.1, we neglected the fluid viscosity and higher-order terms that capture these phenomena to some extent. The full set of nonlinear equations capturing all the phenomena is very complex to analyze. In Sections 11.2.2.1–11.2.2.3, specific terms are treated, focusing on primary and secondary acoustic radiation forces along with acoustic streaming.

11.2.2.1 Primary Radiation Forces

Bodies, deformable or rigid, are subjected to acoustic radiation pressure in an acoustic field due to the waves impinging on their surface. Although we assume that the acoustic field is harmonic, a net force emerges, which is called *acoustic radiation force*. To compute the acoustic radiation force, time averaging is used along with the integration of the acoustic stress tensor $\boldsymbol{\sigma}$ over the surface. The elements of $\boldsymbol{\sigma}$ approximated to the second order are [3]:

$$\boldsymbol{\sigma} = \begin{pmatrix} -\langle p - p_0 \rangle - \rho_0 \langle v_{1x} v_{1x} \rangle & -\rho_0 \langle v_{1x} v_{1y} \rangle & -\rho_0 \langle v_{1x} v_{1z} \rangle \\ -\rho_0 \langle v_{1y} v_{1x} \rangle & -\langle p - p_0 \rangle - \rho_0 \langle v_{1y} v_{1y} \rangle & -\rho_0 \langle v_{1y} v_{1z} \rangle \\ -\rho_0 \langle v_{1z} v_{1x} \rangle & -\rho_0 \langle v_{1z} v_{1y} \rangle & -\langle p - p_0 \rangle - \rho_0 \langle v_{1z} v_{1z} \rangle \end{pmatrix} \tag{11.16}$$

where v_{1x}, v_{1y}, and v_{1z} are the Cartesian components of the first-order velocity field. In the idealized case of a rigid particle, the velocity on the surface nullifies, simplifying the stress tensor to

$$\sigma_{xx} = \sigma_{yy} = \sigma_{zz} = -\langle p - p_0 \rangle, \ \sigma_{xy} = \sigma_{yx} = \sigma_{xz} = \sigma_{zx} = \sigma_{yz} = \sigma_{zy} = 0 \tag{11.17}$$

In the scope of this chapter, only small spherical particles are considered, where their radius R is much smaller than the acoustic wavelength λ and much larger than the viscous boundary layer $\delta = (\nu/\omega)^{1/2}$, where ν is the kinematic viscosity. The acoustic radiation force acting on an arbitrarily shaped body can be computed using numerical methods (e.g. finite element analysis) along with Eq. (11.16). For rigid and nonrigid spherical particles, there are analytical methods to calculate the acoustic radiation force. Solids and liquids in gases can be idealized as rigid if they are much denser than the gas [3]. King [5] computed analytically the acoustic radiation force acting on a rigid particle generated by plane incident standing waves and traveling waves. For the standing wave, where $p_1 = A\sin(kx)e^{-i\omega t}$, the resulting force is only in the x direction and is approximated as

$$F^x_{PRI} = -\frac{5\pi}{6}\frac{A^2kR^3}{\rho_0 c_0^2}\sin(2kx) \tag{11.18}$$

whereas for the traveling wave, where $p_1 = Ae^{[i(kx - \omega t)]}$, the force is approximated as

$$F^x_{PRI} = \frac{11\pi}{18}\frac{A^2k^4R^6}{\rho_0 c_0^2} \tag{11.19}$$

In the former case, the force nullifies every quarter wavelength, which coincides with an equilibrium position. However, only pressure nodal points (i.e. $x = \lambda n/2$, $n \in \mathbb{Z}$) are stable points to which the particle is attracted, and the antinodes (i.e. $x = \lambda(1+n/2)/4$, $n \in \mathbb{Z}$) are unstable. In the latter case, the force is always positive, which means that the particle is pushed away from the source due to the momentum of the incident wave. Notably, the force produced by a standing wave is of order (kR) [3] times larger than the force produced by a traveling wave, since $kR < 1$.

In the case of a compressible particle, such as a droplet or air bubble submerged in liquid, the compressibility and density of the two media are comparable. Therefore, one can no longer assume a rigid sphere, and the velocity components in the acoustic radiation stress tensor must be considered. In the case of a gas bubble submerged in water, we can no longer assume that the frequency of the acoustic wave is much smaller than the sphere's resonance frequency. Yosioka and Kawasima [6] computed analytically the acoustic radiation force acting on a compressible sphere for incident plane standing and traveling waves. The resulting force for the standing wave is only in the x direction and is approximated as

$$F^x_{PRI} = \frac{\pi A^2kR^3}{3}\left(\frac{1}{c_s^2\rho_s} + \frac{2\rho_0 - 5\rho_s}{c_0^2\rho_0(\rho_0 + 2\rho_s)}\right)\sin(2kx) \tag{11.20}$$

where ρ_s is the density of the particle and c_s is the speed of sound in the particle. The equilibrium points are analogous to those of the rigid particle. However, the particle is attracted to either the nodes or antinodes according to the particle and fluid properties (i.e. the term in the parenthesis can be positive or negative).

For an incident traveling wave the resulting force is approximated as

$$F^x_{PRI} = \frac{2\pi A^2 k^4 R^6 \rho_0}{c_0^2\left(\rho_0^2 + 2\rho_s^2\right)} \left(\frac{2(\rho_0 - \rho_s)^2}{9\rho_0^2} + \left(\frac{\rho_s}{\rho_0} - \frac{c_0^2(\rho_0 + 2\rho_s)}{3c_s^2\rho_s} \right)^2 \right) \tag{11.21}$$

As for the rigid particle, the scaling of the forces and the force direction for a pure traveling wave are analogous.

In 1962, Gor'kov [7] derived a method to estimate the forces acting on a deformable small non-resonating particle. His method is valid for all cases, except for a pure traveling plane wave. To use his formula to calculate the acoustic radiation force, the values of the incident wave field variables (i.e., pressure p_1 and particle velocity $\mathbf{v}_1$) at a close neighborhood of the particle, along with the physical properties of the surrounding medium and the particle, must be measured. A potential field can be computed, whose gradient is the acoustic radiation force:

$$\mathbf{F} = -\nabla U, U = 2\pi R^3 \left[\frac{\rho_s c_s^2 - \rho_0 c_0^2}{2\rho_0 c_0^2 \rho_s c_s^2} \langle p_1^2 \rangle - \frac{\rho_0(\rho_s - \rho_0)}{\rho_0 + 2\rho_s} \langle \mathbf{v}_1 \cdot \mathbf{v}_1 \rangle \right] \tag{11.22}$$

Similarly to the results obtained by Yosioka and Kawasima for a compressible particle, the force changes sign according to the properties of the particle and fluid. A more elaborate model considering thermal effects was developed by Karlsen and Bruss [8]. Calculating the acoustic radiation force acting on a gas bubble in a liquid requires further development since bubbles can resonate [2]. Here, the first vibration mode of a spherical bubble is considered, which is a volumetric vibration mode. In this vibration mode, the bubble experiences monopole oscillations. These volumetric oscillations are more important than the bubble's compressibility [3], and the force acting on it can be approximated as: [2, 9]

$$\mathbf{F} \approx -\langle V \nabla p \rangle \tag{11.23}$$

where V is the bubble's instantaneous volume. Since the bubble is subjected to an oscillating pressure field, either standing or traveling or a combination of thereof, its radius also oscillates as

$$R = R_0 + R_\varepsilon(t) \tag{11.24}$$

where R_0 is the bubble's radius at rest. Assuming small acoustic pressure amplitudes, the variations in the radius are also small compared to the equilibrium radius oscillations (i.e. $|R_\varepsilon| \ll R_0$). The linear radial oscillations equation of a bubble is [2, 10, 11]

$$\frac{\partial^2 R_\varepsilon}{\partial t^2} + \omega\delta \frac{\partial R_\varepsilon}{\partial t} + \omega_0^2 R_\varepsilon = -\frac{p_1(x,t)}{\rho_0 R_0} \tag{11.25}$$

where ω is the acoustic field angular frequency, ω_0 is the bubble's natural frequency, and δ is the total damping constant [2, 9, 11]. Assuming harmonic excitation, i.e. $\partial \bullet / \partial t = -i\omega$, the radial oscillations of the bubble are:

$$R_\varepsilon = \frac{p_1(x)}{\rho_0 R_0 \left(\omega^2 - \omega_0^2 + i\omega^2\delta\right)} \tag{11.26}$$

Substituting Eq. (11.26) into Eq. (11.23), and assuming that the pressure gradient can be estimated at the center of the bubble since $R_0 \ll \lambda$ we obtain for a traveling wave field:

$$F^x_{PRI} = \frac{2\pi A^2 \delta R_0}{\rho_0 \omega^2 \left\{ \left[1 - (\omega_0/\omega)^2\right]^2 + \delta^2 \right\}} \tag{11.27}$$

and for a standing wave field:

$$F^x_{PRI} = -\frac{\pi A^2 R_0 \left(1 - (\omega_0/\omega)^2\right)}{\rho_0 \omega^2 \left\{ \left[1 - (\omega_0/\omega)^2\right]^2 + \delta^2 \right\}} \sin(2kx) \tag{11.28}$$

The force is always positive in the traveling wave case, which means that the bubble is pushed away from the source.

The acoustic radiation force acting on a bubble in a standing wave field is often termed the *primary Bjerknes force*. Notice that if only the first natural frequency is considered, then the bubble oscillates in phase when $\omega_0 \gg \omega$ and out of phase when $\omega_0 \ll \omega$. To understand the behavior of a bubble when it is excited at different frequencies, one can write:

$$F^x_{PRI} \propto \begin{cases} \sin(2kx) & \omega_0 > \omega \\ -\sin(2kx) & \omega_0 < \omega \end{cases} \tag{11.29}$$

Equation (11.29) suggests that bubbles that are excited at a frequency below their natural frequency move toward pressure antinodes, while bubbles that are excited above their natural frequency move toward pressure nodes. Notice that Eq. (11.28) is singular when $\omega = \omega_0$, Yosiokaet al. [3, 12] derived a more complete model that is not singular, with which Eq. (11.28) agrees. If higher vibration modes are considered, the bubble dynamics are more complex, and the interested reader is referred to a comprehensive treatment by Doinikov [9] for further details.

11.2.2.2 Secondary Radiation Forces

Particles excited by an incident pressure field apply *secondary acoustic radiation forces* to neighboring particles. These forces arise when compressible particles can resonate, which is particularly relevant for gas bubbles. Acoustic radiation forces between two adjacent bubbles are termed *secondary or mutual Bjerknes forces.*

Let's consider two bubbles whose centers are located at x_1 and x_2 as shown in Figure 11.3. The bubbles can have a different radius at rest, and the distance between their centers is $L = |\mathbf{x}_1 - \mathbf{x}_2|$. The bubbles are subjected to an acoustic field, and as a response, they vibrate and experience primary Bjerknes forces. In this section, only their interaction force is considered. The secondary acoustic radiation force acts along the vector connecting the centers of the bubbles.

Assuming that the bubbles are widely spaced yet their distance is much smaller than the acoustic wavelength (i.e. $\lambda \gg L \gg R_{10} + R_{20}$), where R_{10} and R_{20} are the first and second bubbles' radii at rest, we consider the response of each bubble to the

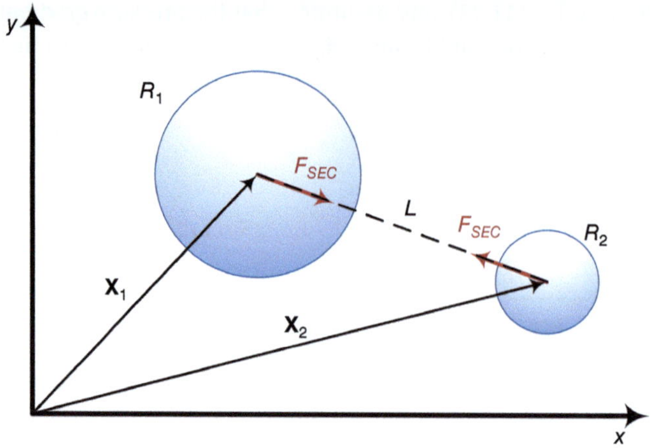

Figure 11.3 The geometry of a bubble pair.

incident wave field to estimate their vibrations according to Eq. (11.25). The pressure field scattered by each bubble can be computed, and the force acting between the bubbles can be computed using Eq. (11.23) as:

$$F_{SEC} = -\langle V_{b1} \nabla p_{b2} \rangle = -\langle V_{b2} \nabla p_{b1} \rangle \tag{11.30}$$

where V_{b1} and V_{b2} are the instantaneous volumes of the first and second bubbles, and P_{b1} and P_{b2} are the scattered pressure fields from the bubbles. The convention is that positive force is attractive while negative force is repulsive. Plugging in the terms for the instantaneous volume and pressure, we obtain [9]

$$F_{SEC} = \frac{2\pi R_{10} R_{20} A^2 \left[\left(1 - \omega_1^2/\omega^2\right)\left(1 - \omega_2^2/\omega^2\right) + \delta_1 \delta_2 \right]}{\rho_0 \omega^2 L^2 \left[\left(1 - \omega_1^2/\omega^2\right)^2 + \delta_1^2 \right] \left[\left(1 - \omega_2^2/\omega^2\right)^2 + \delta_1^2 \right]} \tag{11.31}$$

where $\omega_1, \omega_2, \delta_1$, and δ_2 are the natural frequencies and total damping terms of each bubble. From Eq. (11.31) it can be seen that if the bubbles oscillate in phase (i.e. $\omega_1, \omega_2 < \omega$ or $\omega_1, \omega_2 > \omega$) they attract each other, and if they oscillate out of phase (i.e. $\omega_1 > \omega > \omega_2$ or $\omega_1 < \omega < \omega_2$) they repel each other. This result fails to explain the bubbles' behavior in close proximity. To address this, the mutual interaction between the bubbles can no longer be ignored, as shown by Zabolotskaya [13]. In this case, the coupled equations describing their vibration are:

$$\begin{aligned} \ddot{R}_{\varepsilon 1} + \omega \delta_1 \dot{R}_{\varepsilon 1} + \omega_1^2 R_{\varepsilon 1} + \frac{R_{20}^2 \ddot{R}_{\varepsilon 2}}{R_{10} L} &= -\frac{p_1(\mathbf{x}_1, t)}{\rho_0 R_{10}} \\ \ddot{R}_{\varepsilon 2} + \omega \delta_2 \dot{R}_{\varepsilon 2} + \omega_2^2 R_{\varepsilon 2} + \frac{R_{10}^2 \ddot{R}_{\varepsilon 1}}{R_{20} L} &= -\frac{p_1(\mathbf{x}_2, t)}{\rho_0 R_{20}} \end{aligned} \tag{11.32}$$

This results in an updated equation for the bubble–bubble interaction force:

$$F_{SEC} = \frac{2\pi R_{10} R_{20} A^2}{\rho_0 \omega^2 L^2 \Delta} \left[\left(\frac{\omega_1^2}{\omega^2} - 1 + \frac{R_{10}}{L} \right) \left(\frac{\omega_2^2}{\omega^2} - 1 + \frac{R_{20}}{L} \right) + \delta_1 \delta_2 \right]$$

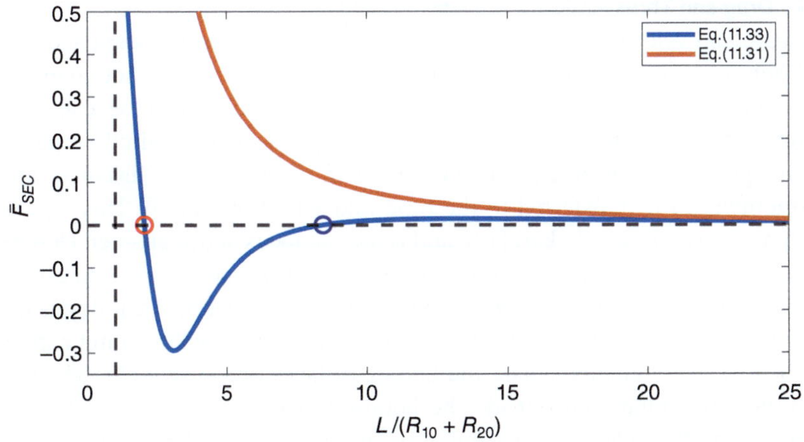

Figure 11.4 The normalized secondary Bjerknes forces as a function of the normalized distance according to two models for bubbles with a natural frequency lower than the acoustic field frequency – $\omega_1, \omega_2 < \omega$.

$$\Delta = \left(\delta_2\left(1-\frac{\omega_1^2}{\omega^2}\right)+\delta_1\left(1-\frac{\omega_2^2}{\omega^2}\right)\right)^2 + \left(\frac{R_{10}R_{20}}{L^2}-\left(1-\frac{\omega_1^2}{\omega^2}\right)\left(1-\frac{\omega_2^2}{\omega^2}\right)+\delta_1\delta_2\right)^2 \tag{11.33}$$

Immediate interpretation of the bubbles' behavior is not clear, but it can be shown that the direction of the force depends on their distance, L, which is not the case for the basic model (i.e. Eq. (11.31)). For example, when the natural frequencies of both bubbles are lower than the excitation frequency, but one is close to it, this can lead to a change in the direction of the force, generating stable and unstable equilibrium distances, as shown in Figure 11.4.

We can approximate the bubbles' natural frequency as [11]

$$\omega_0 = \sqrt{\frac{3\gamma p_0}{\rho_0 R_0^2}-\frac{2\sigma}{\rho_0 R_0^3}} \tag{11.34}$$

where γ is the ratio of specific heats for gas and σ is the surface tension. Figure 11.4 depicts the computed forces according to the simplified model, which is suitable for $\lambda \gg L \gg R_{10}+R_{20}$ (i.e. Eq. (11.31)) and the latter model (i.e. Eq. (11.33)) that considers mutual interaction between the bubbles and is suitable for $\lambda \gg L \sim R_{10}+R_{20}$. The values of the parameters that were used to compute the plotted forces are:

$$\begin{gathered}\rho_0 = 1000\,\text{kg/m}^3,\ \sigma = 0.072\,\text{N/m},\ \gamma = 1.4,\ p_0 = 101.325\,\text{kPa},\ f = 130\,\text{kHz},\\ R_{10} = 25.4\,\mu\text{m},\ f_1 = 128.39\,\text{kHz},\ \delta_1 = 0.10,\\ R_{20} = 30\,\mu\text{m},\ f_2 = 108.82\,\text{kHz},\ \delta_2 = 0.06\end{gathered} \tag{11.35}$$

11.2.2.3 Drag and Thrust-Induced Acoustic Streaming

The pressure and velocity fields vary with time in homogenous quiescent acoustic fields. When the time-averaged velocity does not nullify, mean fluid motion exists, which is termed *acoustic streaming* [3]. Acoustic streaming can be classified according to its driving mechanisms [14]. The energy of high-amplitude acoustic beams in open spaces is dissipated within the fluid over some distance, and pseudo-momentum is transferred to the fluid and sets it in motion. This type of streaming was described by Eckart [15], and is not the focus of this chapter. Herein, the focus is on the acoustic streaming occurring at the microscopic scale near boundaries at ultrasound frequencies. The boundary layer driven streaming is the result of the shear viscosity close to the boundary. The acoustic streaming regime is described in two domains, inside the boundary layer ($<\delta = (\nu/\omega)^{1/2}$), which is termed inner streaming, and outside the boundary layer ($>\delta$), which is termed outer streaming. Other classes of streaming are discussed in detail by Boluriaan and Morris [14].

To compute the acoustic streaming, one needs the governing equations for the fluid, namely the mass and momentum conservation laws without neglecting the viscosity, which is [3, 10]

$$\rho\frac{\partial \mathbf{v}}{\partial t} + \rho(\mathbf{v}\cdot\nabla)\mathbf{v} = -\nabla p + \mu\nabla^2\mathbf{v} + \left(\mu_b + \frac{1}{3}\mu\right)\nabla(\nabla\cdot\mathbf{v}) \tag{11.36}$$

where μ is the shear viscosity and μ_B is the bulk dynamic viscosity. The medium is assumed to be homogenous and quiescent, and the acoustic field parameters are approximated similarly to Eq. (11.5), while considering additional terms that do not oscillate at the same frequency as the first-order terms.

$$p = p_0 + p_1 + p_2 + \cdots, \rho = \rho_0 + \rho_1 + \rho_2 + \cdots, \mathbf{v} = \mathbf{v}_1 + \mathbf{v}_2 + \cdots \tag{11.37}$$

Substituting Eq. (11.37) to Eqs. (11.1) and (11.36), collecting terms of the same order and computing the time average [3, 16, 17], to the zeroth order one obtains (notice that $\langle\partial\bullet/\partial t\rangle = 0$, and $\langle\bullet_1\rangle = 0$):

$$\nabla p_0 = 0 \tag{11.38}$$

This means that the pressure is uniform, as expected. To the first order, one obtains no equation, as the time derivatives and the time average of the first-order terms equal zero. In the second order, the equations become:

$$\begin{aligned}\rho_0\nabla\cdot\langle\mathbf{v}_2\rangle = -\nabla\cdot\langle\rho_1\mathbf{v}_1\rangle, \mu\nabla^2\langle\mathbf{v}_2\rangle + \left(\mu_b + \frac{1}{3}\mu\right)\nabla\nabla\cdot\langle\mathbf{v}_2\rangle - \langle\nabla p_2\rangle \\ = \left\langle\rho_1\frac{\partial\mathbf{v}_1}{\partial t}\right\rangle + \rho_0\langle\mathbf{v}_1\cdot\nabla\mathbf{v}_1\rangle\end{aligned} \tag{11.39}$$

These equations can be solved analytically, but only for select cases. For the rest of the cases, they can be solved numerically [14], using finite element analysis [17, 18] or direct numerical simulation [19].

The two methods that are experimentally used to generate acoustic streaming at the microscale are sharp-edge streaming [20–23] and bubble-induced streaming [24–27]. For both cases, the driving mechanism is viscous energy dissipation in the boundary layer, and for both, obtaining an analytical solution is possible in a variety of cases. For sharp edge streaming, the interested reader is referred to [22], and for a trapped bubble, to [24, 25, 28].

11.3 Acoustic Micromanipulation Techniques

There are three types of forces that are relevant for micromanipulation: primary acoustic radiation force (F_{PRI}), secondary acoustic radiation force (F_{SEC}), and drag or thrust-induced acoustic streaming (F_{ST}). The description of these forces is discussed in Section 11.2. Primary acoustic radiation forces are applied to objects in intense acoustic fields. When the particle size is considerably smaller than the acoustic wavelength ($R \ll \lambda$), this force can be computed using Gor'kov potential (Eq. (11.22)). The sign of the force is determined by the density and compressibility of the particle and fluid, and it may be attracted to either the nodes or antinodes (Figure 11.5a). For air bubbles, the sign of the primary acoustic radiation force, which is exclusively referred to as the primary Bjerknes force, may change as well, but due to a different mechanism (Eq. (11.29)). Bubbles are deformable and can resonate, and according to the ratio of their natural frequency to the acoustic field frequency, they can be attracted to either the nodes or antinodes. The natural frequencies of the bubble highly depend on its radius at rest Eq. (11.34).

Secondary acoustic radiation forces are generated by secondary sources, such as oscillations of bubbles excited by the primary acoustic field (Figure 11.5c). When the excitation frequency is close to one of the bubble natural frequencies, it vibrates with large amplitudes and generates a secondary acoustic field. Through these secondary fields, objects apply secondary acoustic radiation forces to each other. The sign of the secondary acoustic radiation force depends on the natural frequencies of the bubbles as well as the primary excitation frequency (see Figure 11.5c, Eqs. (11.31) and (11.33)). Additional nonlinear acoustic effects include acoustic

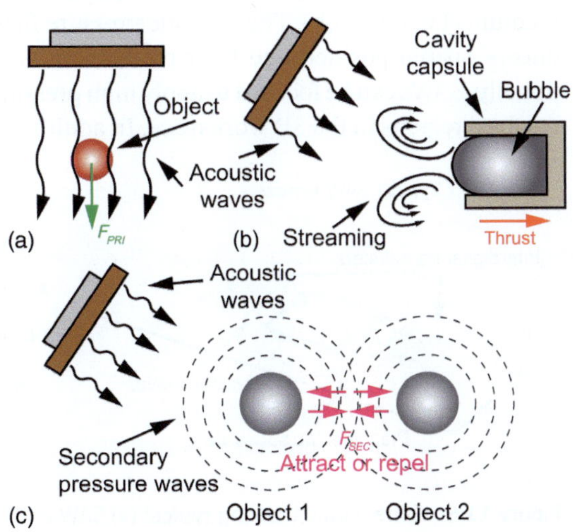

Figure 11.5 Acoustic forces acting on objects submerged in fluid. (a) Primary acoustic radiation force, (b) Acoustic streaming force, and (c) secondary acoustic radiation force.

streaming (see Eq. (11.39) in Section 11.2.2.3), which can be generated around oscillating objects (Figure 11.5b). During the oscillation, there is a momentum exchange between the vibrating object and the surrounding medium in the viscous boundary layer. The acoustic streaming flow near vibrating objects is commonly observed as counter-rotating vortices. We further discuss acoustic streaming to actuate micro/nanorobotic devices in Section 11.4.

11.3.1 Introduction to Acoustic Tweezers

In this section, we describe acoustic micromanipulation techniques that are based on one or a combination of the aforementioned acoustic phenomena. We focus on the manipulation schemes that utilize bulk acoustic waves (BAWs) because they are more suitable for medical applications.

Acoustic tweezers are based on the concept that an acoustic trap (or traps) can be created at a precise and controlled location in space. The trap is usually formed at a focal or nodal point by acoustic radiation forces. Acoustic tweezing devices utilize longitudinal waves in fluids and a combination of longitudinal and transverse waves for on-chip devices to manipulate objects in a controllable fashion.

Since the invention of the "acoustic tweezing" concept [29], numerous experimental platforms have been introduced for the noninvasive manipulation of biological samples [30]. Acoustic tweezers can be divided according to the types of waves they employ: surface acoustic wave (SAW) and BAW (Figure 11.6). SAW tweezers are particularly suited for on-chip manipulation of particles and biological samples [31]. Longitudinal, transverse, and Rayleigh surface waves [32] are generated and propagated along the surface of an elastic solid medium (Figure 11.6a). The energy of the wave's leaks into the fluid medium residing on the surface, and pressure nodes and antinodes are generated. Since the amplitudes of SAWs decay exponentially in the elastic media, the pressure in the coupled media diminishes spatially, allowing manipulation of objects only in proximity to the surface.

BAW tweezers employ longitudinal pressure waves that propagate through the medium (Figure 11.6b). The acoustic pressure field is controlled through the transducers' spatial position and their relative amplitude, phase, and frequency. Additionally, BAWs can be focused to apply high pressure to select objects within spatially resolved regions in the 3D workspace. In addition to serving as the primary acoustic

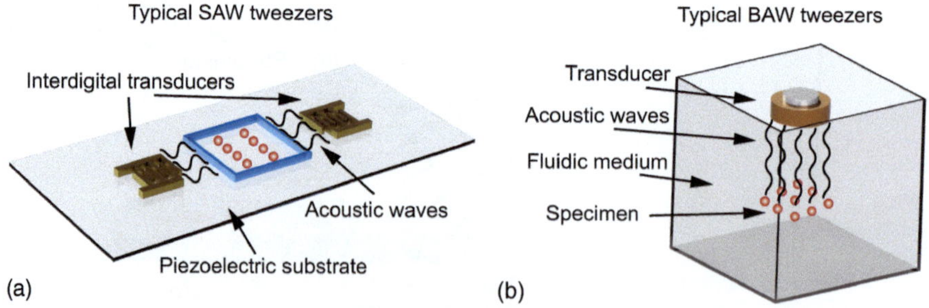

Figure 11.6 Schematics showing typical (a) SAW and (b) BAW tweezers.

source for direct micromanipulation, BAWs can excite structures with entrapped air bubbles [33] or sharp-edged solid features [34] to generate secondary forces. These acoustically excited mechanical microsystems can be used to manipulate particles, cells, and multicellular organisms. The working principles of this actuation scheme and how microrobots can be powered with acoustic waves are described in detail in Section 11.4. In the remaining of Section 11.3, we focus on BAW tweezers and associated micromanipulation techniques because these are better suited for medical applications.

11.3.2 Acoustic Micromanipulation Using Bulk Acoustic Waves

There are two major approaches to BAW tweezers (Figure 11.7). In the first approach, a standing wave field is generated in the medium. Early examples of particle tweezing using standing waves in air date back to the 1930s [35]. It was later quantitatively shown that acoustic waves can manipulate air bubbles in water [36, 37]. Rigorous theoretical analysis followed those seminal experimental demonstrations [5–7, 9, 12, 13, 37]. Due to their ease of use and interesting physics, air bubbles and oil droplets have been extensively used to explore the mechanics of BAW tweezers. Bubbles and droplets are size-controllable and compressible; therefore, all the acoustic forces can be studied and utilized. Bjerknes was one of the first scholars to discuss the behavior of an air bubble in an oscillating media [38], and bubble dynamics were studied by exciting them in standing wave fields [37, 39].

One of the most widely used techniques to generate a standing wave field is to use two identical transducers facing each other so that they generate identical waves propagating in opposite directions (Figure 11.7a). Alternatively, a transducer and a reflector can be placed in such a way that the generated and reflected sound waves interfere to form a standing wave with the desired distribution of nodes and antinodes [40]. In a perfectly collimated system, the pressure nodes are spaced half a wavelength apart. The same is true for the antinodes. Objects in the field move to the

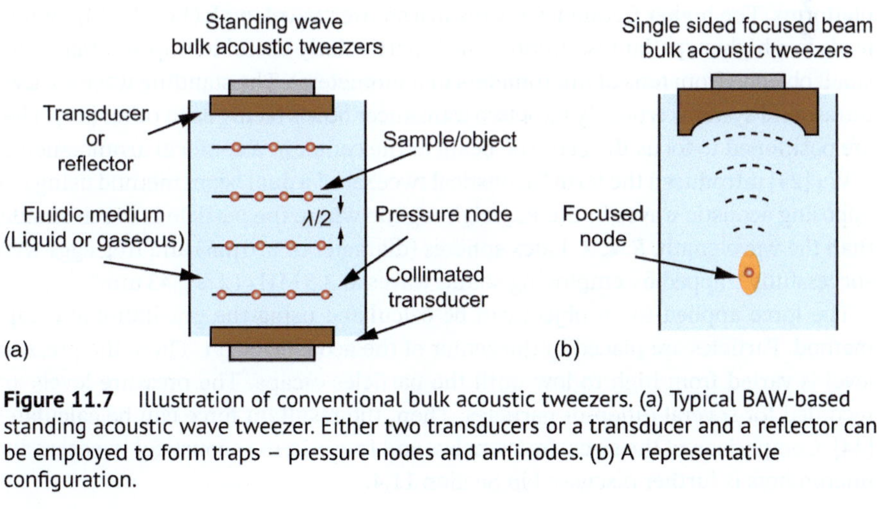

Figure 11.7 Illustration of conventional bulk acoustic tweezers. (a) Typical BAW-based standing acoustic wave tweezer. Either two transducers or a transducer and a reflector can be employed to form traps – pressure nodes and antinodes. (b) A representative configuration.

closest acoustic trap (i.e. either pressure node or antinode), where the displacement is always less than half of a wavelength.

Due to acoustical phenomena that mitigate the wave energy as it propagates in the media (e.g. dispersion and attenuation), the transducers and reflectors are often placed closely, usually several half wavelengths apart. For efficient tweezing, the size of the object must be much smaller than the wavelength wave (i.e. $R \ll \lambda$). If the size of the object is larger, the theoretical equations no longer hold, and the resulting forces diminish. The signal fed to the transducers usually comprises a single harmonic whose amplitude, frequency, and phase are controlled using a signal generator. In most applications, an amplifier is required to drive the acoustic transducers, which are usually made of piezoelectric materials. The deformation of the piezoelectric material with the application of voltage produces the pressure field. The excitation frequencies are set by considering the topology of the generated pressure field; they should also correspond to the resonance frequencies of the transducet for efficiency.

Acoustic tweezing in air is called acoustic levitation, as the acoustic forces counteract gravity, making the trapped particles levitate. The dominant forces in acoustic levitation are the primary acoustic radiation and drag forces. The drag force (F_D) is generated by acoustic streaming. F_{PRI} acts to trap the object in the node or antinode, while F_D makes the tweezing unstable. The F_{PRI} is proportional to the volume of the sample, while F_D is proportional to the radius of the object according to Stoke's law, assuming that $\mathrm{Re} < 1$. When $\mathrm{Re} > 1$, the Schiller–Naumann model can be used to obtain more accurate values [41]. Therefore, F_D/F_{PRI} increases as we scale down the size of the objects. Stability problems occur when the F_D/F_{PRI} is between 0.25 and 0.75 [42, 43]. For values higher than 0.75, the acoustic levitation and trapping become unstable.

Tweezing based on focused ultrasound has been proposed to address the instability problem while considering safety issues. The higher acoustic pressure, P, results in a higher acoustic radiation force ($F_{PRI} \sim P$ [2]). The focused ultrasound transducers usually employ sound waves of higher frequencies compared to standing wave platforms. The higher frequency results in a shorter wavelength (Eq. (11.13)), which leads to a higher spatial resolution. This is particularly useful to trap and maneuver small objects (from tens of micrometers to nanometers). The standing wave-focused ultrasound systems typically have two transducer bowls facing each other [29]. They are positioned to focus the acoustic beam at the center in a confocal arrangement.

Wu [29] introduced the term "acoustical tweezers," a dual beam method using two opposing acoustic waves in the Rayleigh regime where the particle is much smaller than the wavelength, $R \ll \lambda$. Latex spheres (diameter of 270 μm) and frog eggs were successfully trapped by employing sound waves at 3.5 MHz ($\lambda \approx 0.43$ mm).

The force applied to an object can be calculated using the gravitational escape method. Particles are placed in the center of the acoustic beam. Then, the pressure level is varied from high to low until the particles escape. The pressure levels are recorded for several different particles. Then, the resultant force can be calculated [44]. Comparison of the sizes, frequencies, and forces in the acoustical tweezers and microrobots is further discussed in Section 11.4.

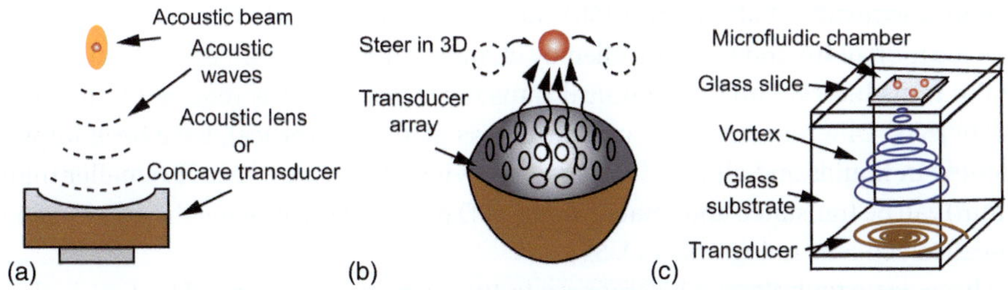

Figure 11.8 Single-sided acoustic tweezers. (a) The acoustic beam is tightly focused using either a transducer with an adjacent acoustic lens or a concave transducer. (b) An array of acoustic transducers is used to 3D-steer a bead in a gaseous media by changing the phase between transducers. (c) An acoustical vortex tweezer is used to trap beads in microchannels. The acoustic waves are transmitted via a glass substrate from the transducer to the microchannel. The 2D steering is achieved by moving the microchannel using an automated stage.

Standing wave bulk acoustic tweezers and standing wave focused ultrasound tweezers require the objects to be surrounded by the transducers, which limits the translation to the clinics. Moreover, standing wave fields consist of multiple equilibrium positions in the workspace that apply forces and trap untargeted objects. The multiplicity of nodes and antinodes leads to the existence of multiple trapping sites that cannot be moved independently. It is ideal to generate forces only on the target object. To resolve these issues, techniques were developed to manipulate particles using transducers from a single side (Figure 11.7b) [45]. These techniques are based on acoustic lenses, transducer arrays, and vortex beams (Figure 11.8).

The radiation force on the particle is exerted through momentum transfer, which is due to reflected and refracted acoustic rays. Refraction is the primary phenomenon causing the particle to move towards the beam's center. Therefore, the acoustic impedance of the particles and surrounding media needs to be matched [46]. The force in the tight focus depends on particle size, and excitation frequency. To generate a tight focus, either the transducers have a curved geometry or an acoustic lens is used with a flat transducer (Figure 11.8a) [47, 48]. One of the important parameters is f-number, $f_{number} = \frac{focal\ distance}{lens\ diameter}$, which is the ratio of focal distance to the lens diameter. The f_{number} is usually chosen less than 1 to keep the focal point tight and limit the energy loss due to attenuation. In addition, to increase the spatial resolution higher frequencies are used. For example, in 126-μm-diameter oleic acid droplets, transducers with a resonance frequency of 30 MHz can be employed [46].

Single-beam levitation by focusing the acoustic rays using either single-sided curved transducers or transducers with acoustic lenses is promising for several applications, such as particle manipulation and drug delivery in vivo. However, their maneuverability (translation and rotation of the traps) is limited. Furthermore, the energy loss, especially when using an acoustic lens, can be high, which lessens its applicability in biologically sensitive media [49].

To address these problems, an array of acoustic transducers can generate multiple acoustic traps and maneuver particles simultaneously (Figure 11.8b). This method employs acoustic radiation forces [50] and acoustic streaming [51] to levitate objects. The location of the traps is controlled by changing the relative phase and amplitude of the transducers. Thus, no physical transducer movement is required to maneuver objects. Different structure arrays, such as plane and vertical, have been implemented in liquids and air [52, 53]. It was shown that particles (usually smaller than 1 mm) can be translated and rotated along a 3D path without moving the arrays using so-called acoustic holograms [50, 54].

These acoustic holograms demonstrate two important features. The first is that single-sided arrays can trap microparticles and levitate them by generating forces of the order of μN, thus counteracting gravity. The second is that the topology of the acoustic field can be tuned on the fly. These allow controlling multiple particles in acoustic fields, thus paving the way for biomedical applications, e.g. microsurgery without physical interference.

Acoustic holograms are promising for several in vivo and in vitro tweezing applications; however, they are challenging to scale down to single cell size. SAW tweezers generate acoustic waves on the surface of piezoelectric substrate (usually $LiNbO_3$), resulting in either grids or lines of nodes and antinodes. This leads to multiplication of trapping areas. Therefore, it is challenging to trap and manipulate single objects (e.g. cells). This challenge can be addressed using acoustical vortex tweezers, which can tightly focus the helical acoustical waves and generate a single trap to maneuver objects on the scale of hundreds of micrometers to tens of micrometers (cell size). The schematic in Figure 11.8c illustrates typical acoustical vortex tweezers [55]. The device is comprised of a piezoelectric substrate at the bottom, a microchip at the top filled with a liquid medium and samples, and a glass substrate bridging the bottom and top subparts. The piezoelectric substrate has a unique crystallography that allows out-of-plane pressure distribution upon electrical signal. The design of spiraling active electrodes forms a vortex and focalizes it inside the microfluidic chip. The electrodes are made of thin layers of metal (usually Ti/Au and Cr/Au). A sinusoidal electrical signal is applied to generate localized deformation around the interdigitated transducers. A bulk acoustical vortex normal to piezoelectric substrate is generated and transmitted to the microchip through the glass in the middle. The working mechanism of the acoustical vortex tweezers is based on Bessel and Hankel beams [56–58]. The contribution of Bessel beams converges to none with the distance. Therefore, Hankel's contribution is the main reason for focal area formation.

The size limit of the maneuverable particles is dictated by the focal area of the vortex in the microchip. The size of the focal area can be controlled by design. For instance, a 4.4 MHz [56] and 47 MHz [55] spiraling transducers can trap 75 and 7 μm particles, respectively. The pressure and, in turn, the resulting force are adjusted by the applied signal. The operation frequency and signal amplitude, along with material properties, are the key parameters to determine the forces. Up to 200 pN were applied to trap single cells and manipulate them along a designated route [55]. One of the main challenges in acoustofluidic applications is joule heating in relatively long experiments. In the acoustical vortex tweezer set-ups, due to the distance between focal points and piezoelectric substrates, joule heating at the microchip is minimal even during long experiments (tens of minutes).

Even though vortex tweezers are currently designed for on-chip applications, they have the potential to be translated to clinics. Current acoustical vortex tweezers are designed to be operated close to the acoustical source. With an ergonomic design of the middle structure (currently glass), the acoustic waves can be transmitted to human organs for selective manipulation of samples.

11.4 Micro/Nanorobotic Devices Actuated by Acoustic Fields

As discussed in Sections 11.2.2.2 and 11.2.2.3, acoustically excited structures can generate steady flows and secondary acoustic radiation forces. In this section, we discuss how robotic devices can be designed, powered, and controlled through these acoustic phenomena. Although proof-of-principle examples have been realized using a variety of transducers and acoustic fields, we anticipate that BAWs will be used in clinical applications. The section starts with relatively simple mobile micromachines and continues with more complex robotic devices that can perform dexterous micromanipulations.

11.4.1 Mobile Acoustic Micromachines

The overall propulsion scheme is based on an actuator that converts acoustic energy into thrust. A gaseous bubble or microfabricated structure (e.g. beams and membranes) can act as such by converting the acoustic waves into mean flow (i.e. acoustic streaming), thus generating thrust for propulsion.

We first describe how a gas bubble entrapped inside a microfabricated cavity can be used as an on-board thruster. Upon acoustic excitation, due to the compressibility of the bubble, the bubble oscillates at the liquid-air interface prescribed by the opening of the cavity. These oscillations are maximized when the sound field frequency coincides with the resonance frequency of the bubble. Then, the bubble's oscillations resemble the corresponding mode shape [26, 59]. As discussed earlier, acoustically excited bubbles experience three different forces: primary acoustic radiation force (F_{PRI}), secondary acoustic radiation force (F_{SEC}), and acoustic streaming force, (F_{ST}). F_{SEC} is due to the interactions among nearby bubbles or the interactions of bubbles with solid boundaries. The thrust stems from the asymmetric design of the capsule, which results in a propulsive force (Figure 11.5b) [25, 27]. F_{PRI} is generally considered to be negligible in the actuation of bubble-based machinery [60].

If the cavity or cavities are engraved on a suspended microstructure, oscillations of the bubble can lead to motion (Figure 11.9a). Using photolithography, sub-mm-sized cylindrical capsules with a single opening were fabricated [62]. The micromachines moved as fast as 45 mm/s[1] (50 body lengths per second). By tuning the design (i.e. shape of the structure, number of bubbles, and their distribution), swimmers capable of turning were realized. Bubble-based micromachines can be selectively actuated

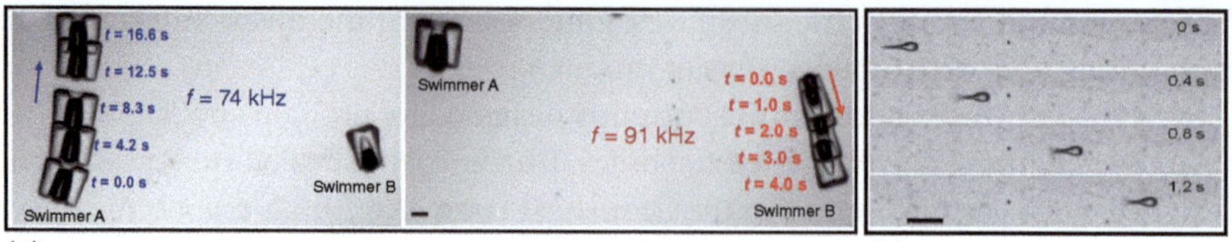

Figure 11.9 Mobile acoustic micromachines based on bubbles and sharp-edge. (a) Swimmers of different microbubble sizes were excited using various excitation frequencies. Selective actuation enabled control of the swimmer's motion. (b) A monolithically fabricated microswimmer with sharp-edges was excited by acoustic waves. Upon actuation, the tail oscillated at a higher amplitude than the head, which resulted in translational motion. Scale bars: 50 μm in (a), 200 μm in (b). Part (a) is Reproduced from Ahmed et al. [61], Springer Nature/CC BY 4.0 Part (b) is Reproduced from Kaynak et al. [23], Royal Society of Chemistry.

if they comprise distinct bubbles (i.e. different volumes and openings) [61]. The natural frequency of the bubble depends on its volume and opening size; therefore, different micromachines can be activated at different frequencies [26, 59]. Considering the typical size of the bubbles used in these studies [27, 61, 63, 64], and the motivation to excite the bubbles at their first natural frequency, the micromachines have been actuated at frequencies ranging from a few hundred kHz to several MHz. These studies have also shown that the micromachines generate high enough forces to allow them to move inside highly viscous and viscoelastic media. This feature is important for in vivo applications considering the rheology of bodily fluids such as blood and mucus. The materials that were used to microfabricate these structures, parylene and PEG, are biocompatible and are widely used for in vivo biomedical applications.

The early prototypes had relatively large air bubbles entrapped in cavities due to the limited resolution of early photolithographic techniques. Recent work has shown that smaller acoustic micromachines can be fabricated with precise dimensions using the two-photon polymerization method [27, 63, 65]. Smaller bubbles have higher resonance frequencies; therefore, 3D nanoprinted micromachines are excited at higher frequencies. For medical applications, this shift in actuation frequencies is desirable, as the smaller micromachines can reach hard-to-reach regions in the body and be controlled by clinically available ultrasound transducers.

In general, when air bubbles are close to rigid substrates, secondary acoustic radiation forces emerge due to mirroring effect (i.e. a rigid boundary acts as a mirror, as if a similar bubble exists on its other side) [2]. This force is always attractive because identical bubbles always attract each other. This attractive force acts in parallel to the thrust generated by the bubble-induced acoustic streaming. As a result, bubble-based micromachines are attracted to walls of containers. These forces can be harnessed for motion control. For example, acoustic micromachines can move along curved surfaces through a surface-slipping motion [66]. This mode of motion could be useful to navigate micromachines inside arteries and the gastrointestinal tract. This last study has recently been extended to locomotion inside viscoelastic fluids. A high shear rate propulsion mechanism was reported, which is a significant step towards applying these microrobots in vivo [67].

Another common method to generate microstreaming employs sharp edge structures. These flexible structures oscillate under acoustic excitation and induce microstreaming at the tip [16]. Exciting the structures at moderate acoustic levels leads to a linear behavior; thus, each tip oscillates at the driving frequency with a proportional amplitude. These parameters, therefore, can be controlled via the input signal generated by an acoustic transducer. The concept of sharp edge streaming was explored to create microscale mixers [68] and pumps [69] within microfluidic channels. Later, rotary systems were developed by designing rotors fitted with extended arms with sharp-edged features [70]. The sharp-edges, along with the compliance of the material, lead to large oscillations at the tip at the designated driving frequency. The micromachines could rotate either clockwise or counterclockwise at distinct frequencies, according to the design. It is reported that the angular speed is proportional to the number of sharp-edge streamers; each is attached to a single arm of a rotor. A 6-arm rotor can span as fast as 1200 RPM. Later, this study was extended to compound machinery by using the rotors inside microfluidic channels to drive microscale turbines and mills [71].

Analogous to bubble-based microsystems, the streaming flow can generate thrust that propels mobile micromachines forward [23]. A micromachine, inspired by sperm cells, with a head and flagellum-like tail was developed (Figure 11.9b). The material properties and structural design maximized the oscillations at the tip of the tail. As a result, the micromachines move forward as fast as 1200 μm/s[1] (~8 body lengths per second). Similar machines were miniaturized further using electrodeposition. In this work [72], although the thickness of the micromachine was fixed throughout the body (<1 μm), the oscillation amplitude due to structural resonance was maximized at the tail. The oscillation at a frequency of approximately 90 kHz generated acoustic streaming and propelled the swimmer in aqueous media.

11.4.2 Soft Robotic Microsystems

The micromachines we discussed so far display simple motions, i.e. translational, and rotational. Full 3D motion has also been realized by controlling multiple bubbles of different sizes on the same device [73]. Compliant mechanisms achieve force and motion control through elastic body deformation. One of the biggest advantages of using compliant mechanisms to create microscale robotic devices is that they can be fabricated as a monolithic structure through 3D printing. Moreover, unlike rigid mechanisms, compliant mechanisms do not suffer from surface forces as the transmission is based on deformation and not friction. Recent work has shown that capsules that entrap bubbles can be connected to flexible beams using two-photon polymerization [27]. The natural frequencies of the entrapped bubbles can be tuned by editing the cavity geometry, e.g. cavity opening diameter and number of openings, for selective actuation [26, 59]. The flexible beams allow continuous motion under various acoustic forces (e.g. F_{ST} and F_{SEC}). A single cavity with three openings that

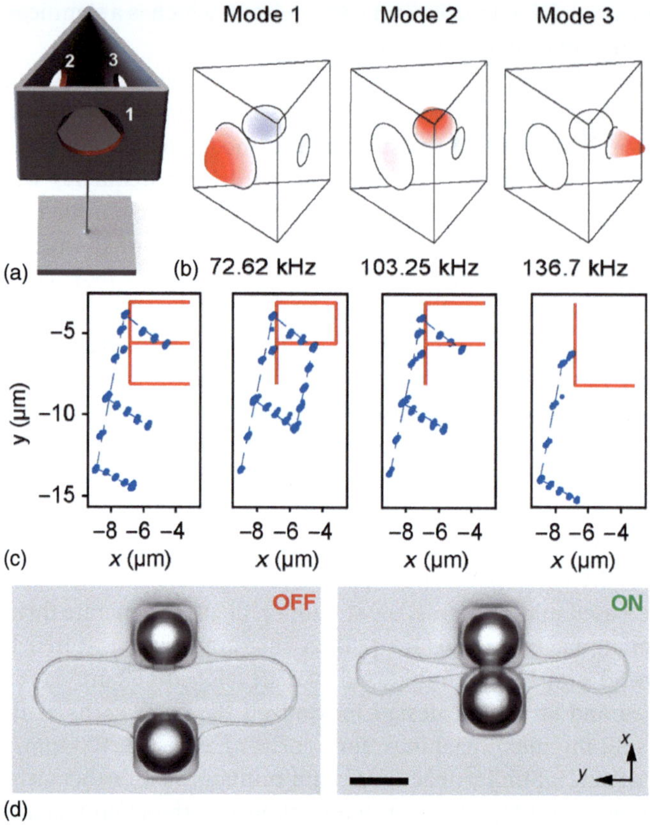

Figure 11.10 3D nanoprinted soft robotic microsystems. (a) The illustration shows a capsule with three distinct orifices. The capsule is attached to a glass substrate via an elastic beam. (b) The direction of the movement of the capsule was controlled via actuation of different modes of the entrapped air bubble. (c) The markers show the tracked position of the capsule while it was moved along paths prescribing the letters "EPFL." (d) Two capsules are connected with a spring, where the top one is fixed to the substrate and the other is free to move. Upon acoustic excitation, the capsules attract each other due to secondary acoustic radiation forces, and the spring is compressed. Scale bar: 50 μm. Source: Reproduced with permission from Kaynak et al. [27], Ann Liebert, Inc.

is linked to the surface with a flexible beam can follow prescribed trajectories due to F_{ST} (Figure 11.10a). Additionally, a linear actuator was developed by linking two cavities facing each other via a flexible spring [27]. The bubbles in the cavities generated F_{SEC} and thus attracted each other resulting in translational motion by deforming the beams. These actuators can serve as building blocks for soft robotic devices with more complex architectures and many degrees of freedom.

An alternative route to creating microrobotic devices is based on the use of sharp-edged hydrogel structures [74]. Monolithic rotary machines and pumps were fabricated inside microfluidic chambers from biocompatible polymers (Figure 11.11). The pumping of fluid was driven by microstreaming generated by 3D conical structures, where the flow velocity reached up to 476 μm/s[1]. Taking

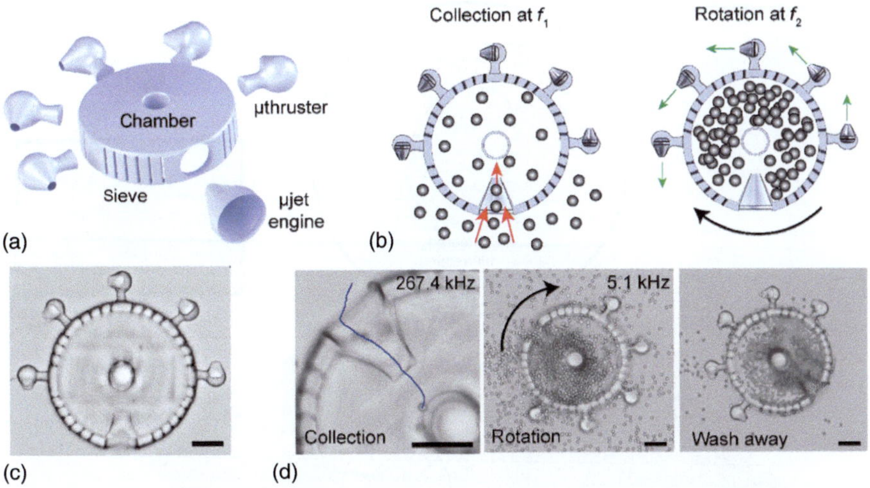

Figure 11.11 An acoustically powered soft robotic device. (a) Schematic of the device highlighting the collection chamber, sieve for filtering, and a μjet engine for pumping media. (b) Illustration showing sample collection and device rotation. (c) A bright field image of the device. (d) Time lapse images showing collection of particles, rotation of the device, and the final state after washing away the suspended particles. Scale bars: 75 μm. Source: Reproduced from Kaynak et al. [74], John Wiley & Sons/CC BY 4.0.

advantage of 3D nanoprinting, a chamber and a sieve were added to the pump to collect samples from the media. As a step forward in the creation of complex functional devices, mass transport and selective actuation of thrusters were combined in a single device (Figure 11.10b). The resulting compound machine could rotate in place and collect microparticles from the surrounding medium in a spatiotemporally controlled fashion. Such untethered biomanipulation tools have the potential to be utilized in vivo as injectable drug delivery devices.

11.5 In Vivo Actuation of Micro/Nanorobotic Devices

Acoustic manipulation technologies can serve as unique tools for fundamental research in life sciences and biomedical applications [75, 76]. Here, we provide examples of how microscale devices can be manipulated inside a living animal. As described in Sections 11.4.1.1 and 11.4.1.2, there are various ways of remotely translating and rotating objects using acoustic transducers, and BAWs are best suited for in vivo applications. So far, the successful demonstrations contain basic trapping and translation of microscale objects. However, the platforms are capable of actuating micro/nanorobotic devices. It is only a matter of time before acoustically powered robotic devices are operated inside living animals.

The first example shows a steerable, vortex-based acoustic trapping beam lifting and moving a 3-mm glass sphere inside the urinary bladder of a live pig [77] (Figure 11.12). A 256-element, focused array with a 12-cm focal distance was operated to generate the vortex beams at around 1.4 MPa pressure. The beam is

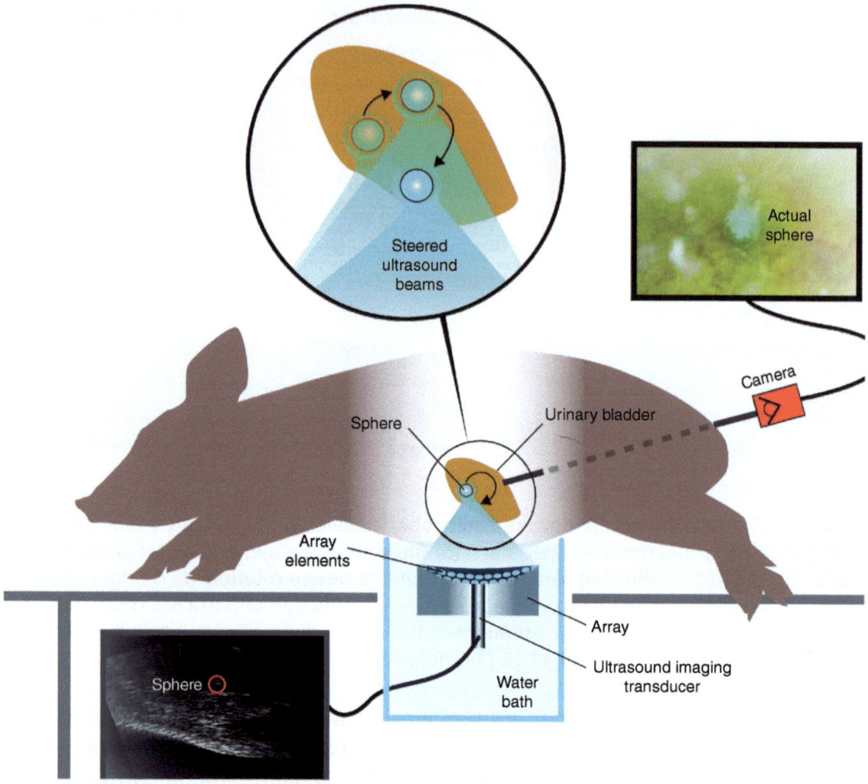

Figure 11.12 Demonstration of in vivo acoustic micromanipulation. A 3-mm sphere is manipulated in a pig bladder using a bulk acoustic tweezer. Schematic of an acoustically powered miniaturized cystoscope. Source: Reproduced with permission from Ghanem et al. [77], Proceedings of the National Academy of Sciences PNAS.

created by altering the phase among the focused array of 256 transducer elements while their amplitude remains constant. The sphere is trapped at the center of the beam generated at 1.5 MHz and moved in a plane transverse to the beam axis. No histological evidence of injury was observed in the pigs following ultrasound exposure, demonstrating the noninvasive nature of the technique.

Scaling down the size of the objects would result in significantly lower primary radiation forces, which is a bottleneck for in vivo applications. On the other hand, as described in Section 11.2.2, significantly higher forces can be generated on microbubbles. Microbubbles have been increasingly used for biomedical applications such as targeted drug delivery. Notably, recent work has demonstrated ultrasound-mediated aggregation and uncaging of drug-loaded liposomes tethered to microbubbles inside the brain vasculature of rats [78]. Using this technique, small molecules are delivered with millimeter-precision without opening the blood-brain barrier at relatively low pressure (around 100–300 kPa). The aggregation of microbubbles, which are 1.5 μm in diameter, is probably due to the secondary Bjerknes forces between the bubbles. After local aggregation at 2.5 MHz, the drugs are uncaged from the microbubbles with a second ultrasound pulse sequence with higher pressure. More recent work has shown that lipid-shelled microbubbles can be concentrated and manipulated for precise positioning within the blood vessels of mice using an acoustic vortex tweezer [79]. The acoustic waves are generated by four transducer elements that are driven by 3-MHz sine waves to produce pressures ranging from 80 to 800 kPa. The wave propagates along the z direction, and phase difference around the axis gradually varies from 0 to 2π radians in each x–y plane. The phase is indeterminate on the axis, which leads to the generation of a null core of zero amplitude, creating a vortex structure with a potential well. The particles near the well experience an inward force and thus can be trapped. In parallel, vortex beams were generated by eight independent transducers driven with the necessary time delays to construct a helicoidal wavefront [80]. The results showed that the trapping beam can operate in 3D after propagating through thick, opaque-to-light, viscoelastic media without suffering from wavefront distortion. It can be sharply focused on localized sites, offering pinpoint control over an individual microbubble even in crowded environments. Finally, a standing wave acoustic tweezer system has been developed where four piezoelectric actuators with a resonance frequency of 4.25 MHz are positioned orthogonally to each other around an off-centered fluidic channel [81]. This platform allows for 2D manipulation of intravenously injected microbubbles inside zebrafish embryos.

As an alternative to primary and secondary radiation forces, acoustic streaming forces can be used to actuate medical devices. To our knowledge, there is only one medical device, a miniaturized urological endoscope that is actuated following this strategy. Surfaces that consist of 2D arrays of microbubbles were fabricated, which oscillate under ultrasound excitation and thereby generate an acoustic streaming force [82]. Mounting such a surface at the tip of a millimeter-scale endoscope realized remotely controlled bending and steering. Bubbles of different sizes are addressed by their unique resonance frequency, so multiple degrees of freedom can readily be incorporated. The flexible endoscope was tested in a rabbit bladder ex vivo.

11.6 Discussion and Outlook

Acoustic technologies are indispensable for medicine. Recent advances in acoustic micromanipulation have shown great potential to provide means for targeted drug delivery and alternative minimally invasive interventions. In this chapter, we summarized the physics behind the widely explored manipulation strategies and elaborated on the most exciting directions by going over proof-of-concept demonstrations. These demonstrations pave the way for the future development of more advanced medical applications and lay the groundwork for translating in vivo acoustic manipulation to human patients. In the coming years, we will surely see the development of more advanced in vivo acoustic manipulation technologies that will transport microscale robots to hard-to-reach tissues and deliver therapies with pinpoint accuracy without posing either risks or side effects or postoperative trauma.

We shall also expect to see remote acoustic steering of guidewires, catheters, and endoscopes for diagnostic and therapeutic operations in near future.

There are a number of factors that define the effectiveness and safety of these manipulation strategies. A broad range of frequencies (from kHz to MHz) have been utilized in acoustic tweezers and microrobotic applications [23, 29, 72]. In the majority of cases, the frequencies dictate the sizes of manipulable objects. As a reference, the speed of sound in water at room temperature is approximately 1500 m/s, and at 100 kHz, the acoustic wavelength is 15 mm. Significant acoustic forces are generated when the size of the sample is much smaller than the wavelength $R \ll \lambda$. In addition, the distance from the acoustic source is important, as the acoustic pressure attenuates with the distance due to dissipative losses [1, 2]. As discussed, solid elastic particles with diameters of 190–390 μm can be trapped in aqueous media, where the resultant trapping force is on the order of 20 nN [49]. To scale down the size of the trapped particles, acoustic waves must be generated at higher frequencies, resulting in a smaller trapping area. Single cells (diameter of ~10 μm) were trapped using a 30 MHz focused transducer [83], where the estimated force goes up to ~100 nN. The studies provide an order of magnitude estimate of the forces that can be applied to microscopic objects through primary acoustic radiation forces. As a comparison, ultrasonic imaging uses frequencies in the range of 2–18 MHz. Acoustic streaming and secondary acoustic radiation forces may provide ways to apply higher forces to microscale objects as they scale more favorably compared to the primary acoustic radiation force. The work summarized in this chapter provides general guidelines on how to harness these secondary forces to power microrobotic devices. The future is vibrant for acoustics resonating with microrobotics.

Acknowledgment

Murat Kaynak is supported by the Swiss National Science Foundation (SNSF) under the postdoc mobility fellowship (Project number: P500PT_211084). Amit Dolev acknowledges the support of the European Research Council under the European Union's Horizon 2020 Marie Sklodovska-Curie Actions Postdoctoral Fellowship (Grant agreement No. 101022448).

References

1 Pierce, A.D. (2019). *Acoustics: An Introduction to Its Physical Principles and Applications*. Springer International Publishing.
2 Leighton, T. (2012). *The Acoustic Bubble*. Academic Press.
3 Hamilton, M.F. and Blackstock, D.T. (1998). *Nonlinear Acoustics*. Academic Press.
4 Blackstock, D.T. (2000). *Fundamentals of Physical Acoustics*. Wiley.
5 King, L.V. (1934). On the acoustic radiation pressure on spheres. *Proceedings of the Royal Society of London Series A* 147: 212–240.

6 Yosioka, K. and Kawasima, Y. (1955). Acoustic radiation pressure on a compressible sphere in a viscous fluid. *Acta Acustica United with Acustica* 5 (7): 167–173.

7 Gor'kov, L. P. (1962). On the forces acting on a small particle in an acoustical field in an ideal fluid. *Soviet Physics Doklady* 6: 773–775.

8 Karlsen, J.T. and Bruus, H. (2015). Forces acting on a small particle in an acoustical field in a thermoviscous fluid. *Physical Review E, Statistical, Nonlinear, and Soft Matter Physics* 92: 043010.

9 Doinikov, A.A. (2005). Bjerknes forces and translational bubble dynamics. In: *Bubble and Particle Dynamics in Acoustic Fields: Modern Trends and Applications*, 95–143. Research Signpost.

10 Lamb, H. (1945). *Hydrodynamics*. Dover Publications.

11 Plesset, M.S. and Prosperetti, A. (1977). Bubble dynamics and cavitation. *Annual Review of Fluid Mechanics* 9: 85–145.

12 Yosioka, K., Kawasima, Y., and Hirano, H. (1955). Acoustic radiation pressure on bubbles and their logarithmic decrement. *Acta Acustica united with Acustica* 5 (6): 173–178.

13 Zabolotskaya, E.A. (1984). Interaction of gas bubbles in a sound field. *Soviet Physics Acoustics* 30: 365–368.

14 Boluriaan, S. and Morris, P.J. (2003). Acoustic streaming: from Rayleigh to today. *International Journal of Aeroacoustics* 2: 255–292.

15 Eckart, C. (1948). Vortices and streams caused by sound waves. *Physics Review* 73: 68–76.

16 Nama, N., Huang, P.H., Huang, T.J., and Costanzo, F. (2014). Investigation of acoustic streaming patterns around oscillating sharp edges. *Lab on a Chip* 14: 2824–2836.

17 Muller, P.B., Barnkob, R., Jensen, M.J.H., and Bruus, H. (2012). A numerical study of microparticle acoustophoresis driven by acoustic radiation forces and streaming-induced drag forces. *Lab on a Chip* 12: 4617–4627.

18 Muller, P.B. and Bruus, H. (2014). Numerical study of thermoviscous effects in ultrasound-induced acoustic streaming in microchannels. *Physical Review E, Statistical, Nonlinear, and Soft Matter Physics* 90: 043016.

19 Behdani, B., Monjezi, S., Zhang, J. et al. (2020). Direct numerical simulation of microbubble streaming in a microfluidic device: the effect of the bubble protrusion depth on the vortex pattern. *Korean Journal of Chemical Engineering* 37: 2117–2123.

20 Ovchinnikov, M., Zhou, J., and Yalamanchili, S. (2014). Acoustic streaming of a sharp edge. *The Journal of the Acoustical Society of America* 136: 22–29.

21 Zhang, C., Guo, X., Royon, L., and Brunet, P. (2020). Unveiling of the mechanisms of acoustic streaming induced by sharp edges. *Physical Review E* 102: 043110.

22 Doinikov, A.A., Gerlt, M.S., Pavlic, A., and Dual, J. (2020). Acoustic streaming produced by sharp-edge structures in microfluidic devices. *Microfluidics and Nanofluidics* 24: 32.

23 Kaynak, M., Ozcelik, A., Nourhani, A. et al. (2017). Acoustic actuation of bioinspired microswimmers. *Lab on a Chip* 17: 395–400.
24 Spelman, T.A. and Lauga, E. (2017). Arbitrary axisymmetric steady streaming: flow, force and propulsion. *Journal of Engineering Mathematics* 105: 31–65.
25 Bertin, N., Spelman, T.A., Stephan, O. et al. (2015). Propulsion of bubble-based acoustic microswimmers. *Physical Review Applied* 4: 064012.
26 Dolev, A., Kaynak, M., and Sakar, M.S. (2022). Dynamics of entrapped microbubbles with multiple openings. *Physics of Fluids* 34: 012012.
27 Kaynak, M., Dolev, A., and Sakar, M.S. (2023). 3D printed acoustically programmable soft microactuators. *Soft Robotics* 10 (2): 246–257.
28 Spelman, T.A., Stephan, O., and Marmottant, P. (2020). Multi-directional bubble generated streaming flows. *Ultrasonics* 102: 106054.
29 Wu, J. (1991). Acoustical tweezers. *The Journal of the Acoustical Society of America* 89: 2140–2143.
30 Ozcelik, A., Rufo, J., Guo, F. et al. (2018). Acoustic tweezers for the life sciences. *Nature Methods* 15: 1021–1028.
31 Ding, X., Lin, S.C.S., Kiraly, B. et al. (2012). On-chip manipulation of single microparticles, cells, and organisms using surface acoustic waves. *Proceedings of the National Academy of Sciences of the United States of America* 109: 11105–11109.
32 Rayleigh, L. (1885). On waves propagated along the plane surface of an elastic solid. *Proceedings of the London Mathematical Society* s1-17: 4–11.
33 Ahmed, D., Ozcelik, A., Bojanala, N. et al. (2016). Rotational manipulation of single cells and organisms using acoustic waves. *Nature Communications* 7: 1–11.
34 Ozcelik, A., Nama, N., Huang, P.H. et al. (2016). Acoustofluidic rotational manipulation of cells and organisms using oscillating solid structures. *Small* 12: 5120–5125.
35 Bücks, K. and Müller, H. (1933). Über einige Beobachtungen an schwingenden Piezoquarzen und ihrem Schallfeld. *Zeitschrift für Physiotherapie* 84: 75–86.
36 Buchanan, R.H., Jameson, G., and Oedjoe, D. (1962). Cyclic migration of bubbles in vertically vibrating liquid columns. *Industrial and Engineering Chemistry Fundamentals* 1: 82–86.
37 Blake, F.G. (1949). Bjerknes forces in stationary sound fields. *The Journal of the Acoustical Society of America* 21: 551.
38 Vilhelm, B. (1906). *Fields of Force.* Columbia University Press.
39 Eller, A. (1968). Force on a bubble in a standing acoustic wave. *The Journal of the Acoustical Society of America* 43: 170–171.
40 Dolev, A., Davis, S., and Bucher, I. (2019). Noncontact dynamic oscillations of acoustically levitated particles by parametric excitation. *Physical Review Applied* 12: 034031.
41 Gu, Y., Chen, C., Mao, Z. et al. (2021). Acoustofluidic centrifuge for nanoparticle enrichment and separation. *Science Advances* 7: eabc0467.

42 Lee, M.C. and Feng, I.A. (1982). Acoustic levitating apparatus for submillimeter samples. *The Review of Scientific Instruments* 53: 854–859.

43 Wu, J. and Du, G. (1990). Acoustic radiation force on a small compressible sphere in a focused beam. *The Journal of the Acoustical Society of America* 87: 997–1003.

44 Hertz, H.M. (1995). Standing-wave acoustic trap for nonintrusive positioning of microparticles. *Journal of Applied Physics* 78: 4845–4849.

45 Silva, G.T. and Baggio, A.L. (2015). Designing single-beam multitrapping acoustical tweezers. *Ultrasonics* 56: 449–455.

46 Lee, J., Teh, S.Y., Lee, A. et al. (2009). Single beam acoustic trapping. *Applied Physics Letters* 95: 073701.

47 Lam, K.H., Hsu, H.S., Li, Y. et al. (2013). Ultrahigh frequency lensless ultrasonic transducers for acoustic tweezers application. *Biotechnology and Bioengineering* 110: 881–886.

48 Cannata, J.M., Ritter, T.A., Chen, W.H. et al. (2003). Design of efficient, broadband single-element (20–80 MHz) ultrasonic transducers for medical imaging applications. *IEEE Transactions on Ultrasonics, Ferroelectrics, and Frequency Control* 50: 1548–1557.

49 Baresch, D., Thomas, J.L., and Marchiano, R. (2016). Observation of a single-beam gradient force acoustical trap for elastic particles: acoustical tweezers. *Physical Review Letters* 116: 024301.

50 Marzo, A., Seah, S.A., Drinkwater, B.W. et al. (2015). Holographic acoustic elements for manipulation of levitated objects. *Nature Communications* 6: 1–7.

51 Li, J., Crivoi, A., Peng, X. et al. (2021). Three dimensional acoustic tweezers with vortex streaming. *Communications on Physics* 4: 1–8.

52 Courtney, C.R.P., Drinkwater, B.W., Demore, C.E.M. et al. (2013). Dexterous manipulation of microparticles using Bessel-function acoustic pressure fields. *Applied Physics Letters* 102: 123508.

53 Seah, S., Drinkwater, B., Carter, T. et al. (2014). Correspondence: dexterous ultrasonic levitation of millimeter-sized objects in air. *IEEE Transactions on Ultrasonics, Ferroelectrics, and Frequency Control* 61: 1233–1236.

54 Melde, K., Kremer, H., Shi, M. et al. (2023). Compact holographic sound fields enable rapid one-step assembly of matter in 3D. *Science Advances* 9 (6): eadf6182.

55 Baudoin, M., Thomas, J.L., Al, S.R. et al. (2020). Spatially selective manipulation of cells with single-beam acoustical tweezers. *Nature Communications* 11: 4244.

56 Baudoin, M., Gerbedoen, J.C., Riaud, A. et al. (2019). Folding a focalized acoustical vortex on a flat holographic transducer: miniaturized selective acoustical tweezers. *Science Advances* 5: eaav1967.

57 Baresch, D., Thomas, J.L., and Marchiano, R. (2013). Spherical vortex beams of high radial degree for enhanced single-beam tweezers. *Journal of Applied Physics* 113 (18): 184901.

58 Baudoin, M. and Thomas, J.L. (2020). Acoustic tweezers for particle and fluid micromanipulation. *Annual Review of Fluid Mechanics* 52: 205–234.

59 Dolev, A. (2023). Entrapped microbubble dynamics. MATLAB central file exchange. https://www.mathworks.com/matlabcentral/fileexchange/123075-entrapped-microbubble-dynamics.

60 Louf, J.F., Bertin, N., Dollet, B. et al. (2018). Hovering microswimmers exhibit ultrafast motion to navigate under acoustic forces. *Advanced Materials Interfaces* 5: 1800425.

61 Ahmed, D. et al. (2015). Selectively manipulable acoustic-powered microswimmers. *Scientific Reports* 5 (6): 1554–1562.

62 Feng, J., Yuan, J., and Cho, S.K. (2015). Micropropulsion by an acoustic bubble for navigating microfluidic spaces. *Lab on a Chip* 15: 1554–1562.

63 Ren, L., Nama, N., McNeill, J.M. et al. (2019). 3D steerable, acoustically powered microswimmers for single-particle manipulation. *Science Advances* 5: eaax3084.

64 McNeill, J.M., Nama, N., Braxton, J.M., and Mallouk, T.E. (2020). Wafer-scale fabrication of micro- to nanoscale bubble swimmers and their fast autonomous propulsion by ultrasound. *ACS Nano* 14: 7520–7528.

65 Sridhar, V., Park, B.W., and Sitti, M. (2018). Light-driven janus hollow mesoporous TiO_2–Au microswimmers. *Advanced Functional Materials* 28: 1704902.

66 Aghakhani, A., Yasa, O., Wrede, P., and Sitti, M. (2020). Acoustically powered surface-slipping mobile microrobots. *Proceedings of the National Academy of Sciences of the United States of America* 117: 3469–3477.

67 Aghakhani, A., Pena-Francesch, A., Bozuyuk, U. et al. (2022). High shear rate propulsion of acoustic microrobots in complex biological fluids. *Science Advances* 8: 5126.

68 Huang, P.H., Xie, Y., Ahmed, D. et al. (2013). An acoustofluidic micromixer based on oscillating sidewall sharp-edges. *Lab on a Chip* 13: 3847–3852.

69 Huang, P.H., Nama, N., Mao, Z. et al. (2014). A reliable and programmable acoustofluidic pump powered by oscillating sharp-edge structures. *Lab on a Chip* 14: 4319–4323.

70 Kaynak, M., Ozcelik, A., Nama, N. et al. (2016). Acoustofluidic actuation of in situ fabricated microrotors. *Lab on a Chip* 16: 3532–3537.

71 Kaynak, M., Ayhan, F., and Sakar, M.S. (2019). Compound micromachines powered by acoustic streaming. In: *2019 International Conference on Robotics and Automation (ICRA)*, 225–230.

72 Ahmed, D., Baasch, T., Jang, B. et al. (2016). Artificial swimmers propelled by acoustically activated flagella. *Nano Letters* 16: 4968–4974.

73 Liu, F.W. and Cho, S.K. (2021). 3-D swimming microdrone powered by acoustic bubbles. *Lab on a Chip* 21: 355–364.

74 Kaynak, M., Dirix, P., and Sakar, M.S. (2020). Addressable acoustic actuation of 3D printed soft robotic microsystems. *Advanced Science* 7 (20): 2001120.

75 Drinkwater, B.W. (2020). A perspective on acoustical tweezers – devices, forces, and biomedical applications. *Applied Physics Letters* 117: 180501.

76 Rufo, J., Zhang, P., Zhong, R. et al. (2022). A sound approach to advancing healthcare systems: the future of biomedical acoustics. *Nature Communications* 13: 3459.

77 Ghanem, M.A., Maxwell, A.D., Wang, Y.N. et al. (2020). Noninvasive acoustic manipulation of objects in a living body. *Proceedings of the National Academy of Sciences of the United States of America* 117: 16848–16855.

78 Ozdas, M.S., Shah, A.S., Johnson, P.M. et al. (2020). Non-invasive molecularly-specific millimeter-resolution manipulation of brain circuits by ultrasound-mediated aggregation and uncaging of drug carriers. *Nature Communications* 11: 4929.

79 Lo, W.C., Fan, C.H., Ho, Y.J. et al. (2021). Tornado-inspired acoustic vortex tweezer for trapping and manipulating microbubbles. *Proceedings of the National Academy of Sciences of the United States of America* 118 (4): e2023188118.

80 Baresch, D. and Garbin, V. (2020). Acoustic trapping of microbubbles in complex environments and controlled payload release. *Proceedings of the National Academy of Sciences of the United States of America* 117: 15490–15496.

81 Jooss, V.M., Bolten, J.S., Huwyler, J., and Ahmed, D. (2022). In vivo acoustic manipulation of microparticles in zebrafish embryos. *Science Advances* 8: eabm2785.

82 Qiu, T., Adams, F., Palagi, S. et al. (2017). Wireless acoustic-surface actuators for miniaturized endoscopes. *ACS Applied Materials & Interfaces* 9: 42536–42543.

83 Ashkenazi, S., Huang, S.W., O'Donnell, M. et al. (2006). High-frequency ultrasound transduction using polymer microring resonators. *Proceedings – IEEE Ultrasonics Symposium* 1: 1056–1059.

Index

Acoustic Technologies in Biology and Medicine, First Edition.
Edited by Adem Ozcelik, Ryan Becker, and Tony Jun Huang.